Advances in Quantum Science II

Taku Onishi
Editor

Advances in Quantum Science II

Artificial Intelligence, Computational and Experimental Approaches

Springer

Editor
Taku Onishi
Graduate School of Engineering
Mie University
Tsu, Japan

Department of Chemistry
University of Oslo
Oslo, Norway

ISBN 978-981-95-9720-8 ISBN 978-981-95-9721-5 (eBook)
https://doi.org/10.1007/978-981-95-9721-5

This Springer imprint is published by the registered company Springer Nature Singapore Pte Ltd.
The registered company address is: 152 Beach Road, #21-01/04 Gateway East, Singapore 189721, Singapore

Preface

Scientific phenomena related to electrons are based on the principle of quantum mechanics. Because electrons are theoretically treated as quantum particles. Recent research findings in quantum science are introduced from the wide aspects of theoretical development, computational method, Artificial Intelligence (AI), machine learning, computation, and experiment. Researchers who deal with electrons are able to find tips to pioneer new researches.

Volume 2 highlights machine learning, and computational and experimental studies for advanced materials, molecules, and biological systems. The use of machine learning is getting popular in quantum science. In Part I, several machine learning techniques are demonstrated for concrete examples, which make them easier to understand. Quantum calculations and dynamical simulations for advance materials (Part II) and molecules (Part III) are illustrated, in combination with the associated experimental results. Methodological developments for computing biological molecules are introduced in Part IV. In Part V, the recent researches in applied physics and physical chemistry are introduced from the point of an experimental approach.

Part I: Machine Learning and Artificial Intelligence

To optimize the Hartree–Fock exchange coefficients in hybrid density functional theory (DFT) methods, machine learning can be applied. In Chap. 1, the optimised HF exchange coefficients to reproduce an experimental bandgap of a semiconducting perovskite are illustrated for different hybrid-DFT methods. In Chap. 2, to express molecular straightness as a mathematical function, molecular curvature has been proposed. The concrete simulation results for several bent molecules are illustrated. In Chap. 3, pharmaceutical properties are predicted through machine learning simulations, where molecular orbital energies are introduced as explanatory variables.

Part II: Quantum Calculations for Advance Materials—Collaboration with Experimental Studies

For iron-containing magnetic systems, Mössbauer spectroscopy has been widely used to investigate the electronic state of iron. In Chap. 4, superconductive properties of iron-containing superconductors are explained through both DFT calculation and Mössbauer spectroscopy. Carbon nanomaterials possess a lot of advantages due to their unique structures. Chapters 5 and 6 review the adsorption mechanism of metals on carbon nanomaterials, and the insertion of nanoribbons into carbon nanotubes, respectively. The development of heat storage materials has been much expected because of the effective use of waste heat energy. In Chap. 7, thermodynamic computations of heat storage materials are illustrated.

Part III: Quantum Calculations for Molecules

Part III focuses on quantum chemical calculations for molecules. In Chap. 8, photodissociation mechanisms in transition metal complexes are investigated using molecular orbital calculations. In Chap. 9, non-covalent interactions are discussed for H_2O-HF and propylene oxide-helium systems.

Part IV: Computational Biology

In Chap. 10, canonical molecular orbital (CMO) computations are applied for investigating functionalities, interactions, and reaction mechanisms of protein molecules. Force fields for molecular dynamics simulation are developed on the basis of quantum chemical calculations in Chap. 11. Readers will gain an understanding of the relationship between quantum mechanics and biological studies.

Part V: Applied Physics and Physical Chemistry

The experimental results are discussed from a quantum mechanical perspective. In Chap. 12, spin-charge interconversion is discussed for a two-dimensional electron gas system (2DEG). A promising heterostructure of (La,Sr)MnO_3/$LaTiO_3$/$SrTiO_3$ is also highlighted. In Chap. 13, quantum effects of hydrogen atom are overviewed. The recent progress of the experimental techniques is also introduced. In Chap. 14, after explaining the electronic structure of porphyrins, supramolecular porphyrin assemblies such as cyclic chlorophyll tetramers, cyclic porphyrin trimers, and a face-to-face porphyrin dimer are illustrated. In Chap. 15, the fundamentals of scanning tunneling microscopy (STM) and tip-enhanced Raman spectroscopy (TERS) are described in the former portion. The experimental results of chiral molecular systems are illustrated in the latter portion. In Chap. 16, the synthesis procedure of inverse opal photonic crystals possessing a three-dimensionally ordered macroporous (3DOM) structure is described. The empirical design guideline to maximize photonic band gap (PBG) is proposed. In Chap. 17, new protein–metal complex hybrid materials are introduced. The structures are investigated from experimental and computational approaches. In Chap. 18, fundamental techniques of oriented molecular beam are explained. Previous stereodynamics studies using oriented molecular beam are also illustrated.

Finally, I would like to thank Dr. Shinichi Koizumi and Springer staffs to produce this book. I hope this book stimulates readers to come up with new scientific ideas.

In the AI era, model construction is getting important

Taku Onishi
A Happy Day 2026

Acknowledgements

The Editor, Taku Onishi, would like to express his special thanks to T.O.

Abbreviations

2D	Two-dimensional
2DEG	Two-dimensional electron gas
3DOM	Three-dimensionally ordered macroporous
AI	Artificial intelligence
CCSD	Coupled cluster singles doubles
CM	Center of mass
CMO	Canonical molecular orbital
CMs	Centers of mass
CNT	Carbon nanotube
DFT	Density functional theory
Eq.	Equation
Eqs.	Equations
ET	Electron transfer
Fig.	Figure
Figs.	Figures
Fml.	Formula
Fmls.	Formulas
GNR	Graphene nanoribbon
HF	Hartree–Fock
MD	Molecular dynamics
PBG	Photonic band gap
QB	Quantum battery
QED	Quantum electrodynamics
STM	Scanning tunneling microscopy
TERS	Tip-enhanced Raman spectroscopy
XPS	X-ray photoelectron spectroscopy

About the Editor

Taku Onishi is a Quantum Scientist who graduated from Department of Chemistry, Osaka University, Japan in 1998. He gained master's degree in 2000 and Ph.D. in 2003, both from Department of Chemistry, Osaka University, Japan. In 2003, he took up academic tenure as Assistant Professor in Graduate School of Engineering, Mie University, Japan. During 2010–2011, he stayed at Centre for Theoretical and Computational Chemistry (CTCC), University of Norway as Excellent Young Researcher Overseas Visit Program, Japan Society for the Promotion of Science. Since 2010, he has been a Guest Researcher of Department of Chemistry, University of Oslo, Norway. In 2021, the title of Guest Professor was given from University of the Ryukyus, Japan. He has served on international scientific activities: a Member of Royal Society of Chemistry; Chair of the Next-Generation Science (NGS) Symposium; a position on the science committee of the International Conference of Computational Methods in Sciences and Engineering (ICCMSE); a member of the editorial board of *Cogent Engineering*, the *Journal of Computational Methods in Sciences and Engineering* (JCMSE). He has reviewed many international proceedings, books, and journals in various research fields of chemistry and physics. For example, *AIP Conference Proceedings*, *Progress in Theoretical Chemistry and Physics*, *Cogent Chemistry*, *Physical Chemistry Chemical Physics*, *Molecular Physics*, *Dalton Transaction*, *The Journal of Physical Chemistry Letters*, *Journal of Computational Chemistry*, *International Journal of Quantum Chemistry*, *Journal of*

Solid State Chemistry, *Solid State Ionics*, *Chemistry of Materials*, *Materials Chemistry and Physics*, *Chemical Engineering Journal*, *Chemical Physics*, *ACS Catalysis*, *Applied Surface Science*, *Chemical Physics*, *Chemical Physics Letters*, *Chirality*, *Applied Surface Science*, *ACS Catalysis*, *Biochimica et Biophysica Acta–Proteins and Proteomics*, *AIP Advances*, *The Journal of Organic Chemistry*, *New Journal of Chemistry*, etc.

About the Corresponding Authors

Taku Onishi (Chaps. 1 and 2) is the Editor of *Advances in Quantum Science I and II*.

Hiroyuki Teramae (Chap. 3) is a full Professor in the Department of Chemistry, Faculty of Science at Josai University. He earned his Doctor of Engineering from Kyoto University in 1984 and joined NTT Basic Research Laboratories in 1985. He served as a Visiting Fellow at University of Colorado Boulder in 1995–1996 and joined Josai University in 2004. His research focuses on computational chemistry and theoretical physical chemistry, including quantum chemical calculations, molecular orbital theory, molecular dynamics simulations, and materials informatics. He has studied electronic structures of polysilanes and properties of biomolecules and nanomaterials (e.g., model DNA) using high-performance computing, and currently applies machine learning to chemistry. He has authored numerous influential publications advancing theoretical approaches in the field.

Dmitrii A. Tayurskii (Chaps. 4 and 5) is a Professor at Kazan Federal University, Russia. He received his Ph.D. in Physics from Kazan State University (1989) and Doctor of Science from Kazan Physical-Technical Institute, Russian Academy of Science (2001). He currently works as a First Vice-Rector for Research at Kazan Federal University. He was invited as a Professor to Kanazawa University, Japan and now he is a Collaborative Professor and a Senior Advisor for the President at this University. Also, during more than 10 years he was invited as a Researcher to the Institute of Physical and Chemical Research RIKEN, Japan and managed comprehensive collaboration between Kazan Federal University and RIKEN. The scope of his scientific interests is quite broad: low-temperature physics, quantum fluids, strongly correlated electron systems, non-extensive thermodynamics, physics of complex systems, computer design of materials, and quantum biology. Together with his graduate students, he derived a theory of superfluid hydrodynamics of quantum liquids in the fractal geometry based on the concept of the fractal Schrödinger equation. He has more than 300 publications in peer-reviewed journals worldwide, six monographs, and seven textbooks. In 2025, he was awarded the honorary title "Honored Scientist of the Russian Federation."

Umedjon Khalilov (Chap. 6) is the Head of the Laboratory of Advanced Materials Modeling at the Institute of Ion-Plasma and Laser Technologies, Academy of Sciences of Uzbekistan. He earned his Ph.D. in Physics/Chemistry from the University of Antwerp in 2013 and was awarded a Doctor of Science (D.Sc.) degree by the Academy of Sciences of Uzbekistan in 2021. He also served as the International Deputy Director of The Belt and Road International Joint Research Center for Advanced Materials at Hebei University of Technology (2024–2028). His research in Computational Materials Science focuses on the synthesis and properties of low-dimensional carbon allotropes, including carbyne and graphene nanoribbons. A two-time recipient of the prestigious FWO grant (2014, 2017), his work utilizes advanced atomistic modeling to bridge fundamental nanoscale phenomena with applications in nanoelectronics and thermal management.

Toyokazu Ishida (Chap. 7) is a Senior Researcher at the National Institute of Advanced Industrial Science and Technology (AIST) in Tsukuba, Japan. After earning his Ph.D. in science from Kyoto University in 2003, he worked at the Institute for Molecular Science (IMS) in Okazaki, Japan, before joining AIST in 2006, where he remains today. From 2011 to 2012, during his academic career, he was seconded to the Ministry of Economy, Trade and Industry (METI) in Japan as a research and development specialist. There, he played a key role in planning and designing next-generation national projects. From 2013 to 2023, he served as the computational science team leader for the national research project on the thermal management materials and technology research (TherMAT). Since 2016, he has also been a Visiting Researcher at the Oak Ridge National Laboratory (ORNL) in the United States, promoting joint research between Japan and the U.S. His primary research areas are theoretical and computational chemistry, particularly the computational analysis of enzymatic reactions and molecular recognition.

Yuzuru Kurosaki (Chap. 8) obtained his B.S. and M.S. degrees from Kyoto University and received his Ph.D. also from Kyoto University under the supervision of Prof. Tokio Yamabe. He then worked as a Postdoctoral Fellow at Japan Atomic Energy Agency (JAEA) under the supervision of Prof. Tetsuo Miyazaki. After the postdoctoral research, he joined Kansai Institute for Photon Science in JAEA. Meanwhile, he stayed in Prof. Herschel Rabitz's laboratory at Princeton University as a visiting scientist for 1 year. Now he is a Senior Principal Researcher at National Institutes for Quantum Science and Technology (QST), which was established from the merger of National Institute of Radiological Sciences (NIRS) with research institutes originally included in JAEA. His research interests have focused on many topics in chemical physics, e.g., quantum tunneling in chemical reactions, photodissociations of transition metal complexes, nonadiabatic transitions between potential energy surfaces, spin-orbit effects on reaction rates, quantum optimal control of physical, chemical processes with ultra-short laser pulses, etc. Recently, he also became interested in negatively charged nitrogen-vacancy (NV) centers in diamonds, which are promising for application in quantum sensors and quantum computing devices. He has so far published about 80 peer-reviewed papers in international journals.

Morihisa Saeki (Chap. 8) obtained his B.S. degree from Waseda University and M.S. degree from The University of Tokyo. He received his Ph.D. from The University of Tokyo under the supervision of Prof. Takashi Nagata (1999). He then worked as a Postdoctoral Fellow at Institute for Molecular Science (IMS) and at National Institute for Materials Science (NIMS). After the postdoctoral research, he worked in Japan Atomic Energy Agency (JAEA). While working in JAEA, he stayed at Helmholtz Zentrum Dresden Rossendorf (HZDR) as a Visiting Scientist for 1 year. Now he is a Senior Principal Researcher at National Institutes for Quantum Science and Technology (QST). His research fields are molecular and cluster sciences using laser spectroscopy, X-ray absorption spectroscopy and quantum chemical calculations, and their application to nuclear energy, precious metal recycling and biology. He has so far published about 70 peer-reviewed papers in international journals.

Federico Palazzetti (Chap. 9) received his Ph.D. in Chemical Sciences from the University of Perugia in 2009. There, he began his research studying the kinetics and dynamics of the gas phase from theoretical and experimental perspectives. To this end, he conducted research at the University of Coimbra, the University of Barcelona, Osaka University, and National Taiwan University. He then became a Research Fellow at the University of Perugia and at the Institute of Inorganic and Plasma Methodologies of the CNR in Bari. From 2014 to 2018, heled a unit at the Scuola Normale Superiore in Pisa, working on a project funded by the Italian government that focused on eco-sustainable cements. From 2015 to 2019, he was appointed directly as an Assistant Professor at the University of Perugia, where he was the principal investigator of a government-funded project on chiral discrimination processes. Since 2019, he has expanded his area of interest to include the theoretical and computational study of biological systems and organometallic complexes. Since 2024, he has been an Associate Professor at the San Raffaele University in Rome.

Toshiyuki Hirano, Ph.D. (Chap. 10) is an Associate Professor at the Tokyo Metropolitan College of Industrial Technology and a Research Fellow at the Institute of Industrial Science, The University of Tokyo. Specializing in computational chemistry and high-performance computing, his research focuses on elucidating the electronic structures of large-scale biological systems. He is a core developer of the ProteinDF program, having contributed significantly to the implementation of the third-generation density functional theory (DFT) method utilizing Cholesky decomposition. He also developed QCLObot, an orchestration software that automates the generation of initial guesses using the quasi-canonical localized orbital (QCLO) method. Furthermore, he is actively working on advanced analytical tools, such as Energy Density Analysis (EDA) and the Molecular Orbital Overlap (MOO) method, to visualize complex molecular interactions. Alongside his research, he is dedicated to fostering the next generation of engineers and scientists through his educational activities at the college.

Fumitoshi Sato, Ph.D. (Chap. 10) is a Professor at the Institute of Industrial Science and the Department of Mechanical Engineering, Graduate School of Engineering, The University of Tokyo. He is a pioneer in computational chemistry for large-scale biological systems and the primary developer of the ProteinDF program. His research focuses on the development and application of canonical molecular orbital (CMO) calculation methods based on hybrid density functional theory (DFT). He achieved the world's first all-electron CMO calculations for proteins such as cytochrome c and the insulin hexamer. His current work integrates these advanced computational methods to elucidate protein reaction mechanisms and support novel enzyme and drug design.

Takefumi Yamashita, Ph.D. (Chap. 11) is an Associate Professor in the Department of Physical Chemistry, School of Pharmacy and Pharmaceutical Sciences, Hoshi University (Tokyo, Japan), and a Project Associate Professor at the Research Center for Advanced Science and Technology (RCAST), The University of Tokyo. His research is based on quantum chemical calculations and molecular dynamics (MD) simulations, with a focus on the modeling, understanding, and design of biomolecules and polymeric materials. He has contributed to extracting dynamical information from spectroscopic observables and to developing and refining empirical force fields such as DREIDING-UT. In the area of drug design, he explores rational antibody engineering, antigen–antibody complex formation, and practical workflows for free energy profiling using atomistic MD simulations. In the area of polymeric materials, he develops multiscale approaches that couple MD simulations with continuum methods to predict the properties of polymer networks. More recently, he has incorporated machine learning into the analysis of reaction dynamics and molecular design.

Shinobu Ohya (Chap. 12) is a Professor in the Department of Electrical Engineering and Information Systems at The University of Tokyo, Japan. He received his B.S. degree in electronic engineering in 2000 and his Ph.D. in engineering in 2005 from The University of Tokyo. After completing his doctorate, he joined the same institution as a research associate and later served as assistant professor and associate professor before being promoted to full professor in 2023. From 2006 to 2010, he was also a Researcher in the JST PRESTO program "Structure and Control of Interfaces," where he conducted pioneering work on semiconductor spin-band engineering and its device applications. He was appointed as a JSPS Overseas Research Fellow in 2011 and carried out collaborative research at the University of California, Santa Barbara, working on perovskite oxides and superconducting quantum computing. His research focuses on spintronics, quantum transport, ferromagnetic semiconductors, oxide heterostructures, and spin–orbit torque devices. He has published extensively in the field and has made steady contributions to the understanding of spin-dependent phenomena and their application to electronic and quantum devices. His work aims to advance fundamental knowledge of quantum materials while exploring pathways toward future low-power and functional device technologies.

Katsuyuki Fukutani (Chap. 13) is a Professor of Surface and Interface Physics at the Institute of Industrial Science, The University of Tokyo. He has also served as Group Leader of the Surface and Interface Group at the Advanced Science Research Center of the Japan Atomic Energy Agency. He received his B.Sc. and D.Sc. in Physics from The University of Tokyo, Japan. His research interests include the structure and dynamics of hydrogen at surfaces and interfaces, the electronic effects of hydrogen on surfaces and thin films, and surface and interface magnetism. By developing experimental techniques such as nuclear reaction analysis combined with channeling for depth-resolved site analysis, resonance-enhanced multiphoton ionization for state-selective detection of molecules, and a spin-polarized hydrogen beam, his group focuses on hydrogen transport across surfaces, quantum effects in hydrogen diffusion, spin and electronic effects in surface reactions, and ortho–para hydrogen conversion. He is a member of several scientific societies and has published more than 250 scientific papers.

Takahiro Ozawa (Chap. 13) is a Research Associate at the Institute of Industrial Science, The University of Tokyo. He received his B.Sc. in Physics from Kyushu University, Japan, and his Ph.D. in Applied Physics from the Graduate School of Engineering, The University of Tokyo, Japan. His research focuses on atomic and electronic transport on surfaces and in thin films. He is interested in nuclear quantum effects and investigates how they influence the dynamics and structures of light elements such as hydrogen and lithium using original experimental approaches that combine ion beam analysis, thermal desorption spectroscopy, and electrical resistivity measurements. He is also interested in modifying physical properties through hydrogen doping. He employs in situ measurement systems that integrate photoemission spectroscopy and optical absorption spectroscopy with low-temperature hydrogenation techniques to elucidate the role of hydrogen in the electronic structures of various materials, including metal alloys and high-entropy oxides, as well as two-dimensional materials such as graphene and germanene grown on thin metal films.

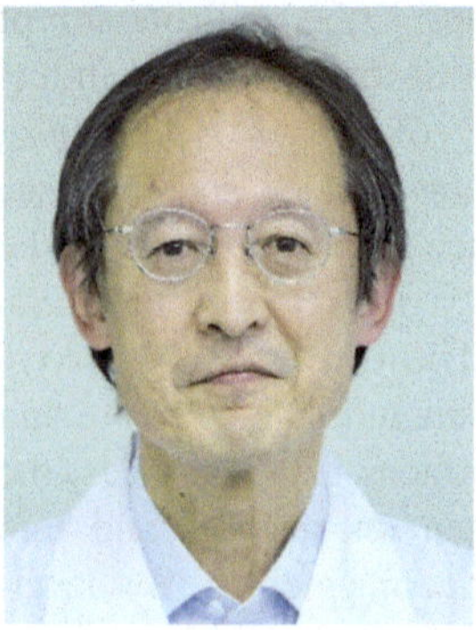

Joe Otsuki (Chap. 14) is currently a Professor of Chemistry at the College of Science and Technology, Nihon University, Japan. He obtained hid Ph.D. from The University of Tokyo in 1991. He worked as an Assistant Professor at the Institute of Industrial Science, The University of Tokyo, until 1999. In 1993, he spent 1 year at Université Louis Pasteur in Strasbourg as a Canon Europa Fellow. He moved to Nihon University in 1999. His research focuses on basic sciences related to supramolecular chemistry, coordination chemistry, and photochemistry. His work is based on synthetic molecules produced in his own laboratory. His recent interests focus on porphyrin assemblies. Well-designed synthetic orphyrin molecules form well-defined assemblies: some mimic the initial steps of photosynthesis, while others act as cage-like containers for chemical entities. He has received several honors, including the Lecture Award at the ISPAC 2017; the Publons Peer Reviewer Awards in 2018 and 2019; the Solar Energy Best Paper Award 2019; and the Sir J. C. Bose Endowment Award 2024. He has been listed among the World's Top 2% scientists since 2023. He served as Chair of the Organizing Committee of the Annual Meeting of the Chemical Society of Japan in 2021, 2024, and 2026.

Yuji Kuwahara, (Chap. 15) who is a Professor of The University of Osaka, received his Ph.D. in Science from Kyoto University in 1988. He subsequently worked as a Researcher at RIKEN, Japan, and in 1996 was appointed Associate Professor at The University of Osaka (formerly Osaka University), where he has continued his research activities. His research focuses on the physical and chemical properties of molecules on solid surfaces, investigated at the molecular and atomic levels. He is internationally recognized for pioneering work using advanced scanning tunneling microscopy (STM) and its innovative extensions, including tip-enhanced Raman spectroscopy (TERS) and STM-induced light emission techniques. He has also played a key role in developing sophisticated experimental systems that integrate these methods for comprehensive surface analysis. His contributions have significantly advanced ultra-high-resolution experimental techniques in nanotechnology, physical chemistry, and surface science. He has authored more than 150 original research articles in leading international journals and has delivered numerous invited lectures worldwide. In recognition of his contributions to international collaborative research with Poland, he received the Medal AMICO from the University of Lodz. His broader achievements in surface science, particularly in research on chiral molecules, were further recognized by the Society Award of the Japan Society of Vacuum and Surface Science in February 2024.

Paweł Krukowski (Chap. 15) is an Associate Professor at the Faculty of Physics and Applied Informatics, University of Lodz. His research focuses on chiral organic molecules on metallic surfaces and low-dimensional materials, with particular emphasis on nanoscale characterization at the single-molecule level. He has been affiliated with the University of Lodz since 2016 and was promoted to Associate Professor in October 2024 after obtaining the habilitation degree (Doctor Habilitatus) in Physical Sciences. He received his Ph.D. in Physics from the University of Lodz for research on stable free radicals and holds an M.Sc. in Experimental Physics, as well as an MBA in IT from the Polish-Japanese Academy of Information Technology in Warsaw. His scientific interests include organic chiral molecules, graphene-based heterostructures, and advanced surface-sensitive techniques such as scanning probe microscopy and tip-enhanced Raman spectroscopy. He has held international research positions as a Specially Appointed Researcher at Osaka University (now The University of Osaka) from 2013 to 2016 in the group of Prof. Yuji Kuwahara, and as a JSPS Postdoctoral Fellow at the Institute for Solid State Physics, The University of Tokyo, from 2010 to 2012 in the group of Prof. Fumio Komori. He has authored more than 45 peer-reviewed publications in leading international journals. His current research addresses stable chiral radicals on metallic substrates, aiming at combined structural, chiral, and chemical identification at the single-molecule level.

Toshihiro Moriga (Chap. 16) is a Professor of Department of Applied Chemistry, Graduate School of Science and Technology for Innovation, Tokushima University, Japan. He graduated from Department of Chemistry, Osaka University, Japan in 1988 and was awarded Dr. Sci. from Osaka University in 1996. He was adopted as an Assistant Professor at Tokushima University in 1991 and had his sabbatical stay at Department of Materials Science and Engineering, Northwestern University, USA from 1996 to 1997, co-working with Profs. T. Mason and K. Poeppelmeier. He has been a Professor at Tokushima University since 2009. His specialized research is based on experimental approaches to enhancing the functionality of wide bandgap oxide/oxynitride semiconductors, particularly phosphors for white LEDs, proton-conducting ceramic fuel cell electrolytes, and visible-light-driven photocatalysts. He is serving as the Chair of the Chugoku-Shikoku Branch of the Ceramic Society of Japan for FY2024–2025. He is highly valued for his contribution to education in the faculty and got the Best Teacher Award in the Faculty of Science and Technology (formerly Faculty of Engineering), Tokushima University in 2005, 2009, 2013, 2017, and 2021. In 2025, he was selected as a Hall of Fame Teacher at Faculty of Science and Technology, Tokushima University.

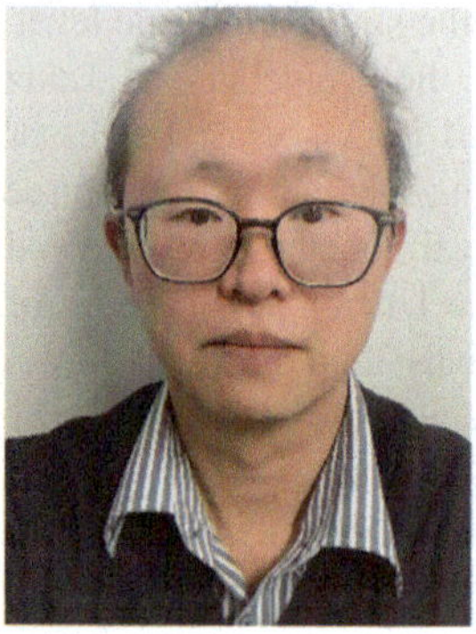

Takashiro Akitsu (born in 1971, Hamamatsu City, Japan) (Chap. 17) is a full Professor of chemistry at Department of Chemistry, Faculty of Science Division II, Tokyo University of Science since 2016. He graduated from Department of Chemistry, Faculty of Science, Osaka University (currently The University of Osaka) and got his Ph.D. degree in 2000 (inorganic coordination chemistry and crystallography). He worked at Institute of Protein Research in the same university (protein crystallography of metalloproteins), Department of Chemistry at Keio University (photo-magnetic functional organic/inorganic hybrid materials), Stanford University (bioinorganic chemistry of metalloproteins) and Tokyo University of Science since 2008. Chiral Schiff base metal complexes and their hybrid systems with functional materials were the main themes of his group. These studies were carried out both experimental and theoretical approaches. He has published more than 200 papers in reputed journals and has been serving as an editorial board member of repute, for example, one of co-Editors of *Acta Crystallographica E* (International Union of Crystallography).

Dock-Chil Che (Chap. 18) received the Ph.D. degree from the Department of Chemistry, Graduate School of Science, The University of Osaka, Osaka, Japan, in 1993. He was a Research Fellow supported by the Japanese Society of Promotion of Science, from 1992 to 1994. From 1995 to 1996, he was a Post-Doctoral Fellow at Institute of Advance Medical Studies, Taiwan. In 2015, he was promoted to the position of Associate Professor with the Department of Chemistry, Graduate School of Science, The University of Osaka, Osaka. His current research interests include reaction dynamics of radical with small molecules, the spectroscopy of small molecules, and the development of the gas analytical technique.

Deok-Chul Ohe [illegible] received the Ph.D. degree from the Department of Chemistry, Graduate School of Science, [illegible] University of [illegible], Japan, in 199[illegible]. He was a Research Fellow supported by the Japanese Society of Promotion of Science from [illegible] [illegible]. From [illegible] with the [illegible]

Contents

Part I
Machine Learning and Artificial Intelligence

Chapter 1
Statistical Optimisation of Hartree-Fock Exchange Coefficient in Hybrid-Density Functional Theory: Supervised Machine Learning by Python Programming

Taku Onishi

Abstract After a scientific terminology *machine learning* became used, regression has been recognised as one of supervised machine learning techniques. In quantum chemical calculations by the use of hybrid-density functional theory (DFT) method, a calculated bandgap of $SrTiO_3$ perovskite varies, depending on the magnitude of the Hartree-Fock (HF) exchange coefficient. In a regression analysis, bandgaps changing HF exchange coefficient are interpolated by a regression function. After determining a regression function, the HF exchange coefficient to reproduce the experimental bandgap is optimised. In other words, the appropriate HF exchange coefficient is statistically evaluated. In this chapter, optimisation results are illustrated for several modified hybrid-DFT methods. We discuss whether bandgaps evaluated by quantum chemical calculations with the optimised HF exchange coefficients correspond to the experimental bandgap or not. Finally, a theoretical direction for the generalised hybrid-DFT method is described.

Keywords Quantum chemical calculation · Hybrid-density functional theory (DFT) · $SrTiO_3$ perovskite · Bandgap · Optimisation · Hartree-Fock (HF) exchange functional · Supervised machine learning · Regression · Coefficient of determination · Python

---

T. Onishi (✉)
Graduate School of Engineering, Mie University, Tsu, Japan
e-mail: onishi.taku@mie-u.ac.jp; taku.onishi@kjemi.uio.no

Department of Chemistry, University of Oslo, Oslo, Norway

T. Onishi (ed.), *Advances in Quantum Science II*,
https://doi.org/10.1007/978-981-95-9721-5_1

1.1 Introduction

1.1.1 Machine Learning—Statistical Approach

Application of machine learning has been much expected in various fields of quantum science. Because many physical quantities can be statistically predicted. Machine learning can be classified into three types.

Classification

1. **Supervised machine learning.**
2. **Unsupervised machine learning.**
3. **Deep learning using neural network.**

Regression is categorised as supervised machine learning. Discrete experimental or calculated data is interpolated by a mathematical function called a regression function. Before the appearance of machine learning, regression was recognised as an independent technique in applied mathematics. Nowadays, regression is becoming regarded as one of machine learning techniques. In a regression analysis, after defining the form of a regression function, the unknown coefficients of a regression function are statistically evaluated.

Clustering is one of unsupervised machine learning techniques. Grouping discrete data is performed on the basis of metric (distance).

In 2024, the noble prizes in chemistry and physics were awarded to artificial intelligence (AI) studies related to *deep learning* [1, 2]. Instead of regression, the relation between input and output data is written in multiple transformations. Coefficients of transformation matrices are statistically optimised using huge amount of data. For example, it is now used for accurate protein structure prediction [3, 4].

1.1.2 Functionals of Hybrid-Density Functional Theory (DFT) Method

From the viewpoint of methodological developments in quantum chemical calculations, the new combination of functionals is optimised for a theoretical model under restricted conditions. In fact, electron-electron interactions within and between atoms depend on electron species (s, p, d, etc. electrons) and surrounding electrons. In hybrid-density functional theory (DFT) method, the contribution of exchange and correlation interactions is parameterised as coefficients of the associated exchange and correlation energies. There is no general procedure to determine the combination of associated functionals on the basis of a calculation model. For simple organic molecules, B3LYP method has been widely recognised as one of a standard combination of exchange and correlation functionals. However, some shortcomings are pointed out in transition metal compounds and organometallics [5]. In an actual manner, after performing several hybrid-DFT calculations, an appropriate combination of functionals is determined in comparison with experimental results.

1.1.3 Hartree-Fock (HF) Exchange Functional in Hybrid-DFT Method

Hybrid-density functional theory (DFT) method can reproduce the correct electronic states for strongly correlated transition metal compounds. Because the localisation effect can be incorporated with the HF exchange functional [6, 7]. Here are introduced two cases when Hartree-Fock (HF) exchange functional has an important role.

Case 1: Effective Exchange Integral in Transition Metal Cluster

In an insulating spin model, strongly correlated Mn-F-Mn cluster, the magnetic interaction between manganese atoms via fluorine atom becomes antiferromagnetic, due to superexchange interaction [6]. The covalent bonding is formed between a manganese atom and a bridge-fluorine atom. Effective exchange integral (J) is a physical quantity to represent a magnetic interaction in a spin system. Depending on the magnitude of the HF exchange coefficient in hybrid-DFT method, the calculated J values varied [7].

Case 2: Bandgap of $SrTiO_3$ Perovskite

$SrTiO_3$ perovskite is a typical strongly correlated perovskite. The energy difference between highest occupied molecular orbital (HOMO) and lowest unoccupied molecular orbital (LUMO), which corresponds to a bandgap, varies in changing the HF exchange coefficient in hybrid-DFT method [8]. Pay attention that a HOMO–LUMO gap and a bandgap are defined in a real space and a momentum space, respectively. Both gaps indicate the energy gap between electron-occupied and electron-unoccupied quantum states. If quantum states in both spaces essentially stand for the same, it can be regarded that a HOMO–LUMO gap is equal to a bandgap. The calculated bandgap varies, depending on the magnitude of the HF exchange coefficient in hybrid-DFT method.

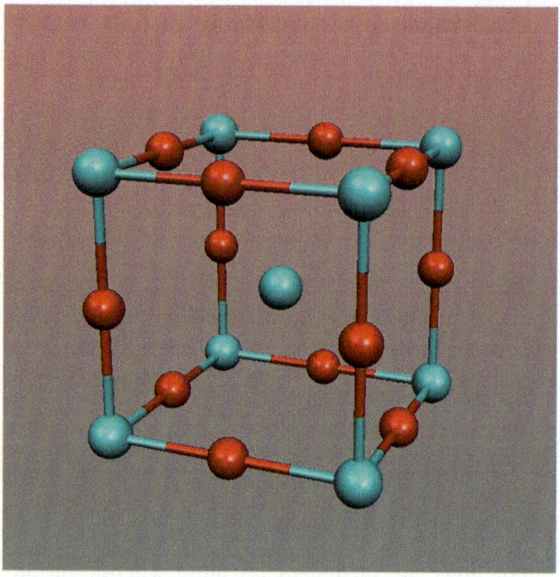

Fig. 1.1 Unit structure of cubic $SrTiO_3$ perovskite (by Molekel): $SrTi_8O_{12}$ model. Edge-atoms and bridge-atoms are titanium and oxygen, respectively. At the cubic centre, strontium is allocated

1.2 Structure and Experimental Bandgap of $SrTiO_3$ Perovskite

The perovskite-type titanium oxide $SrTiO_3$ is a typical cubic perovskite [9] (see Fig. 1.1). The bandgap was experimentally determined to be 3.27 eV [10]. Bandgap is generally estimated as a simple electron transfer energy from occupied quantum state to unoccupied quantum state. Pay attention that an experimental bandgap is not directly related to an energy band in a momentum space.

1.3 Bandgap Evaluation of $SrTiO_3$ Perovskite by Conventional Hybrid-DFT Methods

1.3.1 Setting up Hybrid-DFT Calculation

Quantum chemical calculations are performed with Firefly program package. Basis sets employed here are 6-31G* for oxygen atom and minimal basis set (MINI) [11] for titanium atom and strontium atom. A unit structure of cubic $SrTiO_3$ perovskite, $SrTi_8O_{12}$ model, is used as a cluster model. For a cluster model calculation of a solid, three conditions:

1. **No neutral condition is required.**
2. **No geometry optimisation is performed.**

3. **Experimental interatomic distances and angles are employed.**

are required [6]. Because a cluster model is regarding as not a molecule but a part of solid.

1.3.2 Exchange and Correlation Energies in Hybrid-DFT Method

In a hybrid-DFT method, the exchange and correlation energies are represented by

$$E_{\mathrm{XC}} = C_{\mathrm{X}}^{\mathrm{HF}} E_{\mathrm{X}}^{\mathrm{HF}} + C_{\mathrm{X}}^{\mathrm{Slater}} E_{\mathrm{X}}^{\mathrm{Slater}} + C_{\mathrm{X}}^{\mathrm{Becke}} E_{\mathrm{X}}^{\mathrm{Becke}} + C_{\mathrm{C}}^{\mathrm{VWN}} E_{\mathrm{C}}^{\mathrm{VWN}} + C_{\mathrm{C}}^{\mathrm{LYP}} E_{\mathrm{C}}^{\mathrm{LYP}}, \quad (1.1)$$

where the first, second, third, fourth and fifth terms denote Hartree-Fock (HF) exchange energy, Slater exchange energy, Becke exchange energy [12], Vosko-Wilk-Nusair (VWN) correlation energy [13] and Lee-Yang-Parr (LYP) correlation energy [14], respectively. $C_{\mathrm{X}}^{\mathrm{A}}$ denotes the coefficient of A exchange energy (A = HF, Slater or Becke), while $C_{\mathrm{C}}^{\mathrm{B}}$ denotes the coefficient of B correlation energy (B=VWN or LYP).

In BHHLYP method, Slater exchange energy and VWN correlation energy are not included.

$$E_{\mathrm{XC}}^{\mathrm{BHHLYP}} = 0.5 E_{\mathrm{X}}^{\mathrm{HF}} + 0.5 E_{\mathrm{X}}^{\mathrm{Becke}} + E_{\mathrm{C}}^{\mathrm{LYP}} \quad (1.2)$$

In BLYP and B3LYP methods, the exchange and correlation energies are given by

$$E_{\mathrm{XC}}^{\mathrm{BLYP}} = E_{\mathrm{X}}^{\mathrm{Becke}} + E_{\mathrm{C}}^{\mathrm{LYP}} \quad (1.3)$$

and

$$E_{\mathrm{XC}}^{\mathrm{B3LYP}} = 0.2 E_{\mathrm{X}}^{\mathrm{HF}} + 0.8 E_{\mathrm{X}}^{\mathrm{Slater}} + 0.72 E_{\mathrm{X}}^{\mathrm{Becke}} + 0.19 E_{\mathrm{C}}^{\mathrm{VWN}} + 0.81 E_{\mathrm{C}}^{\mathrm{LYP}}, \quad (1.4)$$

respectively.

1.3.3 Comparing BLYP, B3LYP, BHHLYP and HF Methods

The localisation effect can be appropriately incorporated through modifying the contribution of HF exchange functional. HF method can be regarded as one of hybrid-DFT methods, since coefficients except for HF exchange energy are all zero in Fml. (1.1). Pure DFT method can be also regarded as one of hybrid-DFT methods, since the coefficient of HF exchange energy is zero in Fml. (1.1). Therefore, the effect of HF exchange energy can be investigated in comparison with HF and pure DFT methods.

B3LYP method underestimated the $SrTiO_3$ bandgap (2.16 eV), while BHHLYP overestimated it (4.48 eV). The HF exchange coefficient between 0.2 (B3LYP) and 0.5 (BHHLYP) reproduces the experimental value [8]. In spite of different contribution of exchange and correlation energies in three hybrid methods, the variation of bandgap becomes approximately linear, when the HF exchange coefficient increases. The simple statistical model:

$$\Delta E = k\Delta E_{\mathrm{BHHLYP}}, \tag{1.5}$$

where ΔE, $\Delta E_{\mathrm{BHHLYP}}$ and k denote a revised bandgap, a bandgap calculated by BHHLYP method and a scaling factor, respectively, was proposed. The k value was statistically determined to be 0.73.

1.4 Calculation Procedures

1.4.1 Quantum Chemical Calculation by Modified BHHLYP Method

In Firefly program package [15], the HF exchange coefficient in hybrid-DFT method can be modified within the closed interval [0.0, 1.0]. The exchange and correlation energies of *modified BHHLYP method* is expressed as

$$E_{\mathrm{XC}}^{\mathrm{mod\ BHHLYP}} = C_{\mathrm{X}}^{\mathrm{HF}} E_{\mathrm{X}}^{\mathrm{HF}} + 0.5E_{\mathrm{X}}^{\mathrm{Becke}} + E_{\mathrm{C}}^{\mathrm{LYP}}, \tag{1.6}$$

where $0.0 \le C_{\mathrm{X}}^{\mathrm{HF}} \le 1.0$.

1.4.2 Determining Regression Function—Supervised Machine Learning

Changing the HF exchange coefficient within the closed interval [0.0, 1.0], bandgaps are evaluated. The scatter plot of the HF exchange coefficient versus bandgap is interpolated by a regression function. Herein, a simple linear function, a quadratic polynomial function and an exponential function are employed as regression functions. The applicability of a regression function is assessed by the coefficient of determination (R^2), where $0 \le R^2 \le 1$.

Criteria of Regression Efficiency

If R^2 is close to 1.00, a regression function fits with evaluated bandgaps.

If R^2 is far from 1.00, a regression function deviates from evaluated bandgaps.

In a regression analysis, `curve_fit` function of SymPy library [16] is used. Graphs of regression functions and evaluated bandgaps are plotted with matplotlib library. Python 3 is used as a programming language [17].

1.4.3 How to Optimise Hartree-Fock Exchange Coefficient in Hybrid-DFT Method

In strongly correlated $SrTiO_3$ perovskite, the evaluated bandgap monotonously increases when the HF exchange coefficient increases in hybrid-DFT method. Therefore, from the point of statistical approach rather than methodological approach, we optimise an appropriate HF exchange coefficient to reproduce an experimental bandgap (Fig. 1.2). Here are considered three *modified hybrid-DFT methods* based on BHHLYP, B3LYP and PBE1PW91 methods.

This process is statistical.

Because the appropriate HF exchange coefficient is optimised using a regression function.

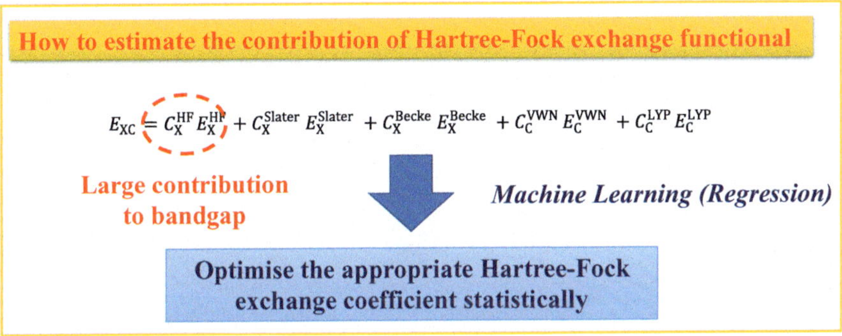

Fig. 1.2 How to optimise the appropriate Hartree-Fock exchange coefficient to reproduce an experimental bandgap

1.5 Results I: Modified BHHLYP Method

1.5.1 Mulliken Charges in Modifying Hartree-Fock Exchange Coefficient

From the viewpoint of quantum computational chemistry, the HF exchange coefficient must be within the closed interval [0.0, 1.0]. In Firefly program package, *the artificial HF exchange coefficient*, which is greater than 1.0, is also possible in modified hybrid-DFT methods.

> ***The idea of varying the HF exchange coefficient in the extended interval is for the purpose of investigating a mathematical behaviour of bandgap variation.***

When the HF exchange coefficient increases, the Mulliken charge of titanium monotonously increases, while that of oxygen monotonously decreases (Fig. 1.3). The Mulliken charge of strontium is almost the same. It shows that localisation effects in titanium and oxygen are enhanced for the large HF exchange coefficient. The formal charges of titanium, oxygen and strontium are +4, −2 and +2, respectively. If the larger HF exchange coefficient is applied in Fml. (1.6), all Mulliken charges are getting closer to the formal charges.

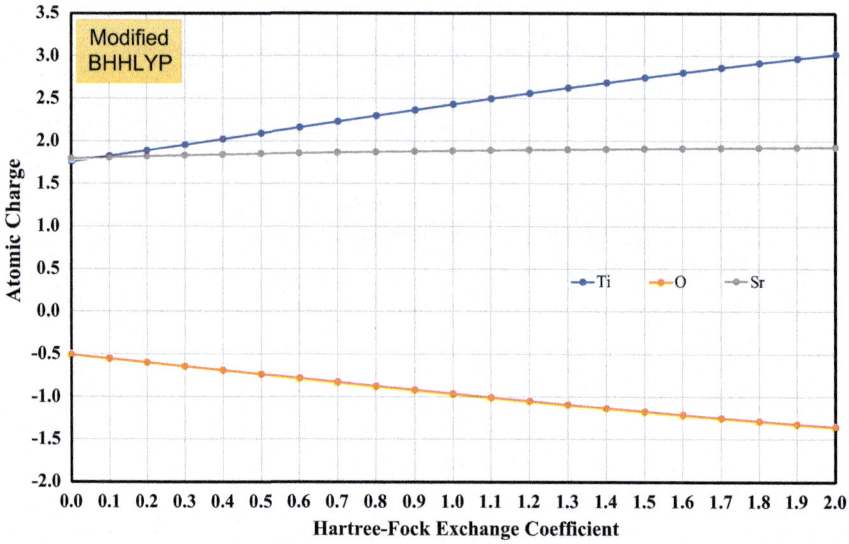

Fig. 1.3 Mulliken charges of titanium, oxygen and strontium, when changing the Hartree-Fock exchange coefficient in the closed interval [0.0, 2.0] ($SrTi_8O_{12}$ model, modified BHHLYP method)

1.5.2 *Orbital Energies in Modifying Hartree-Fock Exchange Coefficient*

Figure 1.4 shows that the orbital energy of HOMO monotonously decreases and that of LUMO monotonously increases, when the HF exchange coefficient increases. A bandgap is evaluated as the orbital energy difference between HOMO and LUMO. As the result, a bandgap monotonously increases as the HF exchange coefficient increases (Table 1.1). It is found that the experimental bandgap of 3.27 eV is reproduced between $C_X^{HF} = 0.3$ and 0.4.

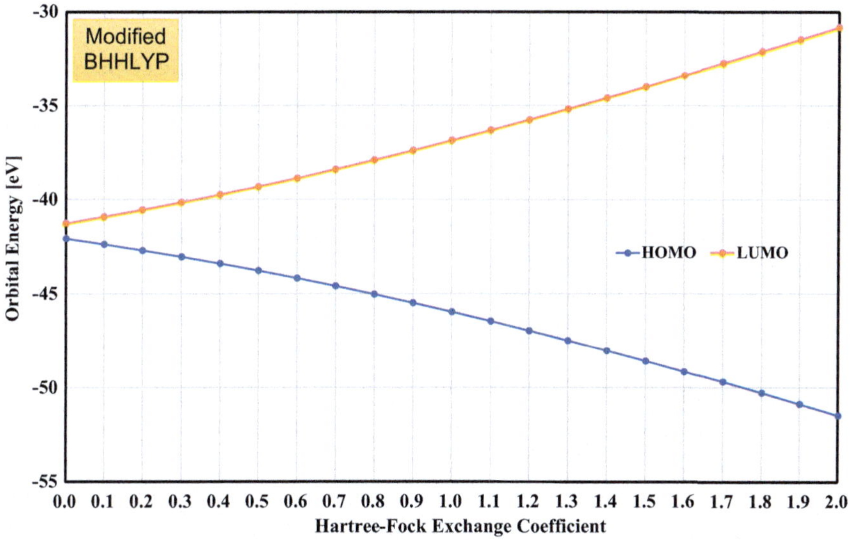

Fig. 1.4 Orbital energies of the highest occupied molecular orbital (HOMO) and the lowest occupied molecular orbital (LUMO), when changing the Hartree-Fock exchange coefficient in the closed interval [0.0, 2.0] ($SrTi_8O_{12}$ model, modified BHHLYP method)

Table 1.1 Bandgaps when changing the Hartree-Fock exchange coefficient in the closed interval [0.0, 2.0] ($SrTi_8O_{12}$ model, modified BHHLYP method)

C_X^{HF}	0.0	0.1	0.2	0.3	0.4	0.5	0.6	0.7	0.8
Bandgap	0.811	1.469	2.161	2.895	3.665	4.476	5.323	6.215	7.146
C_X^{HF}	0.9	1.0	1.1	1.2	1.3	1.4	1.5	1.6	1.7
Bandgap	8.115	9.119	10.158	11.228	12.327	13.453	14.599	15.775	16.966
C_X^{HF}	1.8	1.9	2.0						
Bandgap	18.180	19.415	20.667						

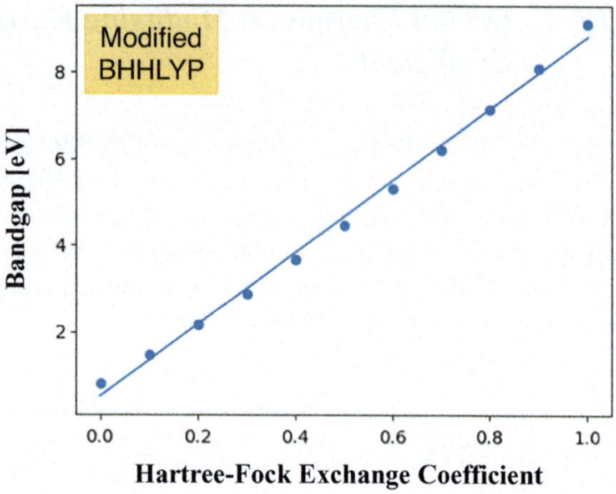

Fig. 1.5 Linear regression function (straight line) and bandgaps (dotted points) calculated by modified BHHLYP method

1.5.3 *Linear Model*

In a linear regression analysis, the discrete points within [0.0, 1.0] are interpolated by the linear function with the form:

$$f_{\text{lin}}^{\text{mod BHHLYP}}(x) = a + bx, \tag{1.7}$$

where the variable x denotes the HF exchange coefficient, and a and b are constant. The returned regression function became

$$f_{\text{lin}}^{\text{mod BHHLYP}}(x) = 0.5187 + 8.307x. \tag{1.8}$$

As the R^2 value was 0.996, it is found that a linear regression model is statistically reasonable. In Fig. 1.5, downward deviations are seen in $C_{\text{X}}^{\text{HF}} < 0.2$ and $C_{\text{X}}^{\text{HF}} > 0.8$, while upward deviations are seen in $0.2 < C_{\text{X}}^{\text{HF}} < 0.8$. The x value to satisfy $f_{\text{lin}}^{\text{mod BHHLYP}}(x) = 3.27$ was 0.331. By performing the quantum chemical calculation by the modified BHHLYP method with $C_{\text{X}}^{\text{HF}} = 0.331$, the bandgap value of 3.13 eV was returned.

1.5.4 *Quadratic Polynomial Model*

To improve upward and downward deviations in a linear model, a quadratic function $f_{\text{qua}}^{\text{mod BHHLYP}}(x)$ was applied. Because a curve connecting discrete points should be convex downward. The form is written as

$$f_{\text{qua}}^{\text{mod BHHLYP}}(x) = a + bx + cx^2, \tag{1.9}$$

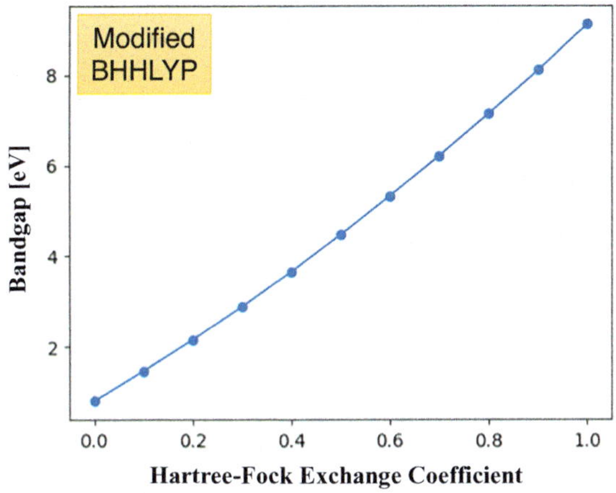

Fig. 1.6 Quadratic polynomial regression function (curve) and bandgaps (dotted points) calculated by modified BHHLYP method

where the variable x denotes the HF exchange coefficient, and a, b and c are constant. The returned regression function became

$$f_{\mathrm{qua}}^{\mathrm{mod\ BHHLYP}}(x) = 0.8130 + 6.346x + 1.961x^2. \tag{1.10}$$

The R^2 value was improved (1.000), in comparison with a linear model (0.996). It is found that Fml. (1.10) fits discrete points well. In Fig. 1.6, you can see that a convex downward function is reproduced. The x value to satisfy $f_{\mathrm{qua}}^{\mathrm{mod\ BHHLYP}}(x) = 3.27$ was 0.349. By performing a quantum chemical calculation by the modified BHHLYP method with $C_{\mathrm{X}}^{\mathrm{HF}} = 0.349$, the bandgap value of 3.27 eV was returned. In a quadratic polynomial model, the experimental bandgap was reproduced better than a linear model.

1.5.5 Exponential Model

The exponential regression function with the form:

$$f_{\mathrm{exp}}^{\mathrm{mod\ BHHLYP}}(x) = ke^{ax}, \tag{1.11}$$

where the variable x denotes the HF exchange coefficient, and k and a are constant, was applied. The returned exponential function became

$$f_{\mathrm{exp}}^{\mathrm{mod\ BHHLYP}}(x) = 1.734e^{-1.719x}. \tag{1.12}$$

Although the convex downward property was reproduced in Fig. 1.7, the R^2 value (0.970) was smaller than that of a quadratic polynomial model.

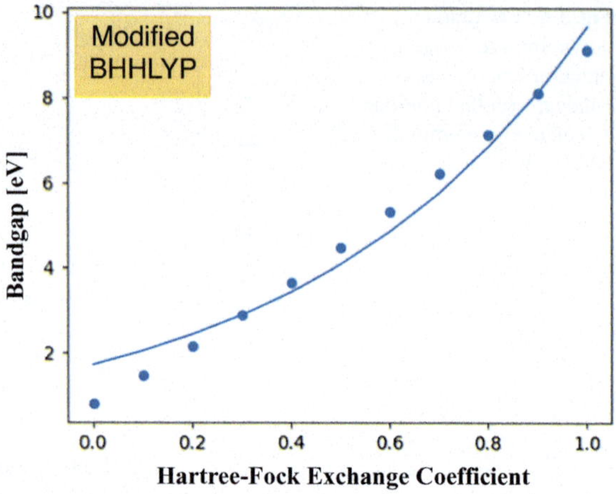

Fig. 1.7 Exponential regression function (curve) and bandgaps (dotted points) calculated by modified BHHLYP method

1.5.6 Artificial HF Exchange Coefficient

In this subsection, the mathematical behaviour in the artificial region of the HF exchange coefficient [1.0, 2.0] is compared with the actual region in quantum computational chemistry [0.0, 1.0]. We discuss whether the inclusion of the artificial region enhances the precision of a statistical optimisation or not.

Including the artificial region, a regression analysis was performed using the same quadratic polynomial regression function: Fml. (1.9). The returned regression function became

$$f_{\text{qua2.0}}^{\text{mod BHHLYP}}(x) = 0.7368 + 6.767x + 1.624x^2. \tag{1.13}$$

The R^2 value was 0.992. In comparison with that of $f_{\text{qua}}^{\text{mod BHHLYP}}(x)$ ($R^2 = 1.000$), the regression precision was degraded.

Table 1.2 shows the energy difference between an interpolated bandgap and a bandgap calculated by a modified BHHLYP method at a specific x value. Note that an interpolated bandgap is evaluated through substituting x value for $f_{\text{qua}}^{\text{mod BHHLYP}}(x)$ or $f_{\text{qua2.0}}^{\text{mod BHHLYP}}(x)$. In $f_{\text{qua}}^{\text{mod BHHLYP}}(x)$, there is only the third decimal misalignment within [0.0, 1.1], while the first decimal or second decimal misalignment is seen over $x > 1.1$. In $f_{\text{qua2.0}}^{\text{mod BHHLYP}}(x)$, though the first decimal misalignment disappeared at all x values, the second decimal misalignment spread (see also Fig. 1.8).

Table 1.2 Energy difference between an interpolated bandgap and a bandgap calculated by modified BHHLYP method at a specific x value. Note that an interpolated bandgap is evaluated through substituting x value for $f_{\text{qua}}^{\text{mod BHHLYP}}(x)$ or $f_{\text{qua2.0}}^{\text{mod BHHLYP}}(x)$

$x\left(C_X^{\text{HF}}\right)$	0.0	0.1	0.2	0.3	0.4	0.5	0.6	0.7	0.8
$f_{\text{qua}}^{\text{mod BHHLYP}}(x)$	0.002	0.002	0.000	0.002	0.000	0.000	0.004	0.001	0.001
$f_{\text{qua2.0}}^{\text{mod BHHLYP}}(x)$	0.074	0.040	0.006	0.017	0.038	0.049	0.058	0.053	0.043
$x\left(C_X^{\text{HF}}\right)$	0.9	1.0	1.1	1.2	1.3	1.4	1.5	1.6	1.7
$f_{\text{qua}}^{\text{mod BHHLYP}}(x)$	0.002	0.001	0.008	0.024	0.049	0.087	0.144	0.211	0.301
$f_{\text{qua2.0}}^{\text{mod BHHLYP}}(x)$	0.027	0.008	0.014	0.033	0.050	0.061	0.060	0.055	0.034
$x\left(C_X^{\text{HF}}\right)$	1.8	1.9	2.0						
$f_{\text{qua}}^{\text{mod BHHLYP}}(x)$	0.408	0.533	0.680						
$f_{\text{qua2.0}}^{\text{mod BHHLYP}}(x)$	0.003	0.039	0.097						

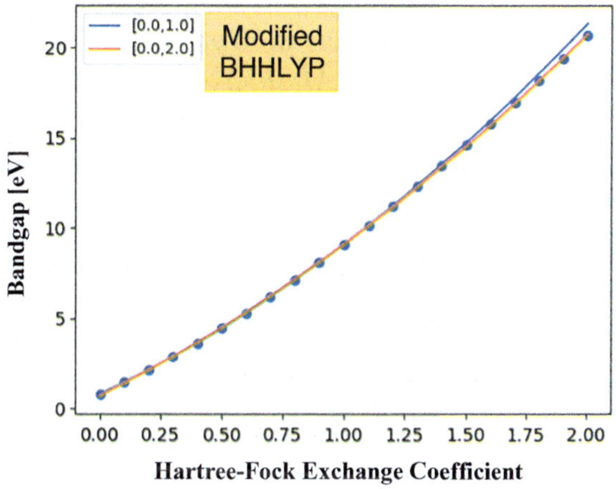

Fig. 1.8 Quadratic polynomial regression functions (curves) and bandgaps (dotted points) calculated by modified BHHLYP methods. Blue and orange lines are for quadratic polynomial regression functions interpolated in [0.0, 1.0] and [0.0, 2.0]

1.6 Results II: Modified B3LYP and PBE1PW91 Methods

1.6.1 Exchange and Correlation Energies in Modified B3LYP and Modified PBE1PW91 Methods

The exchange and correlation energies of B3LYP method [13, 18] are expressed in Fml. (1.4), while those of PBE1PW91 [19–21] method is given by

$$\begin{aligned} E_{\text{XC}}^{\text{PBE1PW91}} &= 0.25E_{\text{X}}^{\text{HF}} + 0.75E_{\text{X}}^{\text{PBE 1996}} \\ &+ E_{\text{C}}^{\text{Perdew 1991 nonlocal}} + E_{\text{C}}^{\text{Perdew-Wang 1991 LDA local}}, \end{aligned} \quad (1.14)$$

where PBE is the abbreviation of Perdew-Burke-Ernzefhof. Note that different exchange and correlation functionals are used in PBE1PW91 method. The exchange and correlation energies of modified B3LYP and modified PBE1PW91 methods are expressed as

$$E_{\mathrm{XC}}^{\mathrm{mod\ B3LYP}} = C_{\mathrm{X}}^{\mathrm{HF}} E_{\mathrm{X}}^{\mathrm{HF}} + 0.8E_{\mathrm{X}}^{\mathrm{Slater}} + 0.72E_{\mathrm{X}}^{\mathrm{Becke}} + 0.19E_{\mathrm{C}}^{\mathrm{VWN}} + 0.81E_{\mathrm{C}}^{\mathrm{LYP}} \quad (1.15)$$

and

$$\begin{aligned} E_{\mathrm{XC}}^{\mathrm{mod\ PBE1PW91}} &= C_{\mathrm{X}}^{\mathrm{HF}} E_{\mathrm{X}}^{\mathrm{HF}} + 0.75E_{\mathrm{X}}^{\mathrm{PBE\ 1996}} \\ + E_{\mathrm{C}}^{\mathrm{Perdew\ 1991\ nonlocal}} &+ E_{\mathrm{C}}^{\mathrm{Perdew-Wang\ 1991\ LDA\ local}}, \end{aligned} \quad (1.16)$$

where $0.0 \leq C_{\mathrm{X}}^{\mathrm{HF}} \leq 1.0$, respectively. As same as modified BHHLYP method, the HF exchange coefficient of $C_{\mathrm{XC}}^{\mathrm{HF}}$ is in the closed interval [0.0, 1.0].

1.6.2 Linear Models in Modified B3LYP and Modified PBE1PW91 Methods

In both modified methods, the experimental bandgap (3.27 eV) can be reproduced between $C_{\mathrm{X}}^{\mathrm{HF}} = 0.3$ and 0.4, as same as modified BHHLYP methods (Table 1.3). However, the orbital energy of HOMO at $C_{\mathrm{X}}^{\mathrm{HF}} = 0.0$ is close to that at $C_{\mathrm{X}}^{\mathrm{HF}} = 0.1$. The same is applied to LUMO. Hence, the closed interval [0.1, 1.0] is considered in a regression analysis. The R^2 values became 0.996 and 0.997 for modified B3LYP and modified PBE1PW91 methods, respectively. For modified B3LYP and modified PBE1PW91 methods, the linear regression functions:

$$f_{\mathrm{lin}}^{\mathrm{mod\ B3PLYP}}(x) = 0.3764 + 8.516x \quad (1.17)$$

and

$$f_{\mathrm{lin}}^{\mathrm{mod\ PBE}}(x) = 0.3806 + 8.523x \quad (1.18)$$

were returned, respectively. In Figs. 1.9 and 1.10, deviations are seen. The HF exchange coefficients to satisfy 3.27 eV were 0.340 (modified B3LYP method) and 0.339 (modified PBE1PW91 method). By performing a quantum chemical calculation by the modified B3LYP method with $C_{\mathrm{X}}^{\mathrm{HF}} = 0.340$, the bandgap value of 3.20 eV was returned. For modified PBE1PW91 method with $C_{\mathrm{X}}^{\mathrm{HF}} = 0.339$, it was 3.19 eV.

Table 1.3 Bandgaps when changing the Hartree–Fock exchange coefficient in the closed interval [0.0, 1.0] (modified B3LYP and modified PBE1PW91 methods)

$x\ (C_X^{HF})$	0.0	0.1	0.2	0.3	0.4	0.5	0.6	0.7	0.8
B3LYP	1.393	1.467	2.161	2.893	3.663	4.474	5.325	6.218	7.151
PBE1PW91	1.410	1.469	2.163	2.895	3.665	4.479	5.333	6.226	7.159
$x\ (C_X^{HF})$	0.9	1.0							
B3LYP	8.120	9.132							
PBE1PW91	8.134	9.143							

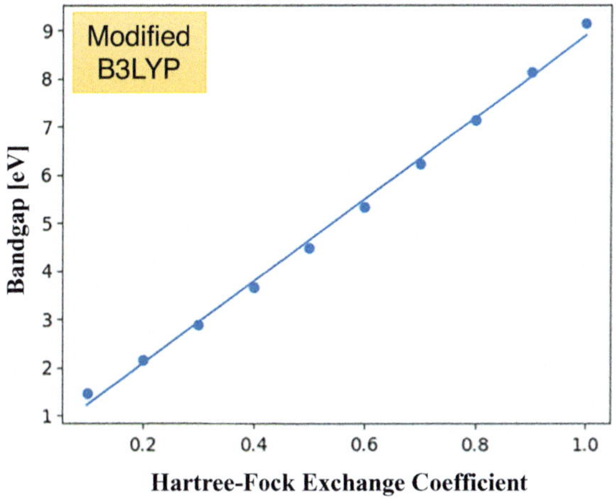

Fig. 1.9 Linear regression function (curve) and bandgaps (dotted points) calculated by modified B3LYP method

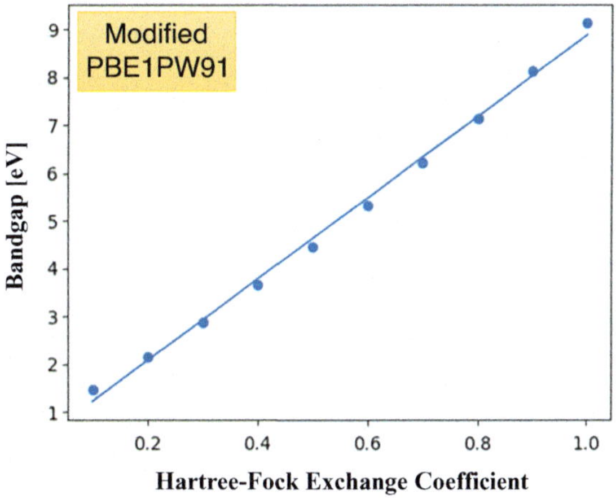

Fig. 1.10 Linear regression function (curve) and bandgaps (dotted points) calculated by modified PBE1PW91 method

1.6.3 Quadratic Polynomial Models in Modified B3LYP and Modified PBE1PW91 Methods

To represent convex downward curves, a quadratic polynomial regression using the same form as Fml. (1.9) was applied for both methods. For modified B3LYP and modified PBE1PW91 methods, the regression functions:

$$f_{\mathrm{qua}}^{\mathrm{mod\ B3PLYP}}(x) = 0.816 + 6.319x + 1.999x^2 \tag{1.19}$$

and

$$f_{\mathrm{qua}}^{\mathrm{mod\ PBE}}(x) = 0.818 + 6.323x + 2.004x^2 \tag{1.20}$$

were returned, respectively. As the R^2 values are both 1.000, it is found that the regression functions are statistically reasonable. The HF exchange coefficients to satisfy 3.27 eV were 0.350 (modified B3LYP method) and 0.349 (modified PBE1PW91 method). Quantum chemical calculations using $C_{\mathrm{X}}^{\mathrm{HF}} = 0.350$ in modified B3LYP method and $C_{\mathrm{X}}^{\mathrm{HF}} = 0.349$ in modified PBE1PW91 method reproduced the experimental bandgap of 3.27 eV.

As shown in Fig. 1.11, the curves of three quadratic polynomial regression functions are almost overlapped. It is found that the HF exchange coefficient is only the parameter to vary the bandgap at the level of rounding to two decimal places.

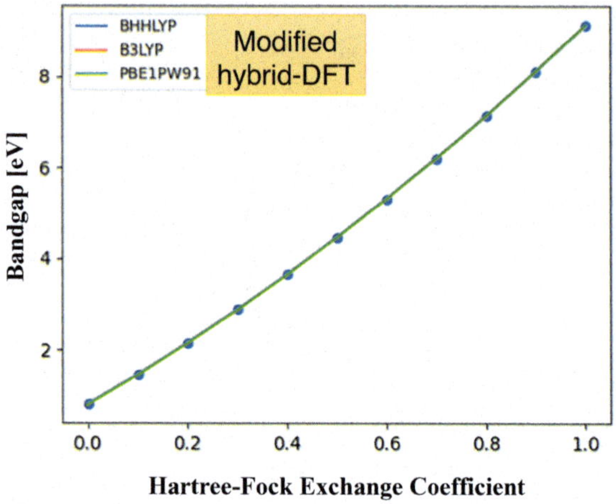

Fig. 1.11 Quadratic polynomial regression functions (curves) and bandgaps (dotted points) calculated by modified hybrid-DFT methods: modified BHHLYP, modified B3LYP and modified PBE1PW91 methods

1.7 Conclusions

In quantum chemical calculations, the experimental bandgap of $SrTiO_3$ perovskite (3.27 eV) was reproduced by modifying the HF exchange coefficient in hybrid-DFT method. The appropriate HF exchange coefficient was statistically optimised with supervised machine learning. In a regression analysis, the scatter plot of HF exchange coefficient versus bandgap calculated by modified hybrid-DFT method was interpolated by regression functions. The candidate regression functions was a linear function, a quadratic polynomial function or an exponential function. From the point of the coefficient of determination (R^2), it was found that a quadratic polynomial function is a statistically appropriate for a regression function. By performing quantum chemical calculations using statistically optimised HF exchange coefficients in modified hybrid-DFT methods, the experimental bandgap was well reproduced.

1.8 Outlook

1.8.1 Road to Generalised Hybrid-DFT Method

In the framework of hybrid-DFT, the exchange and correlation interactions are separated into multiple terms. The contribution of exchange and correlation interactions are changeable, depending on the sorts of molecules or clusters. Quantum interactions between different electrons affect the contribution. For a molecular system consisting s, p and d electrons, the patterns of quantum interactions are below.

Quantum Interaction Patterns

1. **s orbital–s orbital**
2. **s orbital–p orbital**
3. **s orbital–d orbital**
4. **p orbital–d orbital**
5. **d orbital–d orbital**.

In the generalised hybrid-DFT method, ***the combination of exchange and correlation functionals and the corresponding coefficients are determined from a system information***. Basic information is for structural data and sorts of atoms. Further, not only quantum interaction patterns but also a number of each pattern affects the combination of exchange and correlation functionals. Fig. 1.12 shows the schematic of the generalised hybrid-DFT method. In principle, the appropriate combination of exchange and correlation functionals can be fixed before performing a quantum chemical calculation.

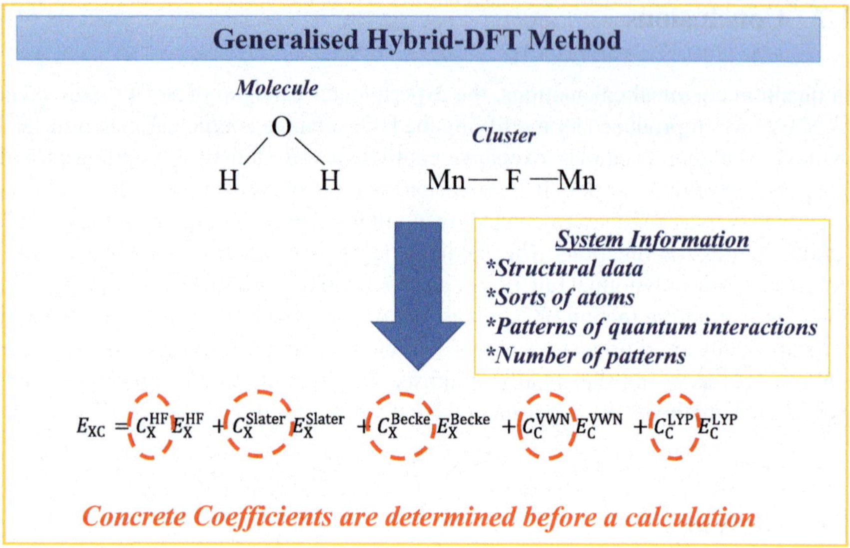

Fig. 1.12 Schematic of the generalised hybrid-DFT method

1.8.2 *Recent Trends of Neural Network Hybrid-DFT Method*

Recently machine learning techniques have explored to determine the combination of exchange and correlation functionals for a specific molecule. As an advanced machine learning technique, deep learning using a neural network is getting popular for the same purpose [22–26]. An input layer and an output layer are connected with intermediate layers. Multiple input and output data are treated as an input layer and an output layer, respectively. To enhance prediction accuracy, it is considered that multiple intermediate layers are favourable in a neural network. The relations between layers are described as transformations. The hybrid-DFT method using deep learning-predicted exchange and correlation functionals is called *neural network hybrid-DFT method*. In quantum chemical calculations, coupled cluster (CC) method is often used as reference data in deep learning. However, CC method often predicts wrong electronic states in strongly correlated systems. Pay attention that the incorrect reference data cause the wrong prediction.

1.8.3 *Machine Learning in Information Engineering: Basic Tool in Artificial Intelligence (AI) Era*

In general, machine learning is categorised in a research area of engineering: information engineering, communication engineering and so on.

As statistical techniques are applied in machine learning, pay attention that scientific phenomena are not substantiated but just predicted.

From the point of solving a mere mathematical problem, formulae and equations employed in machine learning contain no scientific interest. Looking at them from a different perspective, hints to solve scientific problems are hidden in formulae and equations.

In the artificial intelligence (AI) era, researchers must acquire machine learning technique as a piece of basic knowledge. Active algorithm development is much expected.

Acknowledgements This work was partially supported by the Research Council of Norway through its Centres of Excellence scheme, project number 262695.

References

1. Callaway E. DEVELOPERS OF ALPHAFOLD WIN CHEMISTRY NOBEL the prize celebrates artificial-intelligence tools that could revolutionize drug discovery. Nature. 2024;634:525–6.
2. Gibney E, Castelvecchi D. SCOOPED BY MACHINELEARNING PIONEERS John Hopfield's and Geoffrey Hinton's research enabled the development of neural networks. Nature. 2024;634:523–4.
3. Jumper J, Evans R, Pritzel A, Green T, Figurnov M, Ronneberger O, Tunyasuvunakool K, Bates R, Žídek A, Potapenko A, Bridgland A, Meyer C, Kohl SAA, Ballard AJ, Cowie A, Romera-Paredes B, Nikolov S, Jain R, Adler J, Back T, Petersen S, Reiman D, Clancy E, Zielinski E, Steinegger M, Pacholska M, Berghammer T, Bodenstein S, Silver D, Vinyals O, Senior AW, Kavukcuoglu K, Kohli P, Hassabis D. Highly accurate protein structure prediction with AlphaFold. Nature. 2021;596:583–92.
4. Baek M, DiMaio F, Anishchenko I, Dauparas J, Ovchinnikov S, Lee GR, Wang J, Cong Q, Kinch LN, Schaeffer RD, Millán C, Park H, Adams C, Glassman CR, DeGiovanni A, Pereira JH, Rodrigues AV, van Dijk AA, Ebrecht AC, Opperman DJ, Sagmeister T, Buhlheller C, Pavkov-Keller T, Rathinaswamy MK, Dalwadi U, Yip CK, Burke JE, Garcia KC, Grishin NV, Adams PD, Read RJ, Baker D. Accurate prediction of protein structures and interactions using a three-track neural network. Science. 2021;373:871–6.
5. Zhao Y, Truhlar DJ. Density Functionals with broad applicability in chemistry. Accounts of Chem Res. 2008;41:157–67.
6. Onishi T. Quantum computational chemistry - modelling and calculation for functional materials. Springer Nature; 2018.
7. Onishi T. Quantum chemistry in functional inorganic materials. Adv Quant Chem. 2012;64:31–81.
8. Onishi T. The hybrid-DFT study on bandgap estimation for the perovskite-type titanium oxide of $SrTiO_3$. Int J Quant, Chem. 2008;108:2856–61.
9. Kan D, Kanda R, Kanemitsu Y, Shimakawa Y, Takano M, Terashima T, Ishizumi A. Blue luminescence from electron-doped $SrTiO_3$. Appl Phys Lett. 2006;88:191916.
10. Capizzi M, Frova A. Phys Rev Lett. 1970;25:1298–302.
11. Huzinaga S, Andzelm J, Klobukowski M, Radzio-Andzelm E, Sakai Y, Tatewaki H. Gaussian basis sets for molecular calculations. Elsevier; 1984.
12. Becke AD. Chem Phys. 1993;98:1372–7.
13. Vosko SH, Wilk L, Nusair M. Can J Phys. 1980;58:1200–11.

14. Lee C, Yang W, Parr RG. Phys Rev B. 1988;37:785–9.
15. Granovsky AA. Firefly version 8. http://classic.chem.msu.su/gran/firefly/sssindex.html.
16. Meurer A, Smith CP, Paprocki M, Čertík O, Kirpichev SB, Rocklin M, Kumar A, Ivanov S, Moore JK, Singh S, Rathnayake T, Vig S, Granger BE, Muller RP, Bonazzi F, Gupta H, Vats S, Johansson F, Pedregosa F, Curry MJ, Terrel AR, Roučka Š, Saboo A, Fernando I, Kulal S, Cimrman R, Scopatz A. SymPy: symbolic computing in python. Peer J Computer Science. 2017;3:e103.
17. Python 3. https://www.python.org/.
18. Becke AD. J Chem Phys. 1993;98:5648–52.
19. Perdew JP, Wang Y. Phys Rev B. 1992;45:13244.
20. Perdew JP, Chevary JA, Vosko SH, Jackson KA, Pederson MR, Singh DJ, Fiolhais C. Phys Rev B. 1992;46:6671.
21. Adamo C, Barone V. J Chem Phys. 1999;110:6158.
22. Ryczko K, Strubbe DA, Tamblyn I. Deep learning and density-functional theory. Phys Rev A. 2019;100:022512.
23. Kasim MF, Vinko SM. Learning the exchange-correlation functional from nature with fully differentiable density functional theory. Phys Rev Lett. 2021;127:126403.
24. Dick S, Fernandez-Serra M. Machine learning accurate exchange and correlation functionals of the electronic density. Nat Commun. 2020;11:3509.
25. Bogojeski M, Vogt-Maranto L, Tuckerman ME, Müller K-R, Burke K. Quantum chemical accuracy from density functional approximations via machine learning. Nat Commun. 2020;11:5223.
26. Nagai R, Akashi R, Sugino O. Machine-learning-based exchange correlation functional with physical asymptotic constraints. Phys Rev Res. 2022;4:013106.

Chapter 2
Molecular Curvature for Polyethylene Molecules: Mathematical Modelling by Python Programming

Taku Onishi

Abstract For polyethylene molecules, a straight chain structure is intuitively defined from the point of a structural formula in chemistry. If the Cartesian coordinates are used to represent positions of constitute atoms, the connecting line of atomic positions does not correspond to a straight line. In a mathematical sense, it is recognised as a nonlinear molecule. In order to be able to express molecular straightness of a structural formula with a mathematical function, the new physical quantity called *molecular curvature* has been proposed. After separating a molecule into multiple groups, centres of mass (CMs) of groups are used for evaluating molecular curvature. Atomic positions are obtained through geometry optimisation by quantum chemical calculations using hybrid-density functional theory (DFT) method. A regression function, which interpolates CMs, is statistically determined. The formula of molecular curvature can be derived from a regression function. If a local maximum exists in the curve of molecular curvature, it is defined that there exists a bend point. Even if a polynomial function is statistically reasonable as a regression function, the number of bend points is restricted. To solve the problem, a periodic trigonometric function is applied. To avoid the overshooting at terminal groups, in the application of spline interpolation, unnecessary local maxima are removed.

Keywords Molecular structure · Structural formula · Polyethylene molecule · Hybrid-Density Functional Theory (DFT) · Regression function · Centre of mass · Molecular curvature · Straight chain · Molecular bend · Python

---

T. Onishi (✉)
Graduate School of Engineering, Mie University, Tsu, Japan
e-mail: onishi.taku@mie-u.ac.jp; taku.onishi@kjemi.uio.no

Department of Chemistry, University of Oslo, Oslo, Norway

T. Onishi (ed.), *Advances in Quantum Science II*,
https://doi.org/10.1007/978-981-95-9721-5_2

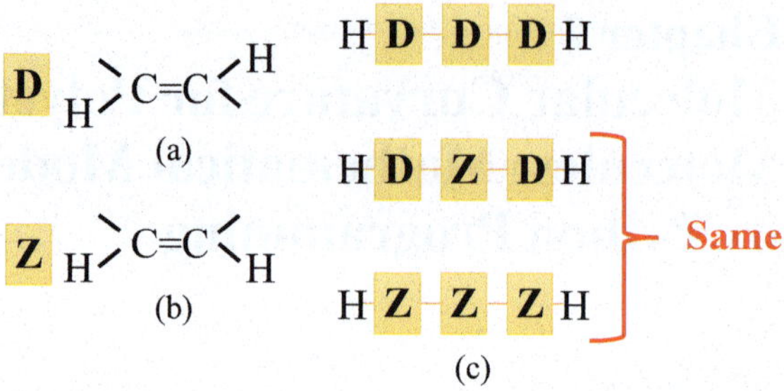

Fig. 2.1 Two types of repeating units in planar polyethylene molecules: **a** diagonal type (D), **b** zigzag type (Z). **c** Selected possible combinations of repeating units for a polyethylene molecule ($n = 3$)

2.1 Introduction

2.1.1 Structures of Polyethylene Molecules

Polyethylene molecules consist of multiple repeating units. For a polyethylene molecule with n repeating units, the molecular structure is written as $H(HC = CH)_nH$. At the left and right terminal repeating units, extra hydrogen atoms are added. Under consideration of bonding orientation, the repeating unit is precisely distinguished into two sorts: D and Z units (Fig. 2.1). If polyethylene is formed only by D unit, the molecular structure is called *straight chain* in chemistry.

For $n = 2$: H-D-Z-H, H-D-D-Z, H-Z-Z-H, all molecular structures become straight chain. Note that D unit must be rotated for H-Z-D-H formation. For $n > 2$, positional flexibility appears. For example, for $n = 3$, there are two types of molecular structures. H-D-D-D-H forms a straight chain, while H-D-Z-D-H is bent. As extra hydrogen atoms are added to left and right terminal repeating units, H-D-Z-D-H corresponds to H-Z-Z-Z-H. In the same reason, H-Z-D-Z-H corresponds to H-D-D-D-H, and H-Z-Z-D-H corresponds to H-Z-Z-Z-H.

2.1.2 Overview of Molecular Curvature

For the purpose of introducing a mathematical concept of *curvature* to a structural formula in chemistry, *molecular curvature* has been proposed. In this chapter, we consider molecules such that all atomic positions can be written in two-dimensional (2D) Cartesian coordinates. The process how to evaluate molecular curvature is as follows:

<u>**How to evaluate molecular curvature**</u>

1. Separate a molecule into groups such that each intermediate group can have a bend point potentially. The centre of mass of each group is utilised as positional coordinates. See a bend point in the next section.

2. Quantum chemical calculations based on molecular orbital theory are performed to evaluate an optimised molecular structure.
3. A mathematical function to interpolate all centres of mass is determined in a regression analysis. As a regression function, a polynomial function or a trigonometric function is used.
4. Using the regression function, derive a formula of molecular curvature for a molecule.

2.2 Calculation Procedures

2.2.1 Grouping

Let us divide a polyethylene molecule into groups. The molecule consists of *n* repeating units and two hydrogen atoms bound with left and right bare single bonds. The left repeating unit bound with an additional hydrogen atom (H-D or H-Z) is the first group. In the intermediate part, each group consists of D or Z. The numbering of groups is performed from the left. The right repeating unit bound with an additional hydrogen atom (D-H or Z-H) is the last group. If the number of repeating units is *n*, the last group becomes the *n*-th group.

Grouping

First group: H-D or H-Z.

Intermediate groups (2nd group, 3rd group, $\cdots$): D or Z.

Last group (*n*-th group): D-H or Z-H.

2.2.2 Molecular Bend in Polyethylene Molecules

In this chapter, we consider planer polyethylene molecules as calculation models. The positions of constituent atoms can be expressed in 2D Cartesian coordinates. For $H(HC=CH)_3H$ ($n = 3$), if three groups are diagonally (in a cis direction) connected, we say that there is *no bend point*. On the other hand, two terminal groups are, in a trans direction, connected with an intermediate group, we say that there is *a bend point* at an intermediate group. For $H(HC=CH)_5H$ ($n = 5$), four types of molecular structures exist. In Fig. 2.2a, as all groups are connected in a cis direction, there is no bend point. In Fig. 2.2b, as the second and fourth groups (–HC3 = C4H– and –HC7 = C8H–) are connected with the third group (–HC5 = C6H–) in a trans direction, there is a bend point at the third group (–HC5 = C6H–). In Fig. 2.2c, there are two bend points at the second and fourth groups (–HC3 = C4H–, –HC7 = C8H–). In

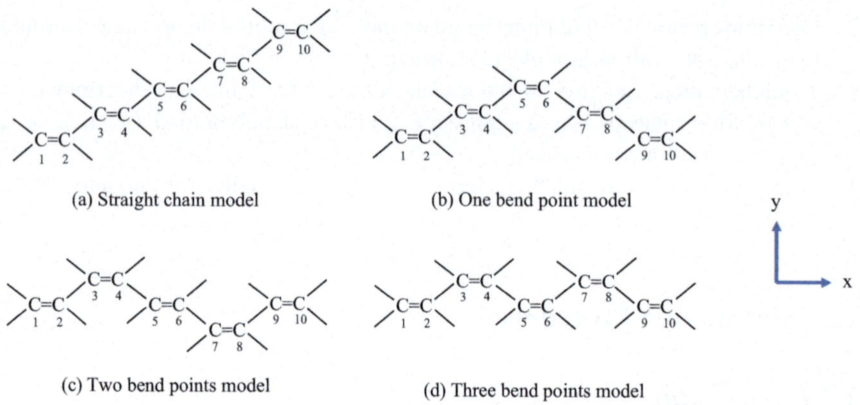

Fig. 2.2 Structural formulae of a planar polyethylene molecule ($n = 5$: $C_{10}H_{12}$). Extra hydrogen atoms are added at bare single bonds. The x and y axes are defined as indicated. The origin is defined for each model

Fig. 2.2d, there are three bend points at the second, third and fourth groups (–HC3 = C4H–, –HC5 = C6H– and –HC7 = C8H–).

2.2.3 *Quantum Chemical Calculations*

To obtain an optimised molecular structure of a planar polyethylene molecule, quantum chemical calculations are performed with Firefly program package [1]. The calculation method employed here is B3LYP method [2, 3], which is one of hybrid-density functional theory (DFT) methods, in combination with 6-31G* basis set. Geometry optimised molecular structures are obtained for four types of molecular conformations (Fig. 2.2a–d).

2.2.4 *Centre of Mas*

In chemistry, the definition of molecular straightness is vogue and intuitive. From the point of a strict mathematical sense, a zigzag structure is formed in a straight chain model (Fig. 2.2a). Taking the molecular volume or electron density into account, the model can be approximately regarded to be straight.

In this chapter, we consider how to treat a straight chain as a linear model in a mathematical sense. As the solution, the centre of mass is introduced to all groups in a polyethylene molecule. If a group consists of n atoms, a centre of mass (CM) of a group: **R** is represented by

$$M\mathbf{R} = \sum_{i=1}^{n} m_i \mathbf{r}_i,$$

where m_i denotes the mass of the i-th atom, $\mathbf{r}_i$ denotes the Cartesian coordinates of the i-th atom and the total mass (M) is given by

$$M = \sum_{i=1}^{n} m_i.$$

2.2.5 *Molecular Curvature in Two-Dimensional (2D) Space*

For a 2D parameterised curve:

$$\boldsymbol{\gamma}(t) = \left[\gamma_1(t), \gamma_2(t)\right],$$

where t is a parameter, *curvature* [4] is defined as the norm of $d^2\boldsymbol{\gamma}(t)/dt^2$:

$$\|\frac{d^2\boldsymbol{\gamma}(x)}{dt^2}\| = \sqrt{\left(\frac{d^2\gamma_1(x)}{dt^2}\right)^2 + \left(\frac{d^2\gamma_2(x)}{dt^2}\right)^2}, \tag{2.1}$$

under the condition such that

$$\|d\boldsymbol{\gamma}(t)/dt\| = 1. \tag{2.2}$$

In the definition of *2D molecular curvature*, the condition (2.2) is removed. If a regression function $f(x)$ can be expressed in a vector form:

$$\boldsymbol{\gamma}(x) = \left[x, f(x)\right],$$

the 2D molecular curvature is defined as

$$\|\frac{d^2\boldsymbol{\gamma}(x)}{dx^2}\| = \sqrt{\left(\frac{d^2 f(x)}{dx^2}\right)^2}. \tag{2.3}$$

2.2.6 Regression Functions

In a regression analysis [5], a regression function connecting discrete points in a 2D space can be statistically determined. The centres of mass (CMs) of groups in a planer polyethylene molecule are used as discrete points. Python 3 [6] is used for describing computation algorithms. In the application of `curve_fit` function in SciPy library [7], a regression function is evaluated.

Polynomial functions and trigonometric functions are applied for regression functions. Let the n-th degree polynomial function be denoted as

$$y = a_n x^n + a_{n-1} x^{n-1} + \cdots + a_2 x^2 + a_1 x + a_0. \tag{2.4}$$

A linear function ($n = 1$) always yields zero molecular curvature. Molecular curvatures for a quadratic polynomial function, a cubic polynomial function and a quartic polynomial function are given by

$$|2a_2|,$$

$$|6a_3 x + 2a_2|$$

and

$$\left|12a_4 x^2 + 6a_3 x + 2a_2\right|,$$

respectively.

For a cosine function:

$$f(x) = N\cos(ax + b),$$

the molecular curvature is given by

$$\sqrt{N^2 a^4 \cos^2(ax + b)}.$$

To assess the validity of a regression function, *the coefficient of determination*, which is denoted as $R^2 (0 \le R^2 \le 1)$, is used. If R^2 is close to 1.0, it is statistically reasonable.

2.3 Molecular Curvatures for Polyethylene Molecules

Cartesian coordinates of CMs are interpolated by both polynomial functions and trigonometric functions. In a planar polyethylene molecule: H(HC = CH)$_5$H (n = 5), molecular curvatures are numerically evaluated for four different molecular structures.

2.3.1 Straight Chain Model

Let us investigate whether the molecular curvature of a straight chain model becomes zero or not. The origin is allocated at the CM of the third group (–HC5 = C6H–). All C = C bonds are parallel to x axis. In a regression analysis, the regression function:

$$y = 0.580x \tag{2.5}$$

was returned. As the R^2 value became 1.000, it is found that the returned regression function is statistically reasonable (see Fig. 2.3). The associated molecular curvature becomes always zero in a defined region. In a straight chain model, if a linear regression function is statistically reasonable, zero molecular curvature is returned in a defined region.

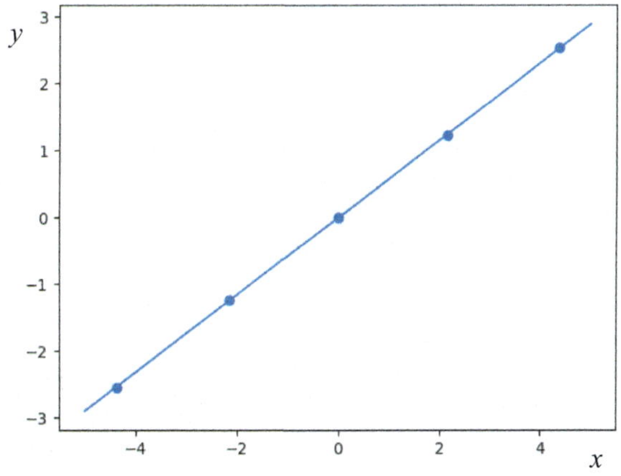

Fig. 2.3 Linear regression function of a straight chain model. Dotted points: centres of mass (CMs)

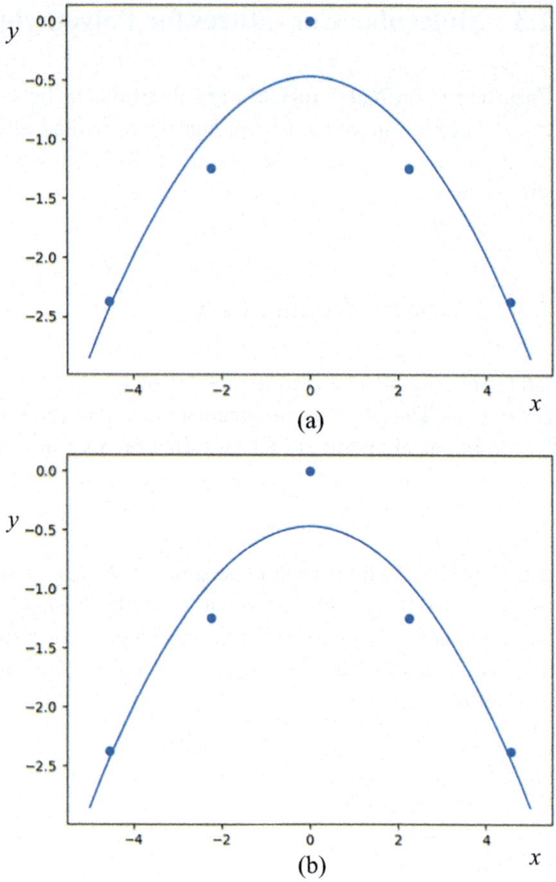

Fig. 2.4 Regression functions of one bend point model with **a** a quadratic polynomial function, **b** a cubic polynomial function. Dotted points: Centres of mass (CMs)

2.3.2 *One Bend Point Model*

Polynomial Regression Functions for One Bend Point Model

In one bend point model, a chain is bent at the third group (–HC5 = C6H–). As same as a straight chain model, the origin is allocated at the CM of the third group (–HC5 = C6H–). The quadratic polynomial regression function:

$$y = -0.465 - 0.0952x^2 \tag{2.6}$$

and the cubic polynomial regression function:

$$y = -0.465 - 0.0952x^2 + 0.00000579x^3 \tag{2.7}$$

Table 2.1 Molecular curvatures of a quartic polynomial regression function at one bend point model

x^*	Group 1 –4.554	Group 2 –2.231	Group 3 0.000	Group 4 2.231	Group 5 4.554
$\Vert\frac{d^2y(x)}{dx^2}\Vert$	1.562	0.071	0.586	0.071	1.562

*For each group, x value corresponds to x coordinate of centre of mass (CM)

were returned in a regression analysis. The cubic polynomial regression function (2.7) approximately corresponds to the quadratic polynomial regression function (2.6), due to approximate zero of a_3 in Fml. (2.4). In both regression functions, there are large deviations from CMs (Fig. 2.4a, b). Further, the same R^2 value (0.893) was returned for both regression functions. As the values deviate from 1.00, it is found that the quadratic and cubic polynomial regression functions are inappropriate for evaluating molecular curvature.

When a quartic polynomial function was used in a regression analysis, the shape of the regression function varied (Fig. 2.5). The returned quartic polynomial regression function became

$$y = -0.293x^2 + 0.00863x^4. \tag{2.8}$$

The R^2 value of 1.00 was returned. Hence, it is found that the regression function is statistically reasonable. The associated molecular curvature is given by

$$\Vert\frac{d^2y(x)}{dx^2}\Vert = \left|0.104x^2 - 0.586\right|. \tag{2.9}$$

In Table 2.1, the molecular curvature increases from the second to third CMs, while it decreases from the third to fourth CMs. There is a local maximum at the third group (–HC5 = C6H–) (Fig. 2.6).

A bend point can be represented by a local maximum of molecular curvature.

From Fml. (2.9), the x coordinate of the bend point can be evaluated to be 0.000. However, there still exists nonlinearity at both terminal groups (H_2C1 = C2H– and –HC9 = $C10H_2$). This is discussed next (trigonometric regression function).

Trigonometric Regression Function for One Bend Point Model

A trigonometric function is next considered as a regression function. For an easy mathematical operation of a periodic function, the Cartesian coordinates of constitute atoms are translated towards positive y direction (+1.183). The returned regression function became

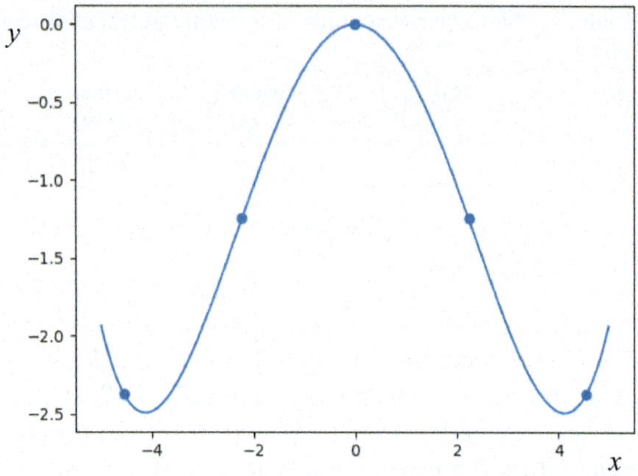

Fig. 2.5 Quartic polynomial regression function of one bend point model. Dotted points: Centres of mass (CMs)

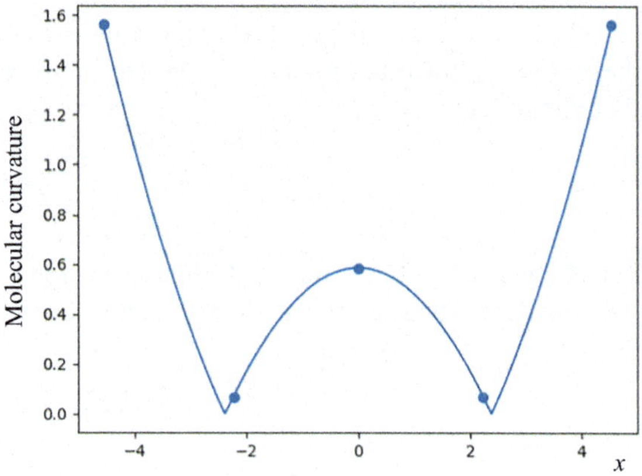

Fig. 2.6 Molecular curvature of one bend point model (a quartic polynomial regression function). Dotted points: Centres of mass (CMs)

$$y = 1.183\cos(0.726x). \tag{2.10}$$

From the obtained R^2 value of 0.9998, it is found that the regression function (2.10) is statistically reasonable. In comparison with Fig. 2.5, the large bends at both terminal groups ($H_2C1 = C2H$– and –$HC9 = C10H_2$) are reduced (Fig. 2.7).

The associated molecular curvature is given by

$$\left\|\frac{d^2\boldsymbol{\gamma}(x)}{dx^2}\right\| = |0.623\cos(0.726x)|. \tag{2.11}$$

In Table 2.2, the molecular curvature increases from the second to third CMs, while it decreases from the third to fourth CMs. The local maximum is given at $x = 0.000$ (Fig. 2.8). It shows that the molecule is bent at the third group (–$HC5 = C6H$–). The x coordinate of the bend point is 0.000. Though two local maxima exist at both

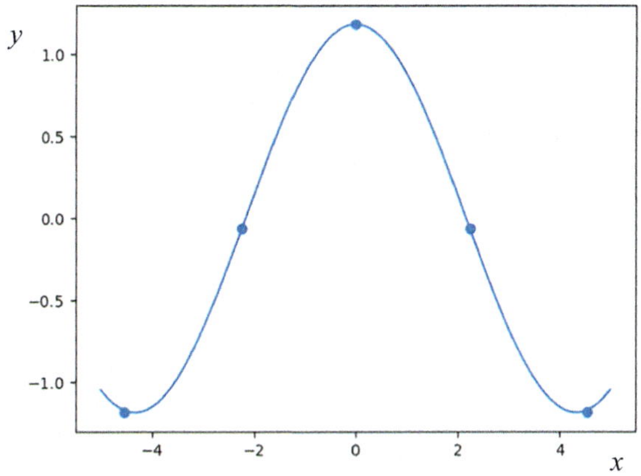

Fig. 2.7 Trigonometric regression function of one bend point model. Dotted points: Centres of mass (CMs)

terminal groups ($H_2C1 = C2H$– and –$HC9 = C10H_2$), there is actually no bend point there. It comes from the overshooting due to application of periodic regression function.

In order to remove the overshooting coming from unnecessary bend points at both terminal groups ($H_2C1 = C2H$– and –$HC9 = C10H_2$), spline interpolation [8, 9] is applied. A linear regression function:

Table 2.2 Molecular curvatures of a trigonometric regression function for one bend point model

x^*	Group 1 –4.554	Group 2 –2.231	Group 3 0.000	Group 4 2.231	Group 5 4.554
$\left\| \frac{d^2y(x)}{dx^2} \right\|$	0.615	0.030	0.623	0.030	0.615

*For each group, x value corresponds to x coordinate of centre of mass (CM)

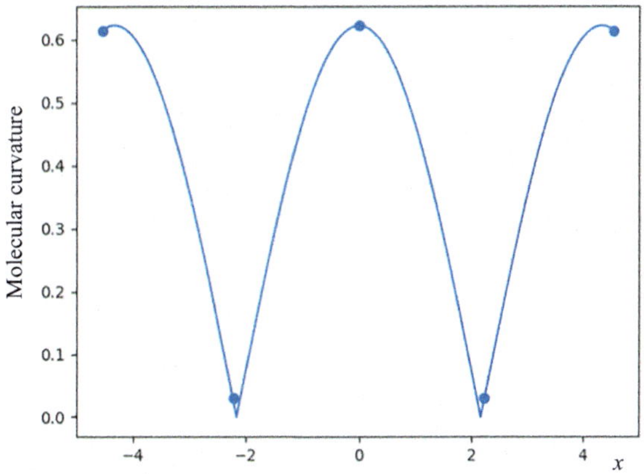

Fig. 2.8 Molecular curvature of one bend point model (a trigonometric regression function). Dotted points: Centres of mass (CMs)

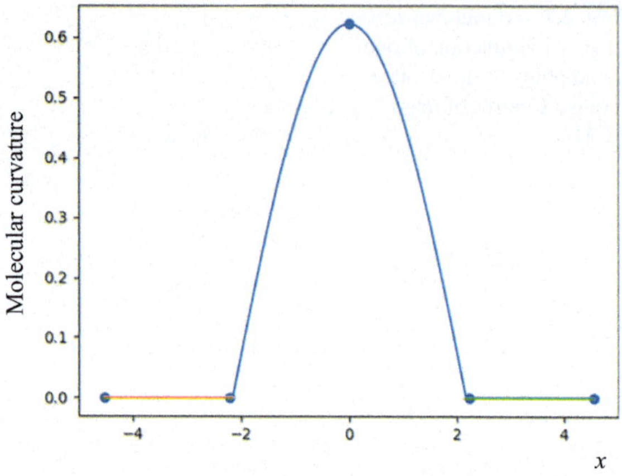

Fig. 2.9 Corrected molecular curvature of one bend point model (after spline interpolation for a trigonometric regression function). Dotted points: Centres of mass (CMs)

$$y = a_1 x + a_0 (x \leq -2.164, x \geq 2.164),$$

where a_1 and a_0 are constant, is applied to satisfy

$$\| \frac{d^2 \gamma(x)}{dx^2} \| = 0 (x \leq -2.164, x \geq 2.164).$$

Note that molecular curvature becomes zero at $x = \pm 2.164$. The corrected molecular curvature is shown in Fig. 2.9.

2.3.3 *Two Bend Points Model*

Polynomial Regression Function for Two Bend Points Model

For the interpolation of the CMs in two bend points model, both a local maximum and a local minimum must be represented in a regression function. At least, the third degree or greater degree polynomial function is required to reproduce them. Applying a cubic polynomial function for a regression function, the cubic polynomial regression function:

$$y = -0.684x + 0.0365x^3 \tag{2.12}$$

was returned. As the R^2 value became 1.000, CMs fit well with the regression function (2.12) (Fig. 2.10).

The associated molecular curvature is given by

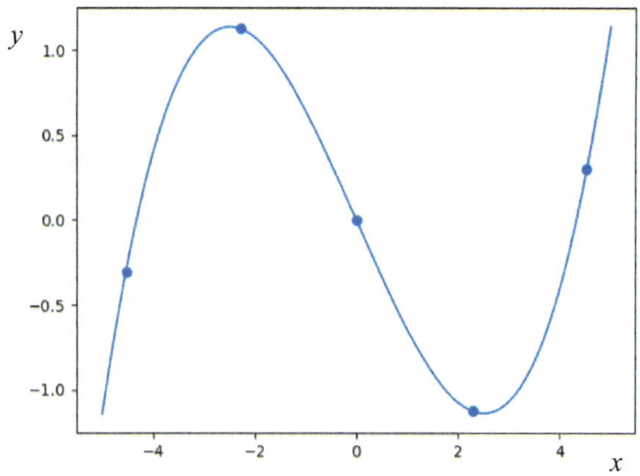

Fig. 2.10 Cubic polynomial regression function of two bend points model. Dotted points: Centres of mass (CMs)

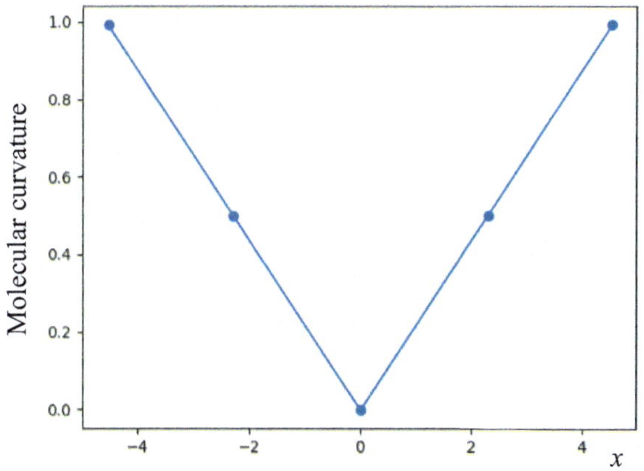

Fig. 2.11 Molecular curvature of two bend points model (a cubic polynomial regression function). Dotted points: Centres of mass (CMs)

$$\|\frac{d^2\boldsymbol{\gamma}(x)}{dx^2}\| = |0.219x|. \tag{2.13}$$

The local maximum is ungiven in Fig. 2.11. Though the use of a cubic polynomial regression function is statistically reasonable for the interpolation of CMs, it is impossible to apply a cubic polynomial regression function for the purpose of evaluating molecular curvature.

Trigonometric Regression Function for Two Bend Points Model

To represent both a local maximum and a local minimum, a trigonometric regression function is applicable for two bend points model. As the third CM is allocated at the origin, the translation of the Cartesian coordinates of constitute atoms is unperformed.

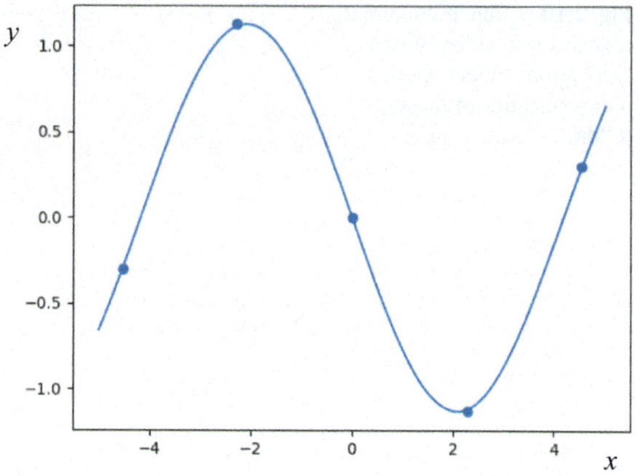

Fig. 2.12 Trigonometric regression function of two bend points model. Dotted points: Centres of mass (CMs)

The returned trigonometric regression function became

$$y = -1.1275 \sin(0.752x). \tag{2.14}$$

It is found that the regression function is statistically reasonable, due to $R^2 = 1.000$ (Fig. 2.12).

The associated molecular curvature is given by

$$\|\frac{d^2y(x)}{dx^2}\| = |0.638 \sin(0.752x)|. \tag{2.15}$$

In Table 2.3, molecular curvature increases from the first to second CMs (from the third to fourth CMs), while it decreases from the second to third CMs (from the fourth to fifth CMs). Figure 2.13 shows the molecular curvature of two bend points model using trigonometric regression function. Two local maxima exist at the second and fourth groups (–HC3 = C4H– and –HC7 = C8H–). It shows that the molecule is bent at the second and fourth groups. The corresponding x coordinates of two bend points are ±2.089. As same as one bend point model using a trigonometric regression function, overshooting is seen at both terminal groups (H_2C1 = C2H– and –HC9 = $C10H_2$).

Table 2.3 Molecular curvatures of a trigonometric regression function for two bend points model

x*	Group 1 –4.535	Group 2 –2.291	Group 3 0.000	Group 4 2.291	Group 5 4.535
$\|\frac{d^2y(x)}{dx^2}\|$	0.170	0.631	0.000	0.631	0.170

*For each group, x value corresponds to the x coordinate of centre of mass

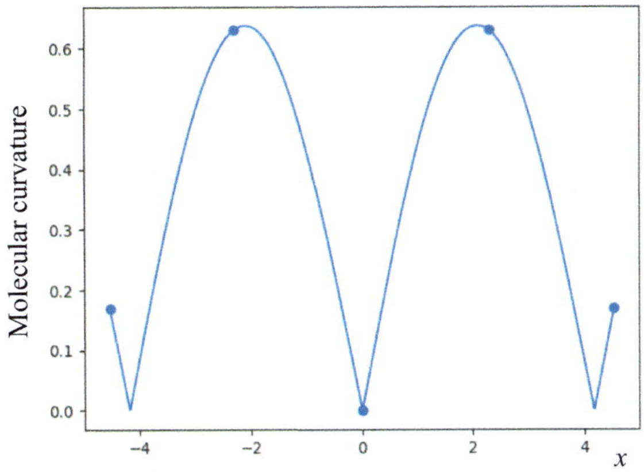

Fig. 2.13 Molecular curvature of two bend points model (a trigonometric regression function). Dotted points: Centres of mass (CMs)

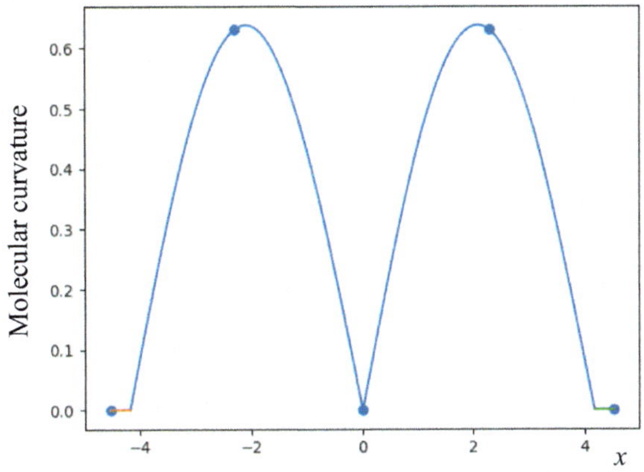

Fig. 2.14 Corrected molecular curvature of two bend points model (after spline interpolation for a trigonometric regression function). Dotted points: Centres of mass (CMs)

To remove the overshooting, spline interpolation was applied. The corrected curve of the molecular curvature is shown in Fig. 2.14.

2.3.4 Three Bend Points Model

To represent three local maxima, a trigonometric regression function is applicable for three bend points model. For an easy mathematical operation of a periodic function, the origin is defined as the point above the third CM (+0.715 in the y direction). The returned trigonometric regression function became

$$y = -0.715 \cos(1.264x). \tag{2.16}$$

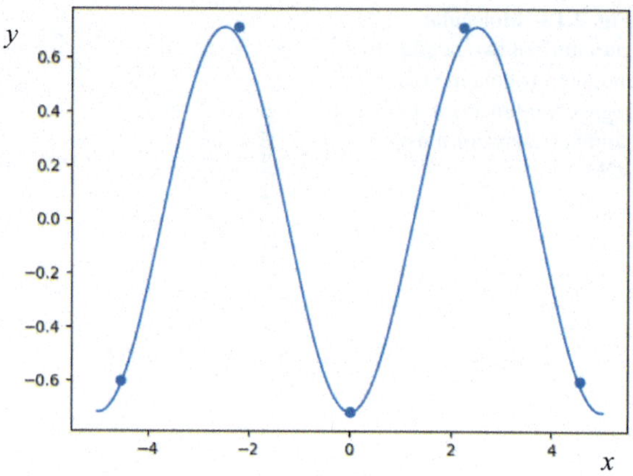

Fig. 2.15 Trigonometric regression function of three bend points model. Dotted points: Centres of mass (CMs)

In Fig. 2.15, two local maxima exist at the second and fourth groups (–HC3 = C4H– and –HC7 = C8H–), while the local minimum exists at the third group (–HC5 = C6H–). Due to the R^2 value of 0.999, it is found that the regression function is statistically reasonable.

The associated molecular curvature is given by

$$\|\frac{d^2\gamma(x)}{dx^2}\| = |1.142\cos(1.264x)|. \tag{2.17}$$

In Fig. 2.16, we can see three local maxima at the second, third and fourth groups (–HC3 = C4H–, –HC5 = C6H– and –HC7 = C8H–). It shows that there are three bend points at the second, third and fourth groups. The x coordinates of three bend points are 0.000, ±2.485. Table 2.4 shows the molecular curvatures at five CMs. Large molecular curvatures are given at the first and fifth CMs. To remove the overshooting at the terminal groups (H_2C1 = C2H– and –HC9 = $C10H_2$), spline interpolation was applied. The corrected curve of the molecular curvature is shown in Fig. 2.17.

2.4 Conclusions

Mathematical modelling is possible in the application of machine learning. In this chapter, mathematical modelling for a structural formula in chemistry is demonstrated.

> ***The new physical quantity molecular curvature, enabled us to express molecular straightness in a mathematical fashion. A bend point, which is expressed in Cartesian coordinates, can be defined.***

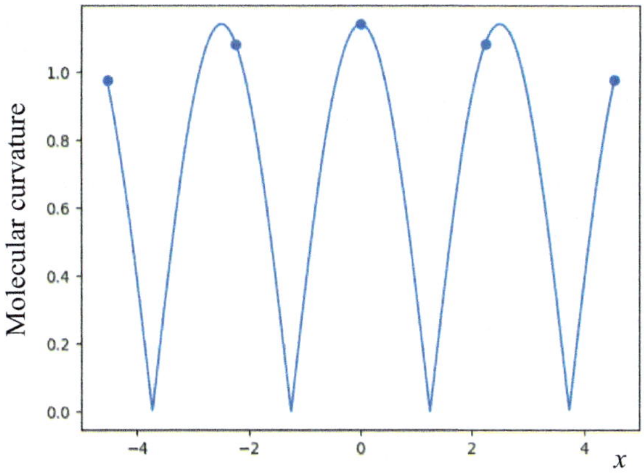

Fig. 2.16 Molecular curvature of three bend points model (a trigonometric function). Dotted points: Centres of mass (CMs)

Table 2.4 Molecular curvatures of a trigonometric regression function for three bend points model

x^*	Group 1 –4.541	Group 2 –2.228	Group 3 0.000	Group 4 2.228	Group 5 4.541
$\lVert \frac{d^2\gamma(x)}{dx^2} \rVert$	0.978	1.082	1.142	1.082	0.978

*For each group, x value corresponds to the x coordinate of centre of mass

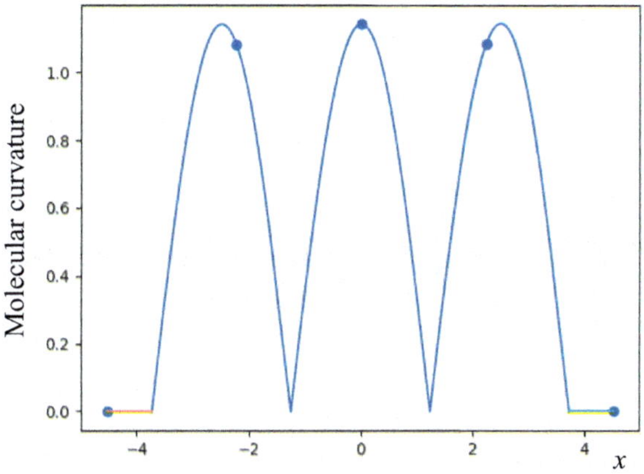

Fig. 2.17 Corrected molecular curvature of three bend points model (after spline interpolation using a trigonometric regression function). Dotted points: Centres of mass (CMs)

Definition 1: Two-Dimensional Molecular Curvature Let a regression function of centres of mass (CMs) for groups in a molecule be denoted as $\boldsymbol{\gamma}(x) = [x, f(x)]$. The associated molecular curvature is given by:

$$\left\|\frac{d^2\gamma(x)}{dx^2}\right\| = \sqrt{\left(\frac{d^2f(x)}{dx^2}\right)^2}.$$

Definition 2: Two-Dimensional Bend Points Let a regression function of centres of mass (CMs) for groups in a molecule be denoted as $\gamma(x) = [x, f(x)]$. The x coordinates of bend points are given as the x value for

$$max\left(\left|\frac{d^2f(x)}{dx^2}\right|\right)$$

in a defined region.

2.4.1 Regression of Centres of Mass

After separating a molecule into multiple groups, centres of mass (CMs) are used to express a molecular position. The effectiveness of the returned regression function is assessed with the coefficient of determination (R^2). If R^2 value ($0.0 \leq R^2 \leq 1.0$) is close to 1.0, it is discriminated to be statistically reasonable. Then, the associated molecular curvature is derived from a regression function.

2.4.2 Sorts of Regression Functions: Polynomial Function, Trigonometric Function

A linear regression function is favourable for straight chain molecules, due to representation of zero molecular curvature in a region. If a polynomial regression function is applied for bent molecules, the number of bend points is restricted. For example, if $n = 2$ and $a_2>0$ in Fml. (2.4), no local maximum is given for a quadratic polynomial regression function. Hence, the number of bend points is zero for such quadratic regression function.

To remove the restriction, a periodic trigonometric function:

$$y = k\sin(ax + b), \tag{2.17}$$

where k, a and b are constant, is used as a regression function. If translated along y axis, the regression function is rewritten as

$$y = k\sin(ax + b) + y_0, \tag{2.18}$$

where y_0 is constant. The molecular curvature of Fml. (2.17) becomes the same with that of Fml. (2.18). It is given by

$$\|\frac{d^2\boldsymbol{\gamma}(x)}{dx^2}\| = \left|ka^2\sin(ax + b)\right|. \tag{2.19}$$

As the result, the translation of the regression function along y axis is negligible in evaluating the associated molecular curvature.

2.4.3 Discrimination Criteria of Molecular Straightness and Bend Point

Molecular straightness and bend points can be discriminated by evaluating molecular curvature in a defined region. If zero molecular curvature is realised in a region, it is discriminated that a region is straight. If local maxima are given in a region, it is discriminated that there exist bend points in a region. The number of local maxima stands for the number of bend points.

Discrimination Criteria

Molecular straightness—zero molecular curvature in a region.

Number of bend points—number of local maxima in a region.

As a special case, if zero molecular curvature is realised by connecting multiple linear regression functions, zero molecular curvature is realised in a region. However, molecular straightness is broken at a connecting point. Basically, zero molecular curvature in a region must be considered for a sole regression function, except for around terminal groups.

The local maximum of a molecular curvature depends on (1) a structural spread in the y direction and (2) bend periodicity (see Fml. (2.19)).

2.4.4 Removal of Overshooting: Spline Interpolation

A boundary trigonometric function suffers from representing molecular straightness in a region. Because nonzero molecular curvature is realised only at a point. Further, even if terminal groups have no bend point, unnecessary local maxima are sometimes given. To overcome the problem, spline interpolation is applied. In spline interpolation, a regression function is generally represented by multiple regression functions.

For molecular curvature, two linear regression functions are applied under or over x values to satisfy

$$\|\frac{d^2\gamma(x)}{dx^2}\| = 0$$

around right and left terminal groups. As the result, no bend point around terminal groups is reflected in molecular curvature. After spline interpolation is performed to molecular curvature, count the number of local maxima.

2.5 Outlook

2.5.1 Effects of Functional Groups on Molecular Curvature

The present molecular curvature can be used for 2D straight chains or 2D bent chains. If functional groups are doped in polyethylene molecules, the corresponding CMs are displaced. If functional groups are written in the three-dimensional (3D) space, the direct usage of the 2D molecular curvature: Fml. (2.3) is impossible. This will pose a challenge.

2.5.2 Three-Dimensional Molecular Curvature

In biological chemistry, protein secondary structure is represented by a three-dimensional (3D) curve. If the concept of secondary structure is applied for polymers, the associated molecular curvature is defined in a 3D space. If a 3D parameterised curve is written as

$$\gamma(x) = \left[x, f(x), g(x)\right],$$

where x is a parameter, *3D molecular curvature* is defined as the norm of $d^2\gamma(x)/dx^2$:

$$\|\frac{d^2\gamma(x)}{dt^2}\| = \sqrt{\left(\frac{d^2f(x)}{dx^2}\right)^2 + \left(\frac{d^2g(x)}{dx^2}\right)^2}.$$

Assessing the applicability of the above formula is in progress.

2.5.3 Artificial Intelligence (AI)-Based Predictions

By the introduction of molecular curvature to a structural formula, it became possible to handle molecular bend as a bend point. Through usage of a bend point for biomacromolecules such as deoxyribonucleic acid (DNA), ribonucleic acid (RNA), enzymes and so on, the statistical processing of a biomolecular structure will become easier and faster. Artificial intelligence (AI)-based predictions using molecular curvature will accelerate the developments of pharmaceuticals. For the purpose, evaluation of molecular curvature for an implicit function is required. For example, a spiral structure of a biomolecule is represented by an implicit function [10].

Acknowledgements This work was partially supported by the Research Council of Norway through its Centres of Excellence scheme, project number 262695.

References

1. Granovsky AA. Firefly version 8, http://classic.chem.msu.su/gran/firefly/index.html.
2. Lee C, Yang W, Parr RG. Development of the Colle-Salvetti correlation-energy formula into a functional of the electron density. Phys Rev B. 1988;37:785–9.
3. Becke AD. Density-functional thermochemistry. III. The role of exact exchange. J Chem Phys. 1993; 98:5648–5652.
4. Pressley A. Elementary differential geometry, 2nd ed. Springer Nature;2000.
5. Raschka S, Liu Y, Mirjalili V. Machine Learning with PyTorch and Scikit-Learn: develop machine learning and deep learning models with Python. Packt Publishing;2022.
6. Python 3, https://www.python.org/.
7. Virtanen P, Gommers R, Oliphant TE, Haberland M, Reddy T, Cournapeau D, Burovski E, Peterson P, Weckesser W, Bright J, van der Walt SJ, Brett M, Wilson J, Millman KJ, Mayorov N, Nelson ARJ, Jones E, Kern R, Larson E, Carey CJ, Polat İ, Feng Y, Moore EW, Plas JV, Laxalde D, Perktold J, Cimrman R, Henriksen I, Quintero EA, Harris CR, Archibald AM, Ribeiro AH, Pedregosa F, van Mulbregt P, and SciPy 1.0 Contributors. SciPy 1.0: fundamental algorithms for scientific computing in python. Nat Methods 2020; 17:261–272.
8. Karpfinger C. Calculus and linear algebra in recipes—terms, phrases and numerous examples in short learning units. Springer Nature;2022.
9. Venkateshan SP, Swaminathan P. Computational methods in engineering. Springer Nature;2023.
10. Onishi T. Mathematical modelling of helical DNA. In: AIP Conference proceedings, in press.

2.5.3 Artificial Intelligence (AI)-Based Predictions

References

Chapter 3
Study on New Quantum Artificial Intelligence Method Predicting Pharmaceutical and Molecular Properties with Machine Learning Models Using Molecular Orbital Energies as Explanatory Variables

Hiroyuki Teramae

Abstract We have proposed new machine learning methods to predict physical properties of chemicals such as pharmaceutical potency, dipole moment, and entropy. The advantage is that only molecular orbital energies are used as explanatory variables in our machine learning method. In this chapter, machine learning predictions: (1) DPPH radical scavenging ability of ferulic acid, (2) octanol/water partition coefficient (log *P*), which is a parameter of Hansch-Fujita quantitative structure-activity relationship (QSAR) calculations, (3) five properties of benzodiazepine anxiolytic: potency of anxiolytics, anti-spastic, muscle relaxant, half-life of concentration ($T_{1/2}$), time to reach maximum body concentration (T_{max}), (4) dipole moments of diatomic and hydrides are introduced. In addition, our developing machine learning method is also mentioned.

Keywords Machine learning · Quantum artificial intelligence (AI) · Qualitative structure-activity relationship · Drug discovery · Molecular orbital energy · Explanatory variable · Regression · Coefficient of determination · Root mean square errors · R language

H. Teramae (✉)
Department of Chemistry and Biological Science, Faculty of Science, Josai University, 1-1 Keyakidai, Sakado, Saitama 350-0295, Japan
e-mail: Teramae@gmail.com

Present Address:
X-Ability Co., Ltd., Ishiwata Building 3rd Floor, Hongo 4-1-5 Bunkyo-ku, Tokyo 113-0033, Japan

T. Onishi (ed.), *Advances in Quantum Science II*,
https://doi.org/10.1007/978-981-95-9721-5_3

3.1 How Artificial Intelligence (AI)-Based Prediction Was Conceived

In recent years, advances in machine learning have made it possible to perform classification and prediction in a variety of fields that were previously impossible. It is expected that machine learning will be applied to many fields in chemistry. Since properties of compounds are diverse, in machine learning, it is difficult to prepare appropriate explanatory variables (also called features or descripters). To our best knowledge, almost no attempt has been made. Though we have worked with molecular orbital methods, the need to apply machine learning in computational chemistry led us to enter the field several years ago.

The quantitative structure-activity relationship (QSAR) method, which started with the Hansch-Fujita method [1], is one of the most important results in chemistry. QSAR method is widely used not only for pharmaceuticals but also for the development of various materials. Especially in recent years, attention has also been paid to QSAR from the viewpoint of minimizing animal experiments.

As shown in Fig. 3.1, the QSAR method divides intermolecular interactions between a ligand (a reaction species of interest) and a receptor (a reaction site) into various interactions such as hydrogen bonding, electrostatic interaction, hydrophobic interaction, etc. As the number of parameters increases, the interactions become complex, the wrong prediction result is often given. This is particularly true for macroscopic quantities in animal experiments. We encounter a question. Is it possible to create a quantitative structure-activity relationship method that can be applied to any macroscopic system by simplifying the interaction part?

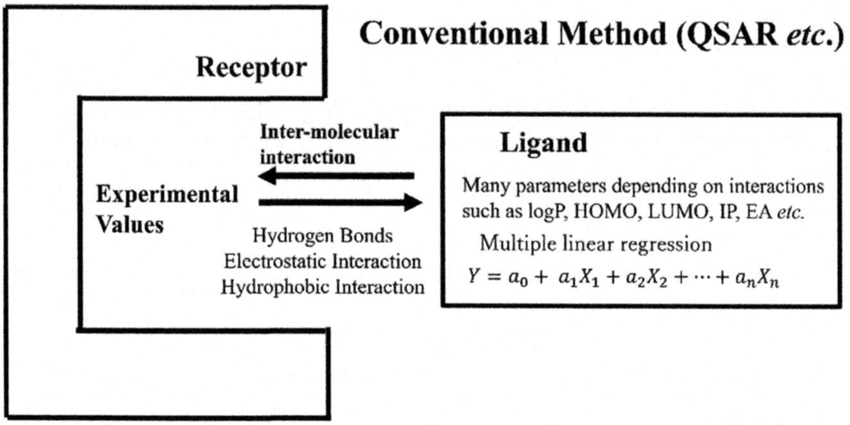

Fig. 3.1 How to understand intermolecular interactions using conventional methods, QSAR, *etc.*

3.1.1 Coefficient of Determination R^2

The coefficient of determination (R^2) is a measure of the explanatory power of the model in regression analysis and is expressed as follows:

$$R^2 = 1 - \frac{SSR}{SST}, \tag{1}$$

where sum of squares of residuals (SSR) is the sum of squares of the difference between the observed and predicted values, and sum of squares of all squares (SST) is the sum of squares of the difference between the observed and its average.

$$SSR = \sum_{i=1}^{n} (y_i - \hat{y}_i)^2 \tag{2}$$

$$SST = \sum_{i=1}^{n} (y_i - \overline{y})^2 \tag{3}$$

By substituting SSR and SST, we have the following:

$$R^2 = 1 - \frac{\sum_{i=1}^{n} (y_i - \hat{y}_i)^2}{\sum_{i=1}^{n} (y_i - \overline{y})^2}, \tag{4}$$

where y_i is the observed value, $\hat{y}_i$ is the predicted value (value from regression model), $\hat{y}$ is the average of observed value. R^2 values range from 0 to 1. If closer to 1, it shows the high explanatory power of the model.

3.1.2 Differences from Quantitative Structure-Activity Relationship (QSAR) Methods and why this Approach Works

In 2008, authors investigated correlations: (i) between anticonvulsant strength and molecular orbital energy, (ii) anxiolytic efficacy and molecular orbital energy. The orbital energy of the next-highest occupied molecular orbital (next-HOMO) was considered. Figure 3.2 shows a dispersion plot of the log values of the anticonvulsant potency of the benzodiazepine minor tranquilizer versus the molecular orbital energy of HOMO-1. See Fig. 3.3 for next-HOMO (HOMO-1). The calculation procedure employed here is Hartree–Fock method with 6-31G(d,p) basis set (HF/6-31G(d, p)). Note that HOMO-1 has the next largest molecular orbital energy after HOMO. It is found that simple regression is applicable by taking logarithm for the anticonvulsant strength. Though the logarithm was not taken for the objective variable in Ref. [2], the logarithm was taken for the variable in Fig. 3.2. As the result, R^2 value becomes 0.801

and improved from 0.663. The weak correlation was observed between anxiolytic efficacy and molecular orbital energy for 17 benzodiazepine-type anxiolytics. In summary, the substituent effect of benzodiazepine-type tranquilizer on anticonvulsant strength could be quantitatively estimated using only the next-HOMO energies. This is the first attempt in the world to estimate structure-activity relationships using only a molecular orbital energy as an explanatory variable.

In the current QSAR method, intermolecular interactions have been subdivided so much. One extreme example is docking simulation, where a three-dimensional structure between ligand and receptor is taken into consideration. However, whether such segmentation is really necessary or not remains unsolved.

Why can we make various predictions based on molecular orbital energy alone? This is closely related to the frontier orbital theory [3], where the interaction is estimated by molecular orbital energies and orbital phases of an electron donor occupied orbital and an electron accepter unoccupied orbital.

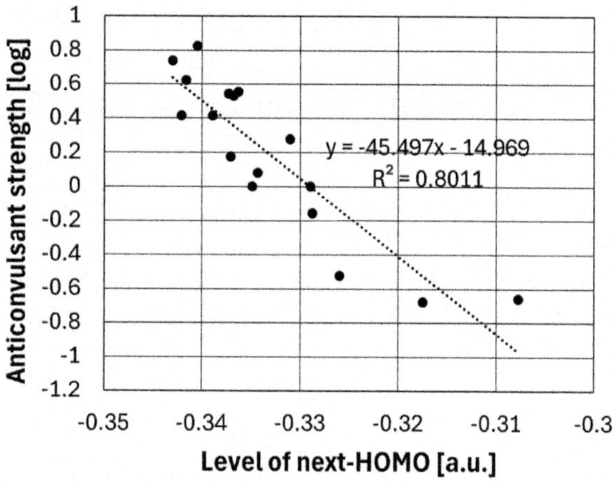

Fig. 3.2 Plot of next-highest occupied molecular orbital (HOMO) versus anticonvulsant strength of benzodiazepine minor tranquilizer. Next-HOMO (HOMO-1) is calculated at Hartree–Fock (HF)/6-31G(d, p) level

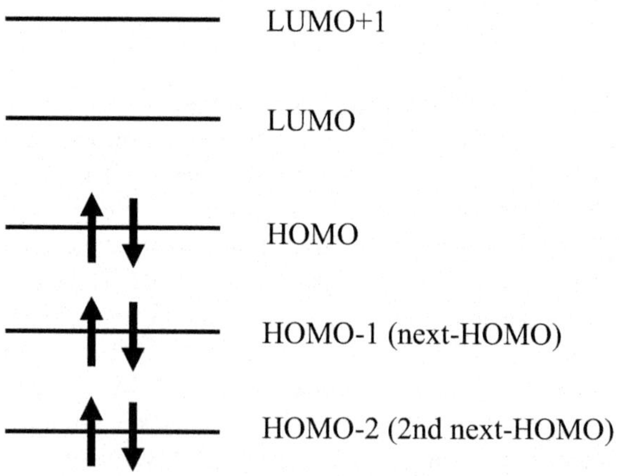

Fig. 3.3 Schematic view of occupied and unoccupied molecular orbitals. HOMO and lowest unoccupied molecular orbital (LUMO) are abbreviations of highest occupied molecular orbital and lowest unoccupied molecular orbital, respectively. Arrows indicate up and down spins of electrons. HOMO-1 and HOMO-2 are the second and third largest occupied molecular orbitals, respectively

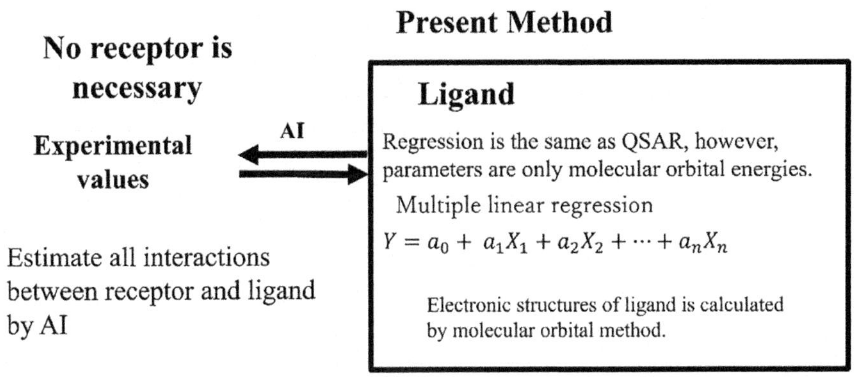

Fig. 3.4 How to understand intermolecular interactions using present new methods

As shown in Fig. 3.1, conventional methods require the decomposition of the molecular interaction between the receptor and ligand. Each interaction has different descriptors and parameters. Therefore, we have considered to stop decomposing molecular interactions into various interactions. We apply the paradigm shift that the molecular interactions are described using single physical quantity as explanatory variable. In our quantum artificial intelligence (AI) technique, molecular orbital energy is applied as explanatory variable. See Fig. 3.4.

3.2 Calculation Methods on Machine Learning

With the progress of our research, we have employed several computational methods, especially in machine learning. Details are explained in this section.

3.2.1 Gaussian for Calculating Molecular Orbital Energies

As mentioned in introduction, it is necessary to execute molecular orbital calculations to obtain molecular orbital energies, which are used as explanatory variables in machine learning. Optimization calculations are performed at the HF/6-31G** level for the molecular structures of interest, and molecular vibrational frequency calculations are performed to confirm that these optimized structures are stable points on the potential energy. Calculations were performed using Gaussian 09 and Gaussian 16, a standard molecular orbital calculation program [4].

3.2.2 R and Caret Package for Machine Learning

In recent years, Python [5] and R [6] are commonly used for machine learning, with the Python language using the scikit-learn package [5] and R using the caret package [7]. In the Sect. 3.1, Python/scikit-learn was used. Except the Sect. 3.1, R/caret was used. Rstudio [8] was also used as a computational development environment. Although Windows, Mac, or Linux can be used as a computer, we used Windows 10/11. This section describes the R/caret package.

R, also called the R language, is a language and environment for statistical analysis and graphics and is an open-source free software. It is a set of functions for creating predictive models. There are various regression and classification methods such as neural networks, support vector machines (SVM), random forests (rf), gradient boosting, etc. By using a caret package, you can easily make predictions by machine learning.

3.2.3 Standardization of Objective and Explanatory Variables

The objective variable (or feature) is first standardized (standardization or z-score normalization) before the model is actually created. Explanatory variables are also standardized to avoid errors. Standardization is a method of transforming the original data into data with a mean of 0 and a standard deviation of 1. It is done by subtracting the mean μ of the feature from each data point x and dividing the result by the standard deviation σ of the feature.

$$z = \frac{x-\mu}{\sigma} \tag{5}$$

Standardization is especially useful when the objective variables have widely varying ranges. In Fig. 3.2, the scales of x- and y-axes are not standardized. In the graphs of prediction vs experiment in the Sect. 3, both scales are standardized. The x-axis plots the original objective variable, while the y-axis plots the objective variable predicted by machine learning.

3.2.4 The K-Fold Cross Validation

The k-fold cross validation is a method for evaluating the performance of a machine learning model. The procedure is as follows: (1) Divide the dataset into k subsets (folds). (2) Choose a test data from k folds. Then, treat the remaining (k-1) folds as training data. (3) Perform training using training data. (4) Measure the performance on the test fold. (5) Repeat (1)–(4) by changing test data k times. Take the average of the k results as the final performance metric. Through the efficient use of the entire dataset, it becomes possible to reduce the risk of overfitting, and stably evaluate the generalization performance of a machine learning model. In R language, the *k*-fold cross validation can be performed by using `train` function of caret package.

3.3 Machine Learning Results Using Molecular Orbital Energies as Explanatory Variables

We introduce our machine-learning predictions on free radical scavenging concentration 50% (IC_{50}) [9–12], octanol/water partition coefficient (log *P*) [13], five properties of benzodiazepines [14], and dipole moments of diatomic and hydride molecules [15, 16].

3.3.1 First Attempt to Predict 2,2-Diphenyl-1-Picryhydrazyl (DPPH) Free Radical Scavenging Ability of Ferulic Acid (FA) and its Derivatives

Oxidative stress is known to be deeply involved in the development and progression of neurodegenerative diseases such as Parkinson's disease and Alzheimer's disease [17] and ischemic brain disorders [18]. In recent years, the neuroprotective effects of antioxidants have been attracting attention with the aim of reducing this oxidative stress [19, 20].

Ferulic acid (FA) is one of the phenolic acid derivatives isolated and structurally determined from *Ferula foetida* in 1886 by Hlasiwetz et al. in Austria [21] and has been shown to exhibit relatively strong antioxidant activity [22] by donating hydrogen to free radicals through its phenolic hydroxyl group.

Recently, attention has been focused on natural antioxidants that are less harmful for health promotion, and FA is expected to be effective. Tomono et al. synthesized FA derivatives with various substituents to search for FA derivatives that are more effective and less toxic than FA, and measured their free radical scavenging ability, and found that the scavenging ability largely depends on the substituent [23].

Therefore, based on the 50% 2,2-diphenyl-1-picryhydrazyl (DPPH) free radical scavenging concentration 50% (inhibitory concentration 50%) IC_{50} obtained from the DPPH free radical scavenging capacity measurement by Xuan et al. we considered the relationship between the IC_{50} and the electronic state of radicals that have had their hydrogen removed from the phenolic hydroxyl group of FA and its derivatives. See Fig. 3.5.

Using the molecular orbital calculation program Gaussian09 [24], structural optimization was performed for FA and its derivatives (FAs) at the restricted Hartree Fock (RHF)/6-31G** level, and for the FA radical with the hydrogen removed from the phenolic hydroxyl group of FA and its derivatives at the unrestricted Hartree

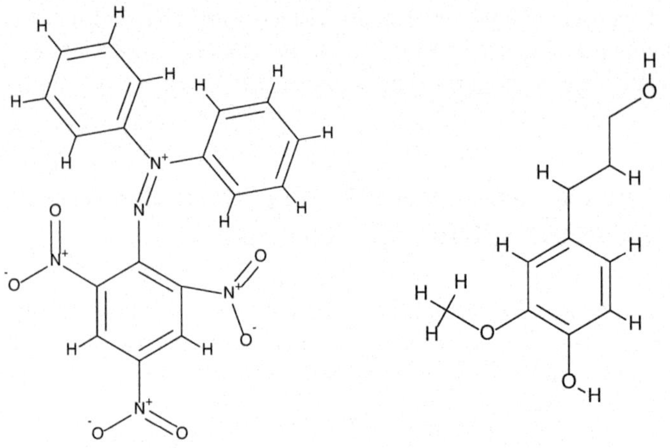

(a) (b)

Fig. 3.5 Molecular structures of **a** Diphenyl-(2,4,6-trinitrophenyl) iminoazanium (DPPH) and **b** Ferulic acid (FA)

Fock (UHF)/6-31G** level. Vibrational calculations were performed on the optimized structure. It was confirmed that it has a stable structure, because it has no imaginary frequency.

First, as we thought that the ease of conversion of individual FAs into radicals might be relevant to the radical scavenging capacity of FAs, the energy differences between the original FAs and the radical species, from which the H radical is removed from the phenolic OH group, were calculated. Table 3.1 summarizes the energy differences between neutral and radical species and the IC_{50} values at the HF/6-31G** level. Note that a higher IC_{50} value indicates a lower radical scavenging capacity. Three antioxidant compounds, coffee acid (CA), ascorbic acid (VC), and Trolox (Trolox), were also included for comparison.

The correlation coefficient of the above results plotted against the $1/IC_{50}$ value became about 0.20, even excluding VC, which has clearly an outlier. Therefore, it can be said that there is no correlation. However, the correlation coefficient is improved to be about 0.47, when the IC_{50} values over 1500 are excluded. Because the measurement was terminated in the middle of the measurement process, meaning that the IC_{50} value is 1500 or higher. Furthermore, if two FA derivatives are also excluded, the correlation coefficient is improved to be 0.65. It is considered that there is a correlation between the energy difference and the IC_{50} value.

Since the difference between the ferulic acid derivatives is only the type of substituent, structural features and density distributions of the molecular framework, which were investigated from the molecular orbital method, are almost the same. In particular, the large difference in radical scavenging ability between FA derivatives **1** and **2** (Fig. 3.6) is only due to the position difference of the hydroxyl group.

Table 3.1 Scavenging concentration 50% (IC_{50}) values and energy differences between neutral and radical species

Compound	IC_{50} (μM)	Energy*1
FA	73	−0.58725
1	93	−0.58464
2	1500	−0.58355
3	99	−0.58732
4	53	−0.58509
5	469	−0.58821
6	Over 1500	−0.58596
7	90	−0.58919
8	161	−0.58970
9	69	−0.58625
10	75	−0.58687
11	82	−0.58721
12	73	−0.57782
13	Over 1500	−0.58405
14	117	−0.58206
15	Over 1500	−0.58095
17	1117	−0.59083
18	398	−0.58667
22	47	−0.58008
24	49	−0.58155
27	755	−0.59008
28	Over 1500	−0.58935
30	202	−0.59009
CA	55	−0.58659
Trolox	52	−0.58024
VC	69	−0.59405

*1 Energy difference between the neutral and radical species

Though the molecular orbital energies of the β-LUMO are significantly different, there is almost no difference in the calculated values. Therefore, in the application of random forest model, we examined the relationship between the energy difference between the α-Singly Occupied Molecular Orbital (SOMO), the β-LUMO, and the radical scavenging ability. It was found that there is a slight correlation.

For individual FA radical species, structural features and density distributions of the molecular framework, which were investigated from the molecular orbital method, are almost the same. The only difference is for substituents. In particular, the large difference in radical scavenging ability between FA derivatives **1** and **2** (Fig. 3.6) cannot be explained. The sole structural difference is the position of the hydroxyl group. Further, there is almost no difference in the calculated values. In

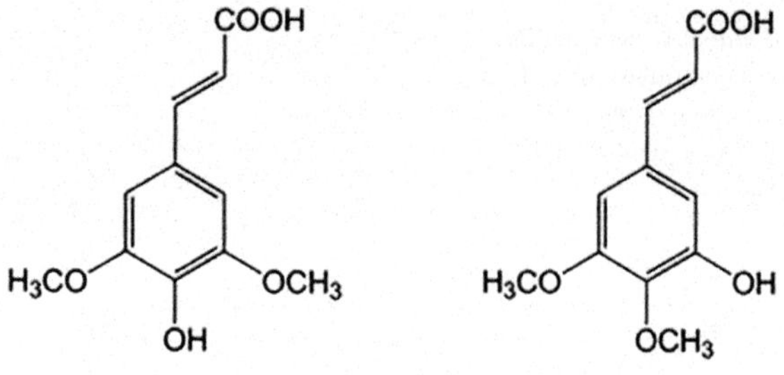

Fig. 3.6 Derivatives of ferulic acid (FA)

both the radical and neutral species, their HOMO and HOMO-1 consist of the doubly degenerate HOMO of benzene. The correlation coefficient between the sum of the molecular orbital energies of the α-SOMO and α-SOMO-1 and the IC_{50} value is 0.46, indicating that there is a slight correlation.

Next, we investigated the relationship between the log IC_{50} and the molecular orbital energies of the radical species: α-SOMO, α-SOMO-1, β-SOMO, α-LUMO, β-LUMO and the neutral species: HOMO-1, HOMO, LUMO, LUMO+1 using a random forest regression method implemented in scikit-learn library. The model was calculated to have a 6% contribution from α-SOMO-1, 24% from α-SOMO, and 45% from α-LUMO, indicating that α-SOMO and α-LUMO have a larger contribution. This is the case that a molecular orbital energy is applied as an explanatory variable, as same as the anticonvulsant strength of benzodiazepine minor tranquilizer.

3.3.2 Second Attempt to Predict DPPH Free Radical Scavenging Ability of Ferulic Acid and its Derivatives

In this subsection, various regression methods implemented in the caret package [6, 7] were employed in machine learning. Instead of Gaussian09 [24], Gaussian16 [4] was used. First, the energy differences between FAs and their radical species were calculated, as the ease of transformation of individual FAs into radicals is considered to be relevant to the radical scavenging ability. For comparison with FAs and the radical species, the antioxidants coffee acid (CA), ascorbic acid (VC), and Trolox (Trolox) are taken into consideration (in total, 26 compounds).

We attempted supervised machine learning in the application of molecular orbital energies as explanatory variables. The relationship between $\log IC_{50}$ and the molecular orbital energies of the radical species: α-SOMO, α-SOMO-1, β-SOMO, α-LUMO, β-LUMO, and those of neutral species: HOMO-1, HOMO, LUMO, and LUMO+1 was analyzed using machine-learning with various sets of explanatory variables.

In general, it is said that molecular orbitals considered here are correlated with each other. Since linear multiple regression (LM) is not suitable for such a case, partial least square (pls) should be used instead. Other regression methods employed here are pls, random forest (rf), radial basis function kernel regularized least squares (krlsRadial), and Bayesian regularized neural network (brnn), a type of neural network method.

Figure 3.7 shows the experimental IC_{50} values, and the predictions obtained by supervised learning using these regression methods. It can be seen that there is a correlation by both approaches. Note that the data values were standardized prior to the calculations, which is why the data values in Fig. 3.7 are between −2 and 2.

The R^2 values of random forest and brnn are more than 0.8, and those of krlsRadial are almost 1, due to overtraining. This indicates that it is highly possible to predict IC_{50} values by using energies and molecular orbital energies from molecular orbital calculations as explanatory variables. It is considered to be applicable to the field of drug design.

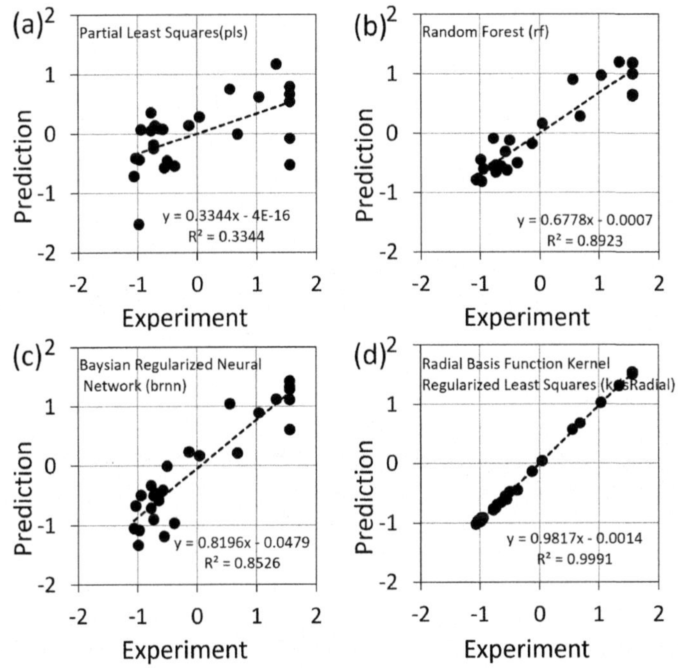

Fig. 3.7 Plots of predicted scavenging concentration 50% (IC_{50}) values versus experimental IC_{50} values: regression methods: **a** pls, **b** rf, **c** brnn, **d** krlsRadial

3.3.3 Third Attempt to Predict DPPH Free Radical Scavenging Ability of Ferulic Acid and its Derivatives

We reported that machine learning is able to predict the anxiolytic and anticonvulsant effects of benzodiazepine (BZP), muscle relaxant properties, for which relationships could not be examined previously, and the pharmacokinetic parameters T_{max} and $T_{1/2}$ [14]. As multiple splits of training and test data were attempted, the best result was adopted because the number of samples was small. Therefore, this subsection shows the results of the recalculation of the antioxidant effect of ferulic acid.

The molecular orbital calculation program Gaussian16 RevB.01 [4] was used to optimize the structure at the RHF/6-31G(d, p) level. Vibrational frequency calculations were performed on the optimized structure. It was confirmed that the structure is stable, because it has no imaginary frequency.

For machine learning calculations, we used the caret package in R [6, 7]. Molecular orbital energies obtained from molecular orbital calculations were used as explanatory variables, up to a maximum of 10 orbitals. The training data consisted of 22 molecules and the test data consisted of 6 molecules. Three-fold validation was used for cross-validation [25]. We have tried several regression methods: xgbLinear (linear gradient boosting), monmlp (monotone multilayer perceptron neural network), rf (random forest), and pls (partial least squares regression).

The partitioning into training and test data was done using random numbers. The root mean square errors (RMSE) of the difference between the predicted and experimental IC_{50} values of the test data were calculated. When the RMSE is minimum, the number of molecular orbital energies is in principle adopted as the optimal number of explanatory variables.

Table 3.2 shows the RMSE between the predicted and experimental values for the test data when 2–10 molecular orbitals were used as explanatory variables. Table 3.3 shows R^2 value for the training data. Figure 3.8 shows the variance of the experimental and predicted values of the log (IC_{50}) values.

XgbLinear gave the minimal RMSE when the number of molecular orbital energies is 2. The coefficient of determination is almost 1 for 2 orbitals or corresponds to 1 for 4, 6, 8, and 10 orbitals. Though monmlp gave the minimal RMSE for 2 orbitals, this minimal value is larger than those of xgbLinear, monmlp, rf, and pls. The coefficient of determination of 1 for 8 and 10 orbitals. Though rf gave the second lowest

Table 3.2 Root mean square errors (RMSE) values of test data for the number of molecular orbital energies. Optimal numbers are marked in blue and RMSE values are expressed in dB [μM]

# of orbitals	2	4	6	8	10
xgbLinear	**0.276**	0.434	0.333	0.672	0.692
monmlp	**0.542**	0.667	0.948	0.860	0.944
rf	0.387	**0.354**	0.374	0.409	0.423
pls	0.568	0.511	**0.468**	0.489	0.489

Table 3.3 Coefficient of determination (R^2) of training set for the number of molecular orbital energies

# of orbitals	2	4	6	8	10
xgbLinear	0.998	1.000	1.000	1.000	1.000
monmlp	0.530	0.888	0.955	1.000	1.000
rf	0.832	0.851	0.851	0.876	0.905
pls	0.333	0.324	0.325	0.344	0.334

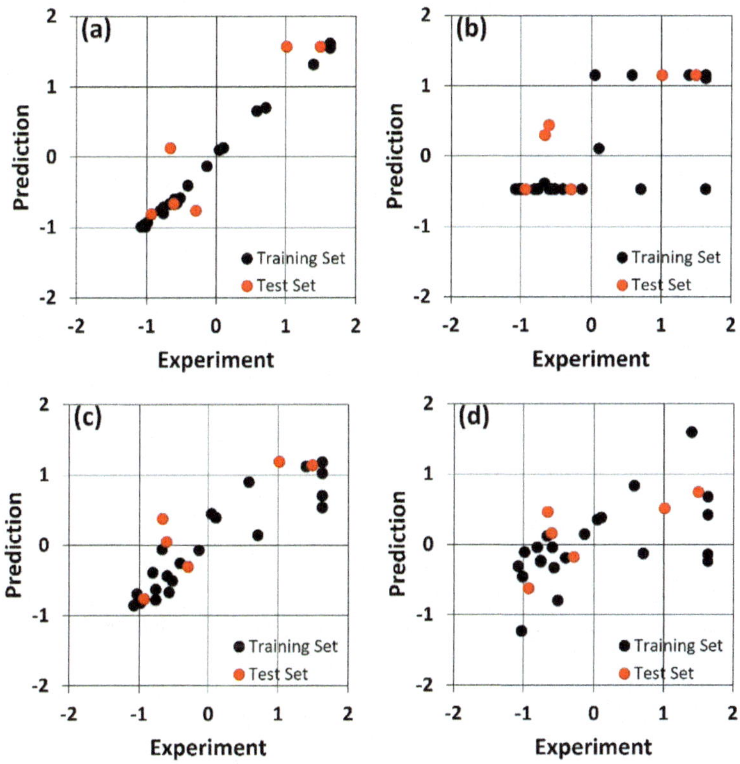

Fig. 3.8 Plots of training and test data of scavenging concentration 50% (IC_{50}) values of ferulic acid derivatives: regression methods **a** xgbLinear, **b** monmlp, **c** rf, **d** pls. Numbers of orbitals are 2 for (**a**) and (**b**), and 4 for (**c**) and (**d**). The black circles denote the training data, and the red circles denote the test data

RMSE, the coefficient of determination is slightly smaller than other methods. See Fig. 3.8. Since pls gave a larger RMSE and a coefficient of determination smaller than 1, pls is considered inappropriate. Finally, it is found that xgbLinear regression method is the best regression method for predicting the radical scavenging potential of ferulic acid.

Thus, machine learning was performed to predict the radical scavenging ability of ferulic acid using one molecular orbital energy of the original ferulic acid derivative as explanatory variables unlike the previous Sect. 3.2. When an appropriate regression method is selected, the combination of molecular orbital calculations and machine learning proves to be a powerful tool for predicting radical scavenging abilities of a wide range of molecules.

3.3.4 Prediction of Octanol/Water Partition Coefficient (Log P) Using Molecular Orbital Energies and Machine Learning

The QSAR method by Hansch-Fujita [1] has long been used in pharmacological activity relationship studies, to estimate structure-activity relationships under consideration of molecular interactions between ligand-receptors. HOMO and LUMO energies are used for descriptor of structure-activity relationship. The octanol/water partition coefficient (log *P*) is the parameter denoting fat solubility.

In our proposed machine learning method, molecular orbital energy is only used as an explanatory variable. The caret package of R [6, 7] was here used for the machine learning calculations. The molecular orbital energies were estimated by molecular orbital calculations. The training data consisted of 16 molecules, while the test data, which is treated as unknown data, consisted of 6 molecules. Data was divided using random numbers. Three-fold-validation was used for gamboost, glm, kknn, krlsRadial, lasso, monmlp, pls, ppr, qrf, ranger, rf, svmLinear, svmRadial, xgbLinear, and xgbTree. The log *P* value can be estimated by molecular orbital energy. The optimal number of explanatory variables correspond to the number of molecular orbital energies when R^2 value between training and test data is the largest. The log P values of 22 molecules, were taken from the experimental values listed in Rekker et al. [26]: atropine, chloramphenicol, chlorothiazide, chlorpromazine, cimetidine, diazepam, diltiazem, diphenhydramine, disopyramide, flufenemic acid, furosemide, haloperidol, imipramine, lidocaine, phenobarbital, phenytoin procainamide, propafenone, propranolol, tetracaine, trimethoprim, and verapamil. However, eight compounds: chlorothiazide, chlorpromazine, diazepam, disopyramide, furosemide, haloperidol, propafenone, and Propranolol, the log *P* values from PubChem [27] were used.

Table 3.4 shows R^2 values of log *P* values for test data, total data and training data, and the minimum RMSE between the predictions of test data and experimental values. Eighteen regression methods are used. The numbers of molecular orbitals range from 2 to 80. Figure 3.9 shows plots of log *P* values for training data and test data by qrf, xgbLinear, xgbTree, and WM methods.

For test data, qrf method combined with HOMO/LUMO orbitals gives the largest R^2 value (0.904), while the smallest RMSE (0.378) is given within other regression methods. The second (0.793) and third (0.778) largest RMSE are given in xgbLinear

Table 3.4 Coefficient of determination (R^2) values of octanol/water partition coefficient (log P) values for test, total, and training data, and the minimum root mean square errors (RMSE) values between the predictions of test data and experimental values

Method	# of orbitals	Test data	Total data	Training data	RMSE
qrf	2	0.904	0.919	0.927	0.378
xgbLinear	4	0.793	0.943	0.999	0.445
xgbTree	10	0.778	0.922	1.000	0.527
WM	10	0.671	0.764	0.797	0.588
rf	2	0.619	0.796	0.867	0.624
HYFIS	14	0.698	0.791	0.893	0.674
ranger	8	0.554	0.806	0.922	0.691
SBC	16	0.530	0.861	0.999	0.699
gamboost	4	0.446	0.659	0.767	0.732
lasso	58	0.489	0.807	0.959	0.735
pls	68	0.385	0.710	0.831	0.745
svmRadial	16	0.464	0.732	0.857	0.760
krlsRadial	24	0.373	0.827	1.000	0.772
kknn	10	0.382	0.616	0.723	0.798
svmLinear	60	0.378	0.776	0.992	0.892
monmlp	64	0.242	0.705	1.000	1.104
ppr	62	0.503	0.132	1.000	2.313
glm	36	0.637	0.385	1.000	4.211

and xgbTree methods, respectively. RMSE values of xgbLinear and xgbTree methods give RMSE values of 0.445 and 0.527, respectively. They are larger than that of qrf method. The WM method gives the fourth largest R^2 value (0.671). As nonlinearity appears in WM method (three types of predictions), further analysis is required.

3.3.5 *Prediction of Five Properties of Benzodiazepine Anxiolytic*

This work was first published in 2022; however, the calculations were not sophisticated, especially in repeating calculations for obtaining the best correlation coefficients with multiple data-splittings. We, therefore, performed the same machine learning with single data-splitting.

As noted in the general introduction, there is a relationship between the potency of the drugs and the Hartree–Fock (HF) molecular orbital energies in predicting five properties of benzodiazepine (BZP) anxiolytics [2]. The conformation analysis was performed using Hamiltonian algorithm [28]. Geometry optimization was

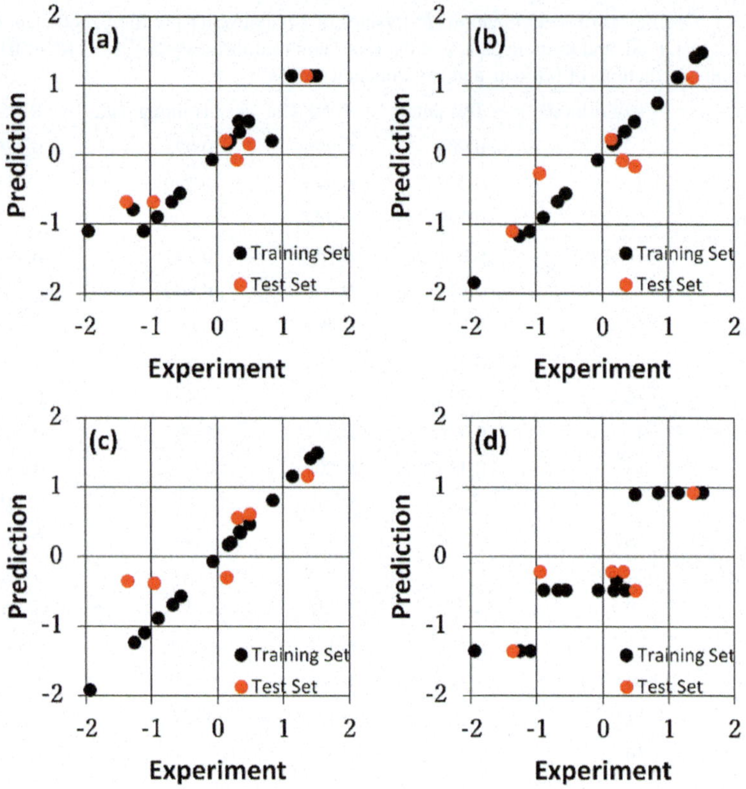

Fig. 3.9 Plots of octanol/water partition coefficient (log *P*) values for training data (black circle) and test data (red circle): regression methods are **a** qrf, **b** xgbLinear, **c** xgbTree, **d** WM. The numbers of orbitals are 2 for (**a**), 4 for (**b**), and 10 for (**c**) and (**d**)

executed for 17 different BZP molecules at the HF/6-31G(d, p) level. The molecular orbital energies were applied as explanatory variables. The objective variables were anxiolytic and anti-spastic intensities, and muscle relaxant intensities (half-life of concentration $T_{1/2}$, time to reach maximum body concentration T_{max}) for seventeen BZP anxiolytics [29]. There was a strong correlation between the molecular orbital energies of next-HOMO (HOMO-1) and the anti-spastic intensity. See Fig. 3.3 for HOMO-1. However, it was found that only one molecular orbital energy of HOMO-1 is inadequate for explanatory variable to estimate correlations with other properties such as anxiolytic and muscle relaxant intensities. If more molecular orbital energies are incorporated as explanatory variables, we thought that predictions of other properties become possible.

Maximum number of molecular orbitals as the explanatory variables is limited to 20. The data split into two groups: (i) for supervised machine learning, (ii) for testing prediction ability. The numbers of data for BZPs are 12 and 5, respectively, except for T_{max}. For T_{max} the data for cloxazolam is not found in the literature. As the data of

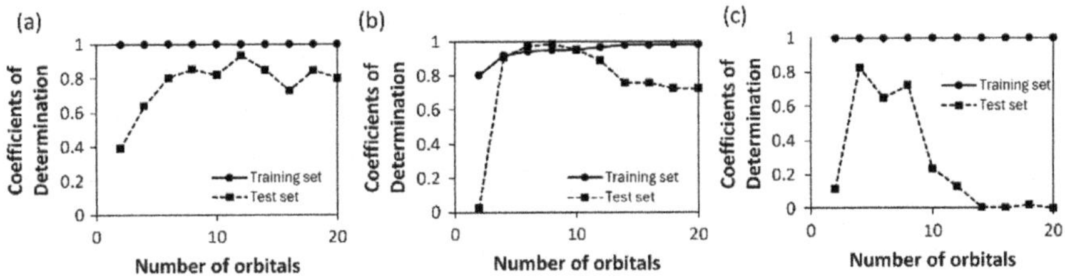

Fig. 3.10 Relation of coefficient of determination (R^2) value and number of molecular orbitals **a** anxiolytic with xgbLinear regression **b** anticonvulsant with bstSm regression, and **c** muscle relaxant intensities with xgbLinear regression

oxazolam is too large, it is excluded to avoid a considerable error. The all regression methods were applied in the machine learning. As the result, the 88 solutions were obtained. The optimal number of molecular orbital energies was determined by the minimal RMSE difference between the predicted and experimental values. Except for the manner, the case of 10 orbital energies was also taken into consideration. The cross validation was performed using the three-fold cross validation for all calculations [6].

Figure 3.10 shows the relations between R^2 value and the number of molecular orbital energies in terms of anxiolytic, anti-spastic, and muscle relaxant intensities of BZP anxiolytics. It is clear that 6 orbitals are sufficient to describe the predictions of these three intensities. When one molecular orbital energy is applied in three cases, R^2 value decreases. It is the reason why sole orbital is insufficient. Increasing the number of molecular orbital energies results in overfitting, as R^2 value decreases.

These results are very reasonable considering these intensities have relations with the HOMO-1 levels. Increasing the number of molecular orbital energies results in the overfitting where the correlation coefficients of the predictions of anxiolytic and muscle relaxant intensities decrease.

Figure 3.11 shows the machine learning of training data and test data for anxiolytic, anti-spastic, and muscle relaxant intensities. The R^2 values of (a) anxiolytic with xgbLinear regression with 14 molecular orbital energies, (b) anticonvulsant with bstSm regression with 4 molecular orbital energies, and (c) muscle relaxant intensities with xgbLinear regression with 4 molecular orbital energies are (a) 0.986, (b) 0.996, and (c) 0.811, respectively. It suggests that strong correlations appear in all cases. Note that in the ideal case, R^2 value becomes 1.0. In all cases, since HOMO-1 (next-HOMO) and HOMO-2 (second next-HOMO) are degenerate in a benzene ring, they are required as explanatory variables. See Fig. 3.3 for HOMO-1 and HOMO-2.

Figures 3.12 and 3.13 show the relations between R^2 value and the number of molecular orbital energies in terms of $T_{1/2}$ and T_{max}, respectively. There are two tendencies: R^2 values of training data are closer to 1.0, if the number of molecular orbital energies is larger than 2; the R^2 values of test data are smaller than those of training data. The maximum values of test data are 0.865 and 1.0 for $T_{1/2}$ and T_{max}, respectively. As the slope in Fig. 3.13b is smaller than 1.0, it shows underestimation of the predicted value.

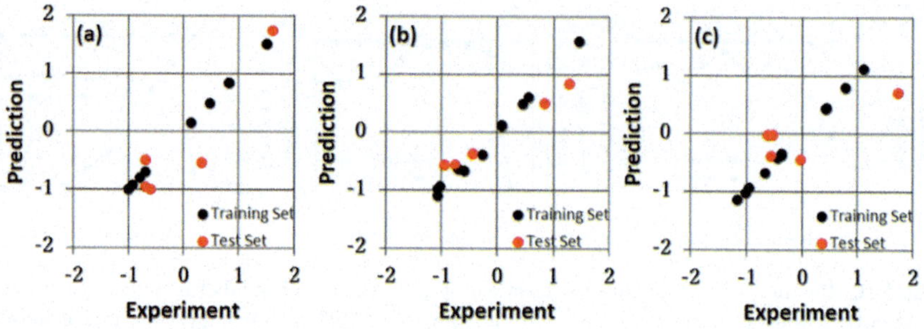

Fig. 3.11 Machine learning results: **a** anxiolytic with xgbLinear regression (14 orbitals), **b** anticonvulsant with bstSm regression (4 orbitals), and **c** muscle relaxant intensities with xgbLinear regression (4 orbitals)

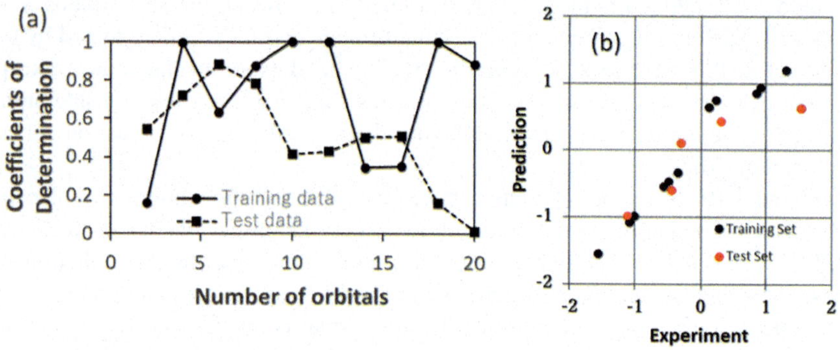

Fig. 3.12 Machine learning results of half-life of concentration in the body ($T_{1/2}$) with neuralnet regression ("neuralnet" is a function name of R/caret): **a** relation between coefficient of determination (R^2) value and number of molecular orbital energies, **b** predictions of training and test data using 6 orbitals

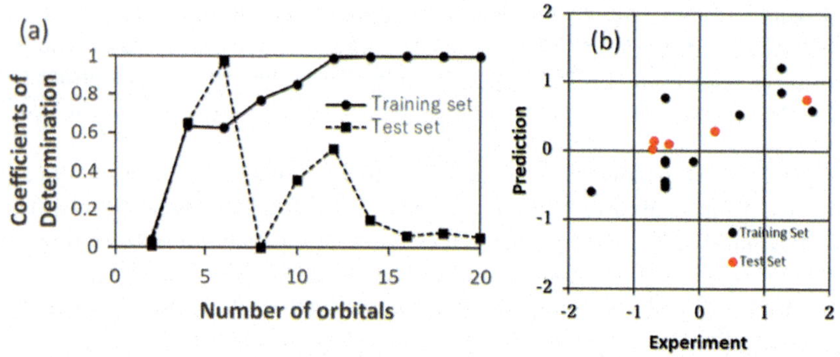

Fig. 3.13 Machine learning results of time to reach maximum body concentration (T_{max}) with xgbTree regression: **a** relation between coefficient of determination (R^2) value and number of molecular orbital energies, **b** predictions of training and test data using 16 orbitals

Finally, it can be concluded that 6 molecular orbital energies are sufficient for predicting three intensities of anxiolytics, anti-spastics, and muscle relaxant. Predicting $T_{1/2}$ and T_{max} is more difficult, since experimental errors are a little bit too large.

3.3.6 Prediction of Dipole Moment of Simple Molecules

This subsection shows the prediction of the dipole moment of a simple molecule such as a diatomic molecule by our machine learning approach using R/caret. Though a dipole moment can be obtained by molecular orbital calculations, the work was performed to verify our approach. In order to examine the dependence of mathematical models, 14 regression models were arbitrarily selected: (1) dnn (deep neural network), (2) glm (least squares regression), (3) HYFIS (hybrid neural fuzzy inference system), (4) knn (k-nearest neighbor method), (5) krlsRadial (normalized radial basis function least squares method), (6) lasso (lasso regression), (7) monmlp (monotone multilayer perceptron neural network), (8) pls (partial least squares regression), (9) ppr (projection tracking regression), (10) rf (random forest), (11) svmLinear (linear support vector machine), (12) svmRadial (radial basis function support vector machine), (13) xgbLinear (linear extreme gradient boosting), (14) xgbTree (tree extreme gradient boosting). The 20 of 28 molecules were selected as training data, while the rest 8 molecules were used as test data. By the use of the relation estimated from training data, the relation of the test data was predicted. The result was compared to experimental values to provide a measure of the accuracy of this machine learning.

Problem of Overfitting

In machine learning research, it is necessary to avoid overlearning in training. Overtraining is the phenomenon in which objective values predicted from training data do not match actual experimental values at all, due to overfitting.

Figure 3.14 shows the training and prediction results for log values of dipole moments using multi-layer perceptron neural network with optional monotonicity constraints (monmlp), a type of neural network. The numbers of orbitals are (a) 2, (b) 4, and (c) 6. At first glance, excellent agreement with experimental values is seen for 6 orbitals. The RMSE values for 2, 4, and 6 orbitals, which are the differences between the predicted values of training data and experimental values of test data, are 0.40, 0.19, and 0.48, respectively. The correlation between training and test results data is poor. This phenomenon is so-called overlearning. The procedure to avoid overlearning must be considered every time.

Since it is possible to increase the number of explanatory variables, it is easy to determine whether overlearning occurs or not. Though the R^2 values of 2 orbitals and 4 orbitals are smaller than 6 orbitals, plots of test data deviates from the linearity of training data. Hence, it can be judged that overlearning occurs in 6 orbitals (Fig. 14c).

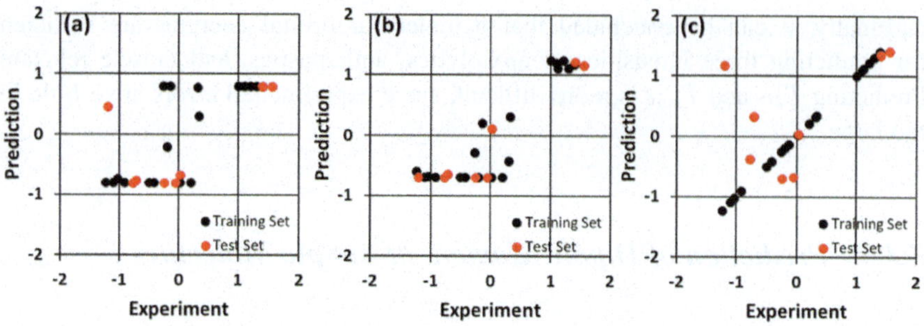

Fig. 3.14 Machine learning results of dipole moments by monmlp regression method. The number of explanatory variables (orbitals) is **a** 2, **b** 4, and **c** 6

The nonlinearity of 2 orbitals is attributed to a simple lack of learning (Fig. 3.14a). Interestingly, in Fig. 3.14b, the small and large values are only predicted to be similar, while only the center part is predicted to be more precise than the two ends. It is necessary to carefully select which regression method and how many orbital energies will give the optimal solution.

Differences Due to Regression Method

Table 3.5 shows the correlation coefficients between the experimental dipole moments and the dipole moments predicted by each regression method on the training set, where the number of explanatory variables (molecular orbital energies) varies from 2 to 10. If the number of molecular orbital energies increases, R^2 value tends to become close to 1.0. The cross validation was performed using the three-fold cross validation for all calculations. As the result, 2-, 3-, 5-, and ten-fold cross validations did not show much difference.

Table 3.7 shows correlation coefficient (R) values, RMSE values, and slopes of the regression lines of the training results and numbers of explanatory variables (molecular orbital energies). Note that the smallest RMSE value and the corresponding number for all regression methods are selected in Table 3.6. For example, in ppr method, the smallest RMSE value and corresponding number are 0.335 and 4, respectively. From the point of RMSE values, ppr and monmlp methods give smaller values in comparison with other methods. However, the corresponding slopes of the regression lines are 0.718 (ppr) and 0.704 (monmlp).

Figure 3.15 shows the machine learning results of training data and test data by the use of (a) ppr, (b) knn, (c) HYFIS, (d) svmRadial, (e) rf, (f) pls, (g) svmLinear, (h) lasso, (i) glm, (j) dnn, (k) xgbLinear, (l) xgbTree, and (m) krlsRadial. The slopes of the regression lines for pls and svmLinear are 0.678 and 0.781, respectively. It shows the deviation of predictions from experimental data. See Fig. 4.12 (f) and (g). In lasso method, not only the smallest slope (0.353) among regression lines but also large RMSE is given. It is seemed that support vector machines such as svmRadial and

Table 3.5 Correlation coefficients between the experimental dipole moments and the dipole moments predicted by each regression method on the training set

Regression method	# of orbitals				
	2	4	6	8	10
dnn	0.646	0.626	0.569	0.672	0.568
glm	0.660	0.719	0.736	0.875	0.912
HYFIS	0.883	0.766	0.769	0.988	0.992
knn	0.694	0.855	0.860	0.769	0.809
krlsRadial	0.924	1.000	0.991	0.874	0.967
lasso	0.659	0.648	0.659	0.655	0.786
monmlp	0.775	0.839	1.000	1.000	1.000
pls	0.608	0.702	0.654	0.839	0.823
ppr	0.791	0.848	0.992	0.951	0.925
rf	0.948	0.966	0.970	0.968	0.965
svmLinear	0.647	0.716	0.708	0.862	0.893
svmRadial	0.877	0.917	0.913	0.911	0.906
xgbLinear	0.999	1.000	1.000	1.000	1.000
xgbTree	0.967	0.978	0.998	1.000	0.999

Table 3.6 Root means square error (RMSE) values between predicted dipole moments and experimental values of divided data (test data)

Regression method	Number of molecular orbital energies				
	2	4	6	8	10
dnn	0.445	0.444	0.442	0.440	0.432
glm	0.431	0.413	0.600	0.447	0.383
HYFIS	0.579	0.221	0.235	0.246	0.256
knn	0.384	0.216	0.257	0.360	0.260
krlsRadial	0.467	0.327	0.262	0.467	0.583
lasso	0.438	0.422	0.410	0.411	0.362
monmlp	0.404	0.187	0.481	0.675	0.314
pls	0.526	0.520	0.446	0.526	0.212
ppr	0.262	0.177	0.689	0.434	0.335
rf	0.514	0.335	0.310	0.331	0.335
svmLinear	0.373	0.438	0.455	0.363	0.301
svmRadial	0.249	0.235	0.222	0.242	0.241
xgbLinear	0.584	0.466	0.219	0.207	0.255
xgbTree	0.484	0.406	0.282	0.291	0.364

Table 3.7 Root means square error (RMSE) values, coefficient of determination (R^2), slope of regression line of training data, and number of molecular orbital energies (MOEs) for several regression methods

Regression method	RMSE	R^2	Slope of regression line	MOEs
ppr	0.177	0.719	0.718	4
monmlp	0.187	0.704	0.704	4
xgbLinear	0.207	1.000	0.981	8
pls	0.212	0.677	0.678	10
knn	0.216	0.731	0.731	4
HYFIS	0.221	0.587	0.559	4
svmRadial	0.222	0.834	0.702	6
krlsRadial	0.262	0.982	0.853	6
xgbTree	0.282	0.996	0.984	6
svmLinear	0.301	0.797	0.781	10
rf	0.310	0.941	0.776	6
lasso	0.362	0.618	0.353	10
glm	0.383	0.832	0.831	10
dnn	0.440	0.452	0.011	8

Table 3.8 Identifier number, coefficients of determination for test data R^2_{test}, test and training data R^2_{total}, training data $R^2_{training}$, RMSE, and assignments of molecular orbital energies using advanced calculation methods

No.	R^2_{test}	R^2_{total}	$R^2_{training}$	RMSE	Regression	Molecular orbital energy used				
16	0.995	0.996	1.000	0.123	xgbTree	H-3	L+3			
16	0.992	0.996	1.000	0.124	kknn	H-3	L+3			
14	0.981	0.994	0.999	0.131	xgbLinear	H-3	L+1			
17	0.984	0.981	0.985	0.246	qrf	H-3	L+4			
293	0.979	0.979	0.987	0.292	svmRadialCost	H-3	H	L+2	L+3	
412	0.956	0.979	0.993	0.240	ppr	H-4	H-2	L+1	L+2	L+3

svmLinear methods are not applicable. The deep neural network (dnn) method shows almost no slope, implying no correlation between explanatory variables. It is due to inadequate training as deep learning. The dnn calculations with more training have been ongoing to obtain the improved results. Though RMSE values for xgbLinear (0.207), xgbTree (0.282) and krlsRadial (0.222) are slightly larger than ppr (0.177) and monmlp (0.187), their slopes of regression lines: xgbLinear (0.981), xgbTree (0.984) and krlsRadial (0.853) are larger than those of ppr (0.718) and nonmlp (0.704). From the points of both RMSE and slope, it is considered that the selections of xgbLinear, xgbTree, and krlsRadial are realistic. In particular, xgbLinear and xgbTree are expected to work well. It should be noted that the selection of regression method depends on the target system. In the selection of a regression method, RMSE value and slopes of a regression line are required as indices.

3.4 Advanced Machine Learning toward Future

3.4.1 Summary of Present Method to Select Molecular Orbitals

Let's summarize how to select molecular orbitals for the use of explanatory variables, which has been developed for the past several years. First, all data are split into training data and test data. The latter data is treated as unknown data to estimate machine learning accuracy. Root mean square errors (RMSE) of test data is used as the reference to determine the optimal number of molecular orbital energies. To reduce calculation patterns, we set the even maximum number of molecular orbital energies. Next, all numbers divided by two are used for the numbers of molecular orbital energies. If the maximum number of molecular orbital energies is 10, numbers of molecular orbital energies become 2, 4, 6, 8, and 10. The selected molecular orbitals are in order from HOMO and LUMO. For example, HOMO and LUMO are selected for 2, while HOMO-1, HOMO, LUMO, and LUMO+1 are selected for 4.

3.4.2 Advanced Method to Select Molecular Orbitals

In our advanced method, the coefficient of determination is for all data including training data and test data, which is denoted as $R^2{}_{total}$. In the previous method, there were two problems: (1) the numbers of molecular orbital energies were usually large to obtain a high $R^2{}_{total}$, (2) near-perfect results have not been obtained so far, even when RMSE values are sufficiently small. It is considered to be why the number of molecular orbital energies are limited. In our advanced method, more combinations of molecular orbitals applied for explanatory variables are taken into consideration.

From 10 molecular orbitals: HOMO-4 to LUMO+4, all combinations patterns are included for the selection of molecular orbitals. If one molecular orbital is selected, there are 10 patterns. However, due to the problem of R language, this case is excluded. We considered all cases of picking out 2, 3, 4, 5, 6, 7, 8, 9, and 10 molecular orbitals. In total, the number of all possible combinations is given by the following:

$$\sum_{n=2}^{10} {}_{10}C_n = \sum_{n=0}^{10} {}_{10}C_n - \sum_{n=0}^{1} {}_{10}C_n = 2^{10} - (1 + 10) = 1013.$$

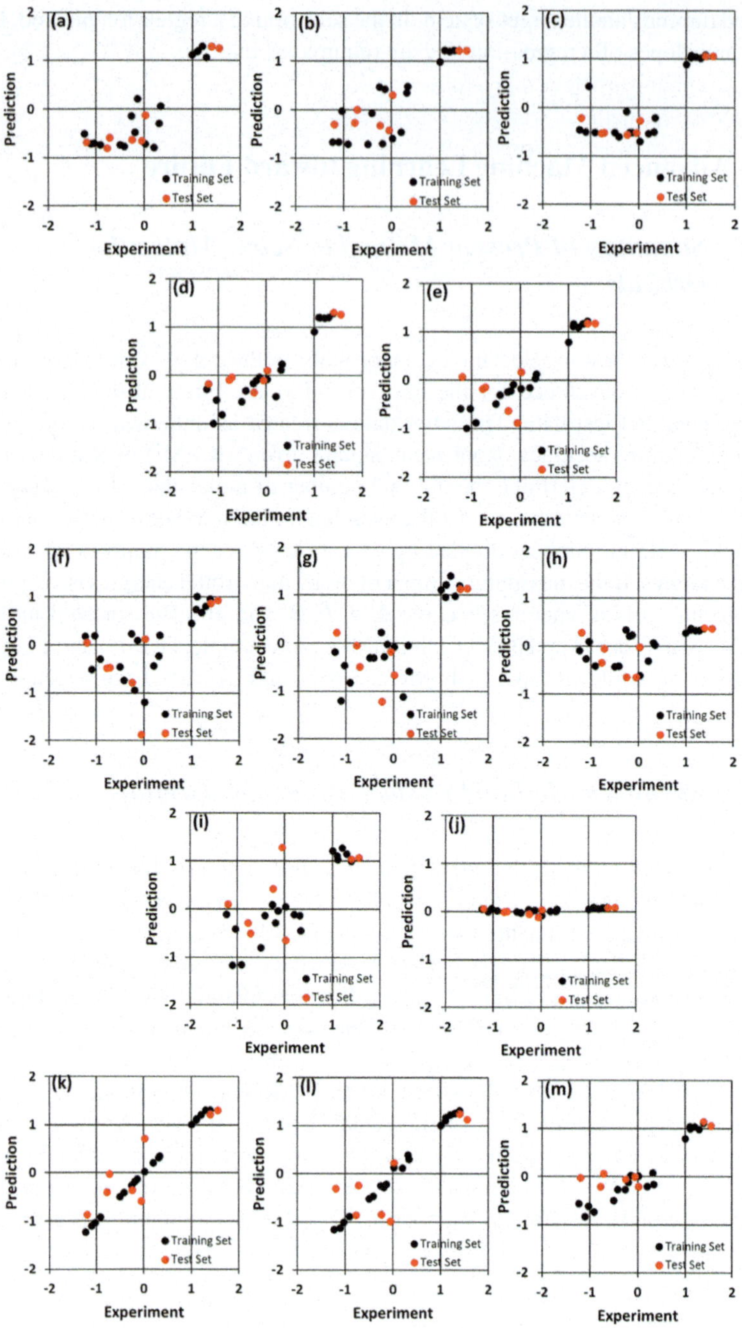

Fig. 3.15 Machine learning results of dipole moments. Regression methods are **a** ppr, **b** knn, **c** HYFIS, **d** svmRadial, **e** rf, **f** pls, **g** svmLinear, **h** lasso, **i** glm, **j** dnn. **k** xgbLinear, **l** xgbTree, and **m** krlsRadial

The training data consisted of 20 molecules, while the test data, treated as unknown data, consisted of 8 molecules. Random numbers were used to divide all data into training data and test data. three-fold cross validation was used for validation. Twenty-one regression methods were used as same as the Sect. 3.6.

Table 3.8 shows the results using the six regression methods in RMSE order. Note that the Identification No. is used to distinguish the 1013 results. It is interesting that HOMO-3 is only included as an occupied orbital except ppr, and LUMO+3 is included as an empty orbital in xgbTree and kknn.

Figure 3.16 shows machine learning results of the best four correlations. As the slopes also became 1.0, the effect of changing combinations of molecular orbital energies is obviously confirmed. In fact, R^2 values of all data (training data + test data) became almost 1.0. It can be considered that the prediction model of dipole moments is perfect. We are investigating the applicability of the advanced method for different predictions.

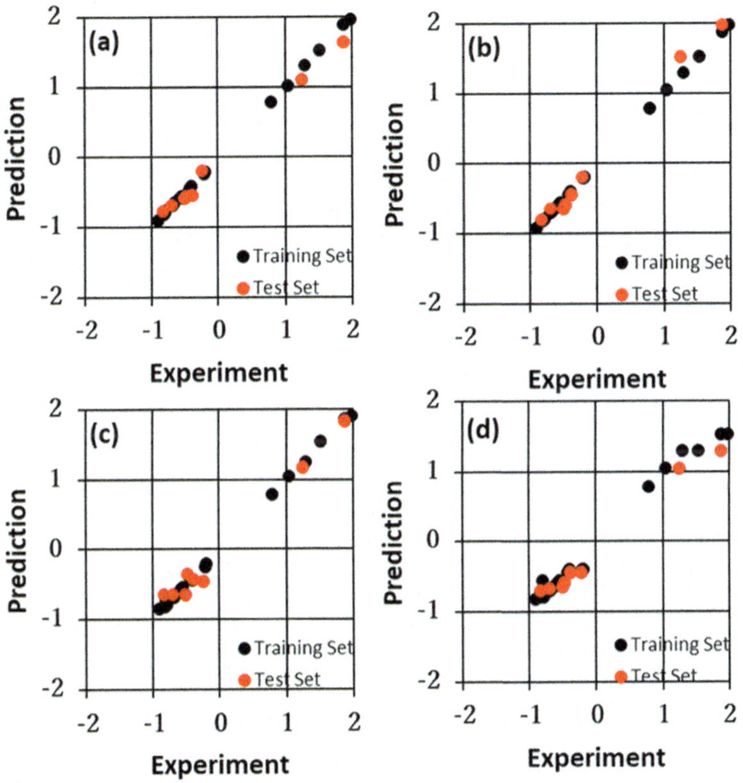

Fig. 3.16 Machine learning results of dipole moments, using advanced machine learning method. Regression methods are **a** xgbTree, **b** kknn, **c** xgbLinear, and **d** qrf

Fig. 3.17 Future prospects of machine learning using molecular orbitals as explanatory variables

QSAR Complex Explanatory Variables
Definition of Many Inter-Molecular Interactions by Classical Mechanics (Complex and Difficult)

⇩

Quantum AI
Development of Machine Learning with Molecular Orbital Energy as Explanatory Variables (Simple and Easy)

⇩

Advanced Quantum AI
Expanding Molecular Orbital Energy as Explanatory Variables (Deep Neural Networks)

⇩

New Application of Drag Discovery or Prediction of Molecular Properties

3.5 Epilog of Machine Learning with Molecular Orbital Energies as Explanatory Variables

In this chapter, our recent works concerning machine learning are introduced. The unique feature is to apply molecular orbital energies as explanatory variables. The procedure is applicable for predicting pharmaceutical potencies and physical properties. We also introduce the new procedure to select molecular orbitals. Predictions of entropy, boiling point, melting point, and anti-HIV activity of alkenyl diarylmethanes are in progress. The works will be presented in the near future. Future prospects of machine learning using molecular orbitals as explanatory variables are summarized in Fig. 3.17.

References

1. Hansch C, Fujita T. J Am Chem Soc. 1964;86:1616–26.
2. Teramae H, Ohtawara K, Ishimoto T, Nagashima U. Bull Chem Soc Jpn. 2008;81:1094–102.
3. Fukui K. Theory of orientation and stereoselection. Springer; 1970.
4. Frisch MJ, Trucks GW, Schlegel HB, Scuseria GE, Robb MA, Cheeseman JR, Scalmani G, Barone V, Petersson GA, Nakatsuji H, Li X, Caricato M, Marenich AV, Bloino J, Janesko BG, Gomperts R, Mennucci B, Hratchian HP, Ortiz JV, Izmaylov AF, Sonnenberg JL, Williams-Young D, Ding F, Lipparini F, Egidi F, Goings J, Peng B, Petrone A, Henderson T, Ranasinghe D, Zakrzewski VG, Gao J, Rega N, Zheng G, Liang W, Hada M, Ehara M, Toyota K, Fukuda R, Hasegawa J, Ishida M, Nakajima T, Honda Y, Kitao O, Nakai H, Vreven T, Throssell K, Montgomery JA Jr, Peralta JE, Ogliaro F, Bearpark MJ, Heyd JJ, Brothers EN, Kudin KN, Staroverov VN, Keith TA, Kobayashi R, Normand J, Raghavachari K, Rendell AP, Burant JC, Iyengar SS, Tomasi J, Cossi M, Millam JM, Klene M, Adamo C, Cammi R, Ochterski JW, Martin RL, Morokuma K, Farkas O, Foresman JB, Fox DJ. Gaussian 16, Revision B.01. Wallingford CT: Gaussian, Inc.; 2016.
5. Pedregosa F, Varoquaux G, Gramfort A, Michel V, Thirion B, Grisel O, Blondel M, Prettenhofer P, Weiss R, Dubourg V, Vanderplas J, Passos A, Cournapeau D, Brucher M, Perrot M, Duchesnay E. Scikit-learn: machine learning in python. J Mach Learn Res. 2011;12:2825.

6. R Core Team. R: a language and environment for statistical computing. Vienna, Austria: R Foundation for Statistical Computing. https://www.R-project.org/.
7. Kuhn M. https://cran.r-project.org/web/packages/caret/vignettes/caret.html.
8. RStudio Team. RStudio: integrated development for R. RStudio. Boston, MA: PBC; 2020. http://www.rstudio.com/.
9. Teramae H, Xuan M, Yamashita T, Takayama J, Okazaki M, Sakamoto T. J Comput Chem Jpn. 2018;17:150.
10. Teramae H, Xuan M, Yamashita T, Takayama J, Okazaki M, Sakamoto T. J Comput Chem Jpn. 2019;18:211.
11. Teramae H, Xuan M, Yamashita T, Takayama J, Okazaki M, Sakamoto T. Proceedings of international symposium on environmental-life science and nanoscales technology 2019 (ISENT2019). University of Yangon; 2020. p. 17–21.
12. Teramae H, Xuan M, Takayama J, Okazaki M, Sakamoto T. J Comput Chem Jpn. 2011;21:103.
13. Teramae H. J Comput Chem Jpn. 2023;22:34.
14. Teramae H, Xuan M, Takayama J, Okazaki M, Sakamoto T. AIP Conference Proceedings. 2002;2611:020007.
15. Teramae H, Matsuo T, Niwatsukino K, Inoue R, Noguchi S, Xuan M, Yamashita T, Takayama J, Okazaki M, Sakamoto T. J Comput Chem Jpn. 2020;19:43.
16. Teramae H. Types of machine learning models and specific applications in chemoinformatics. In: Data collection optimization and analysis methods in chemoinformatics. Tokyo: Gijyutsujouhoukyoukai; 2023. p. 209–17.
17. Yaribeygi H, et al. CNS Neuro Disord Drug Targets. 2018;17:207.
18. Siesjö BK, et al. J Neurotrauma. 1995;12:943.
19. Chang KH, et al. Clin Chim Acta. 2018;485:79.
20. Xuan M, et al. Evid Based Complement Alternat Med. 2015;2015:865986.
21. Hlasiwetz H, Barth L. Chem ZentBlatt. 1866. 37 Book 1(28), 443–445.
22. Kikuzaki H, et al. Journal of agricultural and food. Chemistry. 2002;50:2161.
23. Tomono K, Xuan M, Hayasi K, Kamekawa M, Kani D, Matuzaki H, Takayama J, Sakamoto T, Okazaki M, The 132nd annual meeting of the pharmaceutical society of Japan, March, 2014, Kumamoto, Japan.
24. Frisch MJ, Trucks GW, Schlegel HB, Scuseria GE, Robb MA, Cheeseman JR, Scalmani G, Barone V, Petersson GA, Nakatsuji H, Li X, Caricato M, Marenich A, Bloino J, Janesko BG, Gomperts R, Mennucci B, Hratchian HP, Ortiz JV, Izmaylov AF, Sonnenberg JL, Williams-Young D, Ding F, Lipparini F, Egidi F, Goings J, Peng B, Petrone A, Henderson T, Ranasinghe D, Zakrzewski VG, Gao J, Rega N, Zheng G, Liang W, Hada M, Ehara M, Toyota K, Fukuda R, Hasegawa J, Ishida M, Nakajima T, Honda Y, Kitao O, Nakai H, Vreven T, Throssell K, Montgomery JA Jr, Peralta JE, Ogliaro F, Bearpark M, Heyd JJ, Brothers E, Kudin KN, Staroverov VN, Keith T, Kobayashi R, Normand J, Raghavachari K, Rendell A, Burant JC, Iyengar SS, Tomasi J, Cossi M, Millam JM, Klene M, Adamo C, Cammi R, Ochterski JW, Martin RL, Morokuma K, Farkas O, Foresman JB, Fox DJ. Gaussian 09, Revision C.01. Wallingford CT: Gaussian, Inc.; 2016.
25. Hastie T, Tibshirani R, Friedman J. The elements of statistical learning. 2nd ed. New York: Springer; 2009. p. 241–8.
26. Rekker RF, Laak AMT, Mannhold R. Quant Struct-Act Relationsh. 1993;12:152.
27. PubChem. https://pubchem.ncbi.nlm.nih.gov/.
28. Ohtawara K, Teramae H. Chem Phys Lett. 2004;390:84.
29. Toru M. Psychotropic drug manual. 2nd ed. Tokyo: Igaku Shoin; 2001. p. 188–98.

Part II
Quantum Calculations for Advance Materials—Collaboration with Experimental Studies

Chapter 4
On the Mutual Synergy of ^{57}Fe Mössbauer Spectroscopy Methods and Modern Quantum Chemistry Approaches in the Studies of Iron-Based Superconductors

Airat G. Kiiamov, Vladimir Tsurkan, Dorina Croitori, Hans-Albrecht Krug von Nidda, Alois Loidl, Farit G. Vagizov, Lenar R. Tagirov, and Dmitrii A. Tayurskii

Abstract This chapter demonstrates the powerful synergy between ^{57}Fe Mössbauer spectroscopy and first-principles quantum chemistry calculations for investigating iron-based superconductors, focusing on over-stoichiometric iron telluride FeTe. By combining experimental data with ab initio density functional calculations, we address challenges in interpreting complex Mossbauer spectra resulting from overlapping signals and hyperfine interactions. Traditional two-component models were insufficient, so we developed refined models: a four-group model for $Fe_{1.05}Te$ (isolated plaquettes) and a three-group model for $Fe_{1.125}Te$ (edge-sharing

A. G. Kiiamov · F. G. Vagizov · L. R. Tagirov · D. A. Tayurskii (✉)
Institute of Physics, Kazan Federal University, Kazan, Russia
e-mail: i@dtayursk.ru

A. G. Kiiamov
e-mail: airatphd@kazanfederaluniversity.ru

L. R. Tagirov
e-mail: ltagirov@mail.ru

V. Tsurkan · D. Croitori
Institute of Applied Physics, Moldova State University, Chisinau, Moldova
e-mail: vladimir.tsurkan@physik.uni-augsburg.de

V. Tsurkan · H.-A. Krug von Nidda · A. Loidl
Experimental Physics V, Center for Electronic Correlations and Magnetism, Institute of Physics, University of Augsburg, Augsburg, Germany
e-mail: hans-albrecht.krug@physik.uni-augsburg.de

A. Loidl
e-mail: alois.loidl@physik.uni-augsburg.de

L. R. Tagirov
Zavoisky Physical-Technical Institute, FRC Kazan Scientific Center of RAS, Kazan, Russia

T. Onishi (ed.), *Advances in Quantum Science II*,
https://doi.org/10.1007/978-981-95-9721-5_4

plaquettes). These models show how excess iron perturbs the local electric field gradients and charge densities, with effects extending to the third coordination sphere. The combined approach resolves ambiguity in spectral analysis and links atomic-scale perturbations to bulk properties. This work advances our understanding of stoichiometry-dependent electronic heterogeneity in Fe-based superconductive materials and emphasizes the importance of combining experimental and computational methods for accurate characterization of materials. The results provide a basis for targeted design of quantum materials with tailored superconducting properties.

Keywords Mössbauer spectroscopy · Superconductor · Isomer shift · Quadrupole splitting · Magnetic splitting · Density functional theory (DFT) · Iron telluride · Crystal structure

4.1 Introduction

4.1.1 Superconductor Research in Twentieth Century

The field of superconductivity has a rich and fascinating history that spans over a century. It all began in 1911 when the Dutch physicist Heike Kamerlingh Onnes discovered superconductivity in mercury at temperatures close to absolute zero [1]. This groundbreaking discovery opened up a new area of research and led to the development of various theories to explain the phenomenon.

Initially, most of the research was focused on materials like niobium and lead, which exhibited superconductivity at extremely low temperatures. These superconductors found limited practical applications because of the high cost and complexity of maintaining such low temperatures. However, the discovery of high-temperature superconductors in the late twentieth century revolutionized the field. In 1986, Johannes Georg Bednorz and Karl Alexander Müller discovered a class of copper-oxide-based superconductors, known as cuprates, which exhibit superconductivity at much higher temperatures than previously thought possible [2]. This breakthrough earned them the Nobel Prize in Physics in 1987 and sparked a wave of enthusiasm and renewed interest in superconductivity research.

4.1.2 Iron-Based Superconductors

Despite the promise of cuprates, many challenges remained, including their complex and delicate structure, which made them difficult to work with and limited their practical applications. This is where iron-based superconductors enter the scene. In 2008, these materials, which were discovered by Hideo Hosono's research team at the Tokyo Institute of Technology, marked a significant milestone in the field.

Iron-based superconductors offer a simpler and more robust platform for studying superconductivity. Unlike cuprates, it is easier to synthesize and manipulate them [4].

The emergence of iron-based superconductors expanded the scope of research and opened up new possibilities for understanding the mechanisms underlying superconductivity. These materials have a unique electronic structure and magnetic properties that differ from those of cuprates, providing valuable insights into the complex interplay between electronic, magnetic, and structural properties in superconductors. Moreover, iron-based superconductors exhibit superconductivity at relatively high temperatures; their potential for practical applications has been enhanced [5].

A defining feature of this new class of superconductors is compatible with conventional Mössbauer spectroscopy techniques, which employ the widely-used ^{57}Co gamma-ray source [6]. Unlike cuprates—where this method is fundamentally inapplicable due to the absence of Mössbauer-active nuclei—iron-based superconductors enable direct investigation of their physical properties, particularly within the local environment of iron nuclei. This unique capability provides atomic-level insights into the superconducting mechanisms of iron-based materials.

4.1.3 Mössbauer Spectroscopy

Mössbauer spectroscopy is a powerful scientific technique used to study the properties of atomic nuclei in materials. It appears like a "high-precise microscope" that lets scientists peek into how atoms behave in solids, such as metals, minerals, or even biological samples. By observing how certain atomic nuclei absorb γ-rays, this method reveals details about the chemical and physical environment of those atoms. The method was discovered by Rudolf Mössbauer in 1958, earning him a Nobel Prize. It's special because it can detect incredibly small changes in the energy of gamma rays caused by the environment around the nucleus, such as the chemical bonds or magnetic fields in a material. This technique is unique because it's extremely sensitive to subtle differences in a material's structure, making it a go-to tool in fields like physics, chemistry, and materials science. For example, it can tell scientists whether iron atoms in a mineral are bonded to oxygen or sulfur, or whether a material is magnetic.

Mössbauer spectroscopy has emerged as a powerful and indispensable technique for investigating the local structure of solid-state materials. This method exploits the recoil-free resonance absorption of gamma rays to provide unrivaled atomic-scale resolution of local environments—particularly for iron nuclei—yielding detailed information about electronic structure, magnetic ordering, and lattice dynamics [7]. Due to its unique combination of nuclear-level sensitivity and non-destructive characterization, this method has been established as a transformative tool across condensed matter physics, materials science, and related disciplines [7]. Despite such advantages, Mössbauer spectroscopy can present challenges when it comes to interpreting its results accurately. One of the main difficulties lies in the complexity of the spectra, which were observed especially in materials with multiple phases or having disordered structures. The overlapping of spectral lines, which often makes it hard to discern each line, leads to potential misinterpretations. Additionally, since factors such as hyperfine interactions are depend on external conditions like temperature and pressure, layers of complexity are added for the correct interpretation of the spectra.

Furthermore, during the analysis of Mössbauer spectra, the crystal structure and electronic properties of iron-based superconductors are necessary. Researchers must account for various parameters, including isomer shifts (IS), quadrupole splittings (QS), and hyperfine fields (HS), which can vary significantly depending on the local environment of the nuclei. For a correct evaluation of the various Mössbauer parameters, a thorough knowledge of previous precedents, including numerical modeling, is required. Consequently, while Mössbauer spectroscopy is a powerful tool, the accurate interpretation demands expertise and a comprehensive approach to the data analysis [8].

Principal Experimental Scheme

Mössbauer spectroscopy works by shining gamma rays from a radioactive source onto a sample and measuring how the atomic nuclei in the sample interact with those rays. Let's consider a simplified breakdown of the process:

- The source and the sample: a radioactive source, like ^{57}Co (which decays down to ^{57}Fe), emits gamma rays with a very specific energy. These rays are directed at the sample, which contains atoms of the ^{57}Fe element as an absorber.
- The Doppler effect: for the gamma rays to be absorbed by the sample's nuclei, their energy must match perfectly. To achieve this, scientists move the source back and forth slightly, using a device that shifts the energy of the gamma rays via the Doppler effect. This fine-tunes the gamma rays to the exact energy needed for absorption.
- Measuring the signal: when the gamma rays are absorbed by the sample's nuclei, they produce a "spectrum"—a graph showing how much energy was absorbed at different settings of the source's motion. This spectrum acts like a fingerprint, revealing details about the chemical bonds, oxidation states, or magnetic properties of the atoms in the sample. The key feature of Mössbauer spectroscopy is so-called the "Mössbauer effect," which allows gamma rays to be absorbed with no loss of energy to the surrounding material excitations—the recoilless process. This makes the measurements incredibly precise, as if the atoms are "frozen" in place during the process.

Now continue the exploration of the fundamentals of a matter properties studies by means of Mössbauer spectroscopy through three key concepts: isomer shift, quadrupole, and magnetic splittings, which are essential features of the spectra produced by this method.

Isomer Shift: Probing the Electron Density

The isomer shift is a key feature in Mössbauer spectroscopy that tells scientists about the electron environment around atomic nuclei in a material. The isomer shift is like a slight change in that frequency, caused by how densely electrons are packed around the nucleus.

This shift happens because the energy of the gamma rays absorbed by the nucleus depends on the electron cloud surrounding it. Electrons, especially those in the innermost *s*-orbitals, interact with the nucleus and slightly alter its energy levels. For example, in ^{57}Fe, a common isotope used in Mössbauer spectroscopy, the isomer shift can reveal whether the iron atom is in a different chemical state, like Fe^{2+} or Fe^{3+}, because these states have different electron densities. A denser electron cloud (often in Fe^{2+}) shifts the energy one way, while a less dense one (like Fe^{3+}) shifts it the other way.

In a Mössbauer spectrum, the isomer shift appears as a displacement of the entire spectrum along the energy axis (usually measured in velocity units, like millimeters per second). For non-specialists, this can be a clue to what chemical bonds an atom is involved in, helping scientists identify things like oxidation states or the type of material, such as distinguishing pure iron from rust or other non-oxide iron compounds.

Quadrupole Splitting: The Sensing Electric Fields

Quadrupole splitting occurs when the electric field around a nucleus is not perfectly symmetrical, causing the energy levels of the nucleus to split. The nucleus in Mössbauer-active atoms, such as ^{57}Fe, has a property called a "quadrupole moment," which makes it sensitive to unevenities of electric fields (called electric field gradients—EFG). When the nucleus is in a symmetric environment, like a perfect crystal, no splitting occurs. But in less symmetrical conditions—like when an iron atom is bound to different types of atoms in a molecule—the energy levels of the nucleus split into two. This produces a Mössbauer spectrum with two peaks, called a "doublet," instead of a single peak.

For example, in a mineral, quadrupole splitting can reveal whether iron atoms are in a distorted crystal structure or bound to certain groups, such as oxygen or sulfur. To the non-specialists, quadrupole splitting is like a signal that tells scientists how asymmetric the environment around the nucleus is, providing clues about the structure or chemistry of the material.

Magnetic Splitting: The Detecting Magnetic Fields

Magnetic splitting, also called hyperfine magnetic splitting, occurs when a nucleus is placed in a magnetic field, either from the material itself or from an external source. This effect is caused by the interaction between the magnetic moment of the nucleus and the magnetic field around it. For instance, in magnetic materials like iron or magnetite, the internal magnetic field splits the nuclear energy levels, producing a distinct pattern in the Mössbauer spectrum. The size of the splitting (the distance between the peaks) tells scientists how strong the magnetic field is, while the pattern can indicate whether the material is ferromagnetic (like a permanent magnet) or not.

For non-specialists, magnetic splitting is like a magnet detector: it shows whether the atoms in a material experience a magnetic field and how strong it is. This is incredibly useful for studying magnetic materials used in hard drive technologies or for analyzing minerals in geology to understand Earth's magnetic history.

In summary, the isomer shift, quadrupole and magnetic splittings are the three fundamental features of the Mössbauer spectrum. They act like a fingerprint, revealing details of the electron density, electric field symmetry, and magnetic properties of the atomic nuclei environment. Together, these features make Mössbauer spectroscopy a versatile tool for exploring materials in fields ranging from chemistry and physics to archaeology and planetary science.

4.1.4 **Ab Initio** ***First-Principles Calculations***

This chapter highlights how *ab initio* first-principles calculations greatly enhance the interpretation of Mössbauer spectroscopy data, using iron-based superconductors as a case study. The computational methods provide atomic-scale insights into the electronic and magnetic properties that govern the spectral behavior in these materials. By leveraging *ab initio* modeling, researchers can predict important spectral parameters—such as hyperfine interactions, isomer shifts, and quadrupole splittings—with high precision, establishing stronger correlations between theoretical predictions and experimental observations [9].

The synergy between computational and experimental approaches addresses key challenges in Mössbauer spectroscopy studies of iron-based superconductors. First-principles simulations uncover the quantum mechanical origins of spectral features: overlapping hyperfine lines, spectral difference between structurally similar atomic configurations within the iron sublattice. Due to this integrated methodology, in spectral analysis, qualitative interpretation is transformed into quantitative evaluation, directly linking raw spectral data to the intrinsic electronic and magnetic properties. Ultimately, it advances our understanding of atomic-scale phenomena that dictate both superconducting behavior and the corresponding Mössbauer spectral signatures.

4.2 The Emergence of Iron-Based Superconductors: Iron Telluride Problem

4.2.1 Superconducting LaFePO and FeSe

The story of iron-based superconductors began in 2006. The Hosono research team uncovered superconductivity in an unexpected candidate: LaFePO, a layered compound containing iron, phosphorus, and oxygen [10]. This discovery defied conventional wisdom, as strong magnetic tendencies of iron were for a long time assumed to suppress superconducting behavior. With a modest critical temperature (T_c) of around 4 K, LaFePO represented a tentative but revolutionary step. Its structure—alternating layers of iron-phosphorus (Fe–P) and lanthanum-oxygen (La–O)—hinted at the importance of two-dimensional geometry, a theme that would dominate later research. Though its T_c was low, LaFePO laid the groundwork for a new class of materials, later termed the "1111" family, where lanthanides, iron, pnictogens, and oxygen formed the backbone of unconventional superconductivity [10].

Later in 2008, the narrative took a surprising turn with the discovery of superconductivity in FeSe, a binary iron selenide devoid of arsenic, phosphorus, or oxide layers [11]. Its structure—tetragonal FeSe sheets stacked via weak van der Waals bonds—was strikingly simple, resembling the mineral lead oxide (PbO). Under ambient pressure, FeSe exhibited a T_c of 8 K, but this belied its hidden potential [11]. When compressed, its critical temperature soared to 37 K, while ultrathin films interfaced with substrates like $SrTiO_3$ achieved T_c values exceeding 40 K, even approaching 100 K in some interfacial configurations [12]. The simplicity of the FeSe compound became its strength: it isolated the iron-chalcogenide layer as the essential superconducting unit, free from the complexity of pnictogens or oxide spacers. This revelation redirected focus toward low-dimensional systems and interfacial engineering, while studies of the electronic structure of FeSe uncovered exotic behaviors, such as nematic electronic phases and strong correlation effects.

From LaFePO to FeSe, the evolution of iron-based superconductors illustrates a broader scientific journey—from serendipitous discovery to targeted material design. Early systems relied on intricate doping and layered oxides, but FeSe demonstrated that the essence of superconductivity resided in the iron-chalcogenide planes themselves. These advances not only expanded the palette of high-temperature superconductors but also deepened questions about the universal mechanisms driving unconventional superconductivity, ensuring iron-based materials remain at the forefront of quantum material research.

4.2.2 Superconducting FeTe

Building upon the established superconducting properties of iron selenide (FeSe), the prediction that its isostructural and isovalent counterpart iron telluride (FeTe) would exhibit enhanced superconductivity appeared well-founded from several perspectives [13]. So, both compounds crystallize in the PbO-type structure (space group $P4/nmm$) featuring stacked Fe-X layers (X = Se, Te) with van der Waals gaps between them, while maintaining identical formal oxidation states (Fe^{2+} and X^{2-}). This structural and electronic similarity naturally suggested comparable superconducting behavior. It was hypothesized to enhance the superconducting critical temperature (T_c) in FeTe relative to FeSe. This expectation gained further support from pressure-dependent studies of FeSe, where hydrostatic pressure was shown to increase T_c from 8 K at ambient pressure to 37 K at 8 GPa [14]. Since the Te substitution effectively introduces "chemical pressure", analogous enhancements in superconducting properties were anticipated for the telluride system [15].

However, experimental results contradicted these expectations. Real measurements have shown no superconductivity in iron telluride samples [16]. The significant discrepancy between the predicted and experimentally observed properties of iron telluride prompted intense research focus on this compound, as it potentially indicated fundamental limitations in the theoretical framework.

Thereby, the iron telluride system had emerged as a crucial obstruction to obtain a complete understanding of the physics of iron-based superconductors, owing to remarkable deviations of its properties from theoretical predictions especially in terms of superconductivity [17].

4.3 Crystal Structure of Iron Telluride

One of the first comparative investigations of FeTe and FeSe systems using Mössbauer spectroscopy was reported in the study by Mizuguchi et al. [18]. The work offered an important early contribution to understanding how subtle compositional changes between selenium and tellurium could dramatically alter the electronic behavior. In that study, the Mössbauer spectra of FeSe and FeTe powdered samples showed significantly different line shapes. FeSe displays a simple, symmetric singlet at all temperatures, indicative of a uniform paramagnetic state. In contrast, the FeTe spectrum shows pronounced complexity even at room temperature, with significantly broadened and asymmetric lines that resemble a distorted doublet rather than a pure singlet. Crucially, this spectral complexity in FeTe is related to the establishment of long-range magnetic order. Instead, the anomalous line shape likely reflects intrinsic electronic or structural heterogeneity of the FeTe system.

The asymmetric shape of the iron telluride spectrum had been associated with the presence not only of iron atoms in FeTe layers (Fe1), but also of additional iron atoms traditionally denoted as interstitial iron (Fe2). These excess iron atoms (Fe2) occupy the crystallographic $2c$ position located between the FeTe layers, whereas the iron atoms within the layers themselves reside in the $2a$ position as depicted in Fig. 4.1.

Mössbauer spectra of iron telluride with excess interstitial iron have been widely described in the literature using a two-component model, where each sub-spectrum

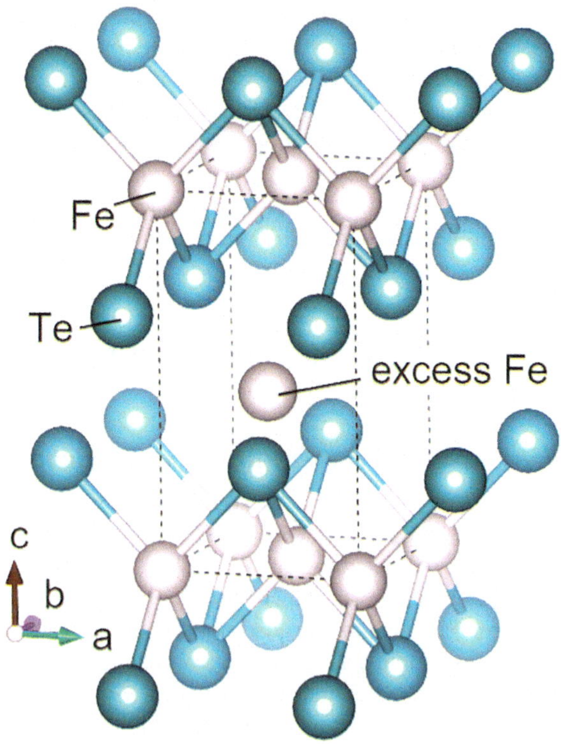

Fig. 4.1 Schematic representation of the iron telluride crystal structure, the labeled excess iron occupying the crystallographic 2*c* position

corresponds to iron atoms occupying the crystallographic 2*a* and 2*c* positions, conventionally labeled as Fe1 and Fe2, respectively [19]. In the present study, we retain the same notation.

4.3.1 The Experimental Configuration and the Geometric Arrangement of the Sample

We employed high-quality single-crystalline iron telluride samples with stoichiometric compositions of $Fe_{1.05}Te$ and $Fe_{1.125}Te$ for our Mössbauer spectroscopy investigations [20, 21]. Single-crystal investigations provide superior structural and electronic characterization compared to powder samples, as demonstrated by at least two reasons:

1. Elimination of grain boundary effects and partially-preferred orientation artifacts;
2. Direct access to anisotropic physical properties through crystallographic orientation control.

It should be emphasized that all absorbers (samples) used for Mössbauer spectroscopy measurements were in the plate form. Moreover, x-ray diffraction analysis confirmed that the crystallographic c-axis was normal to the sample surface in the cases of all examined crystals [20, 21]. The experimental geometry was arranged so that the γ-ray propagation direction was parallel to the sample surface normal. Consequently, the crystallographic c-axis and the γ-ray wave k-vector were mutually parallel. Additional measurements were also performed with a non-zero β-angle between the c-axis and the k-vector.

4.4 Mössbauer Spectroscopy of Single-Crystalline $Fe_{1.05}Te$ Compound

4.4.1 Room-Temperature Measurements

The room-temperature (RT) Mössbauer spectrum of a single-crystalline sample of $Fe_{1.05}Te$ is presented in Fig. 4.2 (linked black dots). The spectrum had shown two separated components, which are called doublet in terms of Mössbauer spectroscopy. The observed asymmetric quadrupole doublet clearly demonstrates the paramagnetic state of iron atoms, as the absence of magnetic splitting combined with quadrupole interaction produces this characteristic spectral signature. The Mössbauer spectrum acquired over a wide velocity range exhibits no additional spectral components that could be attributed to impurity phases—particularly ferromagnetic phases with large hyperfine fields at iron nuclei (see inset, Fig. 4.2).

As a first approximation and in accordance with previous approaches, the room-temperature spectrum of $Fe_{1.05}Te$ was modeled using two doublets, corresponding to iron atoms occupying distinct crystallographic sites (Fe1 and Fe2). The fit (Fig. 4.2, solid lines) yields relative spectral weights of 95% (Fe1) and 5% (Fe2), in excellent agreement with the nominal composition determined by the wavelength-dispersive X-ray electron-probe microanalysis.

The Fe2 sub-spectrum (magenta line) exhibits a symmetric unresolved quadrupole doublet, while the dominant Fe1 doublet (green line) displays an asymmetric peak with line-amplitudes ratio of 3:1.9. Nevertheless, for the case of an ideal single crystal with axial symmetry the expected ratio is 3:1 because the θ angle between the γ-ray propagation vector k and the principal axis of the electric field gradient tensor on the Fe1 nuclei is supposed to be zero due the axial symmetry around the c-axis. Furthermore, the symmetric line shape of the Fe2 sub-spectrum also remains unresolved. The observed deviations arise from different reasons, such as:

1. Sample misalignment—a slight tilt of the crystallographic c-axis away from the surface normal;
2. Impurity phases—minor contributions from secondary phases.

The perpendicular orientation of the crystallographic c-axis with respect to the sample surface as well as the absence of impurity phases were proven by X-ray

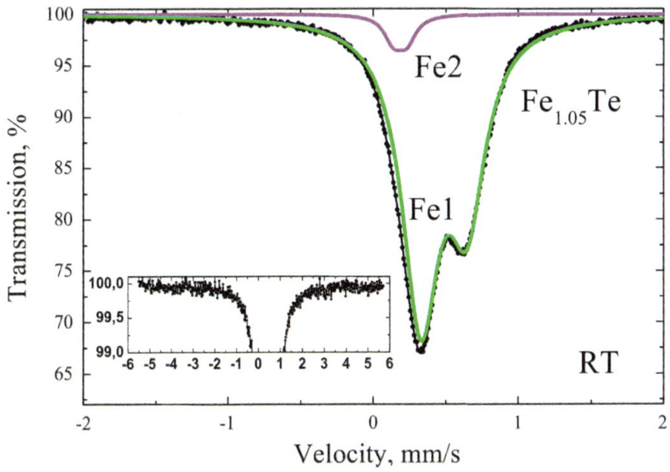

Fig. 4.2 Room-temperature Mössbauer spectrum of $Fe_{1.05}Te$ single-crystalline sample fitted by two doublets (see details in the body text). Inset presents zoomed spectrum obtained in wide velocity range

diffraction analysis and Mössbauer spectroscopy measurements in a wide velocity range. Therefore, we conclude that the hyperfine parameters obtained from this preliminary approximation represent incorrect artifacts due to the model's limitations in our first approach, rather than the system's physical properties. These points on the inadequacy of the simple two-component approach to accurately describe the whole complexity of the intrinsic nature of the compound.

4.4.2 *First-Principles Calculations of $Fe_{1.05}Te$ Compound*

To resolve these complexities in the interpretation of Mössbauer spectroscopy results, we employed *ab initio* density functional theory calculations. The calculations have been performed using the Vienna *ab initio* simulation package integrated into the *MedeA* software complex [23, 24]. We first investigated ideal stoichiometric FeTe (without excess iron) as a reference case. Density functional theory (DFT) calculations using the projector-augmented wave (PAW) method revealed that the principal axis of the electric field gradient tensor at each iron site aligns precisely with the crystallographic *c*-axis, yielding a single quadrupole splitting value of QS$= -0.152$ mm/s. Notably, these results confirm our theoretical supposition that the principal axes of the electric field gradient tensor align with the crystallographic *c*-axis as a direct consequence of the tetragonal symmetry of iron telluride with no excess iron.

In order to model the non-stoichiometric $Fe_{1.05}Te$ system, we employed a $3 \times 3 \times 1$ supercell consisting of 18 Te atoms and 19 Fe atoms, closely matching the

Table 4.1 Electric field gradients—EFG parameters obtained from *ab initio* calculations and fitting of room-temperature Mössbauer experimental data

Sample		Theory			Experiment		
	Site	α (°)	QS (mm/s)	W	α (°)	QS (mm/s)	W
$Fe_{1.05}Te$	Fe2	0.00	0.178	0.05	0.00	0.104	0.05
	Fe1/1	48.89	0.332	0.21	48.53	0.492	0.26
	Fe1/2	32.46	–0.129	0.42	34.70	–0.292	0.46
	Fe1	0.00	–0.163	0.32	0.00	–0.177	0.23
FeTe	Fe1	0.00	–0.152	1.00			

composition of the experimental sample. The presence of excess iron atoms Fe2 perturbs the electric field gradient tensors at adjacent Fe1 sites, leading to changes in both the quadrupole splitting and the α-angle between the EFG principal axis and the crystallographic c-axis (which is parallel with the EFG principal axis of Fe2 atoms) (Fig. 4.3). Based on their EFG signatures, the iron atoms were categorized into four groups: Fe1/1, Fe1/2, Fe1, and Fe2. The excess interstitial iron forms its own distinct group also marked as Fe2. The Fe1/1 and Fe1/2 groups, characterized by non-zero α-angles, reside in two coordination rounds centered around Fe2, with radii of $R_1 = 1.91$Å (Fe1/1) and $R_2 = 4.27$Å (Fe1/2), respectively. Iron atoms located beyond 5Å from Fe2 were classified as Fe1, corresponding to unperturbed lattice sites.

The calculated EFG parameters are summarized in Table 4.1 (under the "Theory" header). Notably, the quadrupole splitting values for iron atoms in the third group (Fe1) and those in ideal stoichiometric FeTe are nearly identical (see Table 1, last two rows). This close agreement indicates that the influence of excess Fe2 on the distant Fe1 atoms is negligible.

Therefore, we can conclude that in $Fe_{1.05}Te$ there are four groups of iron atoms, which have different EFG parameters: three groups consisting of Fe1 iron atoms, the group of Fe2 iron atoms, which contains only excess iron atoms (Fig. 4.3). Although *ab initio* calculations are known to yield only semiquantitative estimates of Mössbauer hyperfine parameters [46–48], they provide valuable complementary information for spectral analysis. Such theoretical results can facilitate the interpretation of complex experimental data, particularly for a multicomponent spectrum. When the room-temperature Mössbauer spectra can be adequately modeled using four distinct doublets, with the hyperfine parameters of each doublet consistent with our computational results. Each doublet in the model is characterized by the isomer shift (IS), quadrupole splitting (QS), the angle θ between the principal axis of the electric field gradient tensor and the γ-radiation propagation direction k-vector, and the partial spectral area (W). As the wave-vector k is parallel to the crystallographic c -axis of the sample, the angles θ and α coincide.

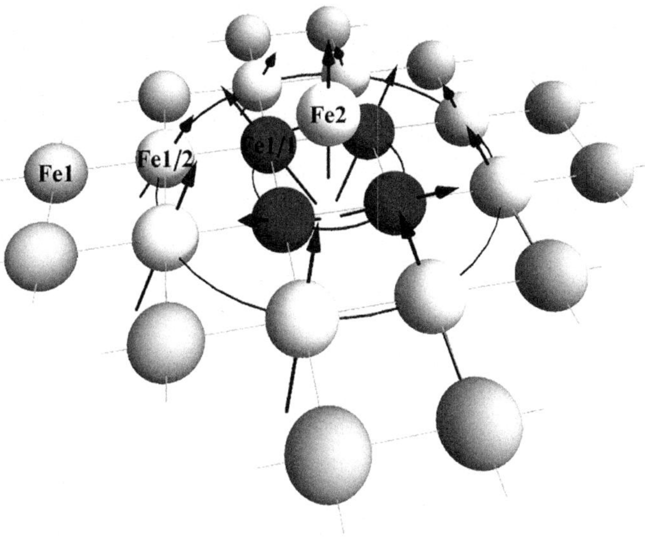

Fig. 4.3 Schematic representation of the groups of iron atoms within the four-group model. Black arrows indicate directions of principal axes of electric field gradients—EFG tensors

Refinement of the Room-Temperature Spectrum of Single-Crystalline $Fe_{1.05}Te$ with Normal Incidence of γ-Radiation

For the fitting procedure, the initial values of QS and α-angle were taken from our *ab initio* calculations (see Table 4.1, "Theory" header), while the initial isomer shift value of IS = 0.45 mm/s was adopted from Ref. [25]. During refinement, the IS, QS, and α values were adjusted slightly to find the best agreement of a model spectrum with the experimental one (Fig. 4.4). The final fitted parameters are summarized in the Table 4.1 columns under the "Experiment" header.

As can be seen from Table 4.1, the EFG parameters obtained from the *ab initio* calculations agree well with the fitted ones: the values of α and W are very close; there are some discrepancies for the QS values which are rather expected within DFT approach [46–48], however, the sign of QS is always correct.

The asymmetric line shapes in the Fe1/1 (violet) and Fe1/2 (magenta) sub-spectra result from the misalignment between the EFG principal axes and the γ-ray propagation k-vector, θ-angles have been obtained as the fitting result. For Fe1 and Fe2 sub-spectra, the QS values are too small (see Table 4.1) to resolve distinct doublets, as they fall below the experimental linewidth (0.097 mm/s, dominated by the natural linewidth).

The isomer shift values for the Fe1/1, Fe1/2, and Fe1 groups are nearly identical. As the distance from the excess Fe2 atom increases, these values converge to 0.452 mm/s, consistent with the reported value of 0.45 mm/s in [25]. In contrast, the IS

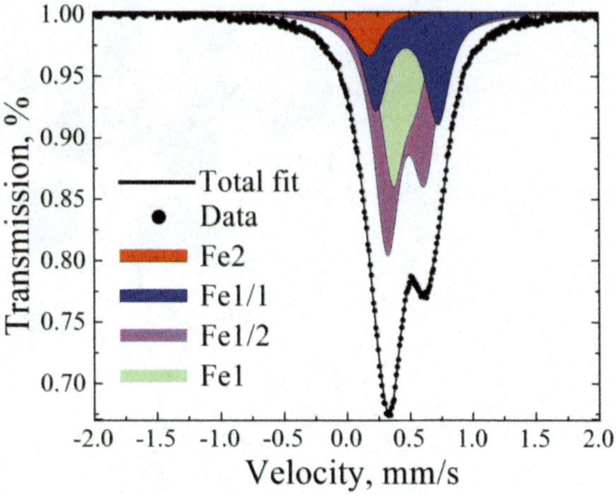

Fig. 4.4 Room-temperature Mössbauer spectrum of single-crystalline $Fe_{1.05}Te$ (black circles) fitted within the four-group model (black solid line). Each sub-spectrum corresponds to one of the groups as it is marked by colors in the legend

for the Fe2 group is nearly three times smaller. This significant difference can be attributed to:

1. Distinct local environments of Fe1 and Fe2 atoms, and consequently;
2. Different valence states—with Fe2 likely in a Fe^{+} state and Fe1 in a Fe^{2+} state.

Refinement of the Room-Temperature Spectrum of Single-Crystalline $Fe_{1.05}Te$ Measured with An Incidence Angle of γ-Radiation of 47°

To validate the refined Mössbauer parameters, we compared the experimental spectrum acquired at a γ-beam incidence angle of $\beta = 47°$ (Fig. 4.5) with a simu-

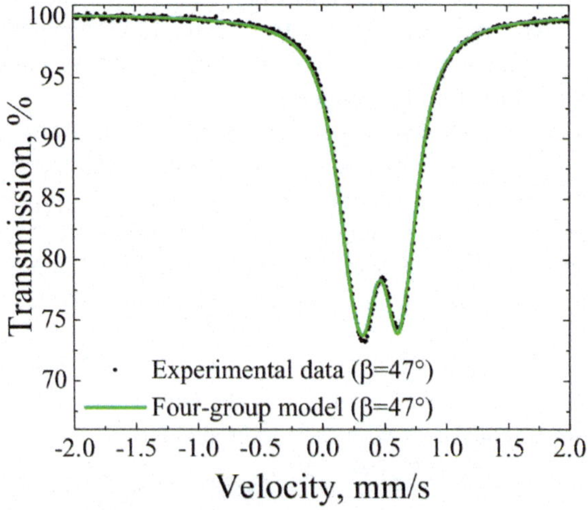

Fig. 4.5 Room-temperature Mössbauer spectrum of single-crystalline $Fe_{1.05}Te$ (black circles) measured at $\beta = 47°$, where β is the angle between the crystallographic c-axis and the γ-photon propagation direction. The green line shows the simulated spectrum calculated using the four-group model with the same geometry ($\beta = 47°$)

lated spectrum using the same angle and EFG parameters (Table 4.1, "Experiment"). Importantly, this comparison did not involve a new fitting procedure—we simply applied the hyperfine parameters obtained from the initial fitting procedure (Table 4.1, "Experiment") to calculate the shape of the spectrum in accordance with the model, the expected spectral shape for the new incidence angle ($\beta = 47°$). The excellent agreement between the experimental and simulated spectra (Fig. 4.5) confirms that the four-group model correctly describes the angular dependence of the Mössbauer spectra in $Fe_{1.05}Te$. This agreement demonstrates that the model, based on our *ab initio* calculations results, accurately represents the intrinsic properties of the compound.

Refinement of the Room-Temperature Spectrum of The Ground Single-Crystal $Fe_{1.05}Te$

To further validate the four-group model, we prepared a powdered sample for analysis, as previous studies have reported similar spectral asymmetry in such form of samples. This approach enables us to systematically test consistency with the model, across all possible orientations of the γ-radiation propagation direction to the crystallographic c-axis of numerous iron telluride crystallites.

Powder sample was derived from a single-crystalline sample by long-time milling. To account for the random grain alignment in the powdered sample, the Mössbauer spectrum was simulated within the four-group model by averaging contributions over all possible values of the β-angle.

This angle defines the orientation between the γ-radiation propagation direction and the crystallographic c-axis of the iron telluride crystallites. The experimentally measured Mössbauer spectrum of the powdered iron telluride sample can be compared with the simulated spectrum in Fig. 4.6. The spectrum retains its characteristic asymmetric doublet shape.

The good agreement between the experimental and simulated spectra proves that the four-group model correctly describes the angular dependence of the Mössbauer spectra in $Fe_{1.05}Te$ for any mutual orientations of the crystallographic axes and γ-radiation propagation direction. It is considered that the observed minor discrepancies arise from preferred crystallites orientation (texture effects, as confirmed by our X-ray diffraction measurements), combined with a potential increase in the Fe^{2+} fraction during powder sample preparation.

4.4.3 Low-Temperature Mössbauer Spectrum of the $Fe_{1.05}Te$ Compound and Its Refinement

Such a differentiation of iron atoms into four groups in the iron telluride structure is obviously expected to manifest also in its magnetic properties. It is known that

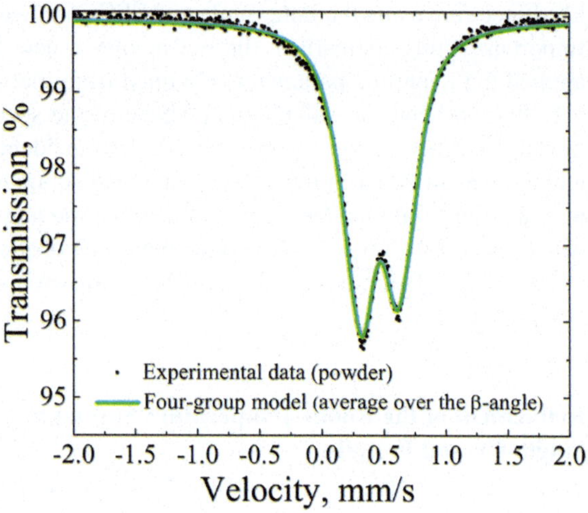

Fig. 4.6 Experimental room-temperature Mössbauer spectrum of the powdered $Fe_{1.05}Te$ sample and the simulated spectrum (green line) within the four-group model (average over the β-angle)

at low temperatures iron telluride is an antiferromagnet [20]. Figure 4.7 displays the Mössbauer spectrum measured at the temperature of 4.2 K for a sample comprising thin flakes exfoliated from a single-crystal ingot, oriented with the (001) cleavage plane parallel to the surface. Low-temperature measurements were carried out with a helium continuous flow cryostat. The spectrum exhibits a complex lineshape with multiple absorption features due to several magnetic sextets. The minimal number of sextets, which allows for describing the low-temperature spectrum accurately, was found to be equal to four. The fit to the low-temperature spectrum is presented in Fig. 4.7 as a black solid line; the hyperfine parameters of the sub-spectra, such as hyperfine field (HS), quadrupole splitting, arbitrary sextet-lines intensities, and partial areas, are presented in Table 4.2.

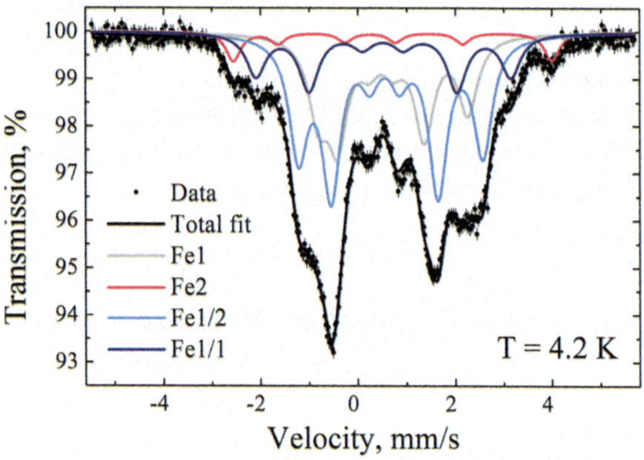

Fig. 4.7 Mössbauer spectrum of $Fe_{1.05}Te$ single-crystalline sample obtained at 4.2 K and fitted by four sextets corresponding to the four-group model of Fe1, Fe2, Fe1/1, and Fe1/2 iron atoms

Table 4.2 The hyperfine parameters of the sub-spectra obtained as a result of the low-temperature Mössbauer spectrum fit (Fig. 4.7). I_i, HS, QS, and W mark, respectively, i-line intensity, hyperfine splitting, quadrupole splitting, and sub-spectra partial area values

Sextet	HS (kOe)	QS (mm/s)	I_1, I_6	I_2, I_5	I_3, I_4	W
Gray	95(2)	0.25	3.0	4.0	1.0	0.31
Pink	202(9)	0.46	3.0	1.2	1.0	0.05
Blue	117(3)	0.12	3.0	4.0	1.0	0.46
Navy blue	162(3)	0.01	3.0	4.0	1.0	0.18

The validity of the four-group model is further corroborated by the four-sextet analysis of the low-temperature Mössbauer spectrum (Fig. 4.7). The relative spectral weights of the partial sextets—0.06, 0.17, 0.46, and 0.31—align closely with the theoretical predictions derived from the four-group model (see Table 4.1, third column under the header "Theory").

Each sextet corresponds to a distinct iron group of the four-group model. Moreover, the sextet attributed to the Fe2 iron group exhibits a large hyperfine field of 202(9) kOe, consistent with a magnetic moment of approximately $2\,\mu_B$ [49–51], while the partial weight of 0.06 observed for that spectrum is very close to a value predicted within the four-group model.

We also note that all the sextets of Fe1/1, Fe1/2, and Fe1 groups display the same characteristic intensity ratio of $3:4:1:1:4:3$, indicating that the magnetic moments of the iron atoms of these groups lie in the sample plane (i.e., perpendicular to the crystal c-axis).

The minor discrepancy between the theoretically predicted and experimentally observed spectral area values arises from slight variations in the Lamb-Mössbauer factors for iron nuclei in different atomic groups. These differences are inherently caused by distinct vibrational properties of the iron atoms across the groups, which directly influence the probability of recoil-free gamma-ray absorption—the fundamental parameter underlying the Mössbauer effect and named as the Lamb-Mössbauer factor. Consequently, we emphasize that all partial spectral areas derived from the experimental data analysis are inherently estimated within a simplification of the mutual equality of Lamb-Mössbauer factor values for the nuclei of iron atoms of each group. Thus, the observed minor discrepancies between the predicted four-group model values and the experimentally fitted sub-spectral areas should not be interpreted as fundamental limitations of the model, but rather as consequences of experimental limitations—particularly the necessary assumption of uniform Lamb-Mössbauer factor values over all iron nuclei. Nevertheless, even within such limitations, the agreement between the model and low-temperature experimental results is fairly good.

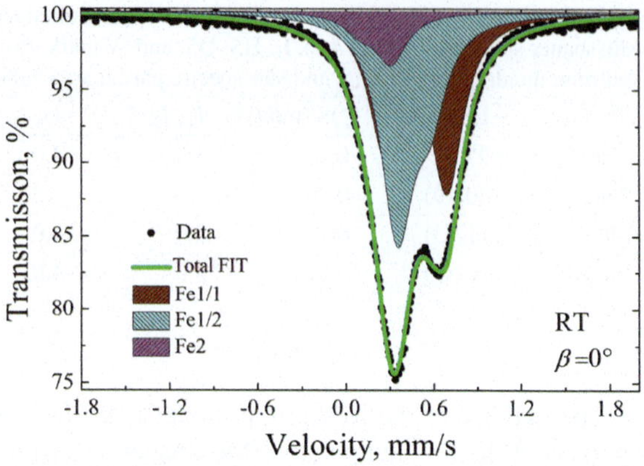

Fig. 4.8 Room-temperature Mössbauer spectrum of single-crystalline $Fe_{1.125}Te$ (black solid symbols) and the spectrum fitted within the *ab initio* results

4.4.4 Summary on Mössbauer Spectroscopy Studies of $Fe_{1.05}Te$ Compound

This section demonstrated that the Mössbauer spectra of $Fe_{1.05}Te$ single-crystalline and powdered samples are accurately described by the four-group model based on our accurate *ab initio* calculations. Within the model, three groups are formed by iron atoms comprising the FeTe layers (Fe1), while the fourth one represents excess interstitial iron (Fe2). This model is validated by a good description of the spectra at both room- and low-temperature spectra, it successfully reproduces the observed angular dependence, and well simulates the spectrum of the powdered sample. The analysis reveals that the over-stoichiometric excess iron atoms Fe2 perturb the charge-density distribution up to the second coordination sphere. Low-temperature measurements further showed that the spin density is also affected, confirming the ability of the model to describe both structural and magnetic modifications induced by the excess iron.

4.5 Mössbauer Spectroscopy of Single-Crystalline $Fe_{1.125}Te$ Compound

Further, we proposed that the validity of the model should also be verified by examining iron telluride with other concentrations of excess iron atoms. In a case of a lower concentration of interstitial iron, the iron atoms would still be distributed among four distinct groups, but only with a higher relative content in the Fe2 group. Thus, it would be more insightful to investigate compounds with an excess iron content greater than 0.05.

Here we report Mössbauer spectroscopy results for highly off-stoichiometric iron telluride $Fe_{1.125}Te$. The spectra were complemented by *ab initio* calculations, to

Table 4.3 Calculated Mössbauer parameters for $Fe_{1.125}Te$ from the *ab initio* simulations

Group	QS	α-angle	W
Fe2	0.11	0	0.11
Fe1/1	0.24	49	0.44
Fe1/2	–0.04	6	0.44

Table 4.4 Mössbauer spectral parameters obtained after the fitting (Fig. 4.8) of the experimental spectrum of single-crystalline $Fe_{1.125}Te$

Group	QS	α-angle	W	IS
Fe2	0.23	0	0.13	0.20
Fe1/1	0.42	46	0.44	0.48
Fe1/2	–0.22	8	0.44	0.46

further validate the four-group model. The Vienna *ab initio* simulation package integrated into the MedeA software complex was employed.

Our calculations for the $Fe_{1.125}Te$ system have shown the appearance of three groups of different iron ions (no Fe1 group appears). These groups have different values of Mössbauer parameters, such as the quadrupole splitting and the α-angle between EFG principal axis and the crystallographic c-axis. One of the groups, the Fe2 group, contains only Fe2 ions. The other two groups accumulate the first and second coordination rings around the excess iron ions Fe2 – Fe1/1 and Fe1/2 groups, respectively. Each group is characterized by unique Mössbauer parameters, including quadrupole splitting and the α-angle between the principal axis of the electric field gradient and the crystallographic c-axis. The relative partial spectrum area of each group has been calculated as a ratio of the number of iron atoms in the group to the total number of atoms in the supercell. The calculated Mössbauer parameters are collected in Table 4.3.

We utilized the *ab initio* results for the analysis of the experimental Mössbauer spectrum as follows:

1. QS and α for the fitting procedure were taken from the *ab initio* calculations (Table 4.3) as an initial guess;
2. the initial value for the isomer shift was taken as IS=0.425 mm/s in accordance with [25];
3. during the fitting procedure, the QS, α-angle, and IS values were varied slightly to better reproduce the experimental spectrum by the proposed model of three groups.

The fitting results are presented in Table 4.4 and displayed in Fig. 4.8.

We also utilized the *ab initio* results for the analysis of the experimental Mössbauer spectrum obtained for the β-angle equal to 54.7°. No any further fitting procedure has been applied—the model spectrum for $\beta = 54.7°$ was obtained using the parameters of Table 4.4 with taking into account the necessary angular dependencies. Figure

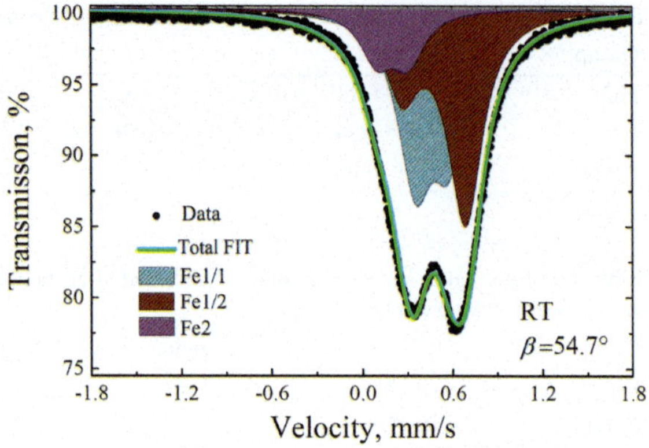

Fig. 4.9 Room-temperature Mössbauer spectrum of single-crystalline $Fe_{1.125}Te$ (black solid symbols) obtained with β-angle value of 54.7° and the spectrum fitted within the *ab initio* results

4.9 represents a comparison of the experimental spectrum and the simulated one. It could be seen that our model is able to describe the angular dependence of the Mössbauer spectra quite well.

The three-group model successfully reproduces the experimental Mössbauer spectra of $Fe_{1.125}Te$ at room temperature, with excellent agreement across varying gamma-ray incidence angles. It could be concluded that the $Fe_{1.125}Te$ system contains three different types of iron ions clearly seen in the Mössbauer spectroscopy experiments. The EFG pattern is generally similar to that of the $Fe_{1.05}Te$ system, where four iron ion groups had been seen.

4.6 Summary on Mössbauer Spectroscopy of Single-Crystalline $Fe_{1+y}Te$ Compounds

The *ab initio* calculations of electric field gradients around interstitial iron (Fe2) atoms revealed a pronounced perturbation in the electron densities of neighboring Fe ions, extending up to the third coordination sphere. The first (Fe1/1) and second (Fe1/2) coordination spheres form together with the central Fe2—a distinct structural motif resembling a plaquette (see Fig. 4.3). While the third coordination sphere remains largely unperturbed, its inclusion into the broader ensemble with Fe2, Fe1/1, and Fe1/2 centers introduces a fourth type of iron site, completing the picture of local electronic heterogeneity.

At low Fe2 concentrations (below 5 at.%), these plaquettes remain well-separated, with negligible overlap, which is due to the low statistical probability of proximal Fe2 interstitials. This regime allowed us to spectroscopically validate the four-group model in $Fe_{1.05}Te$ (Fig. 4.3). However, as the excess interstitial iron Fe2 concentration increases, the plaquettes begin to overlap. At 12.5 at.% interstitial iron—assuming uniform distribution—the third coordination sphere effectively vanishes, leaving no unperturbed regions around Fe2 centers. Our *ab initio* EFG calculations for $Fe_{1.125}Te$

confirm this transition, revealing only three distinct types of iron sites: the interstitial Fe2 ions and the two surviving but modified Fe1/1 and Fe1/2 groups. Notably, the EFG at the Fe1/2 sites is markedly reduced compared to $Fe_{1.05}Te$. It is considered to be due to overlapping influence from neighboring Fe2 centers, which now share common second-sphere Fe1/2 ions.

The case of $Fe_{1.125}Te$ emerges as a critical transition point in the evolution of electronic properties across the $Fe_{1+y}Te$ series. This concentration marks a fundamental shift in the electronic organization of the system:

(i) At concentrations below y = 0.125, the iron telluride forms a distinct "carpet-like" structure composed of well-separated plaquettes. These plaquettes—each consisting of a central Fe2 interstitial surrounded by perturbed Fe1/1 and Fe1/2 coordination spheres—are arranged in a periodic pattern across the Fe1 lattice;

(ii) At precisely y = 0.125, the system reaches a threshold where these plaquettes begin to share edges. The uniform distribution of interstitial iron atoms Fe2 at this concentration eliminates the unperturbed third coordination sphere entirely, causing neighboring plaquettes to interact directly through their Fe1/2 sites. This interaction manifests itself in the notable reduction of electric field gradients at these shared Fe1/2 positions, as revealed by our *ab initio* calculations.

(iii) Beyond y = 0.125, the increasing density of interstitials leads to overlapping plaquettes, forcing the system into a new regime of complex pattern formation. The nature of these emergent structures—likely involving competing interactions between multiple Fe2 centers and their shared coordination spheres—remains an open question that warrants further theoretical and experimental investigation.

This concentration-dependent evolution from isolated plaquettes to interconnected networks provides a unifying framework for understanding the dramatic changes in electronic and magnetic behavior observed across the $Fe_{1+y}Te$ series. The transition at y = 0.125 represents a particularly intriguing case where local perturbations begin to dominate the global electronic structure, offering valuable insights into the interplay between stoichiometry and emergent quantum phenomena in correlated electron systems.

4.7 Conclusion

This study underscores the critical role of integrating Mössbauer spectroscopy with *ab initio* first-principles quantum chemistry calculations to unravel the complex atomic-scale properties of iron-based superconductors, particularly in over-stoichiometric iron telluride systems like $Fe_{1.05}Te$ and $Fe_{1.125}Te$. By combining Mössbauer spectroscopy experimental data with *ab initio* density functional theory calculations, we resolved longstanding challenges in interpreting Mössbauer spectra caused by overlapping hyperfine lines, disordered structures, and hyperfine interactions. Traditional two-component models proved to be inadequate for these systems, prompting the development of refined multi-group frameworks—four distinct iron atom groups for $Fe_{1.05}Te$ and three groups for $Fe_{1.125}Te$. These models revealed how excess interstitial iron perturbs the local electric field gradients and charge densities of neighboring Fe1 sites, extending influence up to the third coordination sphere.

At low interstitial iron concentrations ($Fe_{1.05}Te$), isolated structural plaquettes—comprising a central Fe2 atom and perturbed Fe1 sites—dominate the electronic structure. However, increasing the interstitial iron content ($Fe_{1.125}Te$) triggers edge-sharing between these plaquettes, eliminating unperturbed regions and fostering direct interactions between neighboring plaquettes. This evolution highlights the delicate balance of stoichiometry and emergent quantum phenomena in correlated electron systems. The synergy of experimental and computational approaches not only advances spectral interpretation but also deepens our understanding of the interplay between atomic-scale perturbations and macroscopic superconducting behavior. These insights pave the way for targeted material design, emphasizing the need to account for local electronic heterogeneity in future studies of iron-based superconductors.

So, $Fe_{1+y}Te$ was initially considered a promising superconductor due to its structural similarity to other iron-based superconductors like FeSe, but its persistent antiferromagnetic order and lack of superconductivity puzzled researchers. Our Mössbauer spectroscopy studies have proven that even low concentrations of interstitial iron dramatically alter the electronic structure of $Fe_{1+y}Te$, introducing localized states and enhancing magnetic order, both of which suppress superconductivity. By revealing these effects through precise measurements of isomer shifts, quadrupole splitting, and magnetic splitting, our work highlights the critical role of controlling excess iron to unlock superconductivity in $Fe_{1+y}Te$ and related materials, paving the way for the development of better superconductors for future technologies.

4.8 Outlook

$Fe_{1+y}Te$, an iron-based chalcogenide, was initially viewed as a promising candidate for superconductivity when iron-based superconductors were discovered around 2008. $Fe_{1+y}Te$, with its simple layered structure of iron and tellurium atoms, was expected to follow suit as a parent compound in the $FeTe_{1-x}Se_x$ family, where substituting tellurium (Te) with selenium (Se) increased the superconductivity critical temperature.

The optimism for $Fe_{1+y}Te$ had stemmed from its structural similarity to other iron-based superconductors. Like FeSe, it has a tetragonal crystal structure at room temperature, consisting of iron atoms coordinated by chalcogen atoms (Te or Se) in a layered arrangement. However, despite extensive efforts, pristine $Fe_{1+y}Te$ does not exhibit superconductivity. This unexpected behavior prompted researchers to investigate why $Fe_{1+y}Te$ resists superconductivity, leading to critical insights from Mössbauer spectroscopy.

Our Mössbauer spectroscopy studies have been instrumental in uncovering why $Fe_{1+y}Te$ fails to achieve superconductivity, pinpointing the disruptive role of excess interstitial iron. Mössbauer spectroscopy, which uses gamma rays to probe the local environment of iron-57 nuclei, is uniquely sensitive to the magnetic and electronic properties of iron atoms in the material. Our findings demonstrate that even small amounts of interstitial iron—iron atoms located between the regular FeTe layers—dramatically alter the electronic structure of $Fe_{1+y}Te$, ultimately suppressing superconductivity.

Using ^{57}Fe Mössbauer spectroscopy, we observed that interstitial iron, even at low concentrations (y = 0.05), significantly modifies the electronic environment of iron atoms in $Fe_{1+y}Te$. The Mössbauer spectra at room temperature show an asymmetric doublet, which is attributed to the presence of excess iron in interstitial sites. This doublet reflects variations in the electric field gradient around iron nuclei, indicating that interstitial iron alters the electron density. Specifically, the isomer shift—a measure of electron density at the nucleus—shows that interstitial iron introduces localized electronic states, disrupting the delocalized electron behavior needed for superconductivity.

In contrast, the superconducting FeSe compound exhibits a simpler Mössbauer spectrum of a singlet-line, suggesting a more uniform electronic environment conducive to Cooper pair formation, the basis of superconductivity. Our studies confirm that the excess iron in Fe1+yTe creates an electronic structure dominated by strong correlations and localized magnetic moments, which are incompatible with the itinerant electron states required for superconductivity. This disruption is evident even at low interstitial iron concentrations, highlighting the sensitivity of the electronic structure to these defects.

So, Mössbauer spectroscopy results also reveal that interstitial iron enhances the antiferromagnetic order in $Fe_{1+y}Te$, further suppressing any potential for superconductivity. At low temperatures (e.g., 4.2 K), the Mössbauer spectra of $Fe_{1+y}Te$ display a magnetic sextet, indicating a robust antiferromagnetic order with relatively high hyperfine magnetic fields for iron atoms in the tetrahedral FeTe lattice. Experimental results demonstrate that the interstitial iron not only disrupts the electronic structure but also reinforces the magnetic interactions that compete with superconductivity, effectively "locking" $Fe_{1+y}Te$ into a non-superconducting state.

Our Mössbauer studies provide conclusive evidence that excess iron is a primary reason why $Fe_{1+y}Te$ does not exhibit superconductivity. By systematically analyzing samples with varying amounts of interstitial iron (y), we showed that even small increases in y lead to significant changes in the Mössbauer spectra, including broader hyperfine field distributions and enhanced magnetic splitting. These changes correlate with stronger antiferromagnetic ordering and a more distorted electronic structure, both of which hinder the formation of the superconducting phase.

Acknowledgements The work was supported by the Russian Foundation for Basic Research (RFBR):№ 19-52-12068 and the German Research Foundation (DFG):KR2254/3-1. A. G. Kiiamov and L R. Tagirov acknowledge the support of the Ministry of Science and Higher Education of the Russian Federation (state assignment for the FRC Kazan Scientific Center of RAS). A. G. Kiiamov also thanks Yury Lysogorskiy (ICAMS, Ruhr University, Bochum) for stimulated discussions.

References

1. Onnes K. The resistance of pure mercury at helium temperatures. Commun. Phys Lab Univ Leiden. 1911; 120
2. Bednorz JG, Müller KA. Possible high T_c superconductivity in the Ba-La-Cu-O system. Z Phys B. 1986;64:189–93.
3. Hosono H, et al. Iron-based layered superconductor La[O $_{1-x}F_x$]FeAs (x=0.05-0.12) with T_c=26 K. J Am Chem Soc. 2008; 130:3296–3297
4. Stewart GR. Superconductivity in iron compounds. Rev Mod Phys. 2011;83:1589–652.
5. Ren ZA, et al. Superconductivity at 55 K in iron-based F-doped layered quaternary compound Sm[$O_{1-x}F_x$]FeAs. Chin Phys Lett. 2008;25:2215.
6. Nowik I, et al. Coexistence of ferromagnetism and superconductivity: magnetization and Mössbauer studies of $EuFe_2(As_{1-x}P_x)_2$. J Phys: Condens Matter. 2011;23: 065701.
7. Long GJ, Grandjean F, editors. Mössbauer spectroscopy applied to magnetism and materials science. New York: Springer; 2013.
8. Dyar MD, et al. Mössbauer spectroscopy of earth and planetary materials. Annu Rev Earth Planet Sci. 2006;34:83–125.
9. Zwanziger JW. Computation of Mössbauer isomer shifts from first principles. J Phys: Condens Matter. 2009;21: 195501.
10. Kamihara Y, et al. Iron-based layered superconductor: LaOFeP. J Am Chem Soc. 2006;128:10012–3.
11. Hsu F-C, et al. Superconductivity in the PbO-type structure α-FeSe. Proc Natl Acad Sci USA. 2008;105:14262–4.
12. Ge J-F, et al. Superconductivity above 100 K in single-layer FeSe films on doped $SrTiO_3$. Nature Mater. 2015;14:285–9.
13. Subedi A, et al. Density functional study of FeS, FeSe, and FeTe: Electronic structure, magnetism, phonons, and superconductivity. Phys Rev B. 2008;78: 134514.
14. Margadonna S, et al. Pressure evolution of the low-temperature crystal structure and bonding of the superconductor FeSe ($T_c = 37$ K). Phys Rev B. 2009;80: 064506.
15. Yeh K-W, et al. Tellurium substitution effect on superconductivity of the α-phase iron selenide. Europhys Lett. 2008;84:37002.
16. Li S, et al. First-order magnetic and structural phase transitions in $Fe_{1+y}Se_xTe_{1-x}$. Phys Rev B. 2009;79: 054503.
17. Dai P, et al. Magnetism and its microscopic origin in iron-based high-temperature superconductors. Nature Phys. 2012;8:709–18.
18. Mizuguchi Y, et al. Mössbauer studies on FeSe and FeTe. Physica C. 2010;470:S338-9.
19. Rodriguez EE, et al. Magnetic-crystallographic phase diagram of $Fe_{1+x}Te$. Phys Rev B. 2011;84: 064403.
20. Kiiamov AG, et al. Mössbauer spectroscopy evidence of intrinsic non-stoichiometry in iron telluride single crystals. Ann Phys (Berlin). 2017;529:1600241.
21. Kiiamov AG, et al. Evidence of the Plaquette Structure of $Fe_{1+x}Te$ Iron Telluride: Mössbauer Spectroscopy Study. Phys Status Solidi B. 2019;256:1800698.
22. Rancourt DG. Accurate site populations from Mössbauer spectroscopy. Nucl Instrum Methods Phys Res B. 1989;44:199–210.
23. Kresse G, Furthmüller J. Efficient iterative schemes for ab initio total-energy calculations using a plane-wave basis set. Phys Rev B. 1996;54:11169–86.
24. Petrilli HM, et al. Electric-field-gradient calculations using the projector augmented wave method. Phys Rev B. 1998;57:14690–7.
25. Koteski V, et al. First-principles calculations of tetragonal FeX (X= S, Se, Te): Magnetism, hyperfine-interaction, and bonding. JMMM. 2017;441:769–75.

Chapter 5
Exploring the Adsorption Mechanisms of Heavy Metals and Chemicals on Two-Dimensional Carbon-Based Nanomaterials: Density Functional Theory and Experimental Perspectives

Sadegh Kaviani, Dmitrii A. Tayurskii, Irina Piyanzina, and Oleg V. Nedopekin

Abstract With rapid industrialization and urbanization contributing to the growing contamination of ecosystems, the development of efficient remediation strategies has become increasingly vital. This chapter explores the adsorption mechanisms of heavy metals and chemicals on two-dimensional (2D) carbon-based nanomaterials, integrating insights from density functional theory simulations and experimental studies. Among emerging materials, graphene and its derivatives have shown significant promise due to their high surface area, tunable electronic properties, and exceptional adsorption capacities. We provide an overview of various two-dimensional carbon-based nanomaterials and discuss how structural modifications enhance their adsorption performance. Theoretical perspectives using density functional theory are presented to analyze the electronic structure and adsorption energies, offering a deeper understanding of the adsorption mechanisms at the atomic level. These computational insights are supported by experimental findings that examine critical factors such as pH, temperature, and surface modifications influencing adsorption behavior. By correlating simulation data with experimental results, this chapter highlights the synergistic role of theory and experimental realization in optimizing 2D nanomaterials for environmental applications. Finally, we address current challenges and future directions in the field, emphasizing the need for integrated approaches to design next-generation carbon-based adsorbents for effective environmental remediation.

S. Kaviani
A.B. Nalbandyan Institute of Chemical Physics, 5/2 P. Sevak Street, Yerevan 0014, Armenia

D. A. Tayurskii (✉) · O. V. Nedopekin
Institute of Physics, Kazan Federal University, Kazan, Russia
e-mail: dtayursk@gmail.com

I. Piyanzina
Computational Materials Science Laboratory, Yerevan State University, Yerevan, Armenia

T. Onishi (ed.), *Advances in Quantum Science II*,
https://doi.org/10.1007/978-981-95-9721-5_5

Keywords Adsorption mechanism · Heavy metal · Two-dimensional carbon-based nanomaterial · Density functional theory · Environmental remediation · Chemical contaminant · Graphene derivatives · Surface modification · Adsorption energy · Water treatment

5.1 Introduction

5.1.1 Two-Dimensional (2D) Carbon-Based Nanomaterials

With the rapid expansion of industrialization and urbanization worldwide, ecosystems are increasingly contaminated by pollutants such as heavy metal cations (Pb^{2+}, Hg^{2+}, Cd^{2+}) [1], personal care products (PCPs) [2], and various organic chemicals [3]. These contaminants pose serious risks to environmental health and human well-being by accumulating in water, soil, and living organisms, causing long-term ecological damage and adverse health effects. This urgent issue has driven research toward efficient and sustainable remediation technologies for pollutant removal [4–6]. Among advanced materials, 2D carbon-based nanomaterials have shown exceptional promise [7–9]. Their high specific surface area, excellent electrical and thermal conductivity, and tunable surface chemistry enable remarkable adsorption capacities. Functionalization with surface groups such as hydroxyl (–OH), carboxyl (–COOH), and epoxide moieties, as shown in Fig. 5.1, further enhances their affinity for pollutants, enabling effective adsorption via π–π interactions, electrostatic attraction, and chemical bonding.

2D carbon-based nanomaterials possess unique structural properties, notably their high aspect ratios and extensive surface areas, which make them highly effective adsorbents for environmental remediation. These features provide a large number of active sites that facilitate strong interactions with various contaminants, including heavy metals and organic pollutants such as pesticides, dyes, and pharmaceuticals.

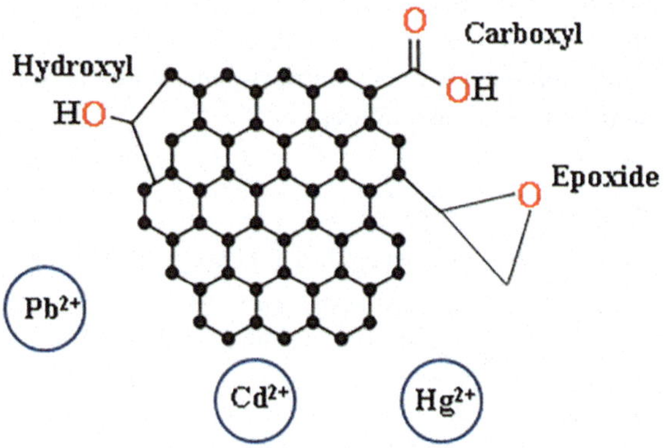

Fig. 5.1 Schematic illustration of a 2D carbon-based nanomaterial functionalized with surface groups (–OH, –COOH, epoxide) for the adsorption of various pollutants, including heavy metals (Pb^{2+}, Hg^{2+}, Cd^{2+}), personal care products (PCPs), and organic chemicals

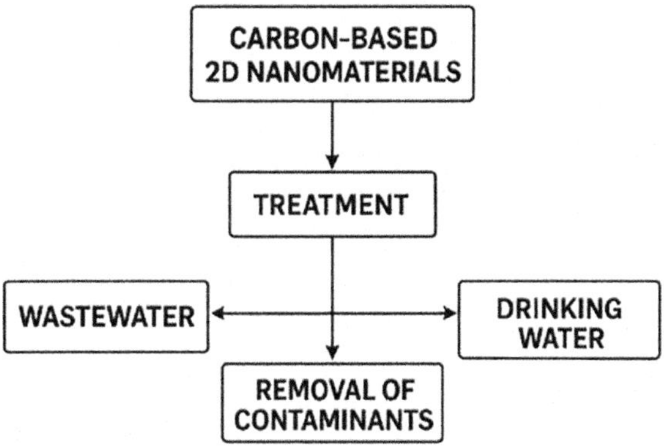

Fig. 5.2 Schematic flowchart illustrating the adsorption of contaminants onto 2D carbon-based nanomaterials

The adsorption process begins as contaminants in aqueous environments come into contact with the nanomaterial surfaces, where physisorption occurs predominantly via van der Waals forces and electrostatic interactions. The large surface area significantly enhances this engagement by offering more binding opportunities. Moreover, the surfaces of these nanomaterials can be chemically functionalized through various modification techniques. Such functionalization introduces specific chemical groups that improve selectivity and binding strength toward targeted contaminants, thereby optimizing the adsorption capacity. Overall, the combination of intrinsic structural advantages and surface engineering enables 2D carbon-based nanomaterials to serve as highly efficient materials for wastewater treatment and the purification of drinking water. This approach represents a significant advancement toward tailored remediation technologies that address the complexities of environmental contamination in modern industrial settings, as illustrated in Fig. 5.2.

5.1.2 Experimental Techniques

Experimental techniques such as adsorption isotherm analyses, kinetic studies, and thermodynamic assessments play a pivotal role in elucidating the interactions between nanomaterials and contaminants, offering critical insights essential for evaluating their efficacy in real-world applications [10–12]. Adsorption isotherm analyses, for instance, involve examining how contaminants are distributed between the solid phase of nanomaterials and the liquid phase, enabling researchers to quantify the capacity and efficiency of these materials in capturing pollutants under varying environmental conditions. This technique helps determine key parameters such as maximum adsorption capacity and the nature of binding sites, facilitating a deeper understanding of the material's performance. Figure 5.3 illustrates typical adsorption isotherms used to analyze the behavior of 2D carbon-based nanomaterials during contaminant removal [13]. To better understand the graph, it is important to note that adsorption isotherms depict the relationship between the amount

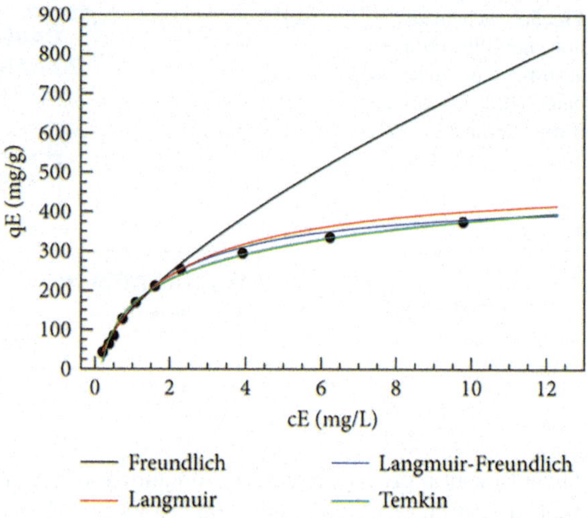

Fig. 5.3 Typical adsorption isotherms illustrating the interaction between contaminants and 2D carbon-based nanomaterials during removal processes. The graphs plot the equilibrium amount of cationic dye adsorbed per unit weight of adsorbent (qE, y-axis) against the equilibrium dye concentration (cE; x-axis) [13]

of contaminant adsorbed onto the nanomaterial and the concentration of contaminant remaining in solution at equilibrium. These curves allow researchers to evaluate adsorption capacity and gain insights into the adsorption mechanism, such as whether it involves monolayer or multilayer coverage and the affinity between contaminant and adsorbent. Different models like Langmuir and Freundlich can be applied to fit the data, providing valuable information on the efficiency and behavior of the nanomaterials under various environmental conditions. Complementing this, kinetic studies explore the rate at which contaminants are removed from solutions by nanomaterials, providing invaluable information about the speed and mechanisms of adsorption processes, which can influence real-time operational effectiveness in environmental remediation applications [14]. Additionally, thermodynamic assessments further enrich this understanding by offering insights into the energy changes associated with adsorption processes, allowing researchers to discern the spontaneity and feasibility of contaminants binding to nanomaterials [15].

5.1.3 Integrated Methodology

Integrating theoretical insights from density functional theory (DFT) simulations with experimental outcomes can significantly advance understanding of adsorption phenomena at the molecular level. DFT enables the calculation of electronic properties, binding energies, and the stability of various adsorption configurations, thereby shedding light on the interactions between contaminants and the nanomaterial surface [16–18]. As illustrated in Fig. 5.4, combining DFT simulations and experimental techniques, such as adsorption isotherms, kinetic studies,

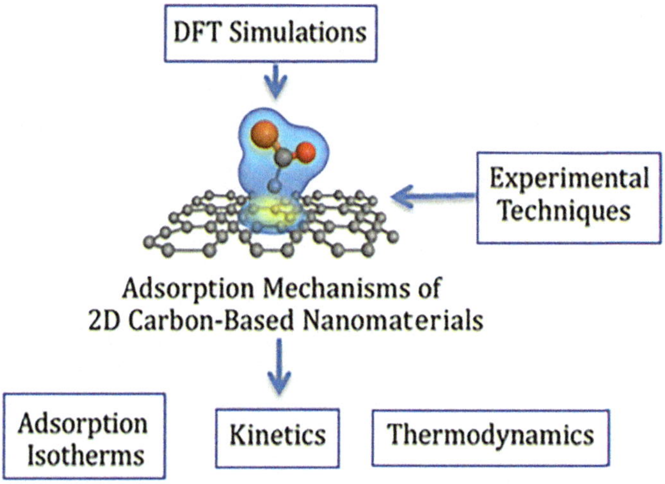

Fig. 5.4 Integration of DFT simulations with experimental techniques to investigate the adsorption mechanisms of 2D carbon-based nanomaterials

and thermodynamic assessments, creates a comprehensive framework for understanding and optimizing the adsorption behavior of 2D carbon-based nanomaterials. This complementary approach not only elucidates the fundamental mechanisms behind adsorption processes but also aids in predicting the performance of these materials under different environmental conditions. An integrated methodology is crucial for designing next-generation 2D carbon-based nanomaterials that effectively address environmental contaminants, ultimately improving remediation strategies and enhancing ecosystem sustainability [27].

In this chapter, we explore the adsorption mechanisms of heavy metals and chemicals on 2D carbon-based nanomaterials through a combination of DFT simulations and experimental studies. The insights gained from both theoretical models and experimental validations are essential for understanding the favorability and mechanisms of adsorption processes. The interplay between theoretical predictions and experimental findings enhances our understanding of how modifications to 2D carbon-based nanomaterials, such as functionalization and structural tuning, can further optimize their adsorption properties. Ultimately, these insights are crucial for designing advanced nanomaterials capable of effectively mitigating the presence of hazardous substances in various ecosystems, thereby contributing significantly to the development of innovative solutions for environmental challenges.

5.2 Basics of Density Functional Theory (DFT) Simulations

5.2.1 *Schrödinger Equation*

The foundation of quantum mechanics and DFT lies in the time-independent Schrödinger equation [20]:

$$\hat{H}\Psi(r_1, r_2, \ldots, r_N) = E\Psi(r_1, r_2, \ldots, r_N). \tag{5.1}$$

Here, $\hat{H}$ is the Hamiltonian operator, Ψ is the many-electron wavefunction, and E is the total energy. Directly solving this equation for many-electron systems is computationally prohibitive, which motivates the use of DFT.

5.2.2 *Kohn–Sham Formalism*

DFT reformulates the many-body problem into a system of non-interacting electrons with the same ground-state density. The Kohn–Sham equation is [21]

$$\left[-\frac{\hbar^2}{2m}\nabla^2 + V_{eff}(r)\right]\Psi_i(r) = \varepsilon_i\Psi_i(r) \tag{5.2}$$

where V_{eff} contains the external potential, Hartree term, and exchange–correlation potential.

5.2.3 *Pseudopotentials*

In DFT, pseudopotentials replace the core electrons and nuclei with an effective potential to simplify calculations. This allows focusing on valence electrons, which are most important for chemical bonding and adsorption phenomena. Modern pseudopotentials are constructed by inverting the Schrödinger equation of free atoms and ensuring that the pseudo-wavefunctions match the true valence wavefunctions beyond a cutoff radius with norm-conservation conditions [22]. This makes the simulations less computationally intensive but still accurate for valence electron behavior.

5.2.4 Functional Selection

The choice of exchange–correlation functional is critical in DFT because it approximates the complex electron–electron interaction. The Local Density Approximation (LDA) assumes uniform electron density and is suitable for systems with slowly varying electron distributions [23]. The Generalized Gradient Approximation (GGA) incorporates density gradients for improved accuracy and is commonly used in adsorption studies [24]. Hybrid functionals combine exact exchange from Hartree–Fock theory with DFT approximations, offering better predictions of electronic properties. Selecting the appropriate functional depends on the system under study and the required accuracy. For adsorption on carbon-based nanomaterials, GGA-type functionals, such as PBE, are frequently preferred because they provide a good balance between computational cost and accuracy [25].

5.3 Basics of Spectroscopic and Microscopic Characterization

5.3.1 Scanning Electron Microscopy (SEM)

SEM is used to obtain high-resolution images of sample surfaces by scanning with a focused electron beam. It provides morphological and topographical information on the nanomaterial surfaces and adsorbed species by detecting secondary electrons emitted from the surface [26].

5.3.2 X-Ray Photoelectron Spectroscopy (XPS)

XPS utilizes X-ray irradiation to eject core electrons from atoms on a sample surface. By measuring the kinetic energy of emitted electrons, it determines elemental composition, chemical states, and electronic environment of surface atoms within a few nanometers depth [27]. It is vital in understanding the chemical binding states involved in adsorption on carbon nanomaterials.

5.3.3 Atomic Force Microscopy (AFM)

AFM is a high-resolution scanning probe technique that maps the surface topography of materials by raster-scanning a sharp tip attached to a cantilever across the sample. The interaction forces between the tip and the sample surface cause deflections in the cantilever, which are detected by a laser beam reflected into a photodetector. AFM can operate in contact, non-contact, or tapping modes, enabling the characterization of nanomaterial morphology, surface roughness, and even mechanical properties with sub-nanometer resolution [28].

5.3.4 Infrared Spectroscopy (IR)

IR spectroscopy measures molecular vibrations by detecting absorption of infrared light at specific wavelengths characteristic of molecular bonds. Techniques like Fourier transform IR (FTIR) identify surface functional groups and adsorbed species by their vibrational signatures, elucidating chemical interactions during adsorption [29].

5.3.5 Ultraviolet–Visible Spectroscopy (UV–Vis)

UV–Vis spectroscopy measures the absorption of ultraviolet and visible light by a sample, which corresponds to electronic transitions in atoms or molecules. It is commonly used to monitor the concentration of adsorbates in solution before and after adsorption, providing quantitative information on adsorption efficiency and kinetics [30]. UV–Vis can also reveal changes in the electronic structure of nanomaterials due to adsorbate interactions, helping to elucidate adsorption mechanisms.

5.4 Overview of Two-Dimensional (2D) Carbon-Based Nanomaterials

This section provides a comprehensive overview of 2D carbon-based nanomaterials, with emphasis on their unique structural characteristics and functional properties. It also highlights emerging trends in their application for environmental remediation, encompassing both theoretical frameworks and practical implementations.

5.4.1 Graphene and Its Derivatives

Graphene, a remarkable single layer of carbon atoms arranged in a 2D lattice, has garnered significant attention from researchers and industry due to its exceptional mechanical, electrical, and thermal properties [31]. This unique carbon allotrope stands out for its extraordinary strength, approximately 200 times greater than that of steel, as well as its impressive flexibility and conductivity [32]. As a result of

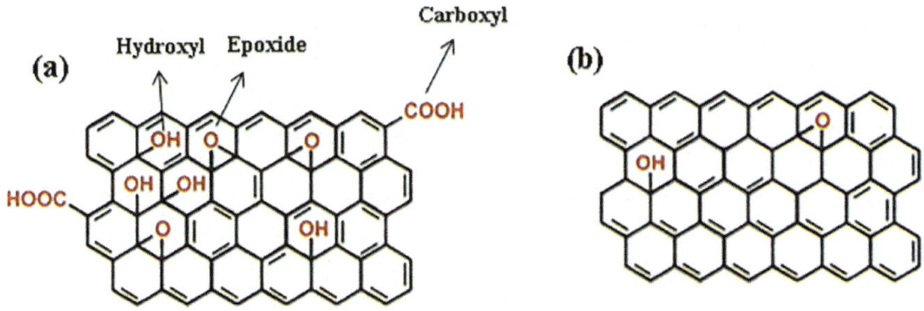

Fig. 5.5 Schematic illustration of **a** graphene oxide (GO) structure, highlighting oxygen-containing functional groups that serve as active binding sites for heavy metal ions, and **b** reduced graphene oxide (rGO) with partially restored sp^2 carbon network and fewer oxygen functionalities

these attributes, graphene has been identified as a superior candidate for applications involving the adsorption and removal of heavy metals from diverse environments, making it particularly valuable for environmental remediation efforts [33–35]. Its distinctive structural features enable it to interact favorably with various contaminants, thereby improving the efficiency of removal processes and contributing to the development of innovative solutions for pollution control and management in contaminated water and soil.

The exploration of graphene derivatives, notably graphene oxide (GO) and reduced graphene oxide (rGO), has further highlighted the importance of chemical modifications in enhancing the adsorption performance of these materials for environmental applications (see Fig. 5.5) [36]. GO is characterized by an abundance of oxygen-containing functional groups (hydroxyl, epoxide, and carboxyl), which significantly enhance hydrophilicity and provide abundant active sites for binding heavy metal ions. In contrast, rGO is obtained through the partial reduction of GO, which removes a substantial portion of the oxygen functionalities, restoring the conjugated π-electron network and improving electrical conductivity [37–40]. This distinction makes GO particularly effective for aqueous-phase adsorption due to its strong hydrophilicity, while rGO offers advantages in applications requiring both adsorption capacity and enhanced conductivity, such as electrochemical or photocatalytic remediation.

5.4.2 Enhancing Adsorption Capabilities

To further improve the adsorption capabilities of graphene and its derivatives, a range of modifications can be employed, leveraging both functionalization techniques and hybridization with other nanomaterials. Functionalization is a crucial strategy that involves either covalent or non-covalent attachment of various chemical groups to the graphene structure, effectively enhancing its interaction with specific contaminants. For example, the introduction of hydroxyl, carboxyl, amine, and epoxide groups can significantly improve electrostatic interactions with heavy metals and organic pollutants, thereby substantially increasing the material uptake capacity [41–44]. This tailored approach allows for the customization of graphene-based adsorbents to target a broader range of contaminants, making them more versatile and efficient in real-world applications.

As illustrated in Fig. 5.6, the figure begins with pristine graphene, followed by three key modification strategies, including doping, functionalization, and hybridization, resulting enhanced adsorption mechanisms. Doping involves introducing heteroatoms such as nitrogen, boron, or sulfur into the graphene lattice, which modifies its electronic properties and creates new active sites for improved interactions with contaminants such as heavy metals. In contrast, functionalization adds chemical groups like hydroxyl, carboxyl, and amine to the graphene surface, enhancing its hydrophilicity and increasing its ability to interact with both organic and inorganic pollutants. Hybridization combines graphene with other nanomaterials to create synergistic effects that further boost adsorption performance. Together, these modification strategies make graphene-based adsorbents more versatile, efficient, and selective for a wider range of contaminants, thereby improving their environmental remediation potential.

Moreover, hybridization with other nanomaterials represents another promising strategy for improving the properties of graphene-based adsorbents [45]. By creating composite materials that integrate graphene or graphene oxide with metal oxides such as titanium dioxide (TiO_2), researchers can harness synergistic effects that result in

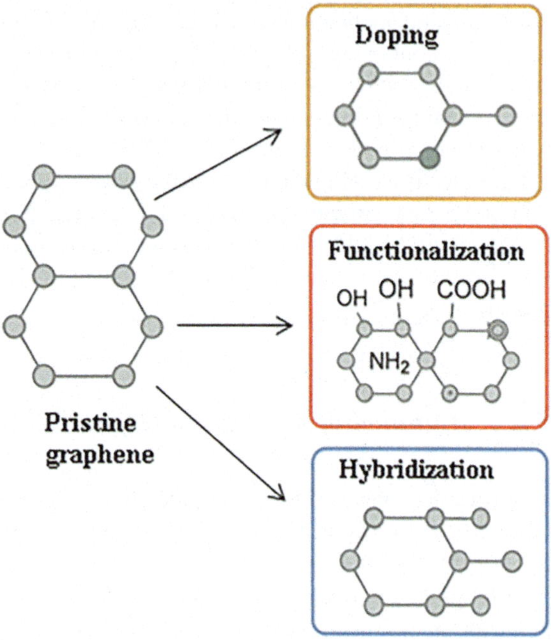

Fig. 5.6 Pristine graphene followed by three main modification strategies, including doping, functionalization, and hybridization

superior stability, selectivity, and overall remediation efficiency (see Fig. 5.7) [46]. In these composites, TiO_2 nanoparticles anchored on the graphene surface act as active photocatalytic sites. Upon light irradiation, electrons in TiO_2 are excited from the valence band (VB) to the conduction band (CB), generating electron–hole pairs. These charge carriers interact with water (H_2O) and hydroxide ions (OH^-) present, producing highly reactive radicals such as hydroxyl radicals (OH·) and superoxide radicals (O_2^-). These radicals are responsible for breaking down organic pollutants into less harmful degradation products. The graphene sheet facilitates electron transfer, reducing the recombination of electron–hole pairs and enhancing photocatalytic efficiency. The inclusion of metal oxides not only enhances the adsorption capacity but also introduces photocatalytic properties, enabling the degradation of organic pollutants under light irradiation and thereby expanding the functional range of these composite materials.

Additionally, rGO stands out as a particularly effective variant of graphene that retains many of the structural characteristics of its parent material while offering a unique balance of electrical conductivity and surface chemistry [47]. The reduction process, which typically involves the partial restoration of the sp^2-hybridized carbon structure in graphene, enhances electron mobility and modifies surface functional groups to optimize interactions with specific contaminants. This unique combination of properties positions rGO as a highly effective adsorbent for targeted applications, enabling enhanced performance in the removal of selected types of contaminants while still maintaining a cost-effective production method [48]. As research continues to explore these modifications, the potential for graphene-based materials

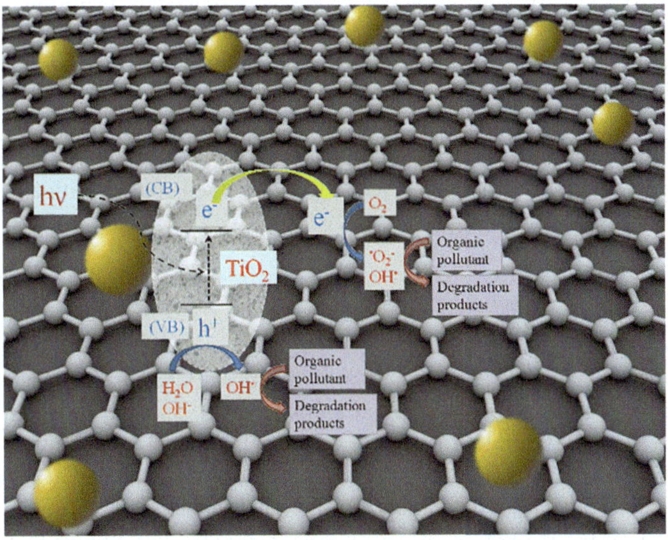

Fig. 5.7 Schematic illustration of the photocatalytic mechanism in TiO_2/graphene nanocomposites. Upon light irradiation, TiO_2 nanoparticles absorb photons, exciting electrons from the valence band (VB) to the conduction band (CB) [46]

in environmental remediation is becoming increasingly promising, paving the way for innovative solutions to address pollution challenges.

5.4.3 Other 2D Carbon-Based Nanomaterials

Beyond graphene, a variety of other noteworthy 2D carbon-based materials warrant discussion, particularly carbon nanotubes (CNTs), carbon nanodots (CNDs), carbon nanoribbons (CNRs), and covalent organic frameworks (COFs). Each of these materials possesses unique properties that enhance their effectiveness as adsorbents for heavy metals and organic pollutants.

Carbon Nanotubes

The CNTs are hollow cylindrical structures composed of carbon atoms arranged in a hexagonal lattice. They exhibit extraordinary mechanical strength, a high aspect ratio, and remarkable electrical conductivity, all of which enhance their performance in environmental applications [49]. The high surface area-to-volume ratio of CNTs provides abundant adsorption sites, while their unique geometry facilitates the diffusion of pollutants toward the adsorbent surface. As illustrated in Fig. 5.8, the tubular structure and surface modification of CNTs play a pivotal role in their adsorption mechanisms for wastewater treatment, which can include physical adsorption driven by van der Waals (vdW) forces and chemical interactions induced through functionalization [50]. Surface modifications introduce various functional groups onto CNTs, which selectively bind heavy metal ions present in contaminated water. This selective binding process is depicted in the flowchart (Fig. 5.8), where heavy metals are adsorbed onto the functionalized CNTs, resulting in the removal of contaminants and the production of purified water. Furthermore, the high aspect ratio of CNTs enhances their dispersal in aqueous solutions, leading to increased contact with contaminants and improved overall adsorption efficiency.

Carbon Nanodots

The CNDs, which are smaller than carbon nanotubes (CNTs) and graphene oxide, are a class of zero-dimensional carbon nanomaterials characterized by their high surface area, tunable photoluminescence, and excellent biocompatibility [51]. These unique properties make CNDs particularly appealing for a variety of applications, including environmental remediation, sensing, and bioimaging [52, 53]. As shown in Fig. 5.9, CNDs can effectively adsorb organic compounds and heavy metal ions such as mercury (Hg^{2+}), nickel (Ni^{2+}), and lead (Pb^{2+}) through complexation, demonstrating their versatility in capturing diverse contaminants. These heavy metal ions are common and hazardous environmental pollutants. Hg^{2+} is a neurotoxin that can

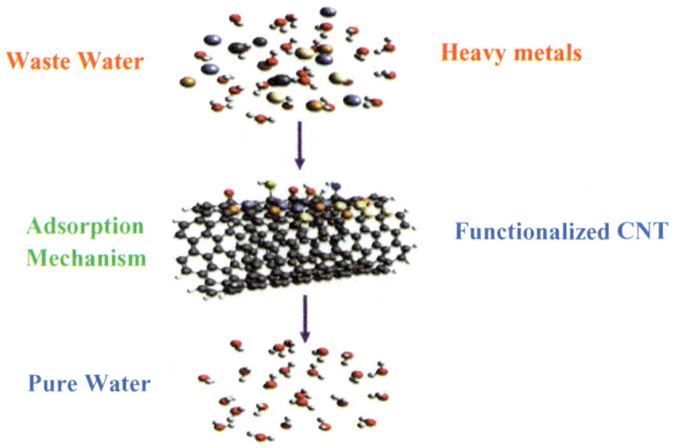

Fig. 5.8 Schematic representation of carbon nanotube (CNT) tubular structure and surface modifications in wastewater treatment [50]

bioaccumulate in aquatic food chains, Ni^{2+} is known for its carcinogenic potential and allergenic effects, and Pb^{2+} is a persistent toxin that affects multiple organ systems, particularly the nervous system. Additionally, the small size of CNDs allows for greater penetration into tight spaces within contaminated matrices, facilitating more comprehensive pollutant remediation. This enhanced accessibility not only increases the efficiency of pollutant uptake but also opens new pathways for developing environmentally friendly technologies aimed at restoring polluted environments and safeguarding public health.

Carbon Nanoribbons

The CNRs have emerged as promising adsorbents for the removal of heavy metals and various chemicals from contaminated environments due to their unique structural properties and high surface area [54]. As depicted in Fig. 5.10, CNRs, which have a two-dimensional (2D) ribbon-like structure composed of sp^2-hybridized carbon atoms, exhibit excellent electrical conductivity and stability, thereby enhancing their interactions with pollutants. Their large surface area provides abundant active sites for adsorption, enabling efficient binding of toxic heavy metals such as lead, cadmium, and mercury, as well as organic pollutants and dyes [55]. Unlike CNTs, which are cylindrical, and carbon nanodots (CNDs), which are zero-dimensional, CNRs have an extended planar morphology that combines aspects of both graphene sheets and CNTs, offering a balance of flexibility, conductivity, and high adsorption capacity. Furthermore, the tunability of their chemical properties through functionalization allows for the optimization of CNRs, enhancing their selectivity and adsorption efficiency. Consequently, CNRs present an effective and sustainable solution for

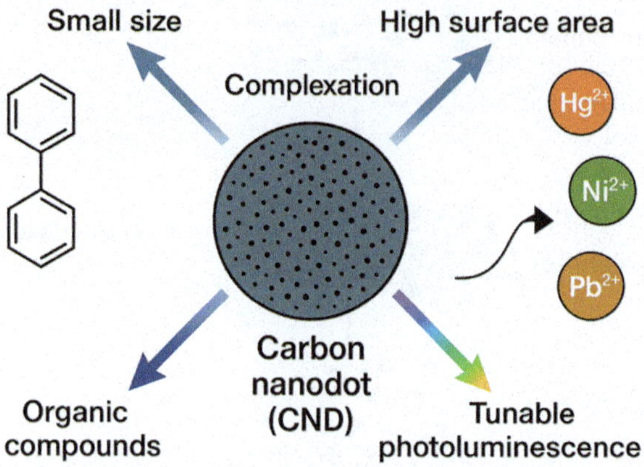

Fig. 5.9 Schematic representation of carbon nanodots (CNDs) adsorbing diverse pollutants from contaminated water. The figure illustrates the adsorption of organic compounds as well as toxic heavy metal ions including mercury (Hg^{2+}), nickel (Ni^{2+}), and lead (Pb^{2+})

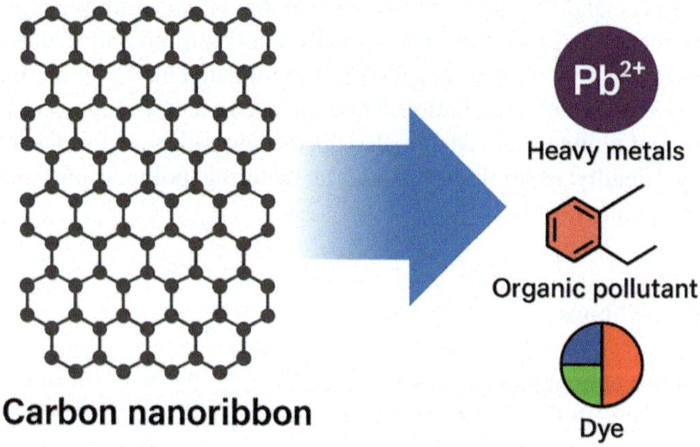

Fig. 5.10 Illustration of carbon nanoribbons (CNRs) adsorbing heavy metals and organic pollutants from contaminated environments

remediation efforts aimed at addressing heavy metal and chemical contamination in water and soil ecosystems.

Covalent Organic Frameworks

COFs have emerged as a significant class of nonporous materials in the field of adsorption, particularly for the removal of heavy metals and hazardous chemicals from contaminated environments [57, 58]. As shown in Fig. 5.11, their unique nanostructure, characterized by highly ordered porous networks, various covalent linkages, and high stability, allows for selective binding and efficient capture of target contaminants, thereby enhancing their adsorption capacity [56]. The stability and functionality of COFs can be customized by altering their chemical composition and pore structure, making them adaptable to various applications in environmental remediation. Moreover, COFs exhibit high surface areas, which facilitate increased interaction with pollutants, while their lightweight nature and ease of synthesis make them economically viable for large-scale applications [59]. As a result, COFs represent a promising solution for mitigating the environmental impacts of heavy metal pollution and toxic chemicals, ultimately contributing to cleaner ecosystems and improved public health.

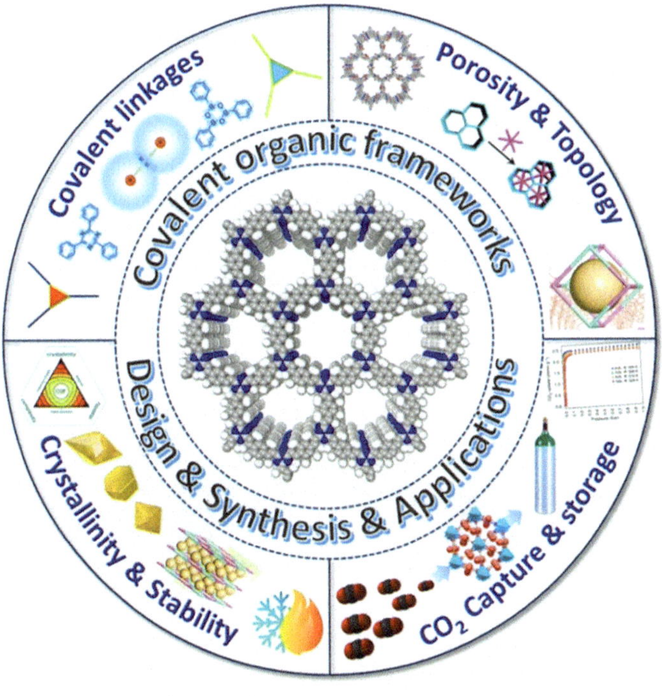

Fig. 5.11 Structural representation of covalent organic frameworks (COFs), highlighting their highly ordered porous networks, diverse covalent linkages, and exceptional stability [56]

5.5 Theoretical Perspectives: DFT Simulations

This section discuss the pivotal role of DFT in elucidating the adsorption mechanisms of heavy metals on 2D carbon-based nanomaterials, with a focus on the key DFT calculations and methods used to predict adsorption behavior at the molecular level.

5.5.1 *Electronic Structure and Adsorption Energies*

DFT has emerged as a powerful computational tool for unraveling the complex interactions between organic compounds and 2D carbon-based nanomaterials [60, 61]. As shown in Fig. 5.12, the electronic structure of these nanomaterials significantly influences their ability to adsorb toxic components, and DFT calculations provide a detailed understanding of these interactions at the quantum level. By employing various functionals, such as the Generalized Gradient Approximation (GGA) or meta-GGA, researchers can obtain accurate electronic density distributions, which are critical for delineating the nature of bonding and charge transfer during the adsorption process [62]. Subsequently, the adsorption energies can be calculated, serving as a quantitative measure of the stability of the metal–nanomaterial complexes. Moreover, DFT enables the investigation of diverse adsorption configurations, elucidating both physisorption and chemisorption pathways (Fig. 5.12), which are crucial for accurately predicting the behavior of harmful compounds in environmental and industrial contexts [63].

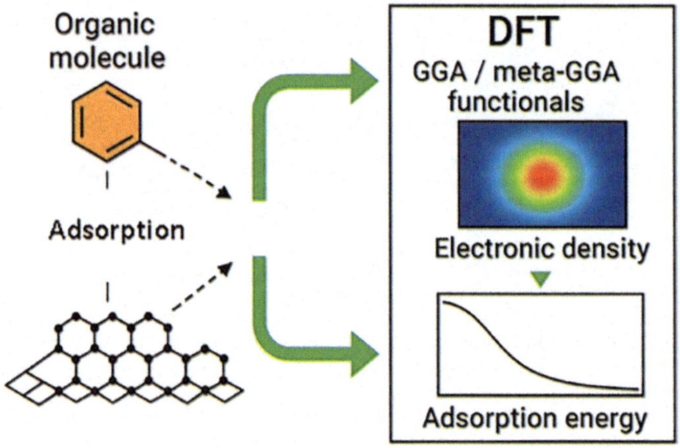

Fig. 5.12 Schematic illustration of the electronic structure of 2D carbon-based nanomaterials and their interactions with toxic components, as revealed through DFT calculations

In a systematic DFT study focused on adsorption phenomena, key parameters such as the position and orientation of the adsorbate relative to the surface of 2D carbon-based nanomaterials are meticulously optimized [64–66]. This optimization is essential for estimating the minimum energy configurations and, consequently, identifying the most favorable adsorption sites. The analysis of the band gap and density of states before and after adsorption further reveals the extent of hybridization between the metal orbitals and the electronic states of the carbon nanomaterials, providing insight into the nature of the interactions [67]. Furthermore, DFT allows for the calculation of charge density distribution upon adsorption, which plays a pivotal role in determining the reactivity and selectivity of these nanomaterials toward various heavy metals. Therefore, an integrated approach combining electronic structure calculations with adsorption energy assessments through DFT can significantly enhance our understanding of the environmental remediation potential of 2D carbon-based nanomaterials in heavy metal removal applications.

5.5.2 Adsorption Mechanisms

Adsorption mechanisms are fundamental to understanding the interaction between heavy metals and 2D carbon nanomaterials, particularly through two primary processes: physisorption and chemisorption [68].

Physisorption

Figure 5.13 visually represents the process of physisorption on the surface of 2D carbon nanomaterials. It depicts adsorbate molecules held close to the surface by weak van der Waals (vdW) forces, emphasizing that these interactions are non-specific and relatively low in energy [69]. The figure also shows various functional groups distributed on the nanomaterial surface, which modify the electrostatic environment. These groups influence the surface charge and create localized areas where heavy metal ions may experience stronger or weaker attraction, contributing to uneven distribution and selective adsorption sites. Additionally, the figure illustrates the role of steric effects by showing how adsorbate molecules of different sizes and shapes orient themselves as they approach the surface. Larger or more complex molecules may find restricted access due to spatial hindrance, while smaller molecules have easier access to active sites. This spatial arrangement affects the overall efficiency and reversibility of physisorption, which is sensitive to environmental conditions such as temperature and pressure. The figure thereby integrates these concepts to provide a comprehensive visual summary of how physical forces, chemical surface modifications, and molecular geometry combine to govern physisorption on 2D carbon nanomaterials.

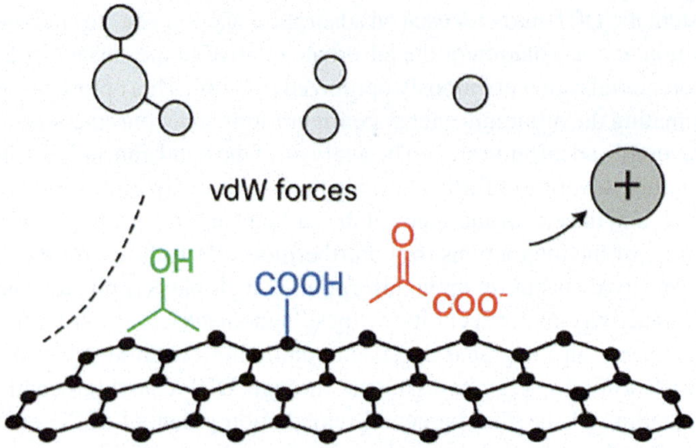

Fig. 5.13 Schematic illustration of the physisorption process on a 2D carbon nanomaterial surface, where adsorbate molecules are held by weak van der Waals (vdW) forces. The image highlights the reversible and low-energy nature of these interactions, as well as the role of surface functional groups in modifying electrostatic interactions and adsorption capacity

Chemisorption

Chemisorption involves the formation of strong chemical bonds between the adsorbate and the surface, resulting in a more permanent attachment that typically requires a higher energy input to reverse [70]. Unlike physisorption, chemisorption is characterized by the specificity of interactions, often involving electron donation from the adsorbent to the adsorbate or vice versa. The presence of active sites on 2D carbon nanomaterials, such as defects or vacancies, significantly enhances this process by creating regions of high reactivity. These sites serve as focal points where the adsorbate forms strong chemical bonds, accompanied by electron transfer that stabilizes the attachment, distinguishing chemisorption from weaker physical forces [71]. In many cases, the process is associated with charge redistribution at the interface, altering the local electronic structure of the adsorbent. This modification can enhance selectivity toward specific toxic compounds, making engineered defect sites particularly valuable for targeted environmental remediation. As illustrated in Fig. 5.14, these active sites act as high-reactivity regions where strong bonds form between the adsorbate and the 2D carbon surface. The schematic shows the electron transfer process that underpins the stability of chemisorption, highlighting the role of defects and vacancies in facilitating these interactions.

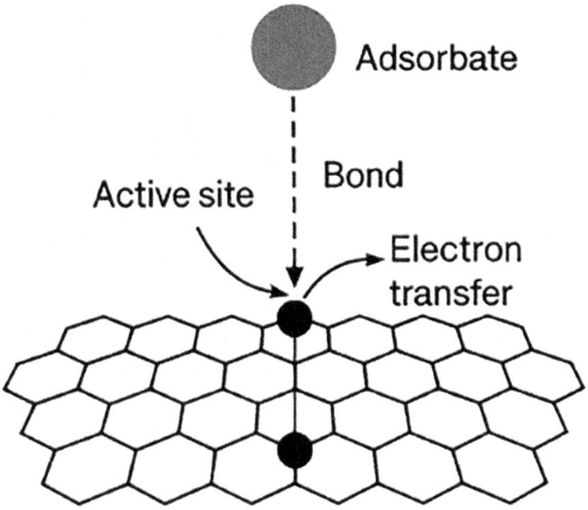

Fig. 5.14 Schematic representation of the chemisorption mechanism on 2D carbon nanomaterials, showing an active site where the adsorbate forms a strong chemical bond through electron transfer, facilitated by defects or vacancies in the structure

5.6 Experimental Validation of DFT Models

This section summarizes the alignment between theoretical predictions and experimental findings, highlighting how empirical data supports and validates DFT models. Figure 5.15 illustrates the key experimental methodologies employed to study adsorption of ibuprofen on doped 2D carbon-based nanosheets [72]. These methodologies include batch adsorption tests, surface characterization techniques, and spectroscopic analyses, all of which provide crucial evidence to complement theoretical models. The main experimental approaches used for validation are summarized in Table 5.1. Batch adsorption tests enable evaluation of adsorption capacity, isotherms, and kinetics by measuring contaminant uptake under controlled conditions. Surface characterization techniques such as scanning electron microscopy (SEM), and X-ray photoelectron spectroscopy (XPS) provide nanoscale visualization of surface morphology and chemical states, revealing interaction sites at the atomic level. Additionally, spectroscopic methods including Fourier transform infrared (FTIR) and UV–Vis spectroscopies confirm the presence of specific functional groups on the nanomaterials and assess their involvement in the adsorption process [73]. Together, the experimental findings from these approaches offer valuable insights that validate the adsorption mechanisms predicted by DFT calculations. By integrating theoretical predictions with empirical data, this comprehensive approach enhances understanding of contaminant–nanomaterial interactions and informs the optimized design of more effective adsorbents for environmental remediation application.

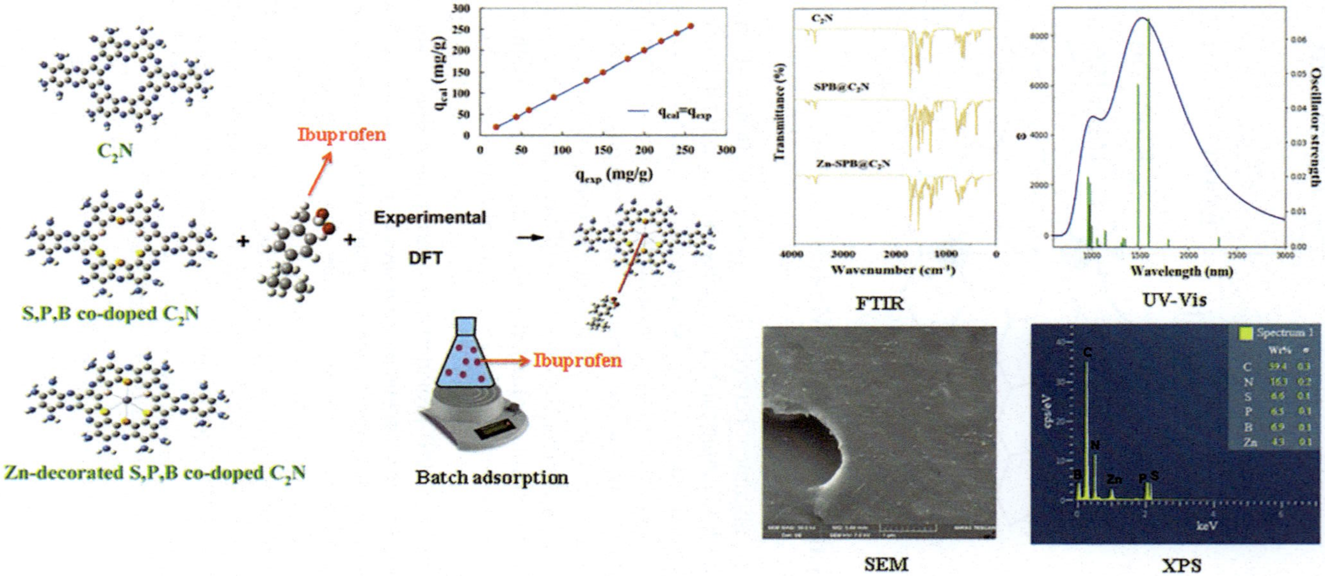

Fig. 5.15 Overview of experimental methodologies for validating DFT models used to study adsorption of ibuprofen on 2D carbon nanomaterials. The figure highlights the synergy between batch adsorption experiments and advanced surface characterization techniques. Batch adsorption tests evaluate ibuprofen uptake capacity, isotherms, and kinetics, providing macroscopic adsorption data. Surface characterization methods, including scanning electron microscopy (SEM) and X-ray photoelectron spectroscopy (XPS), offer nanoscale insights into morphological features and chemical states of Zn-decorated S, P, B co-doped C_2N nanosheets. Complementary spectroscopic analyses, such as Fourier transform infrared (FTIR) and UV–Vis spectroscopies, identify functional groups involved in adsorption interactions. Together, these experimental approaches corroborate the DFT computational predictions, enhancing understanding of adsorption mechanisms for pharmaceutical contaminant removal. Reprinted from Ref. [72], Copyright © 2024, with permission from Elsevier

Table 5.1 Experimental approaches for validation of DFT adsorption models

Experimental approach	Role	Typical techniques/Methods
Batch adsorption test	Evaluate adsorption capacity, isotherms, and kinetics	Adsorption isotherm modeling, kinetics, studies
Surface characterization	Visualize surface morphology and chemical states at atomic level	SEM, XPS
Spectroscopic analysis	Identify functional groups and monitor chemical interactions	FTIR, UV–Vis

5.6.1 Influence of pH and Temperature

The influence of environmental parameters such as pH and temperature on the adsorption efficiency of heavy metals is a critical area of study, as these factors can significantly alter the behavior of adsorbents and the speciation of metals in solution [74–76]. As shown in Fig. 5.16, pH plays a pivotal role by affecting the surface charge of adsorbents, the solubility of metals, and their ionic forms in the aqueous medium [77]. A higher pH can enhance the deprotonation of functional groups on the adsorbent, making more active sites available for metal binding while also converting toxic metal ions into less soluble forms. Experimental studies have demonstrated that optimal pH conditions can dramatically increase the removal rates of metals, showing how specific pH levels can be tailored to enhance adsorption performance [78–80].

The adsorption performance is often quantified in terms of the adsorption capacity (q_e, mg/g), which describes the amount of metal adsorbed per unit mass of adsorbent at equilibrium:

$$q_e = \frac{(C_0 - C_e) \times V}{m} \tag{5.3}$$

where C_0 and C_e (mg/L) are the initial and equilibrium metal ion concentrations, respectively, V (L) is the solution volume, and m (g) is the adsorbent mass. pH variations can significantly influence q_e by altering both the availability of binding sites and the ionic form of the metals.

Similarly, temperature influences adsorption dynamics by affecting molecular interactions and kinetic processes. Elevated temperatures can enhance the diffusion rates of metal ions toward the adsorbent surface and promote the overall adsorption process [75]. Thermodynamic studies often evaluate the temperature dependence of adsorption using the Langmuir equilibrium constant (K), which relates to the adsorption affinity. By plotting Ln K versus 1/T (K^{-1}), the van't Hoff equation can be applied [81]

$$\mathrm{Ln}\,K = -\frac{\Delta H^0}{RT} + \frac{\Delta S^0}{R} \tag{5.4}$$

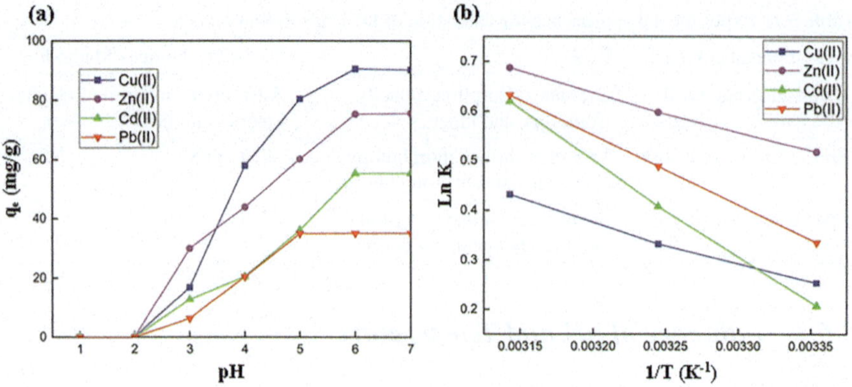

Fig. 5.16 Influence of environmental parameters on the adsorption efficiency of heavy metals by silica functionalized by silica functionalized with pyridin-2-ylmethanol (SiPy) [77]: **a** Effect of pH on the adsorption capacity toward Cu(II), Zn(II), Cd(II), and Pb(II). Adsorption conditions: V = 10 mL, m = 10 mg, optimum concentrations = 180.66×10^{-3}, 140.10×10^{-3}, 99.70×10^{-3}, and 95.63×10^{-3} g L^{-1} for Cu(II), Zn(II), Cd(II), and Pb(II), respectively; t = 25 min at 25 °C. The adsorption capacity increased with pH, reaching maxima at pH 6–7. At low pH, protonation of pyridine and oxygen groups rendered the surface positively charged, repelling metal cations and reducing adsorption. At pH ≥ 8, adsorption efficiency plateaued due to metal hydroxide precipitation. **b** Effect of temperature on metal ion adsorption. Thermodynamic analysis shows negative ΔG°, confirming a spontaneous and favorable process; positive ΔH° indicates endothermic adsorption; and negative ΔS° suggests increased randomness at the adsorbent surface during metal ion binding

where ΔH° (kJ mol^{-1}) is the standard enthalpy change, ΔS° (J/mol K) is the standard entropy change, and R is the gas constant. Positive ΔH° values indicate an endothermic adsorption process, where higher temperatures enhance adsorption efficiency. The Gibbs free energy change (ΔG°) is calculated from following relationship:

$$\Delta G^0 = \Delta H^0 - T\Delta S^0. \tag{5.5}$$

Negative ΔG° values confirm that the adsorption process is spontaneous and thermodynamically favorable under the studied conditions.

Comparison of experimental findings and theoretical insights obtained by DFT reveals a deeper understanding of the adsorption mechanisms [82–84]. DFT calculations can provide detailed information about the electronic structures and energy profiles of both adsorbents and adsorbates under varying environmental conditions. To take pH variation into account, different protonation states of the adsorbent functional groups are modeled, reflecting the deprotonating or protonating that occurs at different pH values [85]. Additionally, the dominant metal species at each pH are considered, as metal ion speciation changes with pH. Adsorption energies (E_{ads}) are then computed for each combination of protonation state and metal species:

$$E_{ads} = E_{complex} - (E_{adsorbent} + E_{metal}). \tag{5.6}$$

This allows identification of the pH conditions that maximize adsorption efficiency.

To account for temperature effects, thermodynamic corrections are applied to the DFT results. After geometry optimization, vibrational frequency calculations provide zero-point energies (ZPE) and thermal corrections to enthalpy and Gibbs free energy. The temperature-dependent adsorption Gibbs free energy ($\Delta G_{ads}(T)$) is calculated as

$$\Delta G_{ads}(T) = \Delta G_{complex}(T) - (\Delta G_{adsorbent}(T) + \Delta G_{metal}(T)) \tag{5.7}$$

By evaluating $\Delta G_{ads}(T)$ at different temperatures, one can determine whether adsorption is endothermic or exothermic and predict how adsorption efficiency changes with temperature. By integrating these computational approaches with experimental results, we can better predict and optimize the conditions necessary for effective heavy metal removal from contaminated environments, ultimately guiding the design of more efficient adsorbents.

5.6.2 *Effect of Co-Existing Ions and Ionic Strength*

In realistic environmental systems, 2D carbon-based nanomaterials interact with complex aqueous matrices containing various ionic species, and the adsorption behavior of heavy metals and organic pollutants is significantly influenced by the presence of co-existing ions such as common alkali and alkaline earth metals (Na^+, K^+, and Ca^{2+}) [86–88]. These co-existing ions often compete with target contaminants for active adsorption sites on the nanomaterial surface, which can suppress the adsorption efficiency and decrease overall adsorption capacity. Additionally, increased ionic strength compresses the electrical double layer surrounding the adsorbent surface, reducing the electrostatic attraction between adsorbent and adsorbate species and generally leading to a decline in adsorption capacity, particularly when ion exchange dominates [89, 90]. However, under certain conditions, higher ionic strength can also facilitate the aggregation of pollutant molecules near the nanomaterial surface, enhancing adsorption via secondary interactions such as hydrogen bonding and π–π stacking, which are particularly relevant for organic molecules. Overall, co-existing ions and ionic strength can exert both inhibitory and facilitative effects on pollutant adsorption to 2D carbon-based nanomaterials, depending on the specific adsorbate and environmental conditions (Fig. 5.17).

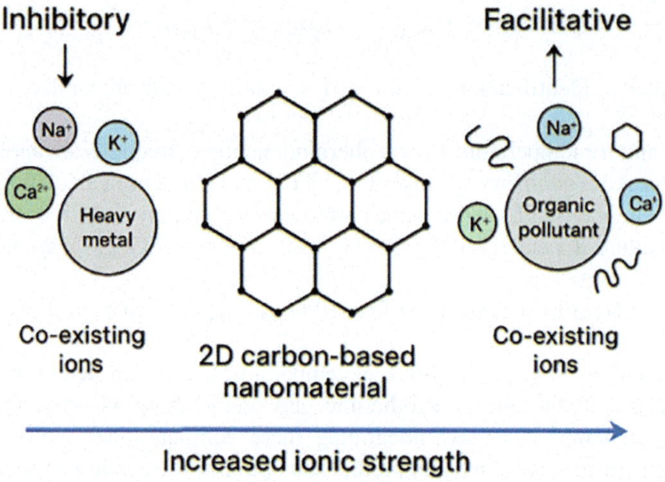

Fig. 5.17 Influence of co-existing ionic species and ionic strength on the adsorption behavior of heavy metals and organic pollutants on 2D carbon-based nanomaterials. Competing ions such as Na^{+}, K^{+}, and Ca^{2+} compete for adsorption sites, while increased ionic strength compresses the electrical double layer, modulating electrostatic interactions

From a theoretical perspective, DFT studies enable the simulation of adsorption behavior in the presence of co-ions, providing valuable insights into changes in electronic distribution, adsorption geometries, and site preferences [91]. Such modeling allows the prediction of how specific ions affect the binding affinity of target pollutants and facilitates the visualization of charge redistribution and orbital interactions under different ionic environments (Fig. 5.18). These simulations can quantify adsorption energies, identify the most favorable binding sites, and reveal the role of competing ions in altering surface reactivity. In particular, projected density of states (PDOS) and charge density difference maps help clarify how ionic species modify orbital overlaps and electron transfer processes [92]. By correlating these atomic-scale insights with experimental adsorption trends, researchers can pinpoint the dominant adsorption mechanisms under realistic environmental condition. Integration of DFT simulations with experimental observations offers a powerful framework for understanding and predicting the influence of co-existing ions and ionic strength on adsorption mechanisms. This combined approach is crucial for the rational design of 2D carbon-based nanomaterials with enhanced selectivity and performance in complex environmental systems.

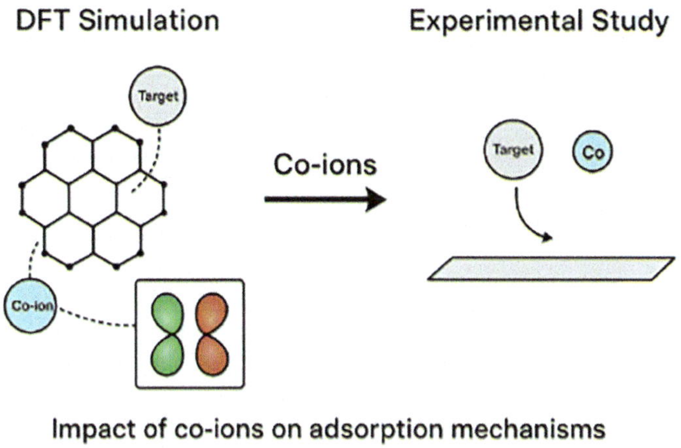

Fig. 5.18 Schematic illustration of the complementary roles of DFT simulations and experimental studies in elucidating the impact of co-ions on adsorption mechanisms in 2D carbon-based nanomaterials

5.7 Challenges and Future Perspectives

The application of 2D carbon-based nanomaterials for adsorption purposes presents several challenges that must be addressed for their effective utilization (Table 5.2). One significant challenge is scalability: while these materials often exhibit remarkable adsorption properties at the laboratory scale, translating this success to industrial-scale applications can be difficult due to production techniques, cost, and material uniformity issues. Regeneration of adsorbents is another critical challenge, as decreased efficiency over time can render the materials less viable for repeated use. Effective regeneration methods need to be developed to restore adsorption capacity without compromising the structural integrity of the nanomaterials. Environmental impacts must also be considered, particularly regarding the fate of these nanomaterials in ecosystems after use; understanding their long-term effects is essential to preventing unintended consequences on wildlife and human health. Furthermore, the integration with molecular-level design, combining DFT and experimental approaches, is necessary to develop novel, effective, and environmentally benign materials tailored for specific pollutants. Finally, regulatory compliance is essential, including the development of guidelines for safe use and adherence to environmental standards to mitigate potential risks. By tackling these challenges and focusing on strategic research directions, the field can move toward the practical application of 2D carbon-based nanomaterials in sustainable environmental practices.

Table 5.2 Challenges in the application of 2D carbon-based nanomaterials for adsorption

Challenge	Description
Scalability	Translating laboratory-scale performance to industrial-scale applications is difficult due to production techniques, cost, and material uniformity issues
Regeneration of adsorbents	Maintaining adsorption efficiency over multiple cycles requires effective regeneration methods that restore capacity without compromising structural integrity
Environmental impact	Potential risks associated with the fate of nanomaterials in ecosystems after use; understanding long-term effects on wildlife and human health is critical
Integration with molecular-level design	Need for combining DFT and experimental methods to design novel, effective, and environmentally benign materials tailored for specific pollutants
Regulatory compliance	Development of guidelines and adherence to environmental standards is essential to ensure safe use and mitigate potential risks

5.8 Conclusion

The exploration of heavy metal and chemical adsorption on 2D carbon-based nanomaterials, through a combination of DFT simulations and experimental validations, highlights an effective approach to addressing contemporary environmental challenges. This multifaceted investigation not only enhances our understanding of the underlying adsorption mechanisms but also reveals the potential for designing advanced materials with specifically tailored properties. Such innovations in material design could significantly improve the efficacy of environmental remediation strategies, offering a promising path forward in the effort to mitigate pollution and restore ecological balance. Moreover, the integration of computational and experimental approaches accelerates the discovery process, allowing researchers to predict material performance more accurately and optimize properties prior to large-scale deployment. This synergy between theory and practice is crucial for translating laboratory successes into real-world environmental applications. As we continue to deepen our insights into these complex interactions, we can leverage this knowledge to develop more efficient, cost-effective, and sustainable solutions for addressing the pressing environmental issues of our time. Ultimately, these advancements will not only help protect natural ecosystems but also contribute to safeguarding public health, promoting global sustainability efforts, and ensuring a healthier planet for future generations.

References

1. Yadav SK, Baruah B, Dutta K, Rai K, Rai A, Rajak R, Gupta A, Misra AK, Wanjari N, Ranjan RK. Exploring the multi-faceted health risks of heavy metals exposure in Sikkim's unique Himalayan ecosystem, India. Environ Monit Assess. 2025;197:1–24. https://doi.org/10.1007/s10661-025-13745-9.
2. Pant M, Singh J, Singhal N, Bisen D. Emerging contaminants in water and wastewater: sources and substances. Springer;2025. p. 331–354.
3. Rauseo J, Spataro F, Patrolecco L. Persistent and emerging organic contaminants in natural environments. Water. 2025;17:436. https://doi.org/10.3390/w17030436.
4. Maddheshiya R, Dhiman S, Gupta H. Emerging contaminants in water and wastewater: sources and substances. Springer;2025. p. 87–105.
5. Osuoha JO, Anyanwu BO, Ejileugha C. Pharmaceuticals and personal care products as emerging contaminants: need for combined treatment strategy. J Hazard Mater Adv. 2023;9: 100206. https://doi.org/10.1016/j.hazadv.2022.100206.
6. Pisciotta JM, Blessing S, Zaybak Z. Biotechnology for environmental sustainability. Springer;2025. p. 507–559.
7. Singh NB, Kundu T, Guin M. Carbon based nanomaterials for drug delivery. Springer;2025. p. 99–127.
8. Swain SK, Biswal A, Patra S, Sahu KM. Carbon based nanomaterials for drug delivery. Springer;2025. p. 1–37.
9. Preethi S, Varghese S, Biswas K, Vijayalakshmi N. Unveiling the properties of layered 2D-based nano-material flexible electronics in biomedical applications: a review. Environ Monit Assess. 2025;197:1–24. https://doi.org/10.1007/s10661-025-13745-9.
10. Alatawi RAS. Electrospun nanofiber chitosan/polyvinyl alcohol loaded with metal organic framework nanofiber for efficient adsorption and removal of industrial dyes from waste water: adsorption isotherm, kinetic, thermodynamic, and optimization via Box-Behnken design. Int J Biol Macromol. 2025;299: 140086. https://doi.org/10.1016/j.ijbiomac.2025.140086.
11. Rajabi M, Keihankhadiv S, Suhas, Tyagi I, Karri RR, Chaudhary M, Mubarak NM, Chaudhary S, Kumar P, Singh P. Comparison and interpretation of isotherm models for the adsorption of dyes, proteins, antibiotics, pesticides and heavy metal ions on different nanomaterials and non-nano materials—a comprehensive review. J Nanostruct Chem. 2023; 13: 43–65. https://doi.org/10.1007/s40097-022-00509-x.
12. Singh V, Ahmed G, Vedika S, Kumar P, Chaturvedi SK, Rai N, Sachchida E, Vamanu A, Kumar A. Toxic heavy metal ions contamination in water and their sustainable reduction by eco-friendly methods: isotherms, thermodynamics and kinetics study. Sci Rep. 2024;14:7595. https://doi.org/10.1038/s41598-024-58061-3.
13. Ederer J, Ecorchard P, Šrámová Slušná M, Tolasz J, Smržová D, Lupínková S, Janoš P, Kooh MRR. A study of methylene blue dye interaction and adsorption by monolayer graphene oxide. Adsorpt Sci Technol. 2022; 7385541. https://doi.org/10.1155/2022/7385541.
14. Bahadi SA, Iddrisu M, Al-Sakkaf MK, Elgzoly MAA, Al-Amrani WA, Ahmed U, Zahid U, Drmosh QA, Onaizi SA. Chemically versus thermally reduced graphene oxide: effects of reduction methods and reducing agents on the adsorption of phenolic compounds from wastewater. Emergent Mater. 2024;7:533–45. https://doi.org/10.1007/s42247-023-00514-y.
15. Abiodun OAO, Oluwaseun O, Oladayo OK, Abayomi O, George AA, Opatola E, Orah RF, Isukuru EJ, Ede IC, Oluwayomi OT. Remediation of heavy metals using biomass-based adsorbents: adsorption kinetics and isotherm models. Clean Technol. 2023;5:934–60. https://doi.org/10.3390/cleantechnol5030047.
16. Selim MM, Tounsi A, Gomaa H, Hu N, Shenashen M. Addressing emerging contaminants in wastewater: insights from adsorption isotherms and adsorbents: A comprehensive review. Alex Eng J. 2024;100:61–71. https://doi.org/10.1016/j.aej.2024.05.022.
17. Gusarov S. Advances in computational methods for modeling photocatalytic reactions: a review of recent developments. Mater. 2024;17:2119. https://doi.org/10.3390/ma17092119.

18. Salahshoori I, Mahdavi S, Moradi Z, Otadi M, Kazemabadi FZ, Nobre MAL, Khonakdar HA, Baghban A, Wang Q, Mohammadi AH. Advancements in molecular simulation for understanding pharmaceutical pollutant adsorption: a state-of-the-art review. J Mol Liq. 2024;410: 125513. https://doi.org/10.1016/j.molliq.2024.125513.
19. Luo J, Yu D, Hristovski KD, Fu K, Shen Y, Westerhoff P, Crittenden JC. Critical review of advances in engineering nanomaterial adsorbents for metal removal and recovery from water: mechanism identification and engineering design. Environ Sci Technol. 2021;55:4287–304. https://doi.org/10.1021/acs.est.0c07936.
20. Schrödinger E. Quantisierung als Eigenwertproblem. Ann Phys. 1926;79:361–76. https://doi.org/10.1002/andp.19263840404.
21. Kohn W, Sham LJ. Self-consistent equations including exchange and correlation effects. Phys Rev. 1965;140:A1133–8. https://doi.org/10.1103/PhysRev.140.A1133.
22. Hamann DR, Schlüter M, Chiang C. Norm-conserving pseudopotentials. Phys Rev Lett. 1979;43:1494–7. https://doi.org/10.1103/PhysRevLett.43.1494.
23. Perdew JP, Zunger A. Self-interaction correction to density-functional approximations for many-electron systems. Phys Rev B. 1981;23:5048–79. https://doi.org/10.1103/PhysRevB.23.5048.
24. Perdew JP, Burke K, Ernzerhof M. Generalized gradient approximation made simple. Phys Rev Lett. 1996;77:3865–8. https://doi.org/10.1103/PhysRevLett.77.3865.
25. Adamo C, Barone V. Toward reliable density functional methods without adjustable parameters: the PBE0 model. J Chem Phys. 1999;110:6158–70. https://doi.org/10.1063/1.478522.
26. Goldstein SJ, Newbury DE, Joy DC, Lyman CE, Echlin P, Lifshin E, Sawyer L, Michael, JR. Scanning electron microscopy and X-ray microanalysis, 3rd ed. New York: Springer;2003. https://doi.org/10.1007/978-1-4615-0215-9.
27. Moulder JF, Stickle WF, Sobol PE, Bomben KD. Handbook of X-ray photoelectron spectroscopy. Eden Prairie: Perkin-Elmer Corporation; 1992.
28. Jalili N, Laxminarayana K. A review of atomic force microscopy imaging systems: application to molecular metrology and biological sciences. Mechatronics. 2004;14(8):907–45. https://doi.org/10.1016/j.mechatronics.2004.04.005.
29. Griffiths P, de Haseth JA. Fourier transform infrared spectrometry, 2nd ed. Hoboken: Wiley; 2007. https://doi.org/10.1002/047010631X.
30. Raza S, Hayat A, Bashir T, Chen C, Shen L, Orooji Y, Lin H. Electrochemistry of 2D-materials for the remediation of environmental pollutants and alternative energy storage/conversion materials and devices: a comprehensive review. Sustain Mater Technol. 2024;40: e00963. https://doi.org/10.1016/j.susmat.2024.e00963.
31. Kaviani S, Izadyar M. First-principles study of the binding affinity of monolayer BC_6N nanosheet: implications for drug delivery. Mater Chem Phys. 2022;276: 125375. https://doi.org/10.1016/j.matchemphys.2021.125375.
32. Zhang F, Yang K, Liu G, Chen Y, Wang M, Li S, Li R. Recent advances on graphene: synthesis, properties, and applications. Compos Part A Appl Sci Manuf. 2022;160: 107051. https://doi.org/10.1016/j.compositesa.2022.107051.
33. Santra S, Ghosh A, Das B, Pal S, Pal S, Adalder A. Beyond the horizons of graphene: xenes for energy applications. RSC Sustain. 2024;2:1631–74. https://doi.org/10.1039/D3SU00445G.
34. Adfar Q, Hussain S, Maktedar SS. Insights into energy and environmental sustainability through photoactive graphene-based advanced materials: perspectives and promises. Nanoscale. 2025;49:2511–650. https://doi.org/10.1039/D4NJ03693J.
35. Baruah A, Newar R, Das S, Kalita N, Nath M, Ghosh P, Chinnam S, Sarma H, Narayan M. Biomedical applications of graphene-based nanomaterials: recent progress, challenges, and prospects in highly sensitive biosensors. Discov Nano. 2024;19:103. https://doi.org/10.1186/s11671-024-04032-6.
36. Sengupta J, Hussain CM. Sensitive and selective detection of heavy metal ions and organic pollutants with graphene-integrated sensing platforms. Nanoscale. 2024;16:14195–212. https://doi.org/10.1039/D4NR00956H.

37. Anegbe B, Ifijen IH, Maliki M, Uwidia IE, Aigbodion AI. Graphene oxide synthesis and applications in emerging contaminant removal: a comprehensive review. Environ Sci Eur. 2024;36:15. https://doi.org/10.1186/s12302-023-00814-4.
38. Deshwal N, Singh MB, Bahadur I, Kaushik N, Kaushik NK, Singh P, Kumari K. A review on recent advancements on removal of harmful metal/metal ions using graphene oxide: experimental and theoretical approaches. Sci Total Environ. 2023;858: 159672. https://doi.org/10.1016/j.scitotenv.2022.159672.
39. Singh S, Naik T, Shehata N, Aguilar-Marcelino L, Dhokne K, Lonare S, Chauhan V, Kumar A, Singh J, Ramamurthy PC. Novel insights into graphene oxide-based adsorbents for remediation of hazardous pollutants from aqueous solutions: a comprehensive review. J Mol Liq. 2023;369: 120821. https://doi.org/10.1016/j.molliq.2022.120821.
40. Sharma A, Sharma S, Yadav S, Arora B, Dutta S, Dixit R, Mehta S, Sharma RK. Covalently functionalized graphene oxide metal complexes: versatile nanocatalysts for organic transformations. Mater Sci Eng B. 2024;310: 117671. https://doi.org/10.1016/j.mseb.2024.117671.
41. Gonçalves MG, Costa VO, Martinez AHG, Régnier BM, Gomes GCB, Zarbin AJG, Orth ES. Functionalization of graphene oxide via epoxide groups: a comprehensive review of synthetic routes and challenges. Front Carbon. 2024;3:1393077. https://doi.org/10.3389/frcrb.2024.1393077.
42. Mokoena LS, Mofokeng JP. Synthesis and characterization of graphene oxide (GO) for the removal of lead ions in water. Carbon Trends. 2024;15: 100339. https://doi.org/10.1016/j.cartre.2024.100339.
43. Lan X, Gao B, Liu Z, Lin P, Liu H, Fu H, Zhang H. Amine-functionalised magnetic graphene-oxide nanosheets for the adsorption and reduction of Cr (VI) in water. Diam Relat Mater. 2025;152: 111891. https://doi.org/10.1016/j.diamond.2024.111891.
44. Zubair M, Syamaladevi R, Ullah A. Challenges and prospects: graphene oxide-based materials for water remediation including metal ions and organic pollutants. Environ Sci Nano. 2024;11:3693–720. https://doi.org/10.1039/D4EN00143E.
45. Farhan A, Rashid EU, Waqas M, Ahmad H, Nawaz S, Munawar J, Rahdar A, Varjani S, Bilal M. Graphene-based nanocomposites and nanohybrids for the abatement of agro-industrial pollutants in aqueous environments. Environ Pollut. 2022;308: 119557. https://doi.org/10.1016/j.envpol.2022.119557.
46. Tayel A, Ramadan AR, El Seoud OA. Titanium dioxide/graphene and titanium dioxide/graphene oxide nanocomposites: synthesis, characterization and photocatalytic applications for water decontamination. Catalysts. 2018;8(11):491. https://doi.org/10.3390/catal8110491.
47. Mombeshora ET, Muchuweni E. Dynamics of reduced graphene oxide: synthesis and structural models. RSC Adv. 2023;13:17633–55. https://doi.org/10.1039/D3RA02098C.
48. Priyadharshini SD, Manikandan S, Kiruthiga R, Rednam U, Babu PS, Subbaiya R, Karmegam N, Kim W, Govarthanan M. Graphene oxide-based nanomaterials for the treatment of pollutants in the aquatic environment: Recent trends and perspectives–a review. Environ Pollut. 2022;306: 119377. https://doi.org/10.1016/j.envpol.2022.119377.
49. Taib MNAM, Khairuddin NFM, Saleh TA. Functionalized carbon nanotubes: synthesis, properties, and application in polymer for flame retardancy—a review. J Therm Anal Calorim. 2025; 1–23. https://doi.org/10.1007/s10973-025-14016-y.
50. Elsehly EM, Chechenin NG. IntechOpen. IntechOpen; 2023. p. 325–339. https://doi.org/10.5772/intechopen.114885.
51. Liu J, Li R, Yang B. Carbon dots: a new type of carbon-based nanomaterial with wide applications. ACS Cent Sci. 2020;6:2179–95. https://doi.org/10.1021/acscentsci.0c01306.
52. Dagdag O, Quadri TW, Daoudi W, Berdimurodov E, Kim H. Carbon dots: recent developments and future perspectives. ACS Publications; 2024. p. 127–145. https://doi.org/10.1021/bk-2024-1465.ch006.
53. Ashraf W, Sharma D, Khan A, Khanuja M. Green carbon dots as adsorbents for removal of toxic chemicals, in green carbon dots: sustainable analytical approaches. ACS Publications; 2025. p. 141–160.

54. Qureshi ZA, Dabash H, Ponnamma D, Abbas MKG. Carbon dots as versatile nanomaterials in sensing and imaging: efficiency and beyond. Heliyon. 2024;10: e31634. https://doi.org/10.1016/j.heliyon.2024.e31634.
55. Nagar V, Sharma V, Kumari P, Jain D, Sharma A, Shenoy SU, Singh A, Awasthi G, Awasthi KK, Sankhla MS. Nanoribbons as advanced nanomaterials for facile detection and efficient removal of heavy metals: a comprehensive review. Int J Environ Sci Technol. 2024;21:8793–802. https://doi.org/10.1007/s13762-024-05559-y.
56. Essalhi M, El-Hassan M, Tayeb A, Al-Qahtani RA, Farooqi AS, Abdelnaby M. Structural design of covalent organic frameworks and their recent advancements in carbon capture applications: a review. Carbon Capture Sci Technol. 2025;14: 100370. https://doi.org/10.1016/j.ccst.2025.100370.
57. Li R, Sun F, Liu Z, Shi Y, He S, Chen J. Research progress and prospect of covalent organic frameworks (COFs) and composites: From synthesis to application in water contaminants. J Environ Chem Eng. 2024;12: 113944. https://doi.org/10.1016/j.jece.2024.113944.
58. Li Y, Wei J, Wang J, Wang Y, Yu P, Chen Y, Zhang Z. Covalent organic frameworks as superior adsorbents for the removal of toxic substances. Chem Soc Rev. 2025;54:2693–725. https://doi.org/10.1039/D4CS00591K.
59. Ge S, Wei K, Peng W, Huang R, Akinlabi E, Xia H, Shahzad MW, Zhang X, Xu B, Ben J, Jiang J. A comprehensive review of covalent organic frameworks (COFs) and their derivatives in environmental pollution control. Chem Soc Rev. 2024;53:11259–302. https://doi.org/10.1039/D4CS00521J.
60. Butera V. Density functional theory methods applied to homogeneous and heterogeneous catalysis: a short review and a practical user guide. Phys Chem Chem Phys. 2023;2:1781–99. https://doi.org/10.1039/D4CP00266K.
61. Rahman MU, Khan S, Khan H, Ali A, Sarwar F. Computational chemistry unveiled: a critical analysis of theoretical coordination chemistry and nanostructured materials. Chem Prod Process Model. 2024;19:473–515. https://doi.org/10.1515/cppm-2024-0001.
62. Saadh MJ, Ajaj Y, Mustafa MA, Yadav A, Bansal P, Kaur H, Logroño PFB, Zainul R, Abbas M, Pandey S. Investigating the potential of graphene and nitrogen carbide nanosheets as efficient sensors for drug detection through theoretical analysis. Inorg Chem Commun. 2024;165: 112572. https://doi.org/10.1016/j.inoche.2024.112572.
63. Zaeem MA, Thomas S, Kavousi S, Zhang N, Mukhopadhyay T, Mahata A. Multiscale computational modeling techniques in study and design of 2D materials: recent advances, challenges, and opportunities. 2D Mater. 2024; 11:042004. https://doi.org/10.1088/2053-1583/ad63b6.
64. Salahshoori I, Wang Q, Nobre MAL, Mohammadi AH, Dawi EA, Khonakdar HA. Molecular simulation-based insights into dye pollutant adsorption: a perspective review. Adv Colloid Interface Sci. 2024;333: 103281. https://doi.org/10.1016/j.cis.2024.103281.
65. Ivanković K, Kern M, Rožman M. Modelling of the adsorption of pharmaceutically active compounds on carbon-based nanomaterials. J Hazard Mater. 2021;414: 125554. https://doi.org/10.1016/j.jhazmat.2021.125554.
66. Kaviani S, Khajavian M, Piyanzina I, Nedopekin OV, Tayurskii DA. Theoretical design of transition metal-doped oxo-triarylmethyl as a disposable platform for adsorption of ibuprofen. J Mol Graph Model. 2024;126: 108647. https://doi.org/10.1016/j.jmgm.2023.108647.
67. Kaviani S, Tayurskii DA, Nedopekin OV, Piyanzina I. DFT insight into Cd^{2+}, Hg^{2+}, Pb^{2+}, Sn^{2+}, As^{3+}, Sb^{3+}, and Cr^{3+} heavy metal ions adsorption onto surface of bowl-like B_{30} nanosheet. Sci Total Environ. 2022;365: 120131. https://doi.org/10.1016/j.molliq.2022.120131.
68. Sheraz N, Shah A, Haleem A, Iftikhar FJ. Comprehensive assessment of carbon-, biomaterial- and inorganic-based adsorbents for the removal of the most hazardous heavy metal ions from wastewater. RSC Adv. 2024;14:11284–310. https://doi.org/10.1039/D4RA00976B.
69. Asif M, Sajid H, Ayub K, Gilani MA, Anwar N, Mahmood T. Therapeutic potential of oxo-triarylmethyl (oxTAM) as a targeted drug delivery system for nitrosourea and fluorouracil anticancer drugs: a first-principles insight. J Mol Graph Model. 2023;122: 108469. https://doi.org/10.1016/j.jmgm.2023.108469.

70. Yang X, Wan Y, Zheng Y, He F, Yu Z, Huang J, Wang H, Ok YS, Jiang Y, Gao B. Surface functional groups of carbon-based adsorbents and their roles in the removal of heavy metals from aqueous solutions: a critical review. Chem Eng J. 2019;366:608–21. https://doi.org/10.1016/j.cej.2019.02.119.
71. Ahmaruzzaman M, Roy S, Singha A, Rtimi S, Aminabhavi TM. Emerging nanotechnologies in adsorption of dyes: a comprehensive review of carbon and metal oxide-based nanomaterials. Adsorption. 2025;31:34. https://doi.org/10.1007/s10450-024-00588-y.
72. Khajavian M, Kaviani S, Piyanzina I, Tayurskii DA, Nedopekin OV. Zn-decorated S, P, B co-doped C_2N nanosheet for ibuprofen adsorption: experimental and density functional theory calculation. Colloids Surf A. 2024;680: 132702. https://doi.org/10.1016/j.colsurfa.2023.132702.
73. Khajavian M, Piyanzina I, Tayurskii DA, Nedopekin OV, Vatanpour V, Ismail S. Phosphonic acid-functionalized g-C_3N_4 nanosheets for chromium (Cr^{6+}) adsorption from aqueous solutions: Experimental insights and density functional theory calculations. Inorg Chem Commun. 2025;174: 113971. https://doi.org/10.1016/j.inoche.2025.113971.
74. Kapnisti M, Lanaras K, Liaska A, Noli F, Samaras P. Removal of cadmium and (radio) toxic metal ions using activated sludge-based sorbents. Sustain Chem Environ. 2025;10: 100230. https://doi.org/10.1016/j.scenv.2025.100230.
75. Klinger C, Reske GD, Vieira Y, Dotto GL, de Castilhos F. Ternary removal of Zn, Ni, and Mn from metal industry wastewater using soybean hulls as adsorbents. Environ Sci Pollut Res. 2025;32:7044–57. https://doi.org/10.1007/s11356-025-36194-w.
76. Wahman AY, Abd El-Hameed K, Markeb AA, El-Said W, El-Maali NA. Validation and optimization of a Prussian blue nanoparticle-based method for efficient detection and removal of lead ions in environmental samples. RSC Adv. 2025; 15:6833–6846. https://doi.org/10.1039/D5RA00199D.
77. Tighadouini S, Radi S, Roby O, Hammoudan I, Saddik R, Garcia Y, Almarhoon ZM, Mabkhot YN. Kinetics, thermodynamics, equilibrium, surface modelling, and atomic absorption analysis of selective Cu(II) removal from aqueous solutions and river water using silica-2-(pyridin-2-ylmethoxy)ethan-1-ol hybrid material. RSC Adv. 2022;12:611–25. https://doi.org/10.1039/D1RA08459J.
78. Yu Y, Feng X, Jian Y, Shen Y, Nie X, Jia W, Wang Q, Qu L, Zhao S. Mutual treatment of algae organic matters and heavy metal ions using a coagulation process: processes and mechanisms. Sep Purif Technol. 2025;362: 131900. https://doi.org/10.1016/j.seppur.2025.131900.
79. Sultana M, Rownok MH, Sabrin M, Rahaman MH, Alam SMN. A review on experimental chemically modified activated carbon to enhance dye and heavy metals adsorption. Clean Eng Technol. 2022;6: 100382. https://doi.org/10.1016/j.clet.2021.100382.
80. Wang L, Wang Y, Ma F, Tankpa V, Bai S, Guo X, Wang X. Mechanisms and reutilization of modified biochar used for removal of heavy metals from wastewater: a review. Sci Total Environ. 2019;668:1298–309. https://doi.org/10.1016/j.scitotenv.2019.03.011.
81. Lima EC, Hosseini-Bandegharaei A, Moreno-Piraján JC, Anastopoulos I. A critical review of the estimation of the thermodynamic parameters on adsorption equilibria: wrong use of equilibrium constant in the Van't Hoff equation for calculation of thermodynamic parameters of adsorption. J Mol Liq. 2019;273:425–34. https://doi.org/10.1016/j.molliq.2018.10.048.
82. Wadhawan S, Jain A, Nayyar J, Mehta SK. Role of nanomaterials as adsorbents in heavy metal ion removal from wastewater: a review. J Water Process Eng. 2020;33: 101038. https://doi.org/10.1016/j.jwpe.2019.101038.
83. Cardoso AO, Bogireddy NKR, Celaya CA, Muñiz J, Agarwal V. Experimental and theoretical approaches to unveil the interaction mechanisms of carbon dots with 4-nitrophenol. J Hazard Mater. 2025;485: 136783. https://doi.org/10.1016/j.jhazmat.2024.136783.
84. Khajavian M, Ismail S. Deep eutectic solvent-modified polyvinyl alcohol/chitosan thin film membrane for dye adsorption: machine learning modeling, experimental, and density functional theory calculations. Int J Biol Macromol. 2025;294: 139479. https://doi.org/10.1016/j.ijbiomac.2025.139479.

85. Zhao C, Zhang J, Zhang W, Yang Y, Guo D, Zhang H, Liu L. Reveal the main factors and adsorption behavior influencing the adsorption of pollutants on natural mineral adsorbents: Based on machine learning modeling and DFT calculation. Sep Purif Technol. 2024;331: 125706. https://doi.org/10.1016/j.seppur.2023.125706.
86. Wen J, Hu X. Metal selectivity and effects of co-existing ions on the removal of Cd, Cu, Ni, and Cr by ZIF-8-EGCG nanoparticles. J Colloid Interface Sci. 2021;589:578–86. https://doi.org/10.1016/j.jcis.2021.01.021.
87. Genua F, Lancellotti I, Leonelli C. Geopolymer-based stabilization of heavy metals, the role of chemical agents in encapsulation and adsorption: review. Polymers. 2025;17:670. https://doi.org/10.3390/polym17050670.
88. Molaey R, Appels L, Yesil H, Tugtas AE, Çalli B. Sustainable heavy metal removal from sewage sludge: a review of bioleaching and other emerging technologies. Sci Total Environ. 2024;955: 177020. https://doi.org/10.1016/j.scitotenv.2024.177020.
89. Zhang Y, Zhu C, Liu F, Yuan Y, Wu H, Li A. Effects of ionic strength on removal of toxic pollutants from aqueous media with multifarious adsorbents: a review. Sci Total Environ. 2019;646:265–79. https://doi.org/10.1016/j.scitotenv.2018.07.279.
90. Tian J, Li G, He W, Tan KB, Sun D, Wei J, Li Q. Insight into the dynamic adsorption behavior of graphene oxide multichannel architecture toward contaminants. Chin J Chem Eng. 2023;53:124–32. https://doi.org/10.1016/j.cjche.2021.12.029.
91. Chérif I, Gassoumi B, Ayachi H, Echabaane M, Caccamo MT, Magazù S, Haj Said A, Taoufik B, Ayachi S. A theoretical and electrochemical impedance spectroscopy study of the adsorption and sensing of selected metal ions by 4-morpholino-7-nitrobenzofuran. Heliyon 2024; 10(5):e26709. https://doi.org/10.1016/j.heliyon.2024.e26709.
92. Mahdi WA, Alhowyan A, Obaidullah AJ. A DFT insight into the potential of cycloparaphenylenes as efficient sensors for detecting Paracetamol. Sci Rep. 2025;15:8150. https://doi.org/10.1038/s41598-025-84263-3.
93. Devarajan Y. Nanomaterials-based wastewater treatment: Addressing challenges and advancing sustainable solutions. BioNanoScience. 2025;15:1–14. https://doi.org/10.1007/s12668-024-01780-8.
94. Raju CV, Cho CH, Rani GM, Manju V, Umapathi R, Huh YS, Park JP. Emerging insights into the use of carbon-based nanomaterials for the electrochemical detection of heavy metal ions. Coord Chem Rev. 2023;476: 214920. https://doi.org/10.1016/j.ccr.2022.214920.
95. Thakur A, Kumar A, Singh A. Adsorptive removal of heavy metals, dyes, and pharmaceuticals: carbon-based nanomaterials in focus. Carbon. 2024;217: 118621. https://doi.org/10.1016/j.carbon.2023.118621.
96. Kaya SI, Yence M, Ozcelikay G, Cetinkaya A, Budak F, Ozkan SA. Two-dimensional materials for environmental applications. Springer;2023. p. 269–297.
97. Suliman MA, Sajid M, Nazal MK, Islam MA. Carbon-based materials as promising sorbents for analytical sample preparation: recent advances and trends in extraction of toxic metal pollutants from various media. TrAC Trends Anal Chem. 2023;167: 117265. https://doi.org/10.1016/j.trac.2023.117265.
98. Dada M, Popoola P. Polymers and two-dimensional nanocomposites. Elsevier;2025. p. 155–179.
99. Yang F, Hu P, Yang F, Hua XJ, Chen B, Gao L, Wang KS. Photocatalytic applications and modification methods of two-dimensional nanomaterials: a review. Tungsten. 2024;6:77–113. https://doi.org/10.1007/s42864-023-00229-x.
100. Balakumar S, Mahesh N, Kamaraj M, Saranya T, Babu PS, Aravind J, Kim W, Govarthanan M. Customized carbon composite nanomaterials for the mitigation of emerging contaminants: a review of recent trends. Carbon Lett. 2024;34:1091–114. https://doi.org/10.1007/s42823-024-00715-3.

Chapter 6
Computational and Experimental Insights into Carbyne and Graphene Nanoribbons Confined in Carbon Nanotubes

Umedjon Khalilov

Abstract One-dimensional (1D) carbon nanostructures, including carbyne and graphene nanoribbons (GNRs), exhibit extraordinary theoretical properties, but their practical utilization has been limited by inherent instability and synthesis challenges. Encapsulation within carbon nanotubes (CNTs) addresses these issues, providing crucial stabilization and creating novel GNR@CNT and carbyne@CNT hybrid systems. This chapter reviews and synthesizes key computational advances that elucidate the structure, properties, and potential of these 1D heterostructures. For carbyne, CNT confinement stabilizes ultra-long chains of thousands of atoms and can suppress the Peierls distortion through charge transfer and radial pressure, potentially inducing a metallic state. For GNRs, encapsulation provides a template for creating atomically precise edges while enhancing mechanical robustness and dramatically increasing thermal conductivity. A central focus is the interplay between the host CNT and the encapsulated structure, where factors such as the tube diameter and chemical termination can be used to tune the electronic, magnetic, and vibrational properties of the composite. These findings highlight promising applications in next-generation nanoelectronics, high-strength lightweight materials, and thermal management systems. By bridging fundamental nanoscale phenomena with practical applications, this work underscores the transformative potential of CNT-confined 1D carbon nanomaterials and outlines the future research directions needed to overcome challenges in synthesis and device integration.

Keywords Carbon nanotubes (CNTs) · Carbyne · Graphene nanoribbons (GNRs) · Encapsulated nanostructures · Quantum confinement · One-dimensional nanostructures · Computational studies

---

U. Khalilov
Arifov Institute of Ion-Plasma and Laser Technologies, Academy of Sciences of the Republic of Uzbekistan, Tashkent 100120, Uzbekistan

U. Khalilov (✉)
Department of Chemistry, University of Antwerp, Antwerp 2020, Belgium
e-mail: umedjon.khalilov@uantwerpen.be

T. Onishi (ed.), *Advances in Quantum Science II*,
https://doi.org/10.1007/978-981-95-9721-5_6

6.1 Introduction

One-dimensional (1D) carbon nanostructures have attracted significant interest within the nanotechnology community due to their extraordinary properties and potential applications. The discovery of novel carbon allotropes, such as fullerenes [1] and carbon nanotubes (CNTs) [2], catalyzed the exploration of lower-dimensional carbonaceous materials. Graphene, a two-dimensional carbon allotrope, further revolutionized the field. However, its intrinsic zero band gap restricts its utility in certain applications. The spatial confinement of carbon into one-dimensional geometries has yielded emergent behaviors. For instance, infilling the hollow core of CNTs with diverse carbonaceous species facilitates the creation of 1D nanowires that exhibit structural and property distinctions from their bulk counterparts [3]. Within such confined volumes, encapsulated atoms or molecules may adopt novel phases, and the properties of the resulting composite can deviate from a linear superposition of the constituents [4]. These advances have spurred the investigation of hybrid nanostructures, which integrate distinct carbon allotropes to engineer novel functionalities.

CNTs are prime examples of 1D carbon nanostructures, known for their remarkable electrical, mechanical, and thermal properties (Fig. 6.1a). Their cylindrical graphitic architecture and nanometer-scale hollow interior make them suitable templates or nanoreactors for synthesizing other 1D materials. Two other noteworthy 1D carbon allotropes are carbyne and graphene nanoribbons (GNRs) (Fig. 6.1b and c). GNRs are narrow, elongated strips of graphene, typically with widths of a few nanometers or less. Their electronic spectrum exhibits a tunable band gap, which helps to mitigate the intrinsic lack of a band gap found in graphene [5]. Carbyne, in contrast, is an infinite linear chain of carbon atoms with sp. hybridization, effectively representing the ultimate one-atom-thick limit of a carbonaceous wire. Although carbyne was theorized in the 1960s, experimental evidence for its stabilized form within multi-walled nanotubes has since been obtained [6]. Both GNRs and carbyne exemplify extreme one-dimensional carbon allotropes, each displaying unique properties, such as size-tunable semiconducting behavior in GNRs and a predicted record-high stiffness for carbyne.

A current frontier in carbon nanoscience involves encapsulating 1D carbonaceous structures within CNTs. This creates integrated hybrid systems such as GNR@CNT (a GNR inside a CNT) and carbyne@CNT (a linear carbon chain inside a CNT) (Fig. 6.1d). The encapsulation concept was first shown with fullerenes inside CNTs, dubbed "carbon peapods", which exhibited many novel electronic characteristics [7]. In 2003, researchers observed one-dimensional carbon chains with over 100 carbon atoms inside multi-walled CNTs [6], offering initial proof of long carbyne chains stabilized by tubular confinement. More recently, progress in chemical and electron-beam techniques has enabled the synthesis of narrow GNRs within single-walled CNTs [8, 9], merging two distinct carbon allotropes into a single nanoscale object. These hybrid structures leverage the protective and templating environment of the CNT. Specifically, the CNT restricts the dimensionality and shields reactive

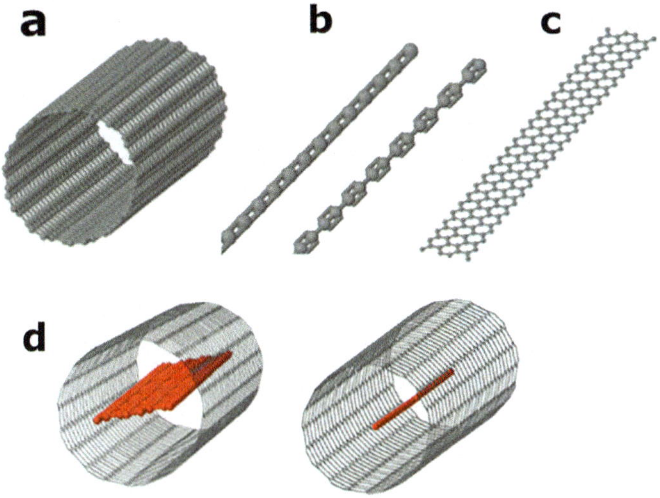

Fig. 6.1 Carbon nanostructures: **a** Carbon nanotube; **b** Carbyne; **c** Graphene nanoribbon; **d** GNR@CNT, and carbyne@CNT

carbonaceous species from the surroundings, while the encapsulated GNR or carbyne lends new properties to the host tube.

The integration of GNRs or carbyne with CNTs is of considerable interest for nanodevice applications. Both CNTs and GNRs, individually, represent promising nanomaterials for next-generation electronics, sensors, and composites [10]. When combined, the resulting nanocomposites may exhibit emergent electrical, mechanical, and even magnetic or catalytic properties that transcend those of the individual components [11]. For instance, a CNT encapsulating a magnetic material can display enhanced magnetism, and a CNT encapsulating a semiconductor ribbon or atomic chain may exhibit modulated conductance. Encapsulated GNRs and carbon chains effectively constitute coaxial heterostructures at the atomic scale, enabling phenomena such as charge transfer, strain engineering, and quantum confinement, which can be tailored through design. Potential applications of these hybrid systems encompass ultra-small electronic interconnects, logic devices with tunable band gaps, nanoscale optical or spintronic devices, high-strength yet lightweight fibers, and advanced thermal management materials.

This chapter focuses on computational studies as a supportive tool for experiments that have explored the structure, synthesis, and properties of carbyne (linear carbon chains) and GNRs encapsulated within CNTs. These computer-based experiments provide critical insights into how confinement within CNTs influences the behavior of these one-dimensional carbon nanostructures, explaining fundamental phenomena such as charge transfer, quantum confinement, and mechanical stability. By leveraging advanced modeling techniques, the aim is to clarify the fundamental mechanisms that underpin the unique properties of these materials and to address the challenges associated with their experimental realization. Furthermore, this chapter will discuss the implications of these computational findings for future research, highlighting how theoretical predictions can guide the design of novel hybrid systems with transformative potential in nanotechnology.

6.2 Carbyne in Carbon Nanotubes (CNTs)

6.2.1 Carbyne@CNT Composite

Carbyne, a carbon allotrope, consists of a one-dimensional chain of carbon atoms with sp. hybridization, unlike the sp^2 bonds in graphene and sp^3 bonds in diamond. In its ideal form, carbyne is an infinitely long linear chain, often called a linear carbon chain (LCC). This structure can be pictured as a string of carbon atoms, each connected to two neighbors. Two main bonding configurations are possible for these chains. One involves an alternating sequence of single and triple bonds (–C ≡ C–), known as a polyyne structure. The other features equal double bonds (=C=C=), termed a cumulene structure. Quantum chemistry predicts that an infinite carbon chain will undergo a Peierls distortion that favors the polyyne form with alternating bond lengths, which opens an energy band gap [12, 13]. Thus, free carbyne is expected to adopt this semiconducting polyyne configuration rather than being a metallic cumulene. Indeed, density-functional calculations find alternating bond lengths of around 1.26 Å and 1.33 Å for the triple and single bonds in carbyne, respectively. The strictly linear structure of carbyne gives it remarkable theoretical properties, including exceptionally high tensile stiffness and strength calculated to surpass those of graphene or carbon nanotubes [14, 15]. On a per-weight basis, carbyne is predicted to be one of the strongest known materials, and its axial Young modulus could also surpass that of sp^2 carbon forms due to the strong sp. σ-bonds along the chain. Carbyne is also expected to have high thermal conductivity along its length, similar to other carbon allotropes, though experimental verification remains challenging.

However, free carbyne is highly chemically reactive and unstable under normal conditions, as long chains readily collapse into graphitic or fullerene carbon or react with other molecules. Until recently, only short chains could be stabilized, typically by capping the chain ends with bulky chemical groups. Encapsulating carbyne inside CNTs offers a method to stabilize much longer chains by isolating them from the external environment and physically preventing reactions or collapse. The CNT acts as a protective sheath that also imposes geometrical constraints. Inside the confined space of a nanotube, a carbon chain remains nearly straight and is forced into an extended configuration. This confinement significantly stabilizes carbyne by preventing instabilities such as bending or cross-linking with neighboring chains. Consequently, the CNT can effectively template an isolated one-dimensional chain.

6.2.2 Synthesis of Carbyne@CNT Systems

Experimental routes to creating carbyne@CNT structures have progressed from early, uncontrolled discoveries to more recent, precise methods. The initial evidence for a carbon chain inside a CNT emerged from high-energy synthesis techniques such as arc discharge. In 2003, Zhao et al. [6] found that the extreme conditions of their arc-discharge process unintentionally produced one-dimensional carbon chains exceeding 10 nm in length (comprising over 100 carbon atoms) inside multi-walled CNTs (MWCNTs). This discovery suggested that the high-temperature, carbon-rich environment facilitates the nucleation of carbyne chains, which subsequently become trapped within the growing shells of the CNTs. While this method first demonstrated the concept, the yield is typically low and the process is uncontrolled. To gain more precision, subsequent research has focused on using the CNT as a nanoreactor for the polymerization of molecular precursors. This common approach involves filling CNTs with small carbon-containing molecules, such as short polyynes (such as $C_{10}H_2$) or other hydrocarbons such as acetylene, and then using thermal treatment to induce them to link into a long, continuous chain [16]. Encapsulation is typically achieved by opening the CNT ends through acid treatment or thermal oxidation before exposing the CNTs to the desired molecules. Upon heating these filled nanotubes to high temperatures (often 800–1000 °C) under vacuum, the precursor molecules polymerize. Evidence for this transformation comes from in situ transmission electron microscopy (TEM) and Raman spectroscopy. For example, Raman spectra show characteristic precursor peaks (around 2000–2200 cm^{-1}) vanishing after high-temperature treatment, while new bands corresponding to longer carbyne chains appear (around 1800–1900 cm^{-1}), a clear sign of coalescence [17]. Using this template-based synthesis, Shi et al. [17] successfully produced ultra-long linear carbon chains (LLCCs) of up to 6000 atoms inside double-walled CNTs (DWCNTs), representing the longest carbyne segments ever observed in a stable state. This method offers significant control, as choosing the precursor and conditions allows one to aim for specific chain lengths and end terminations.

Complementing these experimental efforts, simulation studies have provided crucial insight into the optimal conditions and underlying mechanisms for carbyne growth. Full atomistic molecular dynamics modeling by Deng and Cranford [18], for example, systematically investigated the roles of nanotube diameter and temperature (Fig. 6.2).

Their results revealed that narrow confinement is critical, requiring an inner tube diameter below about 0.9 nm. If the nanotube is too wide, simulations indicate that branched networks or small graphene fragments might form, as entropy and alternative bonding configurations compete with linear chain formation. Conversely, in a sub-nanometer channel, the only viable structure is a one-dimensional line. Likewise, the temperature must be high enough to mobilize carbon species and heal defects (exceeding roughly 500 K) but low enough to prevent sublimation, consistent with the practical temperature ranges of 500–2000 K used in experiments [18]. Computational simulations have played a pivotal role in exploring catalytic growth mechanisms.

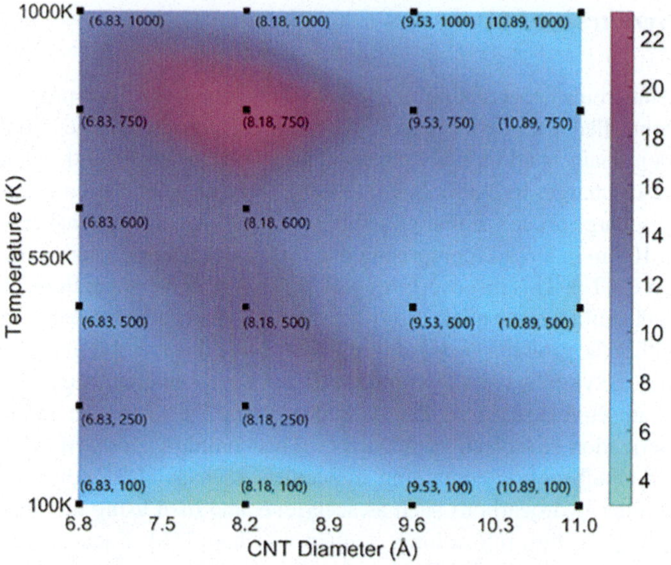

Fig. 6.2 Visualization of maximum carbon chain length as a function of temperature and carbon nanotube diameter. The color scale indicates the chain length, which serves as an indicator of the bond formation rate, and the plot reveals that the peak chain length occurs at lower diameters and higher temperatures. Reprinted from ref. [18]. Copyright 2019, with permission from Elsevier

Research by Khalilov et al. [19] regarding catalyzed systems demonstrates that the nickel catalyst facilitates, among others, the assembly of carbon from small molecular units, incrementally elongating the chain (Fig. 6.3). Crucially, these findings underscore that the choice of feedstock is a determining factor. While pure carbon sources facilitate continuous growth (Fig. 6.3, C), the process is significantly hindered when hydrocarbon species are introduced. This obstruction arises from a competitive interaction where hydrogen atoms passivate the reactive chain ends, effectively terminating growth (Fig. 6.3, C_xH_y). Furthermore, oxygen-containing radicals (such as CO) can deactivate the Ni catalyst through oxidation (Fig. 6.3, C_xO_z), causing chain formation to cease entirely [20]. Consequently, a strictly controlled Chemical Vapor Deposition (CVD) strategy, utilizing a catalyst and a hydrogen-free carbon source, appears to be the most viable pathway for continuous synthesis.

Despite this progress, significant practical challenges remain. There is great interest in using single-walled CNTs (SWCNTs) to allow for more direct probing of the properties of the chain; however, the required ultra-narrow SWCNTs are inherently unstable. Their isolation after synthesis can also require harsh procedures such as ultrasonication and surfactant wrapping, making device integration difficult [21]. Overall, by creating the right conditions within the CNT as both a template and a stabilizing container, carbyne, once purely theoretical, can now be studied as chains of hundreds or thousands of atoms, opening up experimental investigations of its properties.

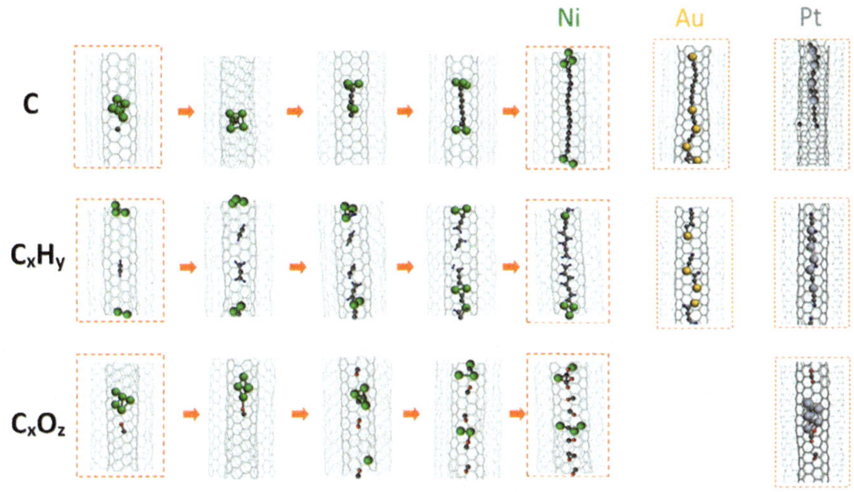

Fig. 6.3 Catalytic growth of carbyne inside a (5,5)@(10,10) double-walled carbon nanotube (DWCNT) using distinct precursors: **a** C, **b** C_XH_Y, and **c** C_XO_Z. The indices (5,5) and (10,10) correspond to the chiralities of the inner and outer walls, respectively. Color coding for atoms: incorporated carbon (red), hydrogen (blue), and metal catalysts Ni (green), Au (yellow), and Pt (gray)

6.2.3 Properties of Carbyne@CNT Structures

When an LCC is encapsulated inside a CNT, the two constituents interact with each other in ways that can significantly alter their individual properties [22–24]. The confinement, charge transfer, and mechanical coupling between the carbyne and the CNT give rise to a variety of new phenomena in these carbyne@CNT systems [16, 25, 26].

Structural and Vibrational Properties

Experimentally, the most significant structural effect of encapsulation is the dramatic stabilization of the carbon chain. The CNT acts as a protective shield, allowing the chain to survive at much higher temperatures and protecting it from chemicals or radiation that would otherwise cause it to break. The vibrational (phonon) properties, which are a key characteristic of these systems, are also directly observed through experiments. Raman spectroscopy typically detects distinct stretching modes for the chain (around 1800–2100 cm^{-1} for polyyne structures), and a key finding is that encapsulation tends to shift these vibrational frequencies and can create multiple peaks due to the interaction between the chain and tube [27].

Computational modeling provides a detailed explanation for these experimental observations. DFT calculations confirm that the lowest energy configuration for the chain is along the central axis of the tube, where it is held in place by a balance of weak van der Waals forces and quantum confinement effects [28]. These studies also show that an optimal CNT diameter exists that minimizes the interaction energy, thus holding the chain most stably. In addition to its stable position, the chain can vibrate or oscillate within the nanotube's hollow core, and the observed shifts in vibrational frequency are attributed to these confinement and interaction effects.

Electronic Properties and Tunability

A key experimental finding is that the electronic properties of encapsulated carbyne are tunable. Work by Shi and colleagues demonstrated that the band gap of the chain can be controlled by selecting the diameter of the host SWCNT [29]. This ability to engineer the electronic structure is explained by a complex interplay of factors revealed through computational studies.

Theoretically, an isolated carbyne chain is predicted to be a semiconductor with a moderate band gap (typically 1–2 eV) because Peierls distortion favors a polyyne state with alternating bonds. Encapsulation fundamentally alters this picture through mechanisms such as charge transfer and orbital hybridization (Fig. 6.4). Ab initio and DFT simulations show that electrons can transfer between the tube and the chain, which can suppress the Peierls distortion, reduce bond-length alternation, and make the chain more cumulene-like [12, 13, 28, 30, 31]. As a result, the combined system can become metallic, a phenomenon predicted for a chain inside a (9,0) SWCNT [18]. The tunability observed experimentally is also theoretically explained. A wider tube allows the chain more freedom to relax into its natural semiconducting state, while a tighter-fitting tube can exert radial pressure, forcing the bonds to become more uniform and pushing the chain toward a metallic state. Furthermore, theoretical work shows that chemical termination is another effective tuning method; for example, hydrogen termination can substantially decrease the band gap of a metal-terminated carbyne chain inside a CNT [19].

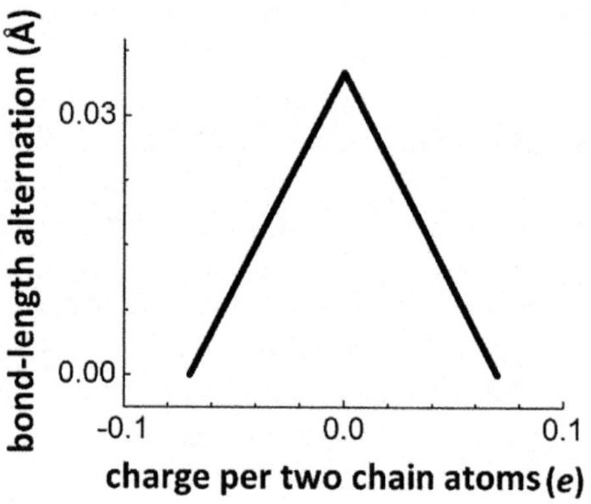

Fig. 6.4 Schematic representation of the relation between bond-length alternation (BLA) and charge transfer (q) in a linear carbon chain. The "triangle-like" dependence illustrates how charge transfer influences the Peierls distortion

Mechanical and Thermal Properties

The understanding of the mechanical and thermal properties of carbyne@CNT composites comes primarily from simulation and theoretical modeling, as direct experimental measurements at this scale are exceptionally challenging.

Mechanically, the coupling between the two components can theoretically enhance the properties of the composite. Due to the exceptional stiffness of carbyne, simulations show that encapsulation can increase system stiffness and enhance the resistance of the CNT to buckling under compression. The effect on tensile strength, however, is complex and depends critically on load transfer efficiency. If the chain is pinned to the tube, it can carry a significant tensile force. Yet, if it can slip freely, its contribution is small, and some simulations even suggest it can decrease the fracture strength by initiating failure [32]. The dynamic properties are also distinct; the encapsulated chain can vibrate such as a guitar string, a behavior that could be exploited in NEMS for ultra-high frequency resonators or mass sensors [33].

Thermally, computational studies show that encapsulation significantly modifies transport properties in a diameter-dependent manner [34, 35]. For smaller diameter CNTs (e.g., in the 5.42–10.85 Å range), LCC encapsulation enhances thermal conductivity by up to 15%. This is because the LCC provides an additional pathway for heat-carrying phonons, and good alignment between the phonon modes of the two components facilitates efficient energy transfer. In larger diameter CNTs (e.g., 13.56 Å), however, the increased distance weakens this phonon coupling, and dissimilar modes lead to increased scattering, which can cause a reduction in thermal conductivity (Fig. 6.5).

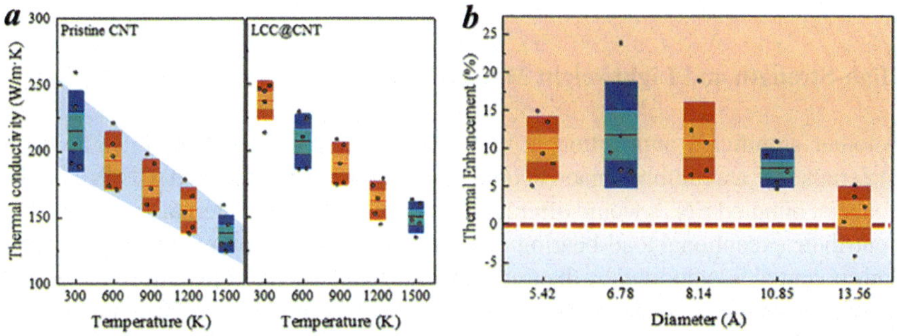

Fig. 6.5 **a** Temperature-dependent thermal conductivity of pure CNTs and CNTs with encapsulated LCCs. **b** The improvement factor of thermal conductivity due to LCC incorporation. Box plots display the mean, standard error (SE), and standard deviation (SD) of the data. The SD measures the dispersion of data points (x_i) relative to the mean (μ) and is calculated as $\sigma = \sqrt{\frac{1}{N}\sum_{i=1}^{N}(x_i-\mu)^2}$. The SE of the mean estimates how precisely the sample mean represents the true population mean and is found by dividing the SD (σ) by the square root of the sample size (n), as $SE = \frac{\sigma}{\sqrt{n}}$. Minimum and maximum SD values are shown in blue, and other SD values in orange. Adapted from ref. [34]. Copyright 2024, with permission from Elsevier

6.2.4 Applications of Carbyne@CNT Composites

The unique characteristics of carbyne@CNT composites, established through both experiment and simulation, enable several innovative applications [36, 37]. While many of these are still in the conceptual stage, they highlight the transformative potential of these hybrid materials.

Ultrathin Nanowires and Nanoelectronic Devices

One of the most exciting prospects is the use of carbyne@CNT as ultrathin nanowires in future electronic circuits. Carbyne itself, being only one atom thick, represents the fundamental limit of interconnect miniaturization. The experimental demonstration that encapsulation within a CNT enhances the stability of carbyne and offers electrical shielding makes this application conceivable. In this design, the CNT can function as either a conductor or a dielectric based on its specific type, allowing for the creation of what could be the world's smallest coaxial cables [36, 37].

Beyond simple wiring, the theoretically predicted and experimentally observed tunability of the electronic properties suggests more advanced functions. The system could act as a nanoscale tunable semiconductor or a switching device if the charge transfer could be modulated, for instance, by gating the nanotube. Furthermore, theoretical models predict that combining a metallic CNT with a semiconducting carbyne (or vice versa) could form a one-dimensional p-n junction or Schottky contact at their interfaces. This could lead to the creation of tiny diodes or rectifiers on the nanoscale.

High-Strength and Lightweight Materials

Another significant application is in creating superior high-strength, lightweight materials by assembling macroscopic fibers from carbyne-filled CNTs. In this concept, bundled CNTs would offer lateral stability while the internal carbyne chains contribute exceptional load-bearing capacity, similar to a reinforced nanoscale fiber. This potential is supported by theoretical work from Gao and colleagues, which indicated that such composite fibers could achieve nearly double the elastic modulus of pure CNT fibers while maintaining a low density [36]. These predictions suggest that such materials would be invaluable in fields such as aerospace or for structural reinforcements, where minimizing weight is paramount.

Current Status and Future Directions

While many of these applications are speculative and depend on overcoming significant fabrication challenges, the demonstrated progress in creating long, stable carbyne chains inside nanotubes is a key enabling step. This research has also expanded fundamental scientific understanding by providing a testbed for studying 1D physics, such as charge-density waves and quantum confinement in a line of atoms. As synthesis and integration techniques improve, researchers anticipate advancing closer to practical devices that harness this ultimate one-dimensional carbon wire.

6.3 Graphene Nanoribbon (GNR) in CNTs

6.3.1 GNRs: From Isolated Strips to Encapsulated Systems

Graphene nanoribbons (GNRs) are narrow strips of graphene, effectively one-dimensional structures since their width is constrained to the nanometer scale, typically less than 10 nm. Conceptualized as being cut from a larger graphene sheet, their properties depend notably on width and the crystallographic orientation of their edges. The two main edge types are armchair (AGNRs) and zigzag (ZGNRs), which exhibit distinct electronic behaviors (Fig. 6.6). A key feature of GNRs is that, unlike two-dimensional graphene, they generally exhibit a non-zero band gap, making them semiconductors.

The edge type dictates the specific electronic properties. For AGNRs, the band gap size depends on the ribbon width, making their electronic characteristics tunable. ZGNRs, in contrast, are predicted to have unique magnetic edge states. The two opposite edges host localized electronic states that tend to antiferromagnetically couple (one edge spin-up, the other spin-down), which opens a small band gap and

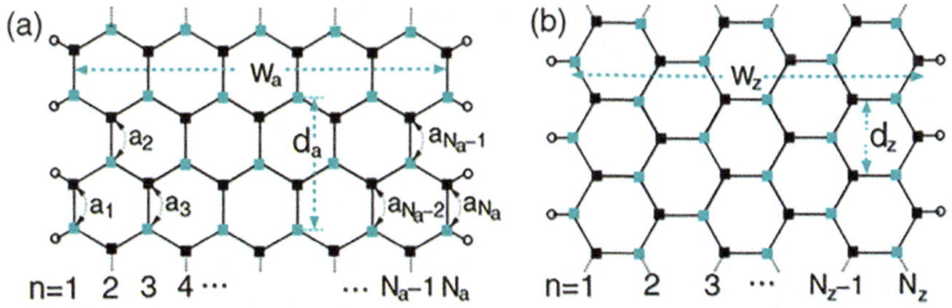

Fig. 6.6 Structural representations of **a** an 11-AGNR and **b** a 6-ZGNR. In both models, hydrogen atoms (open circles) terminate the edge carbon atoms, and black and gray rectangles differentiate the two sublattices of the graphene structure. Panel **a** labels the 1D unit cell dimension (d_a), ribbon breadth (w_a), and the carbon-carbon distance in the n-th dimer row (a_n). Panel **b** indicates the corresponding dimension (d_z) and breadth (w_z) for the ZGNR. The number in each name specifies the ribbon's width; "11" in 11-AGNR refers to the number of carbon dimer rows across the width, while "6" in 6-ZGNR refers to the number of zigzag chains. Reprinted with permission from ref. [38]. Copyright 2006 by American Physical Society

makes the ribbon a spin-polarized semiconductor. This makes ZGNRs interesting for spintronics, while AGNRs are promising for electronics due to their variable band gap.

Despite this potential, realizing these properties in isolated GNRs is hindered by significant challenges related to stability and fabrication. In terms of chemical stability, armchair-edged GNRs are generally more energetically stable than their zigzag-edged counterparts [5]. The zigzag edge, with its unsatisfied bonds, tends to be highly reactive and often requires passivation to prevent it from reconstructing. Furthermore, electronic transport in GNRs is heavily influenced by physical edge roughness. For applications, smooth, well-controlled edges are crucial, but achieving this with conventional top-down fabrication methods such as lithography is difficult and often produces rough edges that scatter charge carriers. These challenges provide a strong motivation for alternative approaches, such as bottom-up synthesis or protective encapsulation.

To overcome these challenges of edge reactivity and structural imperfections, a powerful strategy is to synthesize and protect GNRs within the pristine, confining environment of CNTs. When a GNR resides inside a CNT (forming a GNR@CNT composite), its structure must adapt to the cylindrical interior. The GNR might not remain perfectly flat; it can curve, scroll, or even twist into a helical shape depending on its width relative to the tube diameter [39–42].

This encapsulation approach provides several advantages. GNRs synthesized in situ within CNTs often have perfectly smooth, chemically passivated edges and uniform widths, as the CNT acts as an ideal one-dimensional template. The CNT shell also protects the GNR from the external environment. Consequently, the fundamental electronic and magnetic characteristics of the GNR largely persist, though they can be modulated by the electronic and structural interactions with the host CNT. This creates a robust hybrid material that allows for the study and potential application of otherwise delicate GNR structures.

6.3.2 GNRs Inside CNTs

Templated Nanosynthesis: An Overview

Creating a GNR inside a CNT is a fascinating example of templated nanosynthesis. The general method relies on filling a nanotube with specific molecular precursors and then transforming those precursors into a contiguous graphene ribbon. CNTs, especially SWCNTs, have an inner channel that can accommodate a ribbon of a certain maximum width, typically up to approximately two graphene unit cells across for common SWCNT diameters. Thus, the CNT acts as both a nanoreactor and a template. It defines ribbon width and prevents the ribbon from growing beyond the confines of the tube, a dual role that helps in achieving uniform narrow ribbons.

Experimental Methods

Two independent experimental approaches have demonstrated the feasibility of this process. In one method, electron-beam irradiation was used. Chuvilin and colleagues conducted experiments where a mixture of small, carbon-rich molecules inside SWCNTs was subjected to a focused electron beam in a transmission electron microscope (TEM) [8]. The electron beam knocks out non-carbon atoms and mobilizes carbon species, causing them to rearrange and bond. Remarkably, under prolonged exposure, the molecules were observed to coalesce into a continuous GNR.

In another approach, thermal annealing of filled tubes was employed. Talyzin et al. [9] filled SWCNTs with polyaromatic hydrocarbons (planar carbon molecules) and then heated them to high temperatures. Under confinement and heat, the molecules reacted and fused into a continuous GNR. This process can be viewed as an in situ polymerization of molecular precursors. Confinement plays a critical role; if the same molecules were heated without the CNT, they would likely form bulk graphite or disordered carbon. Inside the nanotube, however, the only configuration allowing carbon atoms to maximize bonding is a narrow graphene strip aligned along the tube axis (Fig. 6.7).

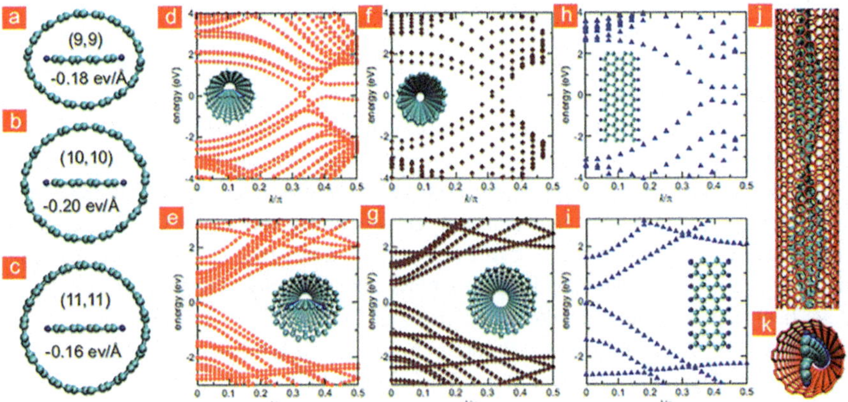

Fig. 6.7 First-principles simulations of the structural and electronic properties of GNRs encapsulated within SWNTs. Structural models show **a–c** ball-and-stick representations of coronene-based GNRs inside armchair SWNTs with their chiral indices and vdW bonding energies, along with **j, k** top and side perspectives of a twisted coronene-based GNR. The calculated electronic band structures compare the composite systems, including **d** a coronene-based GNR within a (10,10) SWNT and **e** a pyrene-based GNR within a (14,0) SWNT, to the band structures of the **f, g** isolated SWNTs and **h** the isolated pyrene-based GNR. Reprinted with permission from ref. [9]. Copyright 2011 American Chemical Society

Synthesis Challenges and Control

While synthesis is achievable, achieving precise control over the final ribbon structure presents key challenges and depends on several interdependent factors.

The Role of the CNT Template. The role of the CNT as a template cannot be overstated. It not only constrains the width of the ribbon but also directs the alignment of precursor molecules. Inside a narrow cylinder, molecules naturally stack and align in certain ways, often in a roughly linear column. When these molecules fuse, they form a ribbon that is automatically straight and narrow. The term "nanoreactor" is often used because reactions that might not occur in bulk can proceed in the confined environment of the nanotube. The CNT can also strain the forming ribbon; as noted in experimental observations, the SWCNT might deform slightly to become oval to accommodate a flat ribbon, and the ribbon itself may twist if needed.

Precursor Selection and Edge Structure Control. The choice of precursor is critical. Ideal precursors are planar polycyclic aromatic hydrocarbons (PAHs) that are slightly smaller than the cross-section of the CNT, such as coronene or perylene derivatives. In contrast, encapsulating and heating spherical fullerenes (such as C_{60}) often results in their collapse into disordered carbon clusters rather than a well-defined ribbon.

A primary challenge is controlling the crystallographic edge structure. The experiments by Chuvilin and Talyzin, for instance, allowed the ribbon to form naturally, which resulted in zigzag-edge ribbons [8, 9]. This outcome likely occurred because the precursors were not pre-shaped for an armchair configuration, and a zigzag form is often the most natural way to tile the cross-section of the tube. If an armchair-edge ribbon is desired, different precursor molecules that bias the formation of armchair edges would be needed. Supporting this, theoretical work by Khalilov and Neyts [43] suggests the choice of hydrocarbon feedstock also influences GNR nucleation/formation, with the presence of hydrogen playing a key role in stabilizing the structures during growth (Fig. 6.8). Experimentally, the majority of reported GNR@CNT have been zigzag-oriented, likely because zigzag ribbons of certain widths fit well in typical SWCNTs and may be favored under certain synthesis conditions.

Advanced Geometries. Another interesting development is the possibility of forming multiple or layered ribbons. In principle, if a CNT is large enough, it might be possible to fit two or more ribbons side-by-side. Indeed, simulations by Khalilov and Neyts [43] also demonstrate the formation of bilayer GNRs inside SWCNTs (Fig. 6.8). In practice, however, what is typically observed in wider tubes is often a single, wider twisted ribbon or a scroll of graphene, rather than multiple separate ribbons. For the narrow cases primarily considered, there is usually one ribbon per tube.

Summary and Outlook for Synthesis

The primary route to GNR@CNT involves encapsulating molecular precursors into SWCNTs, followed by activation using thermal or electron-beam methods. This

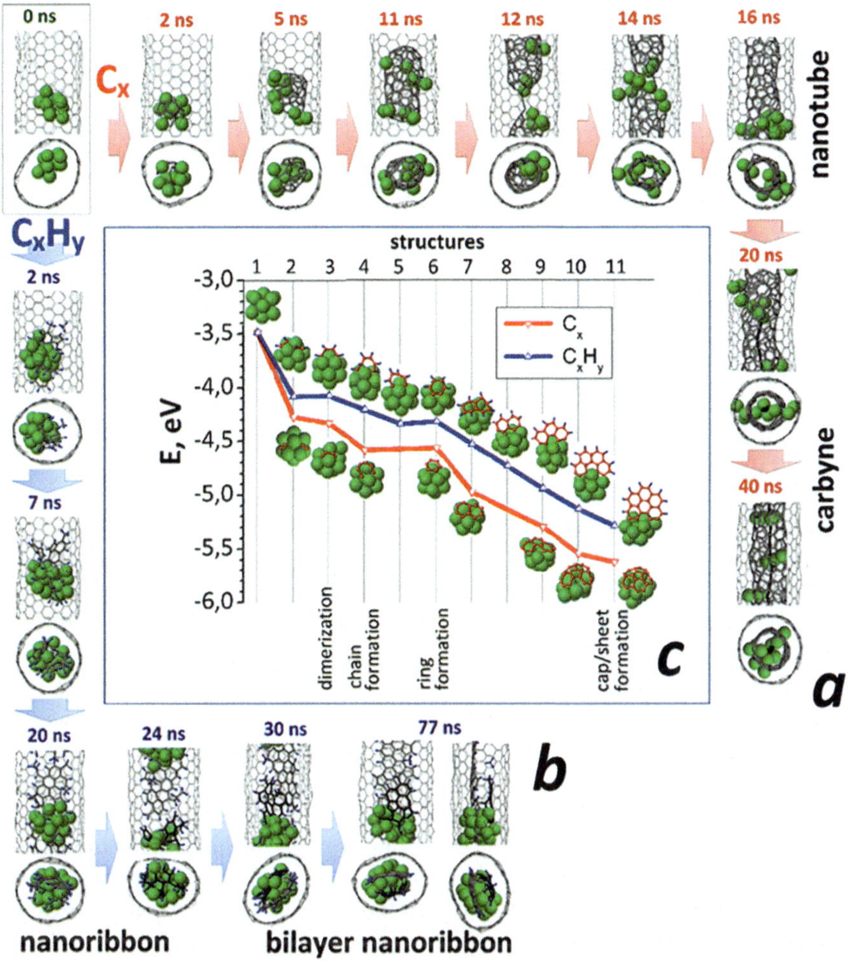

Fig. 6.8 Catalyzed growth of encapsulated carbon nanostructures (ECNs) from different feedstocks **a** Carbon (C_x) and **b** Hydrocarbon (C_xH_y). Ni, C, and H atoms are represented by green, gray/black, and blue, respectively **c** Energies of incipient structures in dimerization, chain, ring, and cap/sheet formation steps of the nucleation stage of tube and graphene growth using C_2 and C_2H_2 feedstocks, respectively

template synthesis yields ribbons with smooth edges and a uniform width defined by the CNT. By using CNTs as nanoreactors, scientists can create atomically precise GNRs that would be difficult to achieve by other means. The encapsulated ribbons are protected by the CNT shell from environmental contamination or further chemical reactions, which is advantageous for stability and subsequent characterization. The trade-off, however, is that the ribbon is confined inside a tube, a situation that might complicate direct electrical contact or device integration. Nevertheless, this encapsulation approach is a powerful demonstration of nanoscale fabrication, achieving atomic-level precision through a self-assembly process that exploits confinement.

6.3.3 Properties of GNR@CNT Systems

When a GNR is encapsulated inside a CNT (forming a GNR@CNT composite), the two components, the ribbon and the tube, interact both structurally and electronically, leading to modified properties for the combined system. Similar to carbyne, confinement and interfacial interactions play a significant role. However, graphene nanoribbons exhibit unique characteristics, as they are two-dimensional sheets constrained in one dimension, unlike the strictly one-dimensional nature of carbyne chains.

Structural and Morphological Interactions

The presence of a GNR inside a CNT can cause distortions in both the ribbon and the tube. As observed experimentally [8, 9], a sufficiently wide ribbon will push the CNT to become oval in cross-section rather than circular. This occurs as the system compromises to accommodate the flatness of the ribbon. The ribbon itself may twist within the CNT, and the degree of this helical twist depends on factors such as ribbon width relative to the tube circumference and the lattice commensurability between the two structures. A smaller lattice constant difference or a narrower ribbon might allow it to lie mostly straight, while a larger difference can cause the ribbon to twist, effectively winding around the tube with a certain pitch. Molecular dynamics simulations support this finding. For example, Li et al. [44] found that multiple GNRs simulated inside a larger SWCNT tended to form a spiral arrangement to maximize van der Waals stacking and accommodate space.

This twist means the edge of the ribbon does not run parallel to the tube axis but at an angle, creating a screw-like morphology. The confinement of the encapsulated ribbon, analogous to a graphene sheet under cylindrical boundary conditions, can also induce ripples or a slight curvature. Despite these deformations, the ribbon and tube typically remain intact and crystalline. This stability arises because their interaction is mainly through van der Waals forces, not covalent bonds. As both are generally graphitic, they do not readily bond with each other, although the edge of the ribbon could potentially bond to the tube wall, an occurrence not widely reported for clean systems. When terminated with hydrogen or another element, the edges of the ribbon become chemically stable and do not bond with the tube, as the inner surface of the tube is essentially chemically inert.

Electronic Properties and Band Structure

The electronic interaction between a GNR and its host CNT can be complex. Each component has its band structure. For example, the GNR has a band gap that can be large for a narrow armchair ribbon or small for a wider or zigzag ribbon, while the CNT might be metallic or semiconducting depending on its chirality. When combined, these two systems can exchange electrons or influence the band structure of one another through electric fields and hybridization [38, 45, 46]. One important consideration is screening, as the CNT walls can partially screen the ribbon from external electric fields and vice versa. If an attempt is made to electrostatically dope the ribbon (for example, by applying a gate voltage), the CNT could shield it unless the gate couples to the entire composite.

Calculations by Lebedeva et al. [47] on a sulfur-terminated zigzag GNR encapsulated within a CNT revealed charge redistribution at the interface. The sulfur termination introduces additional electrons to the ribbon via the lone pairs of sulfur, enabling charge transfer to or from the tube, contingent upon work function alignment. Their DFT study demonstrated that this specific termination profoundly influences the overall electronic structure of the composite. As an illustration, the edge states of a zigzag GNR with sulfur edge atoms can undergo an energy shift due to interaction with the tube. This interaction, stemming from the capacity of both sulfur and the tube to exchange electron density, may lead to a reduction in the magnetic moment at the ZGNR edges relative to an isolated ZGNR. In essence, functional groups at the GNR edge (such as S or H) and the π-π interaction with the CNT will tune the band edges of the ribbon and might induce states in the gap.

Mechanical Properties

Encapsulating a GNR inside a CNT essentially creates a reinforced nanotube [11, 48, 49]. The ribbon provides internal reinforcement, enabling the composite to bear significant loads. Several studies have examined the mechanical behavior of these systems. For instance, molecular dynamics nano-indentation simulations by Fang et al. [11] found that the composite can sustain higher loads than an empty CNT. In their results, the presence of the GNR increased the maximum force the tube could handle by more than 15%, and the composite showed a significant elastic recovery (a "springback" effect approximately 15% greater than the CNT alone), indicating a kind of superelastic behavior. This is likely because the internal ribbon prevents the tube from deforming excessively and helps it snap back into shape.

Another study by Shen and Cheng [50] examined the tensile properties of different helical configurations. They found that a straight GNR provides the greatest mechanical reinforcement, with a much better load-bearing capacity than an empty (10,10) SWNT. Even a 30°-twisted GNR provided noticeable reinforcement, though the benefit from a 60°-twisted GNR was smaller. This is logical because a straighter ribbon carries axial load more effectively. The failure mechanisms also differed.

For example, the composite might fail by the ribbon breaking, the tube breaking, or slippage at the interface, whereas an empty CNT fails by wall fracture.

Thermal Properties

A particularly notable result from Shen and Cheng [50] was that encapsulated GNRs enhance the thermal conductivity of the nanotube system. They calculated that a straight GNR inside a (10,10) CNT roughly doubles the thermal conductivity of the system compared to the CNT alone (about 2.2× higher). Even a 30° helical GNR increased thermal conductivity by approximately 1.5×. The reason is that the GNR provides an additional channel for heat flow. Phonons (quantized lattice vibrations carrying heat) can travel through both the tube and the ribbon. If the ribbon is straight, phonons can propagate relatively unimpeded along it. If the ribbon is helical, transport experiences more scattering, so its contribution is less but still positive. This enhancement makes GNR@CNT composites interesting for applications in thermal management, such as for heat dissipation in microelectronics.

6.3.4 Applications of GNR@CNT Nanostructures

Nanoelectronics and Optoelectronics

The GNR@CNT system holds promise for applications in modern nanoelectronics and optoelectronics. This structure exhibits diverse electronic properties that make it suitable for transistors, memory devices, and optical detectors [5, 22, 51]. The electronic structure of GNR@CNT composites largely reflects a combination of the properties of individual GNRs and CNTs, allowing for the retention of key characteristics such as bandgap tunability and magnetic behavior [9]. Furthermore, the encapsulation of GNRs within CNTs can protect them while preserving their conductive nature, positioning them as candidates for nanoscale interconnects and inductive nanocoils [47]. The ability to control the electronic and magnetic properties of these hybrid structures by tuning the chirality and size of GNRs further expands their potential in logic circuits, switching devices, and photovoltaic applications [10].

Nanoelectromechanical Systems (NEMS)

The enhanced mechanical properties of GNR@CNT composites, especially the observed resilience and superelastic behavior, suggest uses in NEMS. A GNR@CNT could function as a nanoscale spring or resonator that can bend and recover. Its ability to withstand repeated indentation with a springback greater than 15% [11] makes it suitable for devices requiring flexibility and durability, such as flexible nanoswitches or oscillators. The encapsulated ribbon might also dampen or modify the vibration modes of the CNT, which could be useful for controlling resonance frequencies in NEMS sensors.

Summary and Future Outlook for Applications

Overall, the encapsulation of GNRs inside CNTs creates a robust one-dimensional carbon composite with enhanced mechanical strength, significantly improved thermal conductivity, and rich electronic behavior influenced by interfacial interactions. These composites hold promise for a range of applications, including high-strength nanocomposites for mechanical reinforcement, advanced thermal interface materials for cooling electronics, nanoelectronic interconnects or transistors that take advantage of the coaxial geometry, and even spintronic or sensing devices if their unique magnetic or surface properties are harnessed. As research continues, understanding how to precisely control the ribbon-tube interface, including edge termination and relative alignment, will be key to tailoring these properties for specific uses.

6.4 Conclusion and Outlook

Encapsulation of 1D carbon nanostructures, including carbyne and GNRs, within CNTs represents a transformative advance in nanotechnology. This chapter has demonstrated how encapsulation leverages the strengths of both components, where CNTs stabilize otherwise fragile 1D carbon forms while the encapsulated materials impart novel functionalities to the host. These hybrids exhibit enhanced properties compared to their constituents, paving the way for next-generation nanodevices. Carbyne@CNT and GNR@CNT composites share similar outlooks regarding their potential and underlying principles. Encapsulation enables the study of carbon in unprecedented regimes, such as thousand-atom carbyne chains and sub-nanometer GNRs with precise atomic edges. Experimental verification of theoretical predictions has revealed surprising phenomena, including the suppression of Peierls distortion in carbyne and dramatic thermal conductivity enhancements in GNR@CNT systems.

6.4.1 Promising Applications

From an application perspective, these hybrids are promising on multiple fronts. In nanoelectronics, they offer potential solutions to persistent challenges. For instance, the intrinsic lack of a band gap in graphene can be addressed through the use of GNRs. Encapsulating GNRs within carbon nanotubes not only protects these nanoscale semiconductors but also enables their integration into complex architectures. One could envision transistors made from a GNR inside a CNT, where the CNT shields the ribbon and acts as an integrated gate electrode. Similarly, carbon chains inside CNTs could act as potentially the smallest interconnects or logic elements, with the CNT providing stability and a way to modulate the properties of the chain via strain or electrostatic coupling. These systems realize the idea of a "nanocable," a conducting core with an insulating or conducting sheath, for example, a carbyne inside an insulating CNT.

In the realm of advanced materials, the superior mechanical and thermal characteristics of the encapsulated structures can be harnessed. A straightforward outlook is the development of high-strength, lightweight composite fibers where each CNT contains a reinforcing carbon chain or ribbon. Such fibers could be used in aerospace or infrastructure, potentially offering improved performance over existing carbon fibers. The enhanced thermal conductivity of GNR@CNT suggests use of these composites in thermal interface materials or heat spreaders, managing heat in electronics more effectively than traditional CNT films.

There are also exciting prospects in Nanoelectromechanical Systems (NEMS) and sensors. The unique oscillatory modes of a carbon chain inside a CNT (essentially a tunable nanostring) could lead to mass sensors of unprecedented sensitivity. The vibration frequency of the chain would change if an atom adsorbs onto it, and the CNT could transduce that signal. GNR@CNT could similarly function as torsional or flexural oscillators with high Q-factors, given their observed superelastic behavior. With further engineering, NEMS devices such as nanomotors or actuators, where an encapsulated ribbon twists within a tube when a voltage is applied, could potentially be created.

6.4.2 Key Challenges and Future Research Directions

For all this potential, significant challenges and directions for future research remain. One major challenge is controlled synthesis and scalability. While impressive demonstrations have been achieved, producing these materials in large quantities with uniform properties is still difficult. Active research areas include improving filling techniques for CNTs, developing methods to mass-produce ultra-narrow GNRs inside CNTs, potentially through Chemical Vapor Deposition or bulk processing rather than synthesis based on individual TEM. Furthermore, finding ways to extract or contact the encapsulated structures for device integration is crucial. For example, reliably connecting electrodes to a carbyne chain inside a CNT to measure its conductance remains difficult.

Another avenue of research is tailoring the interface by modifying the inner surface of the CNT or the edge chemistry of the GNR. Using different edge terminations, such as hydrogen or sulfur, or doping edge atoms in a GNR by substituting nitrogen, could change how the ribbon and tube exchange charge. This could engineer specific band alignments or magnetic properties. Similarly, introducing periodic perturbations, such as a periodically varying diameter in the CNT or a modulated ribbon width, could create superlattice effects in the 1D electronic structure, leading to new quantum phenomena.

From a theoretical perspective, these systems offer a playground for studying low-dimensional physics and van der Waals heterostructures. Further theoretical work can explore how multiple phases (such as metallic, semiconducting, or magnetic) can coexist or transition in these confined geometries. There is also growing interest in exploring one-dimensional structures beyond carbon inside CNTs, for example, boron nitride ribbons or chains, and even hybrid chains composed of carbon and other elements. The success with carbon might be extended to create other composite one-dimensional materials encapsulated in CNTs, thereby broadening the library of available nanowires.

6.4.3 Concluding Remarks

The combination of GNRs and carbyne with CNTs has established a new class of carbon nanocomposites with extraordinary properties. Although many potential applications remain to be fully explored, the advancements made thus far, from the initial observation of atomic chains in 2003 to the synthesis of long linear chains and precisely defined sub-nanometer ribbons in the 2010s, offer reason for optimism. Future progress in synthesis, characterization, and theoretical modeling is likely to overcome current limitations. As techniques improve, these one-dimensional carbon hybrids may be integrated into real devices, whether in ultra-fast electronics, quantum wires, strong yet lightweight materials, or nanoscale sensors. Continued exploration of these systems not only deepens our fundamental understanding of nanoscale materials science but also fuels innovation by harnessing the exceptional versatility of carbon. The evolving narrative of carbon allotropes continues to deliver surprises, and the exploration of 1D structures confined within CNTs promises many exciting discoveries ahead.

Acknowledgments I gratefully acknowledge financial support from the Agency for Innovative Development of the Republic of Uzbekistan (grant no F-FA-2021-512). I also extend my sincere thanks to my PhD student Kamoliddin Mehmonov.

References

1. Kroto HW, Heath JR, O'Brien SC, Curl RF, Smalley RE. C60: Buckminsterfullerene. Nature. 1985;318:162.
2. Iijima S. Helical microtubules of graphitic carbon. Nature. 1991;354:56.
3. Li H, Sun FW, Liew KM, Liu XF. Stretching behavior of a carbon nanowire encapsulated in a carbon nanotube. ScrA Mater. 2009;60:129.
4. Guo ZX, Ding JW, Xiao Y, Mao YL. Lattice dynamics of carbon chain inside a carbon nanotube. J Phys Chem B. 2006;110:21803.

5. Mandal B, Sarkar S, Pramanik A, Sarkar P. Electronic structure and transport properties of sulfur-passivated graphene nanoribbons. J Appl Phys. 2012;112
6. Zhao X, Ando Y, Liu Y, Jinno M, Suzuki T. Carbon nanowire made of a long linear carbon chain inserted inside a multiwalled carbon nanotube. Phys Rev Lett. 2003;90:187401.
7. Smith BW, Monthioux M, Luzzi DE. Encapsulated C60 in carbon nanotubes. Nature. 1998;396:323.
8. Chuvilin A, Bichoutskaia E, Gimenez-Lopez MC, Chamberlain TW, Rance GA, Kuganathan N, Biskupek J, Kaiser U, Khlobystov AN. Self-assembly of a Sulphur-terminated graphene nanoribbon within a single-walled carbon nanotube. Nature Mater. 2011;10:9.
9. Talyzin AV, Anoshkin IV, Krasheninnikov AV, Nieminen RM, Nasibulin AG, Jiang H, Kauppinen EI. Synthesis of graphene nanoribbons encapsulated in single-walled carbon nanotubes. Nano Lett. 2011;11:4352.
10. Kou L, Tang C, Wehling T, Frauenheim T, Chen C. Emergent properties and trends of a new class of carbon nanocomposites: graphene nanoribbons encapsulated in a carbon nanotube. Nanoscale. 2013;5:3306.
11. Fang TH, Chang WJ, Feng YL. Mechanical characteristics of graphene nanoribbons encapsulated in single-walled carbon nanotubes using molecular dynamics simulations. Appl Surf Sci. 2015;356:221.
12. Rusznyák A, Zólyomi V, Kürti J, Yang S, Kertesz M. Bond-length alternation and charge transfer in a linear carbon chain encapsulated within a single-walled carbon nanotube. Phys Rev B. 2005;72:155420.
13. Zólyomi V, Rusznyák A, Kürti J, Yang S, Kertesz M. Linear carbon chain in the interior of a single walled carbon nanotube. Electron Prop Nov Nanostructures. 2005;786:440.
14. Timoshevskii A, Kotrechko S, Matviychuk Y. Atomic structure and mechanical properties of carbyne. Phys Rev B. 2015;91:245434.
15. Yang S, Kertesz M. Bond length alternation and energy band gap of polyyne. J Phys Chem A. 2006;110:9771.
16. Andrade NF, Aguiar AL, Kim YA, Endo M, Freire PTC, Brunetto G, Galvao DS, Dresselhaus MS, Souza Filho AG. Linear carbon chains under high-pressure conditions. J Phys Chem C. 2015;119:10669.
17. Shi L, et al. Confined linear carbon chains as a route to bulk carbyne. Nature Mater. 2016;15:6.
18. Deng Y, Cranford SW. Mapping temperature and confinement dependence of carbyne formation within carbon nanotubes. Carbon. 2019;141:209.
19. Khalilov U, Vets C, Neyts EC. Catalyzed growth of encapsulated carbyne. Carbon. 2019;153:1.
20. Mehmonov K, Ergasheva A, Yusupov M, Khalilov U. The Effect of Oxygen in the Catalytic Synthesis of Endohedral Carbyne. In: Proceedings of MMIT-2023 (Tashkent, Uzbekistan); 2023.pp. 20–26.
21. Shi L, Yanagi K, Cao K, Kaiser U, Ayala P, Pichler T. Extraction of linear carbon chains unravels the role of the carbon nanotube host. ACS Nano. 2018;12:8477.
22. Berdiyorov GR, Eshonqulov G, Hamoudi H. Electronic transport of CNT-encapsulated carbyne. Comput Mater Sci. 2020;183
23. Cui H, Li Q, Qiu G, Wang J. Carbon-chain inserting effect on electronic behavior of single-walled carbon nanotubes: a density functional theory study. Mrs Communications. 2018;8:189.
24. Fan X, Liu L, Lin J, Shen Z, Kuo J.-L. Density functional theory study of finite carbon chains. ACS Nano. 2009;3:3788.
25. Ferreira RS, Aguiar AL, Alencar RS, San-Miguel A, Souza Filho AG. Flat-to-flat polymerization of single-walled carbon nanotubes under high pressure mediated by carbon chain encapsulation. J Phys Chem C. 2021;125:12857.
26. Freitas A, Azevedo S, Kaschny JR. Structural and electronic properties of linear carbon chains encapsulated by flattened nanotubes. Phys E-Low-Dimens Syst Nanostructures. 2016;84:444.
27. Moura LG, Malard LM, Carneiro MA, Venezuela P, Capaz RB, Nishide D, Achiba Y, Shinohara H, Pimenta MA. Charge transfer and screening effects in polyynes encapsulated inside single-wall carbon nanotubes. Phys Rev B. 2009;80

28. Tapia A, Cab C, Casais-Molina ML, Medina J, Cu T, Canto G. Charge transference and conformational stress influence on the electronic properties of zigzag carbon nanowires. J Nanopart Res. 2019;21
29. Shi L, Senga R, Suenaga K, Kataura H, Saito T, Paz AP, Rubio A, Ayala P, Pichler T. Toward confined carbyne with tailored properties. Nano Lett. 2021;21:1096.
30. Kuwahara R, Kudo Y, Morisato T, Ohno K. Encapsulation of carbon chain molecules; in single-walled carbon nanotubes. J Phys Chem A. 2011;115:5147.
31. Tojo T, Kang CS, Hayashi T, Kim YA. Electronic transport properties of linear carbon chains encapsulated inside single-walled carbon nanotubes. Carbon Lett. 2018;28:60.
32. Faria B, Bernardes CES, Silvestre N, Lopes JNC. C13—a new empirical force field to characterize the mechanical behavior of carbyne chains. Phys Chem Chem Phys. 2020;22:758.
33. Ding D, Zhao Y, Dong S, Yu P, Wang L, Zhao J. The vibration of a linear carbon chain in carbon nanotubes. Materials. 2017;10
34. Zou H, Feng Y, Tang X, Zhang X, Qiu L. Extraordinary thermal conductivity for carbon nanotube encapsulated linear carbon chain. Int J Heat Mass Transf. 2024;227
35. Ganz E, Ganz AB, Yang L-M, Dornfeld M. Carbon nanotube-carbyne composite: a nanoreactor in a quasi-1D liquid state. Comput Mater Sci. 2018;149:409.
36. Gao E, Li R, Baughman RH. Predicted confinement-enhanced stability and extraordinary mechanical properties for carbon nanotube wrapped chains of linear carbon. ACS Nano. 2020;14:17071.
37. Liu R, Zhao Y, Sui C, Sang Y, Hao W, Li J, Wu J, He X, Wang C. Molecular dynamics simulations of Carbyne/carbon nanotube gigahertz oscillators. Comput Mater Sci. 2023;222:112105.
38. Son Y-W, Cohen ML, Louie SG. Energy gaps in graphene nanoribbons. Phys Rev Lett. 2006;97:216803.
39. Jiang Y, Li H, Li Y, Yu H, Liew K, He Y, Liu X. Helical encapsulation of graphene nanoribbon into carbon nanotube. ACS Nano. 2011;5:2126.
40. Li Y, Sun F, Li H. Helical wrapping and insertion of graphene nanoribbon to single-walled carbon nanotube. J Phys Chem C. 2011;115:18459.
41. Xia D, Liu S, Song J, Li Q, Xue Q, Liang C, Dong M. Confined hetero double helix structure induced by graphene nanoribbon. 2D Mater. 2019;6
42. Zhang C, Chen S. Defect- and dopant-controlled carbon nanotubes fabricated by self-assembly of graphene nanoribbons. Nano Res. 2015;8:2988.
43. Khalilov U, Neyts E. Mechanisms of selective nanocarbon synthesis inside carbon nanotubes. Carbon. 2021;171:72.
44. Li Y, Chen W, Ren H, Zhou X, Li H. Multiple helical configuration and quantity threshold of graphene nanoribbons inside a single-walled carbon nanotube. Sci Rep. 2015;5
45. Peng X, Velasquez S. Strain modulated band gap of edge passivated armchair graphene nanoribbons. Appl Phys Lett. 2011;98
46. Pollack A, Alnemrat S, Chamberlain T, Khlobystov A, Hooper J, Osswald S. Electronic property modification of single-walled carbon nanotubes by encapsulation of sulfur-terminated graphene nanoribbons. Small. 2014;10:5077.
47. Lebedeva I, Popov A, Knizhnik A, Khlobystov A, Potapkin B. Chiral graphene nanoribbon inside a carbon nanotube: ab initio study. Nanoscale. 2012;4:4522.
48. Fang T, Chang W, Feng Y, Lu D. Torsional characteristics of graphene nanoribbons encapsulated in single-walled carbon nanotubes. Phys E-Low-Dimens Syst Nanostructures. 2016a;83:263.
49. Fang T, Chang W, Feng Y, Weng C. Tensile fracture of graphene nanoribbons encapsulated in single-walled carbon nanotubes. Acta Mech. 2016b;227:2961.
50. Shen H, Cheng K. Tensile properties and thermal conductivity of graphene nanoribbons encapsulated in single-walled carbon nanotube. Mol Simul. 2012;38:922.
51. Berdiyorov GR, Khalilov U, Hamoudi H, Neyts EC. Effect of chemical modification on electronic transport properties of carbyne. J Comput Electron. 2021;20:848.

Chapter 7
Computer Simulation of Thermal Storage Materials

Toyokazu Ishida

Abstract In our living environment, huge amounts of energy are still being released as waste heat in the industrial, transport, and consumer sectors. Effective management of this reusable waste heat is one of the most important technologies for reducing energy consumption and promoting energy conservation. Among many practical solutions for utilizing such unused waste heat, thermal energy storage is the only way to solve the mismatch of energy use in space and time intervals. Since thermal energy storage is closely related to human activities, many approaches have been proposed to store and utilize the thermal energy released by various social systems. From a material chemistry viewpoint, there are two basic chemical approaches to achieving efficient thermal energy storage: the latent heat storage and the thermochemical heat storage. In Japan, the Ministry of Economy, Trade and Industry (METI) and the New Energy and Industrial Technology Development Organization (NEDO) have conducted a research project on the thermal management of materials and systems (TherMAT) over the last decade (2013∼2023). The author has worked on computational research to find new chemical materials with high-density thermal storage. This chapter introduces the computer simulation study of high-density latent heat storage materials (sugar alcohols) and the computational analysis of the thermochemical heat storage mechanism of alkali earth metal oxide systems.

Keywords Thermal management · Thermal energy storage · Latent heat storage · Sugar alcohols · Thermochemical heat storage · Alkali earth metal oxide · Molecular simulation · First principle calculation

T. Ishida (✉)
Material DX Research Center (MDX), National Institute of Advanced Industrial Science and Technology (AIST), Tsukuba, Japan
e-mail: toyokazu.ishida@aist.go.jp

T. Onishi (ed.), *Advances in Quantum Science II*,
https://doi.org/10.1007/978-981-95-9721-5_7

7.1 Overview of Thermal Management and Thermal Storage Materials

In our daily life and modern society, there is a huge amount of heat energy wasted from various social systems such as industry, transportation, and the housing system [1]. This waste heat energy is generally termed as "*unused heat*," which implies a potential energy resource. According to the NEDO report [1], the majority of unused heat is in the temperature range below 200°C. Therefore, effective handling of this unused heat, especially in the temperature range from medium (200°C) to low (50°C) region, has attracted much attention these days. This waste heat problem is closely connected to global warming and the carbon emission problems. At present, effective management of unused heat is one of the important technologies to reduce energy consumption and promote energy conservation. This technology is generally called thermal (energy) management, and various approaches have been proposed to improve the utilization of unused heat until now [2, 3].

In order to utilize unused heat, it is essential to identify the source of waste heat and to recover heat energy according to the wide temperature range of waste heat. Thermal energy insulation is used for keeping heat energy, while thermal energy storage is employed to store heat energy. These two methodologies are important and fundamental technologies for utilizing *unused waste energy* as *heat energy*. Especially, among possible candidates to deal with waste heat, thermal energy storage (TES) is the key technology in thermal management. In fact, TES is the only practical way to directly store and utilize the thermal energy beyond the wide gap mismatch between energy supply (i.e., heat release) and heat demand [4–6].

In general, heat storage phenomena are classified into three categories according to the TES mechanism (Table 7.1). The simplest method to store heat energy using intrinsic heat capacity is called the sensible heat storage. This mechanism is closely related to our daily life from ancient days, but it has some potential problems, such as a limited temperature range and low heat storage density for use in our living environment. So, if we want to employ more high-density heat storage mechanism, from a chemical and material point of view, there are two major classes of chemicals for achieving TES, which are known as latent heat storage (LHS) materials based on phase change materials (PCM) [7–9] and thermochemical heat storage (TCHS) materials [10–14]. Both materials have their own advantages and are in high demand, and for finding these new materials, basic scientific research is urgently required, especially from industrial society.

Based on the progress and the historical background concerning the thermal management research in Japan, the national research project of "*thermal management materials and technology research*" (TherMAT) [15] has been conducted over the last decade (2013–2023) by METI/NEDO (Fig. 7.1), and the author has conducted computational research with particular focus on the basic TES mechanism of chemical compounds. In this book chapter, below, I will briefly summarize the recent progress of molecular simulation studies on the high-density LHS materials [16, 17]

Table 7.1 Three ways of thermal storage methods

Type	Thermal storage principle
Sensible heat storage	A thermal storage method utilizing the specific heat capacity of the material itself, basically with low storage density
Latent heat storage	A thermal storage method utilizing phase changes in materials, generally offering relatively high heat storage density
Thermochemical heat storage	A heat storage method utilizing reversible chemical reactions, where large exothermic chemical reactions also achieve larger heat storage density

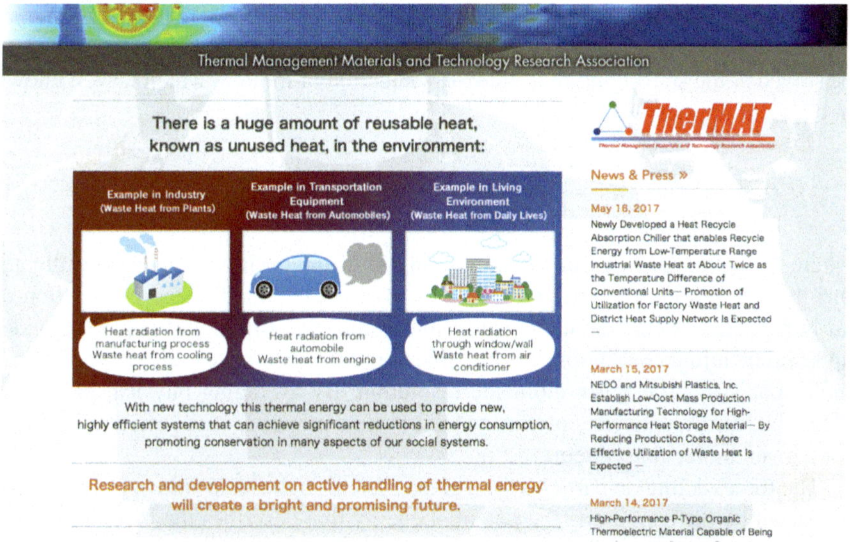

Fig. 7.1 TherMAT project (2013–2023) overview: this picture is excerpted from the project homepage at the time

and the first-principle analysis of the TCHS mechanism for alkali earth metal oxide systems [18], respectively.

7.2 Brief Summary of Computational Methods

Before explaining specific computational results for the thermal storage materials, here we will first briefly introduce an overview of the major computational methods used in this computational research. This part will sever a guidance for understanding the computational research presented in subsequent sections.

7.2.1 Molecular Dynamics (MD) Simulation

Molecular dynamics (MD) simulation is one representative computational tool in molecular simulation. By numerically solving the time evolution of a many-particle system on a computer, one can easily analyze various physical properties as well as structural changes of the target molecular system in an atomistic detail. The mathematical principle of MD simulations is quite simple, and the fundamental equation to be solved at the starting point is Newton's equation of motion, which is well known in classical physics:

$$M_i \frac{\mathrm{d}^2 \mathbf{R}_i(t)}{\mathrm{d}t^2} = -\nabla_i U(\mathbf{R}_1, \mathbf{R}_2, \ldots, \mathbf{R}_N), \tag{7.1}$$

where M_i is the mass of i-th particle (atom), $\mathbf{R}_i$ is the position (nuclear coordinate) of i-th particle (atom), and $U(\mathbf{R}_1, \mathbf{R}_2, \ldots, \mathbf{R}_N)$ is the potential energy function of an N-particle system. By numerically integrating this differential equation under appropriate initial conditions, it is possible to predict the time evolution of the target molecular system at given atomistic resolution. By averaging physical properties along the MD trajectories, one can evaluate system properties and structural changes for a given molecular system.

The forces acting on particles for a given system (right-hand side of Eq. (7.1)) are evaluated by differentiating the potential energy that describes the particle interactions with respect to nuclear coordinates. The reliability of MD simulations is mainly determined by how we evaluate this potential energy function. In principle, if this potential energy could be estimated based on quantum mechanical (QM) principles, more accurate predictions for the target system would be expected. This computational method is generally known as quantum chemical calculation/first-principle calculation.

7.2.2 Quantum Chemical/First-Principle Calculation

The fundamental principles underlying the electronic structure calculations in molecular systems are also remarkably simple. In quantum mechanics, the fundamental equation to be solved is the Schrödinger equation below:

$$H\Psi = E\Psi, \tag{7.2}$$

where H is the Hamiltonian and E is the energy for the target molecular system, Ψ is the wave function that describes the target molecular system. In the atomistic, molecular systems, the Hamiltonian (electron interaction with the nucleus and electrons) in the above equation is precisely defined as follows:

$$\begin{aligned} H = &-\sum_{i=1}^{N}\frac{1}{2}\nabla_i^2 - \sum_{i=1}^{N}\sum_{A=1}^{M}\frac{Z_A}{|\mathbf{r}_i - \mathbf{R}_A|} + \sum_{i=1}^{N}\sum_{j>i}^{N}\frac{1}{|\mathbf{r}_i - \mathbf{r}_j|} \\ &-\sum_{A=1}^{M}\frac{1}{2M_A}\nabla_A^2 + \sum_{A=1}^{M}\sum_{B>A}^{M}\frac{Z_A Z_B}{|\mathbf{R}_A - \mathbf{R}_B|}, \end{aligned} \tag{7.3}$$

where $\mathbf{r}_i$ is the electron coordinate, $\mathbf{R}_A$ is the nuclear coordinate, and Z_A is the nuclear charge. Therefore, once we can correctly solve this equation, we can, in principle, obtain any desired physical properties for given molecular systems. In an actual calculation, by introducing an electron orbital model that describes the molecular electronic structure, the electronic wave function is determined by an approximate wave function expanding the Gaussian basis functions. This computational method is known as the molecular orbital (MO) calculation, which is a representative quantum chemical calculation method.

In the MO theory, the fundamental variable in the Schrödinger equation is the number of electrons inside given molecules. Consequently, for macromolecular systems such as proteins or solids, these variables become extremely large, and computational cost increases significantly in proportion to the number of electrons. In this case, however, by replacing the electron interactions based on the electron orbital models with the electron density representing the spatial distribution of electrons and reformulating the fundamental equation (Kohn-Sham equation), it is known that the electronic structure of molecular and material systems can be calculated from the electron density information. This computational method is generally known as the density functional theory (DFT) method and is now widely used as a representative first-principles calculation method mainly for the computations in condensed matter systems such as solids.

7.2.3 Molecular Mechanics (MM) Calculation

Based on quantum mechanical principles, it is possible to accurately describe the molecular interaction potentials. However, computational cost increases significantly in proportion to the size of the target system. Therefore, even at present, direct application of quantum mechanical principles to complex macromolecular systems such as proteins or molecular liquids is yet impractical. In such cases, fortunately, it is possible to evaluate physical/chemical properties and structural changes of a molecular system with sufficient accuracy by representing molecular interactions using empirical functions with calibrated parameters. This approach is known as the molecular mechanics (MM) method, and typically, the particle interactions within a molecular system (usually called a force field) are described by assuming the following empirical potential functions:

$$E_{total} = \sum_{bonds} k_r(r - r_{eq})^2 + \sum_{angles} k_\theta(\theta - \theta_{eq})^2 + \sum_{dihedrals} \frac{1}{2}V_n[1 + cos(n\phi - \gamma)] + \sum_{i,j}(\frac{A_{ij}}{\mathbf{R}_{ij}^{12}} - \frac{B_{ij}}{\mathbf{R}_{ij}^{6}}) + \sum_{i,j}\frac{Z_i Z_j}{\epsilon \mathbf{R}_{ij}}, \tag{7.4}$$

In this equation, the first three terms describe bonding interactions, including bond-stretch (r), bending (θ), torsion (ϕ, dihedral angle), and the last two terms are non-bonded interactions, including van der Waals and electrostatic interactions. k_r, k_θ, V_n, A_{ij}, and B_{ij} are empirical parameters which are adjusted to reproduce appropriate molecular properties.

7.2.4 Thermodynamic Calculation

Unlike the atomistic-scale computational methods described above, equilibrium thermodynamic calculations utilizing a thermodynamic database are also frequently employed when we focus on macroscopic physical quantities such as the thermodynamic properties of materials. The key essence of this methodology lies in defining the Gibbs energy (G) of a pure substance as a function of temperature (T), based on thermodynamic measurements of pure substances under the standard condition (25 °C, 1 atom):

$$G = a + bT + cT \ln T + dT^2 + eT^3 + \frac{f}{T} + \cdots, +G_{press} \tag{7.5}$$

where a, b, c, d, e, f are adjustable parameters to reproduce experimental or theoretical values, and G_{press} is the pressure correction factor. In the thermodynamic databases, basically, thermodynamic quantities are defined as functions of temperature as shown in Eq. (7.5). Parameters are optimized to reproduce measured or theoretical values. Utilizing this Gibbs energy function enables the computation of thermodynamic quantities such as material composition at thermodynamic equilibrium, phase diagram, and transition temperature.

7.2.5 *Monte Carlo Simulation*

In the MD simulations described above, the time evolution of the particle system is estimated according to Newton's equation of motion (Eq. (7.1)), which enables the calculation of physical/chemical properties and structural changes based on the trajectories of particles. On the other hand, when focusing on time-independent equilibrium properties, one can generate structures randomly without solving the equations of motion. By calculating the energy of generated structures using quantum mechanical or molecular mechanics methods, and then evaluating energy order (or ranking) among the generated structures, one can describe the structural transitions of the target molecular system over a broader energy space. For example, when using the Monte Carlo algorithm in Metropolis sampling, it becomes possible to sample the target system with a probability proportional to the Boltzmann factor at a given temperature, and one can analyze the statistical properties for a given structure ensemble. This type of computational method is generally known as Monte Carlo simulation (sampling).

7.3 Molecular Dynamics Simulation of Latent Heat Storage (LHS) Materials

7.3.1 *Modeling and Simulation of Natural Sugar Alcohols*

As the name implies, the LHS utilizes the latent heat released during the phase change process (solid-liquid or liquid-gas) of a material. The method has the advantage of keeping a constant operating temperature during the phase transition process (Fig. 7.2) and has a larger heat storage density than the sensible heat storage.

Since solid-liquid phase transition is a fundamental property of all materials, in principle, most chemicals and materials can be potential candidates for the LHS application. At present, however, the materials used in the heat storage technology are very limited: water and paraffin are the two most common LHS materials [19]. The aim of our computational research in the TherMAT project is to explore chemical compounds which can achieve high-density thermal storage far beyond existing LHS materials. Our research strategy in the first stage is to clarify fundamental thermal properties of existing materials and to elucidate the molecular factors that govern the high-density thermal storage. And then, we propose a molecular design guideline that achieves a larger thermal storage density to produce a new thermal storage material.

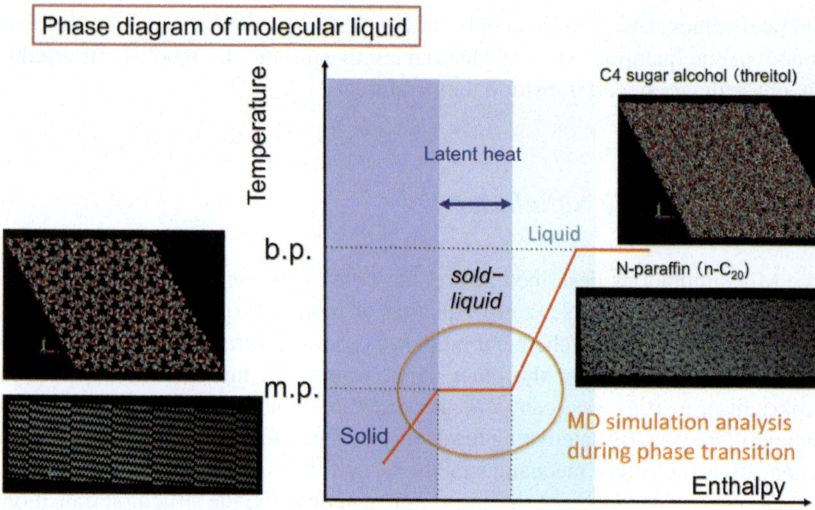

Fig. 7.2 Phase diagram of solid-liquid phase transition process in molecular liquids: solid/liquid states of C4 sugar alcohol and C20 paraffin are shown as typical examples

Computational strategy for finding novel latent heat storage materials based on molecular simulations

1. Based on detailed investigations of available physical and chemical property data, we select possible candidate materials capable of large thermal storage density.
2. Focusing on natural sugar alcohols to elucidate molecular factors that governs thermal storage density, we perform MD simulations on solid-liquid phase transition processes
3. Based on a series of computational analysis, we computationally designed artificial sugar alcohol molecules and predicted their crystal structures at the same time.
4. Systematic MD simulation analyses of solid-liquid phase transition process for the newly designed molecules, we predict novel thermal properties for the designed molecules.

FlowChart 7.1 Computational strategy for finding novel latent heat storage materials based on molecular simulations

This computational strategy for finding new LHS materials is summarized in the following Flow Chart 7.1. For this purpose, in our project, we focused on a natural chemical compound family, sugar alcohols [20, 21].

In nature, there exist some sugar alcohol compounds that have four to six carbons in their backbone structures. For example, C4 alcohol of erythritol is a typical example that has been tested as an LHS material in the past, since erythritol has a relatively large latent heat ($\sim$ 0.3 MJ/kg). In the first step of our computational study, we have carried out systematic molecular dynamics (MD) simulations of the solid-liquid phase transition processes for C4 $\sim$ C6 sugar alcohols and attempted to elucidate the structure-function relationship between the molecular geometries and thermo-physical properties of these natural compounds. We particularly focused on the C6

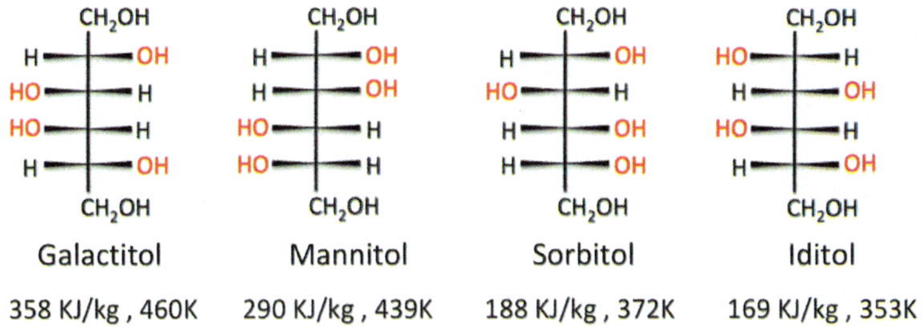

Fig. 7.3 Molecular geometries and thermochemical/physical properties (e.g., enthalpies of fusion or thermal storage density, and melting points) of C6 sugar alcohol isomers

sugar alcohols since the C6 series has four structural isomers as shown in Fig. 7.3. It is experimentally well known that small geometrical differences inside the molecules reflect large thermo-physical differences in the melting point as well as the heat storage density. These interesting chemical phenomena are a great challenge for computational research at the first step.

As for the technical issue in our computational analysis, we employed classical MD simulation [22], and the force field parameters for sugar alcohols have been tested and calibrated through a series of MD simulations [23–25]. We focused mainly on the melting process of phase transition (solid-liquid phase change in Fig. 7.2), and the initial lattice geometries were constructed based on the experimental X-ray structures. During MD simulations, we carefully identified the melting point of the system by monitoring the system temperature, potential energy, and volume fluctuation. On the basis of detailed structural as well as energetic analysis for the simulation cells, we finally determined the enthalpy changes of phase transition (as shown in Fig. 7.2, in this case, this corresponds to the heat storage density) and melting temperature.

In essence, a series of MD simulations for the C6 sugar alcohol isomers well reproduced the melting points and the enthalpies of fusion in comparison with available experimental data, and this confirmed our computational strategy for investigating natural sugar alcohols. Based on these simulation results, the next step is that we have conducted a detailed structural and energetic analysis during the solid-liquid phase transition process. At this stage, we briefly summarize the important points clarified in the simulations.

Important Points in the Molecular Dynamics Simulations

(1) Structural view point

Energy decomposition analysis of molecular interaction energies revealed that the main energetic origin is the intermolecular electrostatic interaction between sugar alcohols and the intra-molecular electrostatic repulsion inside one molecular geometry: a molecular system with large electrostatic stabilization forms a stable solid state and releases a large amount of energy through the melting process, while a molecular system with large electrostatic repulsion does not form a stable solid lattice geometry.

(2) Energetic origin

The major origin of the enthalpies of fusion is a loss of the hydrogen bond network during the solid-liquid phase transition: in all four C6 sugar isomers, on average, one hydrogen bond disappears when the solid structure becomes a random liquid structure.

In short, a series of MD simulations of C6 sugar alcohols clarified that the most significant element that determines the heat storage density between different isomers is the structural stability determined by the electrostatic interactions: a molecular system that forms a stable solid state will release more thermal energy as an enthalpy of fusion [16].

7.3.2 Modeling and Simulation of Artificial (Designed) Alcohols

In the second stage, we have conducted computational molecular design and functional prediction based on the systematic MD analyses mentioned above. In the TherMAT project, we have also developed the thermal energy materials database (PropertiesDB Web) and made it available for public usage [26]. On the basis of detailed computational analysis of naturally occurring sugar alcohols as well as available thermochemical and physical data for existing natural products, we have proposed three simple molecular design guidelines that accomplish high-density thermal storage as follows.

Molecular Design Guidelines

(1) Align the molecular backbone (skeleton) linearly to enhance the intermolecular interaction energy gain.
(2) Properly arrange the direction of hydroxyl groups to avoid unfavorable intramolecular electrostatic repulsion.
(3) Design the carbon backbone structure with an even number of carbon atoms.

Based on these three simple molecular design guidelines, we have performed computational molecular design of artificial sugar alcohols on the computer and then predicted thermochemical and physical properties through a series of MD simulations of the solid-liquid phase transition process. In our computational molecular design, we have proposed the highly symmetrical molecular geometries which resemble a C6 mannitol-like structure (Fig. 7.4 left).

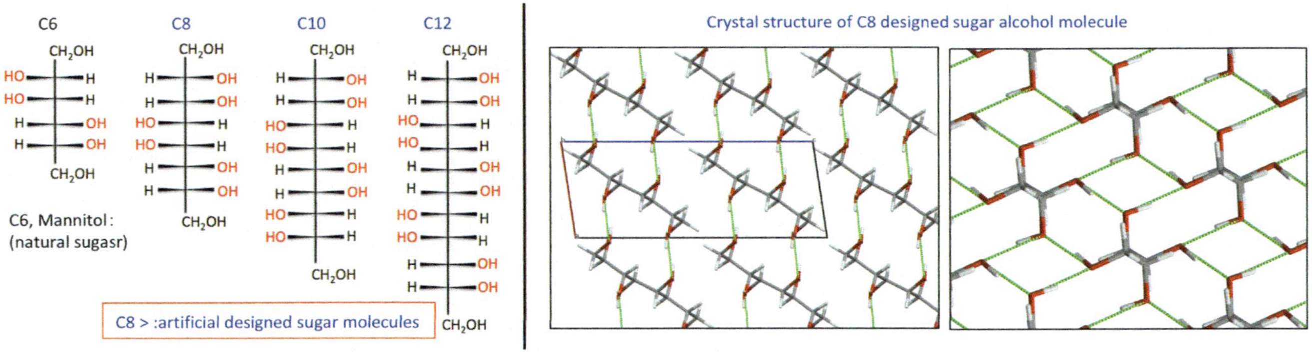

Fig. 7.4 Molecular geometries of artificial sugar alcohols (left). For reference, the C6 natural alcohol of mannitol is included along with the C8, C10, and C12 designed molecules. Predicted crystal structures of the C8 designed sugar alcohol (right). The figures show the structures in two directions: parallel and perpendicular to the carbon backbone axis

Computational procedure of organic crystal structure prediction

1. Using the molecular structure of C6 mannitol as a natural sugar template, generate the structure of a designed molecule with an extended carbon skeleton
2. Based on the organic crystal structure database, explore the crystal structures of designed molecules within representative eight space groups by Monte Carlo sampling
3. For each generated space group, perform structure optimizations at the MM energy level, and rank the energies of stable structures relative to the crystal phase
4. After selecting candidate structures at the MM level, then QM energy corrections are performed on the MM refined structures to determine the final stable structure

FlowChart 7.2 Computational strategy for finding novel latent heat storage materials based on molecular simulations

For predicting thermochemical and physical properties of these artificial molecules, it is necessary to determine a reliable crystal structure in the solid phase. Actually, we have predicted reasonable crystal structures by introducing our original modeling strategy. We briefly summarize the essence of our crystal structure prediction [17] in the following Flow Chart 7.2. At first, we reduced the space group of organic crystals for exploring candidates on the basis of known organic crystal database, and then we selected the eight representative space groups ($P1$, $P\bar{1}$, $P2_1$, $C2$, $P2_1/c$, $C2/c$, $P2_12_12_1$, and $Pbca$. These space groups are known to cover over 80 % of organic molecular crystals [27]). By performing a Monte-Carlo-like (random) sampling of the cell parameters in the given space group, and by adding QM-level energy corrections for the selected crystal structure candidates, we finally determined the most reasonable crystal structure for the designed molecule.

As shown in Fig. 7.4 (right), in the predicted crystal structure, all the hydroxyl groups in the designed molecule form tight hydrogen bonds with the surrounding nearest molecules, and these hydroxyl groups are properly aligned to avoid intramolecular electrostatic repulsion, which implies a very stable solid phase crystal structure. For a series of predicted crystal structures (C8 $\sim$ C20 artificial sugar alcohols), we have performed systematic MD simulations for the solid-liquid phase transition process and have analyzed the details of geometrical as well as energetic changes around the melting points. From a series of MD simulation analyses, we have clarified several interesting molecular properties. The most important point is that the heat storage density of artificial sugar alcohols increases as the carbon number increases, but the carbon number has an optimal value: the maximum density of $\sim$ 0.5 MJ/kg is achieved only when we carefully design the C12 molecular geometry. If we elongate the carbon backbone more than the C12 molecule, geometrical strain (or stress) inside a molecule destabilizes the crystal structure, and consequently, this results in a loss of the heat storage density (Fig. 7.5).

As a summary of LHS materials, the major findings clarified in our systematic computational studies are that for organic polar molecules, we can achieve an ideal heat storage density larger than that of known existing materials if we carefully

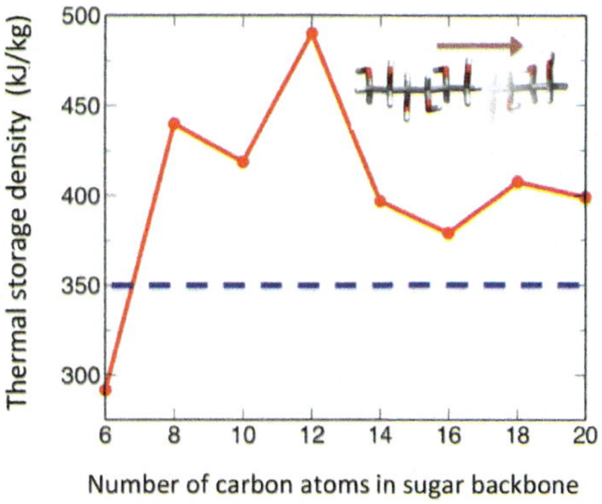

Fig. 7.5 The structure-function relationship between molecular geometries and thermal storage densities in the designed molecular systems

design their molecular structure. The upper limit of LHS is expected to be at most $\sim$ 0.5 MJ/kg if we assume the naturally occurring organic compound as an initial structural template [17].

7.4 First Principle Calculation of ThermoChemical Heat Storage (TCHS) Materials

7.4.1 Chemical Heat Storage

As described above, our computational molecular design revealed that the heat storage density of latent heat is at most 0.5 MJ/kg, even with careful design of the molecular structures. If one needs materials with higher heat storage density, typically with the heat storage density exceeding 1 MJ/kg, TCHS is currently the most promising candidate. The basic concept of TCHS is quite simple: TCHS technology utilizes heat exchange through a reversible chemical reaction (exothermic and endothermic process, in Fig. 7.6).

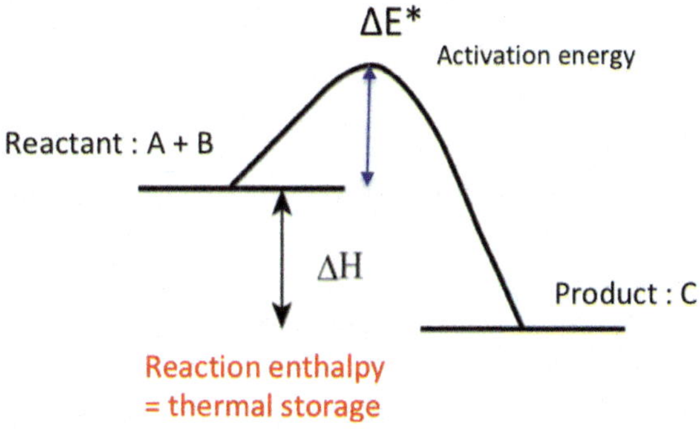

Fig. 7.6 Schematic diagram of thermochemical heat storage (TCHS) process: the picture illustrates the relationship between the activation energy and the reaction heat (ΔH, heat storage density in this case) for a reversible chemical reaction of $A + B \rightleftarrows C + \Delta H$

Table 7.2 Promising reaction candidate for thermochemical heat storage [12, 13]

Type	Thermochemical reaction[a]
Carbonates	$MgO + CO_2 \rightleftarrows MgCO_3 + \Delta H$
	$CaO + CO_2 \rightleftarrows CaCO_3 + \Delta H$
	$SrO + CO_2 \rightleftarrows SrCO_3 + \Delta H$
Hydroxides	$MgO + H_2O \rightleftarrows Mg(OH)_2 + \Delta H$
	$CaO + H_2O \rightleftarrows Ca(OH)_2 + \Delta H$

[a] ΔH means the enthalpy change of each thermochemical reaction

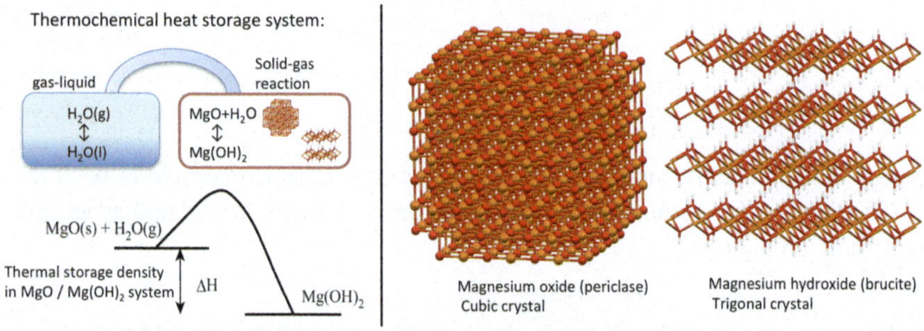

Fig. 7.7 Schematic picture of the typical thermochemical heat storage system of MgO hydration (and $Mg(OH)_2$ dehydration) reaction (left). Crystal structures of ideal MgO and $Mg(OH)_2$ lattices (right)

Due to its simple principle, there are lots of potential candidates for chemical reactions to be applied for TCHS [10–14] (Table 7.2).

However, due to practical constraints in applying it to social systems (e.g., compatibility of environmental impact, profitability of energy issues, handling costs of chemicals, etc.), TCHS still remains a developing technology in the current field of thermal management. Precise handling of complex chemical reaction processes (e.g., controlling the activation step) is a particularly challenging issue. Among many chemical reaction candidates, a typical and well-known thermochemical reaction is the hydration of alkaline earth metal oxide and the dehydration of alkaline earth metal hydroxide [12, 13], and the basic concept of the TCHS system of this type of reaction is presented in Fig. 7.7.

7.4.2 Case Study of MgO/Mg(OH)$_2$ System

We consider magnesium to be the typical candidate of alkaline earth metals in oxides and hydroxides. In this case, the hydration and dehydration reactions are defined as follows, where ΔH indicates the enthalpy of the chemical reaction and equals the thermal storage density in this reaction:

$$MgO + H_2O \rightleftarrows Mg(OH)_2 + \Delta H \tag{7.6}$$

Under the high-temperature condition, $Mg(OH)_2$ (magnesium hydroxide, mineral name *brucite*) dehydration proceeds and $Mg(OH)_2$ decomposes into MgO (magnesium oxide, mineral name *periclase*) and H_2O, and on the contrary, high-temperature steam accelerates the hydration reaction of MgO and results in the product $Mg(OH)_2$ [28] (Fig. 7.7). This reversible reaction is a simple one, but the chemical reaction proceeds slowly below the transition temperature (or turning temperature) in this reversible process. In the research field of thermal energy storage, the transition temperature of the reversible reaction is generally assumed to be the equilibrium condition at which the Gibbs free energy (G) of Eq. (7.7) becomes zero.

$$\Delta G = \Delta H - T\Delta S, \tag{7.7}$$

where H is the enthalpy and S is the entropy of a given chemical reaction.

Currently, one difficulty for applying this chemical reaction to the TCHS is how to efficiently/effectively control the reversible chemical reactions and how to reduce the transition temperature [29–32]. In contrast to the recent progress in TCHS research to overcome this limitation, an elementary, atomistic process inside the TCHS mechanism is not well understood.

7.4.3 First Principle Calculations for MgO/$Mg(OH)_2$ System

To the best of our knowledge, there is no detailed computational investigation for the TCHS mechanism in the MgO/$Mg(OH)_2$ system in Eq. (7.6) [33–36]. Therefore, in the TherMAT project, in order to overcome the practical difficulties for installing a TCHS system into the social system, we have investigated an elementary reaction step on the ideal MgO surface. As mentioned above, the details of this reversible chemical reaction remain elusive at present, and we do not know what the activation factor for this chemical reaction is or what amount of energy is required to activate the chemical reaction. An important finding concerning the molecular mechanism is that the hydration reaction of MgO requires sufficiently wet conditions (as opposed to ambient atmospheric vapor conditions) [37–39]. Therefore, in the first step, we focus only on the hydration mechanism on the ideal MgO (001) surface (Fig. 7.8), and we calculate two fundamental parameters which characterize the thermal energy storage (the reaction heat and the transition temperature) and validate our computational strategy for application to the MgO/$Mg(OH)_2$ system [18].

Reaction Heat

We first calculated the electronic energies of both reactants (MgO and H_2O) and product ($Mg(OH)_2$), and evaluated the reaction heat according to the chemical reac-

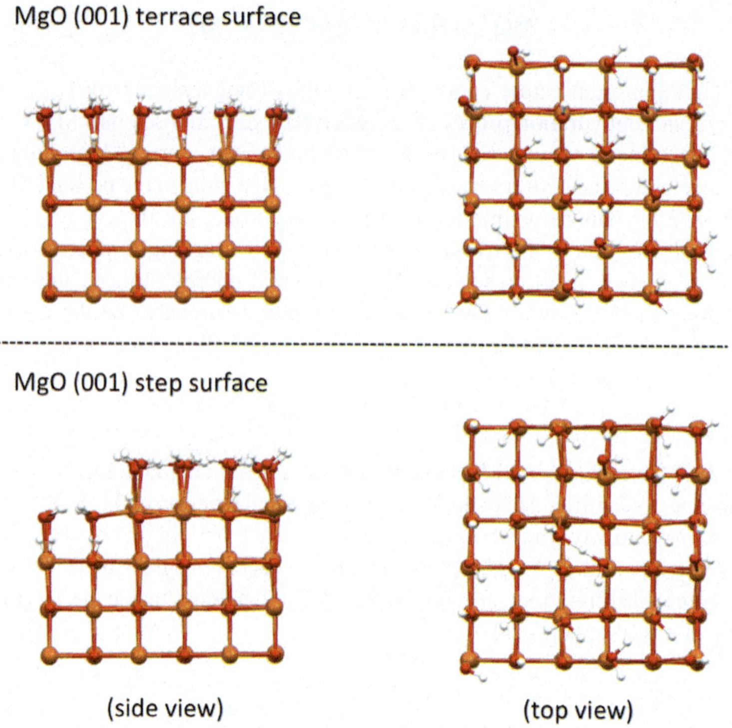

Fig. 7.8 Molecular models of the initial hydration step on the ideal MgO(001) surface: the terrace (top) and step (bottom) plane models targeted in these calculations are shown from the side (left) and above (right), respectively

tion of Eq. (7.6). In this calculation, we employed the first-principles calculation by using the Quantum Espresso package [40, 41], and assumed the ideal solid-state crystal geometry based on experimental information. As described in the computational section, the electronic properties of the reactant and product in the solid state are well described by first-principle calculations, and we evaluated the reaction energetics according to Eq. (7.6). In the case of the hydration reaction of MgO, the reaction heat (ΔH in Eq. (7.6)) is reported to be 78 $\sim$ 81 kJ/mol in the literature [12, 32, 38]. When we compared current computational results with available experimental data, our most accurate computational result of 78.7 kJ/mol well reproduced the thermodynamic parameter for this hydration system.

Transition Temperature

Next, we evaluated the free energy change in the reversible chemical reaction of Eq. (7.6). Figure 7.9 represents the computational free energy change. Additionally, for reference, the Gibbs free energy profiles evaluated by typical thermodynamic

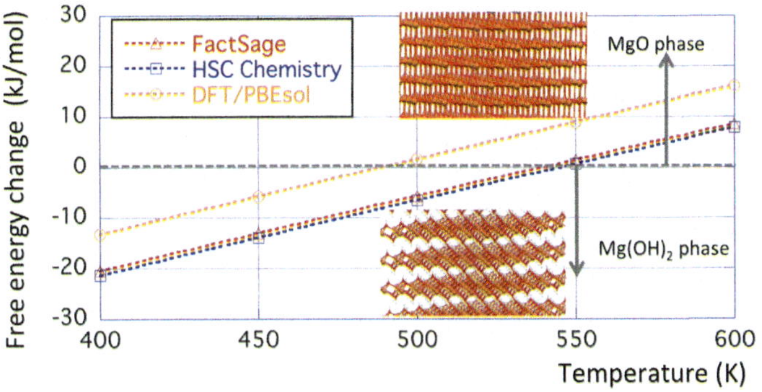

Fig. 7.9 The reaction free energy changes in the MgO hydration and $Mg(OH)_2$ dehydration reaction defined in Eq. (7.1)

calculations (FactSage [42] and HSC Chemistry [43]) are also included in the same figure. In the field of TCHS, the transition (or turning) temperature of the thermo-chemical reaction is assumed to be the net zero energy point of the Gibbs energy change [12, 32, 38]. From these figures, the experimental (or thermodynamic) transition temperature is approximately $\sim$ 530 K, while our computational estimate is $\sim$ 490 K. Compared to the calculation of the reaction heat, the transition temperature shows a slight deviation between the standard thermodynamic data and the *ab initio* calculation; however, if we consider the present computational conditions (here we considered only an ideal lattice condition for both the reactant and product states), these results provide a reasonable estimate for describing the potential energy profile for the hydration reaction described below.

Potential Energy Profile

Based on these numerical estimations, we explored the potential energy profiles of an initial hydration step on the ideal MgO (001) surface by constructing an appropriate 2D periodic interfacial system. Under an appropriate assumption of the hydration process described below, both the terrace and step geometries have been considered in our surface model (Fig. 7.8). To reasonably model the forward step chemical reaction in Eq. (7.6), we assumed that the initial hydration consists of several elementary chemical processes on the MgO lattice surface [18, 34]:

Surface Model Construction

(a) Hydroxylation and Protonation

Hydroxylation and protonation at surface atoms on the ideal MgO (001) surface layer: all water molecules on the first layer of MgO are completely dissociated, and hydroxyl ions are attached to Mg atoms, while protons are attached to oxygen sites.

(b) Translation

Translation (translocation) of Mg-OH groups from the first layer MgO surface: Once activated by high temperature, surface Mg-OH groups are translocated into the surface and interact with other chemical species located on the surface.

(c) Diffusion

Diffusion of Mg atoms (from the surface of the second layer), followed by penetration of water molecules (from the aqueous phase on the surface): after the translocation of Mg-OH units from the first layer, unsaturated Mg atoms inside the second layer move (shift) upwards to the surface regions induced by aqueous water molecules, and remaining vacancies are filled by diffused water molecules from the surface region.

(d) Reorganization

Reorganization of surface Mg-OH species through interactions with other surface atoms: after diffusion of Mg groups on the surface, these chemical species form a specific structural pattern similar to the $Mg(OH)_2$ lattice component.

(e) Lattice growth

Lattice growth of the $Mg(OH)_2$ structural motif on the MgO surface: after diffusion of several Mg-OH / Mg groups from the surface, these chemical moieties form a stable chemical structure, and this $Mg(OH)_2$ nucleation grows in accordance with surface diffusion reactions.

Based on the assumption of the above five elementary reaction steps, we finally constructed a reasonable atomistic chemical model and calculated potential energy profiles by appropriately defining the reaction coordinate in each step. To clearly explain the reaction mechanism on the solid surface, we introduced here a 2D potential energy diagram defined by two reaction coordinates: one is the total dislocation distance of Mg atoms operated in each reaction step, and the other is the sum of the coordination numbers of each Mg atom. Under this definition, the potential energy profile of these surface hydration reactions is described in Fig. 7.10. As can be clearly seen in this figure, an energetically large activation step is accompanied by a large change in the coordination number of the reactive Mg atoms, and this situation is exactly the same for two different surface topologies (step and terrace).

This 2D potential profile clearly tells us that (1) the activation energy is highly dependent on the surface geometry of MgO, (2) while an atomistic motion in this rate-determining process (an extraction of Mg atoms from the inner layer to the surface layer) is exactly the same in both surface geometries (step and terrace). In summary, from this computational analysis, we can understand that stepped geometries are more reactive than the ideal terrace plane, and effective interaction of surface atoms with aqueous solvent molecules will lower the activation energy barrier and accelerate the thermochemical reaction process.

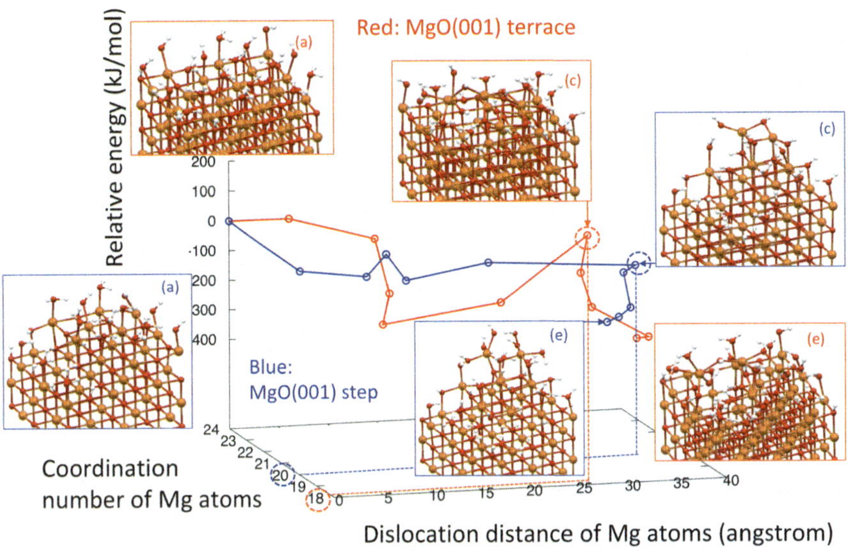

Fig. 7.10 2D potential energy profiles of two MgO(001) surface geometry models along the two selected reaction coordinates: the x-axis indicates the total dislocation distance of the selected four Mg atoms, and the y-axis represents the sum of the coordination numbers of the selected four Mg atoms. The red graph and chemical structures in the red boxes show the hydration reaction on the MgO(001) terrace geometry, while the blue graph and structures in the blue boxes show the hydration reaction on the MgO(001) step geometry. Alphabets within each panel of the structure correspond to the surface model construction process described in the text

7.5 Summary and Outlook: Computational Material Exploration of Thermal Storage Materials

In this book chapter, the author summarizes the essence of our computational investigation for exploring high-density thermal storage materials. As for LHS, by selecting a naturally occurring sugar alcohol structure as a molecular template, we have designed an artificial sugar alcohol molecule and examined its thermochemical and physical properties. Our calculation clearly demonstrated that a designed molecule potentially has a large thermal storage density if we carefully design organic crystal structures, and the upper limit of its latent heat storage density is about ~ 0.5 MJ/kg. If one needs materials with much larger thermal storage densities, the TCHS is the only available candidate. In the first step, we focused on a typical TCHS material of MgO /$Mg(OH)_2$ system, and we have clarified the rate-determining process for the surface hydration reaction of MgO. Our calculations also suggested a simple guideline for the effective handling of this typical thermochemical reaction system: the reactant MgO sample should be finely crushed into a coarse powder form to increase the roughness of the solid surface, and these coarse particles should be formed into porous pellets, or these sample particles should be supported inside a suitable porous medium to enhance an effective surface contact between the coarse-grained MgO sample and the hot steam as the working fluid.

For a complete picture of the thermochemical reaction, the reverse step of the reversible reaction (the dehydration process) is also an important research topic [44, 45]. Currently, the molecular mechanism of this reverse dehydration reaction remains unclear. A more detailed understanding of the reaction mechanism and identification of activation factors are anticipated in the next computational study, and indeed, this is our next research target.

Acknowledgements This work was supported by the New Energy and Industrial Technology Development Organization (NEDO) and the Future Pioneering Projects of the Japanese Ministry of Economy, Trade and Industry (METI).

References

1. Report of waste heat from industrial departments, METI/NEDO TherMAT project, in Japanese (2019). https://www.nedo.go.jp/content/100957934.pdf (accessed May 12, 2026)
2. Dinçer I, Rosen MA. Thermal energy storage: systems and applications, 3rd Edition; 2021. John Wiley & Sons, Inc.
3. Lee HS. Thermal Design: Heat Sinks, Thermoelectrics. Heat pipes: Compact Heat Exchangers, and Solar Cells; John Wiley & Sons Inc; 2010.
4. Alva G, Lin Y, Fang G. An overview of thermal energy storage systems. Energy. 2018;144:341–78.
5. Sadeghi G. Energy storage on demand: Thermal energy storage development, materials, design, and integration challenges. Energy Storage Materials. 2022;46:192–222.
6. Khan MI, Asfand F, Al-Ghamdi SG. Progress in research and technological advancements of thermal energy storage systems for concentrated solar power. J Energy Storage. 2022;55: 105860.
7. Sharma SD, Sagara K. Latent Heat Storage Materials and Systems: A Review Int. J Green Energy. 2005;2:1–56.
8. Khudhair AM, Farid MM. A review on energy conservation in building applications with thermal storage by latent heat using phase change materials Energ. Convers Manage. 2004;45:263–75.
9. Kenisarin MM. Thermophysical properties of some organic phase change materials for latent heat storage. A review Sol Energy. 2014;107:553–75.
10. Aydin D, Casey SP, Riffat S. The latest advancements on thermochemical heat storage systems. Renewable Sustainable Energy Rev. 2015;41:356–67.
11. Yan T, Wang RZ, Li TX, Wang LW, Fred IT. A review of promising candidate reactions for chemical heat storage. Renewable Sustainable Energy Rev. 2015;43:13–31.
12. André L, Abanades S, Flamant G. Screening of thermochemical systems based on solid-gas reversible reactions for high temperature solar thermal energy storage. Renewable Sustainable Energy Rev. 2016;64:703–15.
13. Sunku Prasad J, Muthukumar P, Desai F, Basu DN, Rahman MM. A critical review of high-temperature reversible thermochemical energy storage systems. Appl Energy. 2019;254: 113733.
14. Carrillo AJ, González-Aguilar J, Romero M, Coronado JM. Solar Energy on Demand: A Review on High Temperature Thermochemical Heat Storage Systems and Materials. Chem Rev. 2019;119:4777–816.

15. Thermal Management Materials and Technology Research (TherMAT), METI/NEDO national project, in Japanese (2013–2023). https://www.nedo.go.jp/activities/ZZJP_100097.html (accessed May 12, 2026)
16. Inagaki T, Ishida T. Computational Analysis of Sugar Alcohols as Phase-Change Material: Insight into the Molecular Mechanism of Thermal Energy Storage. J Phys Chem C. 2016;120:7903–15.
17. Inagaki T, Ishida T. Computational Design of Non-natural Sugar Alcohols to Increase Thermal Storage Density: Beyond Existing Organic Phase Change Materials. J Am Chem Soc. 2016;138:11810–9.
18. Ishida T, Ishimura K. Chemical heat storage mechanism in alkaline earth metal oxide: ab initio modeling of the initial hydration reaction on MgO(001) surface. J Phys Chem C. 2024;128:7397–407.
19. Sharma A, Tyagi VV, Chen CR, Buddhi D. Review on thermal energy storage with phase change materials and applications Renew. Sust Energ Rev. 2009;13:318–45.
20. Sole A, Neumann H, Niedermaier S, Martorell I, Schossig P, Cabeza LF. Stability of sugar alcohols as PCM for thermal energy storage. Sol Energ Mat Sol Cells. 2014;126:125–34.
21. Kenisarin M, Mahkamov K. Solar energy storage using phase change materials. Renew Sust Energ Rev. 2007;11:1913–65.
22. Pronk S, Pall S, Schulz R, Larsson P, Bjelkmar P, Apostolov R, Shirts MR, Smith JC, Kasson PM, van der Spoel D, Hess B, Lindahl E. GROMACS 4.5: a high-throughput and highly parallel open source molecular simulation toolkit. Bioinformatics. 2013;29:845–854.
23. Damm W, Frontera A, Tirado-Rives J, Jorgensen WL. OPLS all-atom force field for carbohydrates. J Comput Chem. 1997;18:1955–70.
24. Alavi S, Thompson DL. Simulations of melting of polyatomic solids and nanoparticles. Mol Simul. 2006;32:999–1015.
25. de Waard H, Amani A, Kendrick J, Hinrichs WLJ, Frijlink HW, Anwar J. Evaluation and optimization of a force field for crystalline forms of mannitol and sorbitol. J Phys Chem B. 2010;114:429–36.
26. PropertiesDB Web, METI/NEDO national project, in Japanese (2022). https://www.nedo.go.jp/news/press/AA5_101504.html (accessed May 12, 2026)
27. Belsky VK, Zorkil PM. Distribution of organic homomolecular crystals by chiral types and structural classes. Acta Cryst. 1977;A33:1004–6.
28. Ervin G. Solar Heat Storage Using Chemical Reactions. J Solid State Chem. 1977;22:51–61.
29. Kato Y, Yamashita N, Kobayashi K, Yoshizawa Y. KINETIC STUDY OF THE HYDRATION OF MAGNESIUM OXIDE FOR A CHEMICAL HEAT PUMP. Appl Therm Eng. 1996;16:853–62.
30. Kato Y, Sasaki Y, Yoshizawa Y. Magnesium oxide/water chemical heat pump to enhance energy utilization of a cogeneration system. Energy. 2005;30:2144–55.
31. Ryu J, Takahashi R, Hirao N, Kato Y. Effect of Transition Metal Mixing on Reactivities of Magnesium Oxide for Chemical Heat Pump. J Chem Eng Jpn. 2007;40:1281–6.
32. Myagmarjav O, Zamengo M, Ryu J, Kato Y. Energy density enhancement of chemical heat storage material for magnesium oxide/water chemical heat pump. Appl Therm Eng. 2015;91:377–86.
33. Nair NN, Bredow T, Jug K. Molecular Dynamics Implementation in MSINDO: Study of Silicon Clusters. J Comput Chem B. 2004;25:1255–63.
34. Jug K, Heidberg B, Bredow T. Cyclic Cluster Study on the Formation of Brucite from Periclase and Water. J Phys Chem C. 2007;111:13103–8.
35. Jug K, Heidberg B, Bredow T. Molecular Dynamics Study of Water Adsorption Structures on the MgO(100) Surface. J Phys Chem C. 2007;111:6846–51.
36. Jug K, Heidberg B, Bredow T. Cyclic cluster study of water adsorption structures on the MgO(100) surface. Surf Sci. 2007;601:1529–35.
37. Ishitobi H, Hirao N, Ryu J, Kato Y. Evaluation of Heat Output Densities of Lithium Chloride-Modified Magnesium Hydroxide for Thermochemical Energy Storage. Ind Eng Chem Res. 2013;52:5321–5.

38. Shkatulov A, Krieger T, Zaikovskii V, Chesalov Y, Aristov Y. Doping Magnesium Hydroxide with Sodium Nitrate: A New Approach to Tune the Dehydration Reactivity of Heat-Storage Materials. ACS Appl Mater Interfaces. 2014;6:19966–77.
39. Kurosawa R, Takeuchi M, Ryu J. Fourier-Transform Infrared Analysis of the Dehydration Mechanism of $Mg(OH)_2$ and Chemically Modified $Mg(OH)_2$. J Phys Chem C. 2021;125:5559–71.
40. Giannozzi P, Baroni S, Bonini N, Calandra M, Car R, Cavazzoni C, Ceresoli D, Chiarotti GL, Cococcioni M, Dabo I, et al. QUANTUM ESPRESSO: a modular and open-source software project for quantum simulations of materials. J. Phys.: Condens. Matter.2009;21:395502.
41. Giannozzi P, Andreussi O, Brumme T, Bunau O, Buongiorno Nardelli M, Calandra M, Car R, Cavazzoni C, Ceresoli D, Cococcioni M, et al. Advanced capabilities for materials modelling with Quantum ESPRESSO. J. Phys.: Condens. Matter. 2017;29:465901
42. Bale CW, Bélisle E, Chartrand P, Decterov SA, Eriksson G, Gheribi AE, Hack K, Jung IH, Kang YB, Melançon J, et al. FactSage Thermochemical Software and Databases, 2010–2016. Calphad. 2016;54:35–53. https://factsage.com (accessed May 12, 2026)
43. Roine A. HSC Chemistry [Software], Metso, Pori 2023. https://www.hsc-chemistry.com (accessed May 12, 2026)
44. Churakov SV, Iannuzzi M, Parrinello M. Ab Initio Study of Dehydroxylation-Carbonation Reaction on Brucite Surface. J Phys Chem B. 2004;108:11567–74.
45. Pimminger H, Habler G, Freiberger N, Abart R. Evolution of nanostructure and specific surface area during thermally driven dehydration of $Mg(OH)_2$. Phys Chem Minerals. 2016;43:59–68.

Part III
Quantum Calculations for Molecules

Chapter 8
Theoretical Study on Photodissociations of Palladium Chloride Complexes

Yuzuru Kurosaki and Morihisa Saeki

Abstract In this chapter, we review our recent theoretical study on photodissociations of transition metal complexes. We have been interested in laser-induced particle formation (LIPF) of palladium (Pd) and aim to reveal the early stage of LIPF. To this end, photodissociations of the $PdCl_4{}^{2-}$ complex are investigated, because this complex was confirmed to be the existence form of Pd in solution by extended X-ray absorption fine structure (EXAFS). Using models of both single-Cl atom and double-Cl atoms losses, photodissociations of $PdCl_4{}^{2-}$ are examined with the ab initio and density functional theory (DFT) methods. Not only photodissociations in the gas phase but in aqueous solution are investigated. Solvent effects are considered using the polarizable continuum model (PCM). From our calculation results, the single-Cl atom loss model predicts that $PdCl_4{}^{2-}$ photodissociation results in the $PdCl_3{}^{-}$ and Cl^{-} formation both in the gas phase and in water. The double-Cl atom loss model predicts that the photodissociation leads to the $PdCl_2{}^{-}$, Cl, and Cl^{-} formation in the gas phase. However, it does not give a reasonable prediction for photodissociation in water. It is considered that time-dependent DFT and the PCM are insufficient to describe excited electronic states affected by solvation.

Keywords Laser-induced particle formation · Transition metal complex · Palladium tetrachloride anion · Ab initio multi-reference configuration interaction (MRCI) method · Potential energy surface · Absorption spectrum · Photodissociation · Density-functional theory · Solvent effect · Polarized continuum model

---

Y. Kurosaki (✉) · M. Saeki
Department of Advanced Functional Materials Research, Takasaki Institute for Advanced Quantum Science, National Institutes for Quantum Science and Technology, Takasaki, Gunma 370-1292, Japan
e-mail: kurosaki.yuzuru@qst.go.jp

M. Saeki
e-mail: saeki.morihisa@qst.go.jp

T. Onishi (ed.), *Advances in Quantum Science II*,
https://doi.org/10.1007/978-981-95-9721-5_8

8.1 Introduction

In this chapter, our recent theoretical studies on photochemistry of transition metal complexes (TMCs) are introduced, focusing on the initial processes right after a TMC absorbs light. TMCs have attracted intense interests of chemists and physicists because they have unique electronic properties, and they can be widely applied in catalyst materials. Due to the d electrons in transition metals, electronic excited states of TMCs have properties that are different from those of other molecules including no transition metal; it is of note that the electronic excited states called ligand-to-metal charge transfer (LMCT) states can be realized in TMCs upon ultraviolet (UV) irradiation, because the d electrons are relatively low in energy. Since LMCT states have characteristic features in reactivity, a new trend making good use of them in synthetic techniques has arisen [1]. It is also of note that large spin-orbit coupling makes it possible that TMCs easily transfer between a low-spin and a high-spin states through intersystem crossing [2], which can extend the possibility for the design of new functional materials.

We have been interested in separation and recovery of platinum-group elements (PGEs) from the radioactive waste generated from nuclear-related facilities [3–7]. Figure 8.1 illustrates the concept of our new technique called laser-induced particle formation (LIPF), which enables efficient separation and recovery of PGEs from nuclear spent fuel. A brief description of LIPF is given as follows: The radioactive waste contains the PGEs, such as Pd, Rh, and Ru, together with actinides (Ac), lanthanides (Ln), and transition elements (Mo, Zr, and so on). When the radioactive waste is dissolved in a mixture of acid aqueous solution and alcohol (RCH_2OH), the PG ions have a charge transfer (CT) band in the UV region. Irradiation of UV laser into the radioactive-waste solution induces excitation of the CT bands of PG ions, which is followed by photoreduction of the PG ions to the neutrals. At this time, the UV-laser irradiation does not reduce the other ions (Ac, Ln, Mo, Zr), the CT bands of which exist out of the UV region. The PG neutrals spontaneously form PG particles in the solution because of disappearance of electronic repulsion among the PG ions by the photoreduction and is separated from the Ac, Ln, Mo, and Zr ions. Finally, the PG particles are recovered from the solution of radioactive waste by filtration, precipitation, and centrifugation.

Photochemical separation of metals using lasers was first proposed by Donohue [6–8] in the late 1970s. As far as we know, almost no study has been done in this field of research since then, except for a single report of the PGE separation with laser-induced reactions [9]. This situation means that no significant progress has been made for many years in the technique using laser-induced separation of metals; this might occur because no one has come up with an idea making the best use of photochemical reactions in element separations. Under such situations, a review on LIPF [10] was published in 2009, which prompted us to start the project of using LIPF as a tool for the separation and recovery of PGEs. Our major achievements using LIPF so far are given as follows: First, we [4] succeeded in the separation and recovery of palladium (Pd) from a solution including molybdenum (Mo) using LIPF.

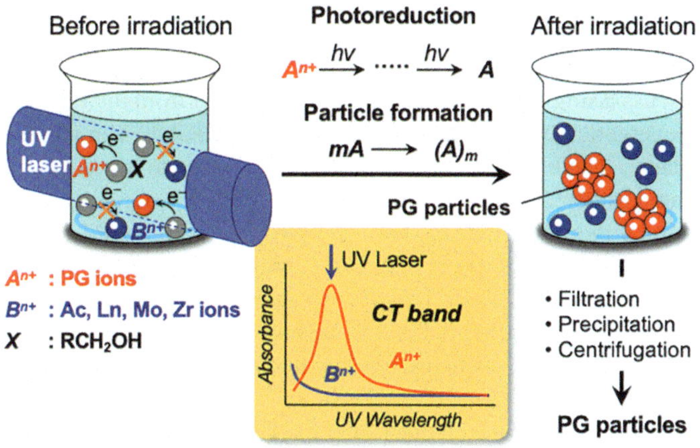

Fig. 8.1 Concept of laser-induced particle formation (LIPF)-based platinum-group (PG) separation. Adapted from Ref. [4] with permission from Elsevier (Copyright 2015)

Next, we [11, 12] found that LIPF can be successfully employed to determine the concentration of radio palladium ^{107}Pd in nuclear spent fuel, thus evaluating radiation safety of the extracted Pd. Recently, we [5] also studied reaction kinetics of LIPF in aqueous/EtOH solution containing Pd using transmission electron microscopy (TEM) and dispersive X-ray absorption fine structure (DXAFS) measurements; as a result, the DXAFS data indicated that the irradiation of the UV pulse induced the reduction of an Pd complex in the solution and the TEM observation revealed the formation of Pd particles with considerable size.

It was confirmed that in aqueous/EtOH solution, a palladium atom exists as one of the constituent atoms of the complex of palladium tetrachloride anion ($PdCl_4{}^{2-}$) [5]. In this chapter, the anion is taken as a typical example of TMCs and its photodissociation dynamics is elaborated. The present LIPF for Pd particle formation proceeds through the reduction of Pd and is roughly expressed as:

$$\mathrm{Pd(II)Cl_4{}^{2-}} \rightarrow \mathrm{Pd(0)} + 2\mathrm{Cl}^- + 2\mathrm{Cl}, \tag{8.1}$$

$$\mathrm{nPd(0)} \rightarrow \mathrm{Pd(0)_n nuclei}, \tag{8.2}$$

where in the parentheses are given the valence number of Pd. LIPF is a muti-step process, and the detailed mechanism of each step remains unexplained from the microscopic viewpoint. Since LIPF is initiated with a UV irradiation, the complex should be electronically excited in the earliest stage, but what follows next is unknown. In this chapter, according to our recent work so far [13–17], early processes right after UV light absorption in $PdCl_4{}^{2-}$ are discussed.

8.2 A Single-Cl Atom Loss Photodissociation

First, we discuss a single-Cl atom loss photodissociation from the $PdCl_4^{2-}$ complex anion. The energetically most stable structure of the $PdCl_4^{2-}$ complex is known to be square planar. It belongs to point group D_{4h} [18]. The bond lengths of the complex were estimated to be 2.29 Å from model fitting of EXAFS oscillations [5]. Using a one-dimensional model system shown in Fig. 8.2, where three Pd-Cl bond lengths are frozen to be the EXAFS value, potential energy curves (PECs) are calculated as a function of a single Pd-Cl bond length (R) that is changed from 1.8 to 6.0 Å. The geometrical symmetry of the model system belongs to point group C_{2v}; the electronic state symmetries are thus classified according to the four irreducible representations: A_1, B_1, B_2, and A_2.

The calculation procedure of the PECs for $PdCl_4^{2-}$ using the multi-reference configuration interaction (MRCI) method is as follows. First, the closed-shell spin-restricted Hartree–Fock (RHF) wave functions are calculated. Then, using the obtained RHF molecular orbitals, complete-active-space self-consistent field (CASSCF) calculations [19, 20] are carried out. Figure 8.3 shows the active space chosen in the present CASSCF calculation, which consists of the 4 s, 4p, and 4d orbitals of Pd and the 3 s and 3p orbitals of Cl, including the 48 electrons distributed in these 25 orbitals. The inner orbitals, the 1 s, 2 s, and 2p of Cl, are always doubly occupied but optimized in the CASSCF calculations. The CASSCF calculation employed here is called CASSCF (48, 25). Finally, the internally-contracted MRCI [21–23] wave functions are calculated using the CASSCF (48, 25) wave functions as reference, where electrons are correlated through single and double excitations but the inner electrons are frozen.

In the present MRCI calculations, the lowest six states are calculated for each irreducible representation (A_1, B_1, B_2, and A_2) and for each spin multiplicity (singlet and triplet). Thus, a total of 48 MRCI energies are obtained at each chosen point of R. In order to know the intensity of electronic transitions, transition dipole moments (TDMs) between the ground and excited states are estimated at the square planar geometry (R = 2.29 Å) using the MRCI wave functions. In all calculations, the basis set employed for Pd is aug-cc-pVTZ-PP [24] which includes the relativistic

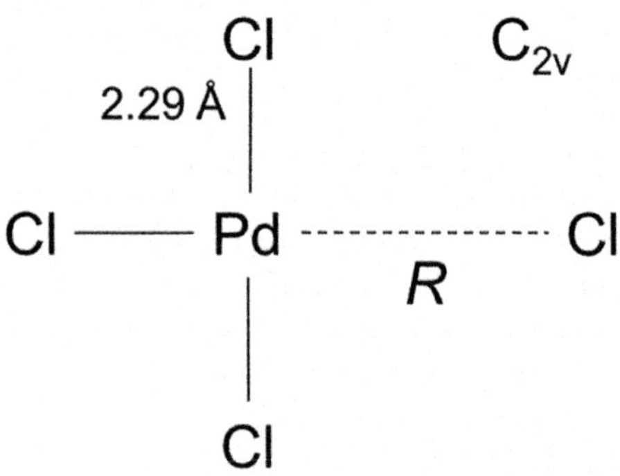

Fig. 8.2 One-dimensional model system used to obtain MRCI PECs for single-Cl atom loss photodissociation from the $PdCl_4^{2-}$ complex anion

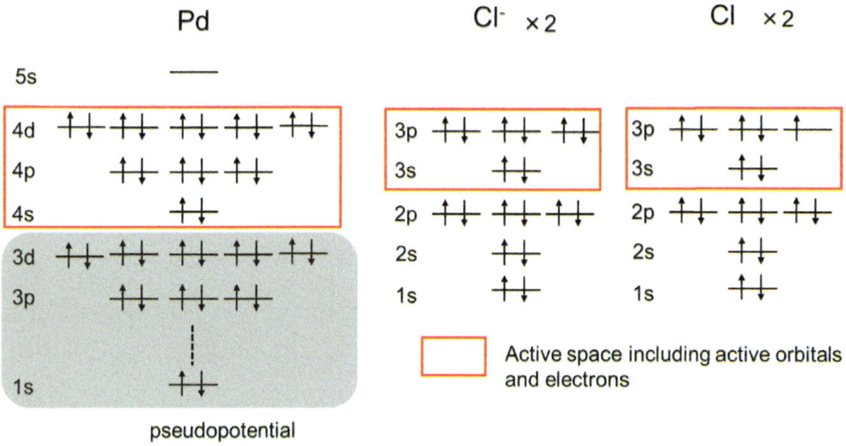

Fig. 8.3 Active space chosen for the $PdCl_4^{2-}$ complete-active-space self-consistent field (CASSCF) calculation. It includes 48 electrons distributed in 25 orbitals

pseudopotential; the basis set employed for Cl is conventional aug-cc-pVTZ [25]. All ab initio calculations are carried out using the MOLPRO program package, version 2019.2, designed by Werner and Knowles [26, 27].

The singlet and triplet MRCI PECs for each irreducible representation are shown in Fig. 8.4a and b respectively. The singlet PECs, Fig. 8.4a, can be divided into three groups (lower, middle, and upper groups) and the triplet PECs, Fig. 8.4b, into two groups (lower and upper groups). The lower group of singlet PECs includes only the ground-state $^1A_1$ PEC, which has a minimum at $R \sim 2.3$ Å and a barrier of ~0.029 hartree (0.79 eV) at $R \sim 4.0$ Å. The middle group of singlet PECs lies ~0.08 hartree higher than the lower group and includes the lowest $^1B_1$, $^1B_2$, and $^1A_2$ PECs and the second lowest $^1A_1$ PEC. Other energetically higher PECs all belong to the upper group, and most of them are of repulsive character. Figure 8.4b shows the triplet MRCI PECs; the lower group includes four PECs of all symmetries, $^3A_1$ $^3B_1$, $^3B_2$, and $^3A_2$, with shallow minima and small barriers, and the upper group includes all other PECs that are of repulsive character.

Before going on to analyzing the MRCI wave functions, a generalization of the HF spin orbital and the occupation number (0 or 1) is described in this paragraph. Given $\Phi(x_1, x_2, \ldots x_N)$ is a wave function for an N-electron system, then the probability of finding an electron in $d\boldsymbol{x}_1$ at $\boldsymbol{x}_1$, without depending on the positions of the other electrons, is given by

$$\rho(\boldsymbol{x}_1) = N \int d\boldsymbol{x}_2 d\boldsymbol{x}_3 \cdots d\boldsymbol{x}_N |\Phi(\boldsymbol{x}_1, \boldsymbol{x}_2, \ldots \boldsymbol{x}_N)|^2 \tag{8.3}$$

In general, $\rho(\boldsymbol{x}_1)$ can be expanded with the orthonormal basis of one-electron function χ:

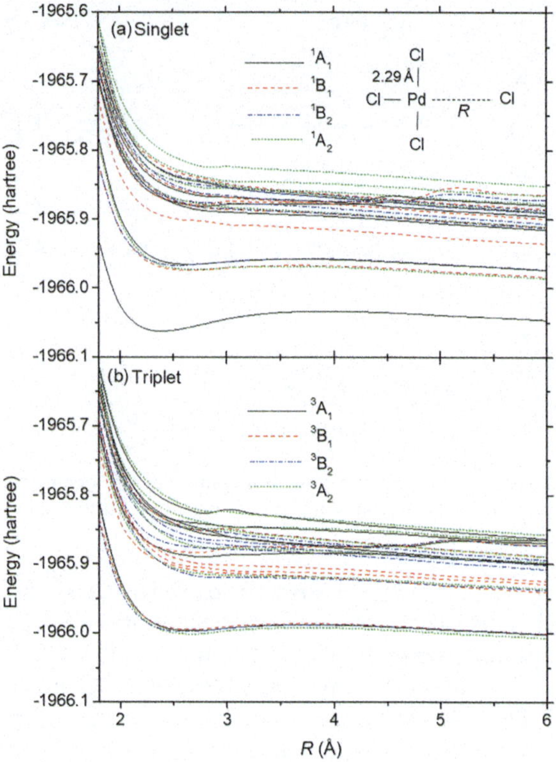

Fig. 8.4 Six MRCI PECs for each irreducible representations, A_1, B_1, B_2, and A_2, as a function of R : **a** spin-singlet and **b** spin-triplet states. Adapted from Ref. [13] with permission from Elsevier (Copyright 2020)

$$\rho(\boldsymbol{x}_1) = \sum_{i,j} \gamma_{ij} \chi_i(\boldsymbol{x}_1) \chi_j^*(\boldsymbol{x}_1). \tag{8.4}$$

When Φ is a HF wave function that is a single determinant consisting of the HF spin orbitals ϕ_i, substituting Φ into Eq. (8.3) results in a diagonal form of $\rho(\boldsymbol{x}_1)$:

$$\rho(x_1) = \sum_i \lambda_i |\phi_i(x_1)|^2. \tag{8.5}$$

$$\lambda_i = \begin{cases} 1, & \phi_i \textit{ is occupied} \\ 0, & \phi_i \textit{ is unoccupied} \end{cases}.$$

When Φ is a general wave function that is of a multi-configuration character like MRCI wave functions, $\rho(\boldsymbol{x}_1)$ has a non-diagonal form:

$$\rho(\boldsymbol{x}_1) = \sum_{i,j} \eta_{ij} \phi_i(\boldsymbol{x}_1) \phi_j^*(\boldsymbol{x}_1). \tag{8.6}$$

However, since η_{ij} is an element of a Hermitian matrix $\boldsymbol{\eta}$, there is a unitary transformation $\mathbf{U}$ that diagonalizes the matrix:

$$\boldsymbol{\zeta} = \mathbf{U}^{\dagger}\boldsymbol{\eta}\mathbf{U}, \tag{8.7}$$

where $\boldsymbol{\zeta}$ is a diagonal matrix, i.e.,

$$\zeta_{ij} = \delta_{ij}\zeta_j. \tag{8.8}$$

Thus, $\rho(\boldsymbol{x}_1)$ is expressed in a diagonal form, in terms of the transformed basis $\varphi_i(\boldsymbol{x}_1)$

$$\rho(\boldsymbol{x}_1) = \sum_i \zeta_i | \varphi_i(\boldsymbol{x}_1)|^2, \tag{8.9}$$

where ζ_i is non-integer, and φ_i is transformed from ϕ_i:

$$\varphi_i(\boldsymbol{x}_1) = \sum_k \phi_k(\boldsymbol{x}_1)U_{ki}. \tag{8.10}$$

Equation (8.9) has a diagonal form similar to the HF case, Eq. (8.4), where ζ_i is the occupation number and φ_i is the natural orbital. Equation (8.9) can be regarded as a generalization of the HF picture that each electron in a molecular system is moving independently in its orbital. The generalization changes the integer occupation number λ_i of the HF orbital ϕ_i, 0, or 1, into the non-integer occupation number ζ_i of the natural orbital φ_i.

Table 8.1 shows MRCI total energies and the occupation numbers (ONs) of the natural orbitals (NOs) [28] that have large coefficients of the Pd 4d atomic orbitals (AOs) (given in parentheses) and other AOs for D_{4h} $PdCl_4^{2-}$ (R = 2.29 Å). As described, the ONs and the coefficients of AOs are obtained by diagonalizing $\rho(\boldsymbol{x}_1)$, Eq. (8.6). Table 8.1 consists of two parts; the upper part shows the lowest-lying states of each symmetry plus the first excited 2^1A_1 state, while the lower part shows the excited states that have nonzero TDMs with the ground 1^1A_1 state. From the ONs of the NOs in the table the electronic structure of the ground 1^1A_1 ($^1A_{1g}$), state can be characterized by vacant $4d_{x^2-y^2}$. The electronic structure of the first excited 2^1A_1 ($^1B_{1g}$) state can be characterized by singly occupied $4d_{z^2}$ and singly occupied $4d_{x^2-y^2}$, and this state can be formed from the ground state through a d-d transition from $4d_{z^2}$ to $4d_{x^2-y^2}$. The electronic configuration of the lowest triplet 1^3A_1 ($^3B_{1g}$) state is the same as that of 2^1A_1 except for spin state. Other states can also be formed from the ground state through d-d transitions; the lowest 1^1B_1 ($^1A_{2g}$) and 1^3B_1 ($^3A_{2g}$) states can be formed through a d-d transition from $4d_{xy}$ to $4d_{x^2-y^2}$, the lowest 1^1B_2 (^1E_g) and 1^3B_2 (^3E_g) states from $4d_{xz}$ to $4d_{x^2-y^2}$, and the lowest 1^1A_2 (^1E_g) and 1^3A_2 (^3E_g) states from $4d_{yz}$ to $4d_{x^2-y^2}$.

The lower part in Table 8.1 shows the excited states that have large TDM with the ground 1^1A_1 ($^1A_{1g}$) state. As shown, they have two NOs whose ONs are close to

Table 8.1 MRCI total energies and ONs of the NOs that have large coefficients (given in parentheses) of the Pd 4d AOs for the D_{4h} $PdCl_4^{2-}$ complex ($R = 2.29$ Å). Adapted from Ref. [13] with permission from Elsevier (Copyright 2020)

State[a]	Energy (hartree)	Orbital					
		Pd $4d_{z^2}$	Pd $4d_{xz}$	Pd $4d_{yz}$	Pd $4d_{xy}$	Pd $4d_{x^2-y^2}$	Other AO
1^1A_1 ($^1A_{1g}$)	−1966.06117	1.975 (0.915)	1.977 (0.916)	1.977 (0.916)	1.980 (0.821)	0.063 (−0.841)	
2^1A_1 ($^1B_{1g}$)	−1965.95362	1.008 (0.944)	1.974 (0.945)	1.974 (0.945)	1.981 (0.826)	1.011 (−0.853)	
1^3A_1 ($^3B_{1g}$)	−1965.98411	0.988 (0.947)	1.983 (0.840)	1.983 (0.838)	1.981 (0.802)	1.001 (−0.886)	
1^1B_1 ($^1A_{2g}$)	−1965.96344	1.981 (0.953)	1.983 (0.765)	1.983 (0.763)	1.001 (0.957)	1.010 (−0.842)	
1^3B_1 ($^3A_{2g}$)	−1965.98327	1.973 (0.955)	1.969 (0.963)	1.969 (0.963)	1.017 (0.961)	1.019 (−0.867)	
1^1B_2 (^1E_g)	−1965.95964	1.982 (0.975)	1.015 (0.944)	1.977 (0.910)	1.977 (0.940)	1.016 (−0.858)	
1^3B_2 (^3E_g)	−1965.98870	1.982 (0.974)	1.007 (0.957)	1.978 (0.858)	1.978 (0.908)	1.008 (−0.886)	
1^1A_2 (^1E_g)	−1965.95937	1.982 (0.975)	1.977 (0.912)	1.015 (0.945)	1.977 (0.941)	1.016 (−0.858)	
1^3A_2 (^3E_g)	−1965.98843	1.982 (0.974)	1.978 (0.858)	1.006 (0.958)	1.978 (0.908)	1.008 (−0.886)	
Excited state							
3^1A_1 (^1E_u)	−1965.86081	1.979 (−0.596)	1.980 (0.762)	1.980 (0.753)	1.980 (0.733)	1.005 (−1.013)	1.012[b]
3^1B_1 (^1E_u)	−1965.86153	1979 (0.588)	1.980 (0.745)	1.980 (0.759)	1.980 (0.741)	1.005 (−1.013)	1.012[c]
6^1B_1 (^1E_u)	−1965.81110	1.941 (0.905)	1.980 (0.730)	1.979 (0.825)	1.975 (0.747)	1.042 (−1.008)	1.049[d]
2^1B_2 ($^1A_{2u}$)	−1965.85654	1.979 (0.872)	1.981 (0.828)	1.981 (0.816)	1.980 (0.866)	1.009 (−1.010)	1.016[e]

[a]The number before the symbol denotes the state level; here, 1 is the ground state, 2 is the first excited state, etc. In parentheses are given the irreducible representations of D_{4h}. The listed four excited states are those that have relatively largeTDMs with the ground 1^1A_{1g} state

[b]The NO is contributed mainly by Cl1 3px (0.377) + Cl2 3px (0.808) + Cl3 3px (0.808) + Cl4 3px (0.377). Cl1 and Cl4 are at diagonal vertexes and so are Cl2 and Cl3

[c]The NO is contributed mainly by Cl1 3py (0.585) + Cl2 3py (0.504) + Cl3 3py (0.504) + Cl4 3py (0.585). The numbering of Cl atoms is the same as above

[d]The NO is contributed mainly by Cl1 3py (−0.313) + Cl2 3py (0.811) + Cl3 3py (0.811) + Cl4 3py (−0.313). The numbering of Cl atoms is the same as above

[e]The NO is contributed mainly by Cl1 3pz (−0.486) + Cl2 3pz (0. 658) + Cl3 3pz (0.658) + Cl4 3pz (−0.486). The numbering of Cl atoms is the same as above

one; one NO is mostly contributed by the Pd $4d_{x^2-y^2}$ AO and the other by the Cl 3p AOs. It is thus explained that these excited states are formed from the ground 1^1A_1 ($^1A_{1g}$) state through an electronic transition from the Cl 3p AOs to the Pd $4d_{x^2-y^2}$ AO, which corresponds to a LMCT.

Table 8.2 shows TDMs calculated at the MRCI level for D_{4h} $PdCl_4^{2-}$ ($R = 2.29$ Å). The LMCT-excited states are only listed. Note that other TDMs become zero due to symmetry. The table also gives calculated oscillator strengths (f) defined by

$$f = (2/3)\Delta E_{fi} |\langle \psi_f | \boldsymbol{\mu} | \psi_i \rangle|^2, \tag{8.11}$$

where ΔE_{fi} is the transition energy between electronic eigenstates ψ_i and ψ_f, and $\boldsymbol{\mu}$ is the electronic dipole operator. Although a discrepancy is found in the absolute value of f between calculation and experiment, the MRCI method reproduces the order of the magnitude of the experimental f. Because of the large TDMs or f's, it is considered that ground-state D_{4h} $PdCl_4^{2-}$ is electronically excited to either or both of the lower and upper ^1E_u states when it absorbs a UV light. In the LIPF experiment, the UV laser at 266 nm (4.66 eV) is used [5]. Hence, the electronic excitation to the lower ^1E_u state is most probable. The lower ^1E_u state corresponds to the 3^1A_1 and 3^1B_1 states in C_{2v} symmetry, and, as shown in (Fig.8.4a), the 3^1A_1 and 3^1B_1 PECs belong to the upper group and have a repulsive character. Thus, a plausible scenario for early-stage photodissociation within the single-Cl atom loss model is considered as follows: Ground-state D_{4h} $PdCl_4^{2-}$ is electronically excited to the lower ^1E_u state upon UV irradiation at 266 nm, followed by a Pd-Cl bond elongation on the repulsive 3^1A_1 and/or 3^1B_1 PECs leading to the $PdCl_3^-$ + Cl^- channel; the atomic species

Table 8.2 TDM $<\psi_f \mid \mu_\lambda \mid \psi_i>$ and f calculated at the MRCI level for D_{4h} $PdCl_4^{2-}$ (R = 2.29 Å), where ψ_i is the ground 1^1A_1 ($^1A_{1g}$) state. Adapted from Ref. [13] with permission from Elsevier (Copyright 2020)

ψ_f	ΔE^a (eV)	λ^b	TDM (a.u.)	f	
				Calc.	Exp. [29]
3^1A_1 (^1E_u)	5.452	x	1.292	0.423^c	0.17
3^1B_1 (^1E_u)	5.432	y	−1.227		
6^1B_1 (^1E_u)	6.805	y	−1.859	1.152^d	0.511
2^1B_2 ($^1A_{2u}$)	5.568	z	0.027	1.014×10^{-4}	7.2×10^{-3}

[a]Transition energy between the ψ_f and $\psi_i$states

[b]x: the axis passing through Pd and leaving Cl; y: the axis in the molecular plane and normal to x; z: the axis normal to the molecular plane

[c]Since the 3^1A_1 and 3^1B_1 states in C_{2v} symmetry should be converged to the degenerate ^1E_u state in D_{4h} symmetry, this value is calculated as the sum of f for each C_{2v} state

[d]There is another $^1A_1$ state that ψ_f t should be converged to the degenerate ^1E_u state in D_{4h} symmetry, together with the 6^1B_1 state; the $^1A_1$ state is, however, higher than the 6^1A_1 state and was not obtained in the present MRCI calculation; this value is thus twice the value of f for the 6^1B_1 state

generated as a product is confirmed to be the Cl^- anion by the Mulliken population analysis (MPA).

In conclusion, using the one-dimensional single-Cl atom loss model, the MRCI calculations predict that the irradiation of a UV laser to $PdCl_4^{2-}$ at 266 nm excites the complex to repulsive PECs, leading to the following dissociation:

$$PdCl_4^{2-} \rightarrow PdCl_3^- + Cl^- \qquad (8.12)$$

8.3 Double-Cl Atoms Loss Photodissociation

In this section, we discuss double-Cl atoms loss from the complex using another model system shown in Fig. 8.5, where r_1 is defined as the distance between Pd and the middle point of leaving two Cls and r_2 the internuclear distance between leaving two Cls. The model system keeps C_{2v} symmetry, under fixing the Pd-Cl bond lengths (2.3602 Å) and the Cl-Pd-Cl angle (90°). Energy values are calculated using the MRCI method at grid points generated by selecting 15 points of r_1 ($1.5 \le r_1 \le 5.0$ Å) and 12 points of r_2 ($1.7 \le r_2 \le 4.0$ Å). The obtained energies are interpolated using the spline function, and contour maps are drawn for electronic states.

The MRCI calculations are carried out using the same active space as the single-Cl atom loss case. First, the RHF wave functions are obtained, and then CASSCF (48, 25) calculations are performed. Using the CASSCF (48, 25) wave functions as reference, MRCI wave functions are calculated for the lowest six spin singlet states for two irreducible representations, $^1A_1$ and $^1B_2$. Only the spin singlet states are discussed here, because excitations from the ground singlet state (1^1A_1) to higher spin states are forbidden, and intersystem crossings between different excited states having different spin multiplicities are considered to be negligible. Only the $^1A_1$ and $^1B_2$ states are chosen, because at the equilibrium square planar geometry for $PdCl_4^{2-}$, TDM between 4^1A_1 and 1^1A_1 and that between 3^1B_2 and 1^1A_1 are large, but TDMs between other states and 1^1A_1 are zero. Thus, a total of 12 MRCI energies are calculated at the chosen grid points in the r_1 and r_2 coordinates. The employed

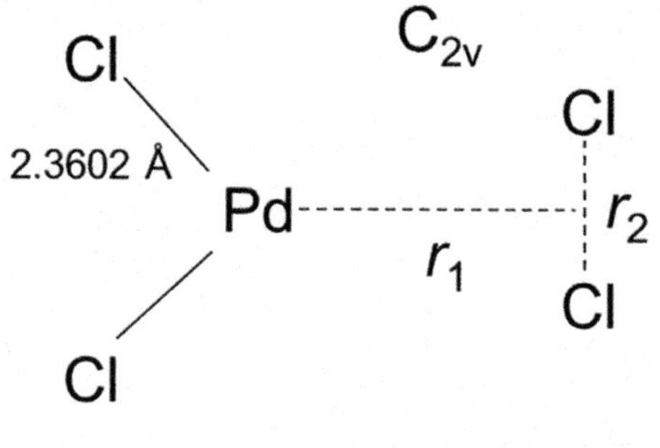

Fig. 8.5 Two-dimensional model system used to obtain MRCI PESs for double-Cl atom losses photodissociation from the $PdCl_4^{2-}$ complex anion

basis sets and the computational program are the same as the single-Cl loss case: aug-cc-pVTZ-PP (Pd); aug-cc-pVTZ (Cl); MOLPRO, version 2019.2.

It is found that the 3^1A_1, 3^1B_1, and 6^1B_1 states in the single-Cl atom loss model, which have large TDMs with the ground electronic state, correspond, respectively, to the 4^1A_1, 3^1B_2, and 6^1B_2 states in the double-Cl losses model. Figure 8.6a shows the ground-state 1^1A_1 potential energy surface (PES) plotted as functions of r_1 and r_2; energy values are given in hartree relative to the energy of the equilibrium D_{4h} geometry. There is a potential well at around $(r_1, r_2) = (1.7, 3.3\ \text{Å})$, which corresponds to the equilibrium geometry. A broken line shows the minimum energy path (MEP) connecting the potential well and dissociation products through a potential barrier of about 0.09 hartree (2.5 eV). Since the barrier height is large, the dissociation hardly occurs by heating on the ground-state PEC. The MEP in the product region shows that the internuclear distance between the leaving two Cl atoms is about 2.6 Å. This suggests that the MEP leads to the Cl_2^- anion formation because a previous calculation [30] predicted that the equilibrium bond length of the Cl_2^- anion molecule is 2.61 Å. It is found from the MPA that the atomic charge of each leaving Cl atom is −0.5, indicating that the Cl_2^- anion is formed as a product.

As mentioned above, the 4^1A_1 and 3^1B_2 PESs correlate with the states that have large TDMs with the ground 1^1A_1 state at the equilibrium D_{4h} geometry. Therefore,

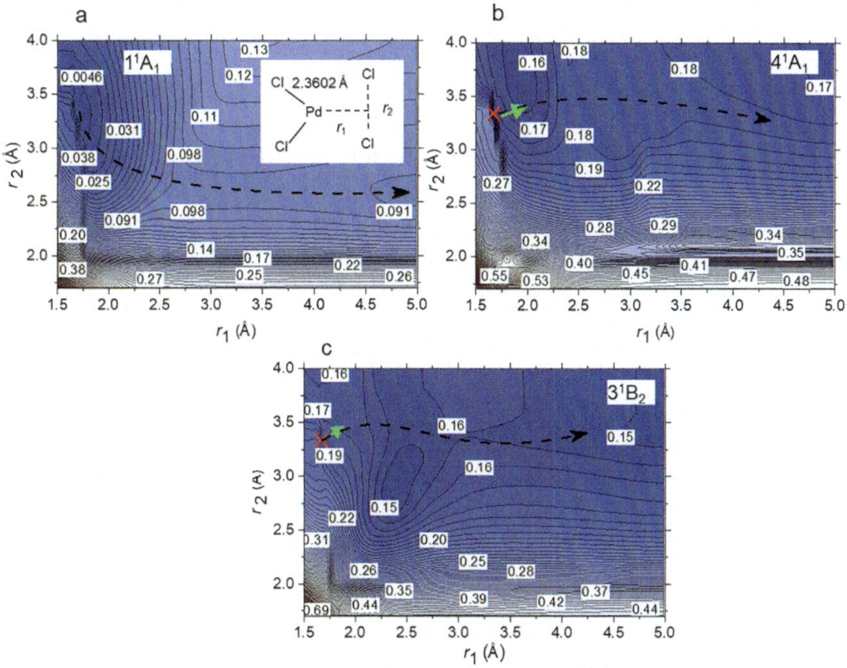

Fig. 8.6 Contour plots of PESs as functions of r_1 and r_2 : **a** the ground 1^1A_1 state; **b** the 3^1A_1 state; **c** the 3^1B_1 state. Energy values are given in hartree, relative to the energy of the equilibrium D_{4h} geometry. Adapted from Ref. [14] with permission from Elsevier (Copyright 2021)

it is considered that the 4^1A_1 and 3^1B_2 PESs play key roles in the photodissociation of the double-Cl atoms loss model. Here, we focus on these two PESs. Figure 8.6b shows the 4^1A_1 PES plotted as functions of r_1 and r_2, where energy values are given in hartree relative to the energy of the equilibrium D_{4h} geometry. If the electronic state is excited to the 4^1A_1 state upon UV irradiation at 266 nm, the wave packet for nuclear motion is created, according to the Franck-Condon (FC) principle, at around a cross which denotes the point of the equilibrium geometry on the 1^1A_1 PES. It is seen from Fig. 8.6b that the slope around the cross on the 4^1A_1 PES is steep, and there is a shallow potential well at around $(r_1, r_2) = (2.0, 3.7\ \text{Å})$. It is hence considered that the created wave packet around the cross goes down the slope, as indicated by a green bold arrow, heading to the well bottom and passing through it. Then, the wave packet spreads over the area, $r_1 > 2.5$ Å and $r_2 > 3.0$ Å, of the 4^1A_1 PES, because this area is very flat and slightly lower in energy than the FC point indicated by the cross, while the other area is energetically very high. A broken curve shows the expected "trajectory" of the wave packet, which finally leads to the generation of two Cl atoms. The MPA predicts that the charge of each leaving Cl atom is −0.5, meaning that the leaving two Cl atoms are neutral Cl and anionic Cl^-. Next, the photodissociation pathway on the 3^1B_2 PES is examined. Figure 8.6c shows the 3^1B_2 PES as functions of r_1 and r_2, where the FC point is denoted again by a cross. As the 3^1B_2 PES looks quite similar to the 4^1A_1 PES, it is considered that the same dissociation scenario holds in this case. A broken curve shows the expected "trajectory" of the wave packet generated at the FC point. It is confirmed from the MPA that the leaving two Cl atoms are Cl and Cl^-.

In conclusion, using the two-dimensional double-Cl atoms loss model, the MRCI calculations predict that the irradiation of a UV laser to $PdCl_4{}^{2-}$ at 266 nm excites the complex to repulsive PESs, leading to the following dissociation:

$$PdCl_4{}^{2-} \rightarrow PdCl_2{}^- + Cl + Cl^-. \tag{8.13}$$

8.4 Solvent Effects

So far we have theoretically investigated the early stage of the $PdCl_4{}^{2-}$ photodissociation using both the single-Cl atom loss and double-Cl atoms loss models. In the calculations, solvent effects were not taken into consideration despite the fact that all steps in LIPF proceed in an aqueous solution. In this section, solvent effects on the photodissociation are discussed to know how the effects change the calculation results in the gas phase. Although it is desirable to estimate solvent effects at the MRCI level, this is not available in the current commercial quantum-chemical software. We hence choose the time-dependent density functional theory (TDDFT) [31] method; TDDFT methods are less accurate compared with MRCI, but have the following practical advantages: (i) both ground and excited states can be calculated at a lower cost; (ii) solvent effects on both ground and excited states can be estimated combined with the polarizable continuum model (PCM) [32]. In order to assess what exchange correlation functionals can most properly reproduce the electronic states of the present $PdCl_4^{2-}$ system in aqueous solution, the absorption spectra of $PdCl_4^{2-}$ in water (dielectric constant is 78.3553) are compared with the experimental result, where two peaks, which correspond to excitations to the lower and upper ^1E_u states, are observed. Thirteen exchange correlation functionals including hybrid and local functionals are tested for the calculations of absorption spectra using the TDDFT method combined with the PCM. In comparison with the experiment, it is found that among the tested functionals, the M11 functional [33] gave the best accordance with the experiment. Therefore, the M11 functional is chosen for the DFT calculations. The employed basis set for Pd is SDD [34], where the core orbitals are substituted by the quasirelativistic pseudopotential, and that for Cl is 6–311 + G(d,p). All DFT calculations here are performed using the Gaussian 16 program package [35].

Table 8.3 lists transition energies (ΔE_1 and ΔE_2) and oscillator strengths (f_1 and f_2) of D_{4h} $PdCl_4^{2-}$ for excitations from the ground state to the lower and upper ^1E_u states, respectively. Also given in the table are dissociation energies (ΔE_{diss}'s) for a Cl^- loss from $PdCl_4^{2-}$, Eq. (8.12), on the ground-state 1^1A_1 PEC. The M11 DFT values are obtained at two Pd-Cl bond lengths (R_{PdCl}), 2.29 and 2.3815 Å, the latter of which is the optimized value in water using the B3LYP method [36] within the PCM. Comparing the MRCI and DFT results at $R_{PdCl} = 2.29$ Å in the gas phase, the MRCI values for ΔE_1 and ΔE_2 are slightly larger than the M11 values. The M11 and MRCI ΔE_{diss}'s in the gas phase agree well with each other and they are all negative, because electrostatic repulsion is dominant in the interaction between the product anions $PdCl_3^-$ and Cl^- at large R in the gas phase. As mentioned, the M11 values for ΔE_1 and ΔE_2 in water and ethanol agree well with experimental ones, which are shown in the table.

The M11 functional is found to reproduce the experimental absorption spectrum for $PdCl_4^{2-}$ in water. It is unclear, however, whether the same functional yields reasonable gas-phase PECs and PESs, in comparison with the MRCI method. It is found that the M11 PECs for single-Cl atom loss in the gas phase (not shown here) can be divided again into the lower, middle, and upper groups, as same as the MRCI

Table 8.3 ΔE_1 and ΔE_2 and f_1 and f_2 for transitions to the lower and upper ^1E_u states for $PdCl_4^{2-}$ and ΔE_{dis} for $PdCl_4^{2-} \rightarrow PdCl_3^- + Cl^-$ on the ground 1^1A_1 state PEC. Adapted from Ref. [16] with permission from Elsevier (Copyright 2024)

Method	Solvent	R_{PdCl} (Å)	ΔE_1 (eV)	f_1	ΔE_2 (eV)	f_2	ΔE_{diss} (eV)
MRCI	Gas	2.29	5.452	0.423	6.805	1.152	−1.459[a]
M11	Gas	2.29	4.736	0.050	6.282	0.649	−1.361
	Gas	2.3815	4.399	0.150	5.759	0.518	−1.061
	Water	2.3815	4.515	0.155	5.676	0.571	0.994
	Ethanol	2.3815	4.503	0.157	5.662	0.576	0.951
Exp [29]			4.429	0.17	5.577	0.511	

[a]The MRCI value of ΔE_{dis} were obtained for the total $PdCl_4^{2-}$ system at $R = 10^3$ Å

PECs given in Fig. 8.4a. There is a difference, however, between the M11 and MRCI PECs in the upper group; the MRCI PECs are all repulsive as shown in Fig. 8.4a, while there is a well in most of the M11 PECs around the equilibrium geometry. We also calculated PESs for double-Cl atoms loss in the gas phase using the M11 TDDFT method (also not shown here). The result shows that although the M11 ground-state 1^1A_1 PES is consistent with the MRCI ground 1^1A_1 PES, (Fig. 8.6a), the M11 excited-state PESs do not accord with the MRCI 4^1A_1 and 3^1B_2 PESs, Fig. 8.6b and c. We hence discuss solvent effects on double-Cl atoms loss only for the ground-state 1^1A_1 PES obtained with the M11 TDDFT within the PCM, because the M11 excited-state PESs are considered to be less reliable.

Figure 8.7 shows the PECs for single-Cl atom loss from $PdCl_4^{2-}$ in water, calculated with the same model shown in Fig. 8.2. The PECs are calculated as a function of R with the other Pd-Cl bond lengths fixed at 2.3603 Å, which is the geometrically optimized value for $PdCl_4^{2-}$ in water with the M11 functional. The system symmetry is kept in C_{2v}, and the electronic states are classified into four irreducible representations, $^1A_1$, $^1B_1$, $^1B_2$, and $^1A_2$. It is seen that the PECs in water can roughly be divided into three groups (lower, middle, and upper groups), which is similar to the gas-phase case. There is a significant difference, however, between the PECs in the gas phase and those in water at large R in the asymptotic region; the PECs in the gas phase are repulsive in the asymptotic region, Fig. 8.4a, while those in water are attractive and the gradients become zero asymptotically as R increases, Fig. 8.7. This is because the anion-anion electrostatic repulsion is shielded by solvent molecules. The MRCI calculations revealed that if $PdCl_4^{2-}$ in the gas phase is electronically excited to the lower and upper ^1E_u states, then it can lose a single-Cl atom spontaneously. One wonders whether the same is true for $PdCl_4^{2-}$ in aqueous solution. Two vertical arrows in Fig. 8.7 show the two electronic transitions: the short arrow corresponding to the transition to the lower ^1E_u state and the long arrow to the transition to the upper ^1E_u state. The lower ^1E_u PEC around the FC point has a very shallow well, and its asymptote at large R is energetically comparable with the FC point. If the system is slightly vibrationally excited and has small kinetic energy along the leaving Cl direction, then the excitation to this PEC leads to a single-Cl atom loss. The upper ^1E_u PEC around the FC point is slightly repulsive, and the excitation to this PEC leads to a spontaneous single-Cl atom loss. It is thus expected from the present calculation that in the photodissociation of $PdCl_4^{2-}$ in water, both excitations to the lower and upper ^1E_u PECs lead to single-Cl atom loss at a high probability.

Table 8.3 also lists the DFT results obtained at $R_{PdCl} = 2.3815$ Å in water and in ethanol (the dielectric constant is 24.852) for comparison. Comparing with the gas-phase values at the same R_{PdCl}, in water and ethanol, it is seen that ΔE_1 is estimated slightly larger, while ΔE_2 is smaller. The ΔE_1 and ΔE_2 estimated in water are slightly larger by about 0.01 eV than those in ethanol, due to the difference in dielectric constant. ΔE_{diss} for the ground 1^1A_1 PEC in water and ethanol are estimated to be about 1 eV positive, i.e., the single-Cl atom loss is endoergic in water and ethanol, because electrostatic repulsion is shielded by solvation, as mentioned.

As for double-Cl atoms loss from $PdCl_4^{2-}$, the ground-state PES correlates with the $PdCl_2^- + Cl_2^-$ channel in the gas phase, Fig. 8.6a. However, this is not the case in

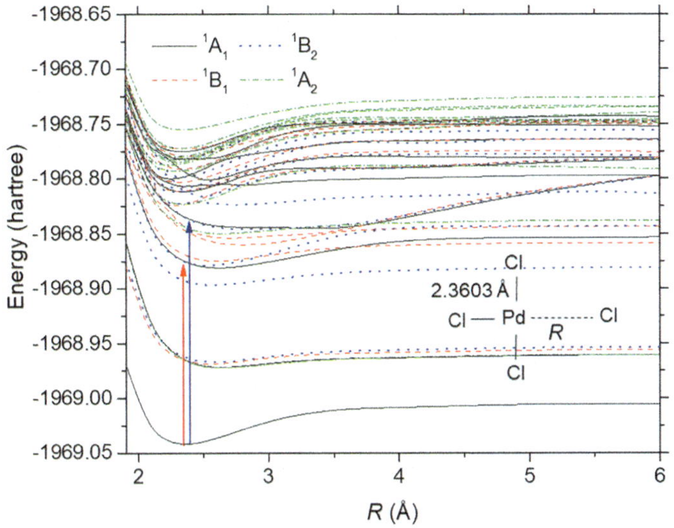

Fig. 8.7 M11 TDDFT PECs for single-Cl loss from $PdCl_4^{2-}$ in water as a function of R. Adapted from Ref. [16] with permission from Elsevier (Copyright 2024)

water. The asymptotic energies in water are calculated for the $PdCl_2^- + Cl_2^-$, $PdCl_2^- + Cl + Cl^-$, and $PdCl_2 + 2Cl^-$ channels using the M11 DFT method within the PCM. It is found as a result that the $PdCl_2^- + 2Cl^-$ channel is energetically lowest and correlates with $PdCl_4^{2-}$ on the ground-state PES. Spin-singlet ground-state PES is calculated using the M11 TDDFT method as functions of internal coordinates r_1 and r_2 defined as the same in Fig. 8.5. Energy values are obtained at grid points generated by selecting 15 points of r_1 ($1.4 \leq r_1 \leq 5.0$ Å) and 17 points of r_2 ($1.7 \leq r_2 \leq 4.0$ Å). The calculated values are interpolated using the spline function to draw a PES contour map. Figure 8.8 shows the ground-state 1^1A_1 PES for double-Cl atoms loss from $PdCl_4^{2-}$ in water; contour labels are given in hartree relative to D_{4h} $PdCl_4^{2-}$. There is a deep well at around $(r_1, r_2) = (1.7, 3.3$ Å), corresponding to the equilibrium D_{4h} geometry of $PdCl_4^{2-}$. In contrast to the gas-phase PES shown in Fig. 8.6a, there is no minimum energy pathway to dissociation on the ground-state PES in water. From the PES, it is considered that the two leaving Cl atoms move apart from one another; they are both the Cl^- anion, which is confirmed by the MPA.

8.5 Summary

In this chapter we review our recent theoretical study using both MRCI and DFT methods to understand early processes in LIPF of Pd. The $PdCl_4^{2-}$ complex is focused on and its photodissociations are examined, because a previous EXAFS observation revealed that the $PdCl_4^{2-}$ complex is the existence form of Pd in a HCl solution. Two

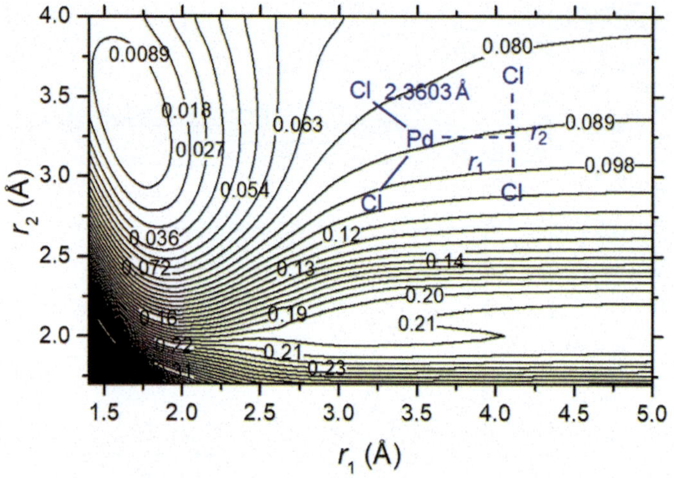

Fig. 8.8 M11 TDDFT ground-state PES for double-Cl atoms loss from $PdCl_4^{2-}$ in water as functions of r_1 and r_2. Adapted from Ref. [16] with permission from Elsevier (Copyright 2024)

dissociation channels, single-Cl atom and double-Cl atoms losses from $PdCl_4^{2-}$ are investigated.

A single-Cl atom loss can be summarized as follows:

(i) Excited-state MRCI PECs at the FC point in the gas phase are repulsive, which causes spontaneous single-Cl atom loss to generate $PdCl_3^- + Cl^-$.
(ii) The TDDFT M11 PECs in water are similar to the MRCI PECs in the gas phase, but a difference is found in the excited-state PECs. The excited-state M11 PECs are not so steep as the excited-state MRCI PECs around the FC point, though excitations to the M11 PECs in water are expected to lead to a single-Cl atom loss to generate $PdCl_3^- + Cl^-$. This is because the asymptotic point at large R is energetically comparable with the FC point.

Double-Cl atoms loss can be summarized as follows:

(i) Two excited-state MRCI PESs correlating with the lower degenerate ^1E_u state at D_{4h} symmetry reveal that such excitations in the gas phase cause spontaneous dissociation leading to the $PdCl_2^- + Cl + Cl^-$ channel.

(ii) Comparing the MRCI PESs with the M11 PESs in the gas phase, both the ground-state PESs agree with each other, though no agreement is seen between the MRCI and M11 excited-state PESs. This suggests that the M11 TDDFT method is less reliable for describing excited PES, and hence, only the ground-state PES is discussed for double-Cl atoms loss in water.

(iii) The ground-state M11 PES in water reveals that D_{4h} $PdCl_4^{2-}$ correlates asymptotically with the $PdCl_2 + 2Cl^-$ channel, while in the gas phase, it correlates with the $PdCl_2^- + Cl_2^-$ channel.

Since most photochemical processes of TMCs proceed in solution, accurate estimation of solvent effects on excited-state PESs is the key to understand TMC photodissociation dynamics. Currently, the TDDFT method plus the PCM is the best practical choice for considering solvent effects on excited-state PESs. However, the combined method does not always provide reliable outcomes, when excited-state solvations are investigated. There are many challenges that should be overcome in the study of TMC photodissociations. We are still working on this project aiming for better understanding TMC photochemistry and LIPF mechanisms.

References

1. Julia F. Ligand-to-metal charge transfer (LMCT) photochemistry at 3d-metal complexes: an emerging tool for sustainable organic synthesis. ChemCatChem. 2022;14:e202200916.
2. Penfold TJ, Gindensperger E, Daniel C, Marian CM. Spin-vibronic mechanism for intersystem crossing. Chem Rev. 2018;118:6975–7025.
3. Saeki M. Laser-induced particle formation: its applications to precious metal recovery from spent nuclear fuel and fundamental studies. In: Ishikawa Y, Nakamura T, Saeki M, Sato T, Sugiyama T, Wada H, Yatsuhashi T, editors. High-energy chemistry and processing in liquids. Singapore: Springer Singapore; 2022. p. 33–55.
4. Saeki M, Taguchi T, Nakashima N, Ohba H. Wet separation between palladium (II) and molybdenum (IV) ions by using laser-induced particle formation: enhancement of recovery efficiency of palladium by laser condition. J Photochem Photobiol A. 2015;299:189–93.
5. Saeki M, Matsumura D, Yomogida T, Taguchi T, Tsuji T, Saitoh H, Ohba H. In situ time-resolved XAFS studies on laser-induced particle formation of palladium metal in an aqueous/EtOH solution. J Phys Chem C. 2019;123:817–24.
6. Donohue T. Photochemical separation of metals in aqueous solution. Chem Phys Lett. 1977;48:119–21.
7. Donohue T. Photochemical separation of europium from lanthanide mixtures in aqueous solution. J Chem Phys. 1977;67:5402–4.
8. Donohue T. Photochemical separation of cerium from rare earth mixtures in aqueous solution. Chem Phys Lett. 1979;61:601–4.
9. Song K, Cha H, Lee J, Choi JS, Lee YI, Choi KS. Extraction of palladium metal from aqueous solution of palladium chloride by laser-induced photochemistry. Microchem J. 2001;68:121–6.
10. Sakamoto M, Fujistuka M, Majima T. Light as a construction tool of metal nanoparticles: synthesis and mechanism. J Photochem Photobiol C. 2009;10:33–56.
11. Asai S, Yomogida T, Saeki M, Ohba H, Hanzawa Y, Horita T, Kitatsuji A. Determination of ^{107}Pd in Pd recovered by laser-induced photoreduction for inductively coupled plasma mass spectrometry. Anal Chem. 2016;88:12227–33.
12. Asai S, Ohta M, Yomogida T, Saeki M, Ohba H, Hanzawa Y, Horita T, Kitatsuji Y. Determination of ^{107}Pd in Pd purified by selective precipitation from spent nuclear fuel by laser ablation ICP-MS. Anal Bioanal Chem. 2019;411:973–83.
13. Kurosaki Y, Nakanishi R, Saeki M, Ohba H. Ab initio MRCI study on potential energy curves for a single cl loss from the palladium tetrachloride anion $PdCl_4{}^{2-}$. Chem Phys Lett. 2020;746:137288.
14. Kurosaki Y, Nakanishi R, Saeki M, Ohba H. Ab initio MRCI study on potential energy surfaces for double cl loss from the palladium tetrachloride anion $PdCl_4{}^{2-}$. Chem Phys Lett. 2021;764:138247.
15. Kurosaki Y, Nakanishi R, Saeki M, Ohba H. Reaction pathways for palladium(I) reduction in laser-induced particle formation of Pd: an ab initio molecular orbital study. Chem Phys. 2023;569:111857.

16. Anzai A, Kurosaki Y, Saeki M, Muraoka A. A DFT study on photodissociation of the palladium tetrachloride anion $PdCl_4^{2-}$ in solution. Chem Phys. 2024;579:112182.
17. Koyama M, Kurosaki Y, Muramatsu S, Saeki M, Inokuchi Y. Study on photodissociation spectra and decay pathways of gas-phase $PdCl_3^-$ and $PdCl_2^-$ anions by electrospray ionization mass spectrometry and MRCI calculation. Chem Phys Lett. 2024;857:141736.
18. Nakamoto K. Infrared and Raman spectra of inorganic and coordination compounds, part a: theory and applications in inorganic chemistry. Hoboken, New Jersey: John Wiley & Sons; 2009.
19. Werner H-J, Knowles PJ. A second order multiconfiguration SCF procedure with optimum convergence. J Chem Phys. 1985;82:5053–63.
20. Knowles PJ, Werner H-J. An efficient second-order MC SCF method for long configuration expansions. Chem Phys Lett. 1985;115:259–67.
21. Werner H-J, Knowles PJ. An efficient internally contracted multiconfiguration–reference configuration interaction method. J Chem Phys. 1988;89:5803–14.
22. Knowles PJ, Werner H-J. An efficient method for the evaluation of coupling coefficients in configuration interaction calculations. Chem Phys Lett. 1988;145:514–22.
23. Shamasundar KR, Knizia G, Werner H-J. A new internally contracted multi-reference configuration interaction method. J Chem Phys. 2011;135:054101.
24. Peterson KA, Figgen D, Dolg M, Stoll H. Energy-consistent relativistic pseudopotentials and correlation consistent basis sets for the 4d elements Y–Pd. J Chem Phys. 2007;126:124101.
25. Woon DE, Dunning TH Jr. Gaussian basis sets for use in correlated molecular calculations. III. The atoms aluminum through argon. J Chem Phys. 1993;98:1358–71.
26. Werner H-J, Knowles PJ, Knizia G, Manby FR, Schütz M. WIREs Comput Mol Sci. 2012;2:242–53. https://doi.org/10.1002/wcms.82.
27. Werner H-J, Knowles PJ et al. MOLPRO, version 2019.2, a package of ab initio programs. https://www.molpro.net.
28. Szabo A, Ostlund NS. Modern quantum chemistry. New York: Macmillan; 1982. p. 252.
29. Rush RM, Martin DS Jr, Legrand RG. Polarized crystalspectra of potassium tetrachloropalladate(II) and potassium tetrabromopalladate(II). Inorg Chem. 1975;14:2543–50.
30. Mondal CK, Das KK, Maji K, Bhattacharyya SP. Photoionization dynamics from Cl_2^-: a one-electron model. Chem Phys Lett. 2004;394:215–9.
31. Bauernschmitt R, Ahlrichs R. Treatment of electronic excitations within the adiabatic approximation of time dependent density functional theory. Chem Phys Lett. 1996;256:454–64.
32. Tomasi J, Mennucci B, Cammi R. Quantum mechanical montinuum solvation modelsm. Chem Rev. 2005;105:2999–3094.
33. Peverati R, Truhlar DG. Improving the accuracy of hybrid meta-GGA density functionals by range separation. J Phys Chem Lett. 2011;2:2810–7.
34. Andrae D, Haeussermann U, Dolg M, Stoll H, Preuss H. Energy-adjusted ab initio pseudopotentials for the 2nd and 3rd row transition-elements. Theor Chem Accounts. 1990;77:123–41.
35. Frisch MJ, et al. Gaussian 16, Revision B.01. Wallingford CT: Gaussian, Inc.; 2016.
36. Becke AD. Density-functional thermochemistry. III. The role of exact exchange. J Chem Phys. 1993;98:5648–52.

Chapter 9
Complementary Approaches to the Study of Non-covalent Interactions in Van Der Waals Clusters: Theory and Experiments

Concetta Caglioti and Federico Palazzetti

Abstract Non-covalent interactions are ubiquitous across all states of matter and multiple contexts. They occur in the simplest nanoaggregates as well as in nanostructured systems involving biologically relevant macromolecules, such as proteins and nucleic acids. These interactions play a crucial role in catalysis and supramolecular chemistry. Understanding these interactions is essential for interpreting numerous phenomena, since a detailed analysis of their nature—which depends on the species involved, their distance, and their mutual orientation—allows us to identify the functional form that best describes the observed behavior. However, the computational approach of these interactions is not always straightforward. High-level electronic-structure methods are often required, and their computational cost increases rapidly with system dimensionality. In this chapter, we focus on interactions in van der Waals clusters, that is, systems composed of two atoms or molecules bound by non-covalent interactions. After reviewing the quantum methods used to calculate interaction energies in such systems, we will present two case studies. These examples show how interaction energies can be evaluated and how analytical potential energy surfaces can be constructed using two different methodological approaches.

Keywords Interaction energy · Dispersion forces · Induction effects · Hydrogen bonding · Coupled Cluster Singles and Doubles (CCSD) · Basis Set Superposition Error (BSSE) · Counterpoise correction · Complete Basis Set (CBS) extrapolation · Density Functional Theory (DFT) · Machine learning

---

C. Caglioti · F. Palazzetti (✉)
Università Telematica San Raffaele, Rome, Italy
e-mail: federico.palazzetti@uniroma5.it

C. Caglioti
e-mail: concetta.caglioti@uniroma5.it

T. Onishi (ed.), *Advances in Quantum Science II*,
https://doi.org/10.1007/978-981-95-9721-5_9

9.1 Introduction

9.1.1 Intermolecular Forces: Foundations and Applications

Intermolecular forces, also referred to as non-covalent interactions, play a pivotal role in biochemistry [1], catalysis [2], materials chemistry [3], nanotechnology [4], and medicinal chemistry [5]. Their accurate evaluation remains a central challenge in theoretical and computational chemistry (see for example [6]). These forces, predominantly electrostatic in nature, act between atoms or molecules. Other types of interactions—gravitational, magnetic, and nuclear—can generally be neglected [7]. Gravitational forces are orders of magnitude weaker than electrostatic ones [8]. Magnetic interactions involving electrons are similarly weak and occur only between species with spin multiplicity greater than one, i.e., molecules possessing unpaired electrons. Nuclear magnetic interactions arise only between nuclei with non-zero spin, which is relatively uncommon; even then, their energies fall well below typical intermolecular force magnitudes [9]. Nuclear forces, due to their extremely short range, are also considered negligible [10].

Despite being several orders of magnitude weaker than covalent bond energies, intermolecular interactions are essential for understanding the behavior of systems across chemistry, physics, and biology. They govern transport phenomena in real fluids [11], cohesion in condensed phases [12], the alignment and responsiveness of liquid crystals [13], enzyme–substrate recognition in biochemical and pharmacological contexts [14], and surface interactions in heterogeneous catalysis [15].

9.1.2 Classification of Intermolecular Forces

Intermolecular interactions are commonly categorized as weak or strong. Among the weak interactions are dipole–dipole forces, which arise between polar molecules with permanent dipole moments, and induced dipole–dipole forces, which occur when a polar molecule distorts the electron cloud of a neighboring nonpolar molecule [16, 17]. This distortion, known as induction, results from the electric field exerted by surrounding species and leads to attractive forces. Because the fields of nearby molecules can either reinforce or cancel each other, induction is highly non-additive.

Dispersion forces (also known as London forces) originate from transient fluctuations in electron distribution [18]. These are inherently quantum mechanical phenomena with no classical analog. As electrons fluctuate, they create instantaneous dipoles that correlate across molecules, favoring low-energy configurations and suppressing high-energy ones. This correlation lowers the average energy and strengthens as molecules approach, resulting in attraction.

Directional interactions include hydrogen bonding, which occurs when a hydrogen atom covalently bound to a highly electronegative atom interacts with another electronegative atom, either within the same molecule or across different molecules. A related but less common interaction is the halogen bond, formed when a halogen atom bonded to an electronegative partner engages with an electron-donating atom [19].

Ion–ion interactions are Coulombic in nature, occurring between two static charges [20]. Their strength is directly proportional to the product of the charges and inversely proportional to the square of the distance between them. From a quantum mechanical perspective, these interactions represent the direct coupling between static charge distributions and are described by first-order perturbation theory. They are strictly pairwise additive and can be either attractive or repulsive.

We also report resonance and exchange interactions. The former one typically arises when at least one molecule is in a degenerate (often excited) state, or when identical molecules differ in their electronic state. These interactions are characteristic of open-shell systems [21]. The exchange interaction, which derives from the Pauli exclusion principle, is a repulsive component of intermolecular interactions. According to this principle, two electrons with the same spin cannot occupy the same quantum state. When two molecules approach each other and their electron orbitals overlap, the electron wave function's symmetry imposes a repulsion that is quantum mechanical, not electrostatic, in origin. This effect is particularly relevant at short distances, when the overlap of the orbitals is significant. It contributes to the shape of the system's potential energy surface (PES). Unlike electrostatic repulsion, the exchange interaction depends on the spatial configuration and spin of the electrons rather than charge.

The study of non-covalent interactions in catalysis has brought particular interactions between the catalyst and the substrate, such as CH/π...HC/π and CH...O/N, back into the spotlight. These interactions are important in asymmetric hydrogenation, particularly in sustainable catalytic systems based on Earth-abundant transition metals [23].

9.1.3 From Quantum Chemical Calculations to Machine Learning Methods

From a computational and theoretical standpoint, the study of non-covalent interactions has long posed significant challenges, prompting sustained efforts from the scientific community (see for example [23–25]). Non-covalent interactions are much weaker than covalent interactions (typically <10 kcal/mol), meaning that even minor errors or approximations in the calculation can lead to significant discrepancies, sometimes overturning predictions about the stability of certain configurations. Some forces, such as dispersion forces, derive from long-range electronic correlations, which are inherently non-local and non-additive. In addition, interaction energies are highly sensitive to molecular geometry and relative orientation, and strongly influenced by the choice of basis set. Moreover, errors such as the basis set superposition error (BSSE) arise from wave function overlap and require appropriate corrections. To achieve accurate results, high-level theoretical methods such as CCSD(T) are required. The acronym CCSD(T) stands for Coupled Cluster Singles and Doubles with perturbative Triples; it is a quantum chemistry method that includes single and double electron excitations exactly, plus triple excitations treated perturbatively. It is widely regarded as the "gold standard" for highly accurate molecular energy calculations. However, these approaches are computationally demanding, as their cost scales exponentially with the number of electrons. Consequently, even systems composed of relatively few atoms can become computationally prohibitive.

Calculations are performed in a wide range of contexts and various strategies have been implemented to make them increasingly accurate. Non-covalent interactions have been studied ab initio in large systems such as supramolecular complexes, molecular crystals, and other soft materials. Strategies based on first principles have therefore been developed, adopting methods such as Quantum Monte Carlo, symmetry-adapted perturbation theory (SAPT), non-canonical coupled cluster techniques, and random phase approximation approaches [26].

In recent decades, density functional theory (DFT) has become a widely used tool for analyzing intermolecular interactions, offering a more computationally efficient alternative to ab initio methods. Significant progress has been made in developing DFT approaches that can accurately describe London dispersion forces. These advances include formulating new functionals and introducing specific corrections to existing ones. Current DFT methods can estimate interaction energies with an accuracy of about 0.5 kcal/mol for small van der Waals complexes containing up to about 20 atoms. However, the accuracy decreases for larger systems with more than 100 atoms, with deviations reaching 3–5 kcal/mol compared to reference values obtained with ab initio methods. It should be noted that uncertainties associated with benchmarks increase with system size and that error profiles vary significantly between DFT methods without showing a systematic trend. Consequently, nanoscale van der Waals complexes are currently one of the most active frontiers in developing DFT techniques to model non-covalent interactions (see for example [27]).

In recent years, the development of machine learning algorithms has involved implementing DFT-related techniques to balance computational cost and accuracy. To this end, reference databases have been refined to evaluate and improve DFT accuracy. Techniques such as general regression neural networks (GRNN) and support vector machines (SVM) have been employed to correct interaction energies derived from DFT, offering promising prospects for enhancing predictive performance [28].

9.1.4 Van Der Waals Clusters

This chapter focuses on the study of interaction energy in van der Waals clusters, which are aggregates of particles (atoms or molecules) that form stable structures due to van der Waals forces. These forces include dispersion, dipole–dipole, and induced dipole interactions. Though weaker than covalent or ionic bonds, van der Waals clusters can remain stable in the gas phase, especially at low temperatures and pressures where thermal fluctuations are minimized. We derive potential energy surfaces (PES), which describe the interaction energy of van der Waals clusters as a function of the system's coordinates. The first case study examines the H_2O–HF system, for which a grid of single-point interaction energies was computed using ab initio methods [29]. These data points are then fitted using a spherical harmonics expansion to obtain an analytical representation of the PES. The second case study involves the propylene oxide–He system, for which the PES is modeled as an improved version of the Lennard–Jones potential [30]. In this case, the parameters are derived from experimental data and further refined through ab initio calculations. In both cases, constructing an analytical potential energy surface is essential for enabling classical and quantum treatments of collisional processes.

9.2 The Interaction Energy Evaluation in Quantum Chemistry

In the context of the Born–Oppenheimer approximation, the intermolecular interaction energy is defined as the difference in energy between system A + B in a given configuration and the energy of fragments A and B when they are separated by an infinite distance. When the distance between the monomers is sufficiently large, internal interactions can be neglected, and the interaction energy depends on the distance between the monomers and their mutual orientation. These interaction energies are very small compared to the energy of the individual monomers, making them difficult to calculate using ab initio methods. To obtain meaningful results, one must resort to advanced theories and extensive bases, which increase computational cost. This problem can be addressed in two ways: (i) directly calculating the energy of the dimer for each geometry and subtracting the energies of the individual monomers or (ii) calculating the interaction energy using perturbation theory as an extension of the Rayleigh-Schrödinger theory (see for example [31, 32]).

The first method, known as the supramolecular approach, is the most common. However, since the energy is very small, calculations with a high level of theory are required. Consider two fragments (or monomers), A and B, which can be atoms or molecules, with energy contents E_A and E_B, respectively. The energy of a monomer is always defined relative to a reference state in which its constituent particles (electrons, protons, and neutrons) are separated by an infinite distance. In this hypothetical situation, the particles no longer interact with one another, so the total energy of the system is set to zero. When the particles come together to form a bound system, the attractive forces between the nuclei and electrons dominate while the repulsive forces between like charges counterbalance them. The resulting equilibrium configuration has a lower energy than the separated particles. This means that the calculated energy of the atom or molecule is negative with respect to zero reference. The more negative the energy, the more stable the system.

The interaction energy is then given by the following:

$$E_{\text{interaction}} = E_{AB} - (E_A + E_B)$$

E_{AB} represents the energy of the supramolecular complex AB, which is the sum of the energies of the two monomers plus the interaction energy. This approach has several advantages: it is easy to implement; it allows one to obtain a multidimensional surface of the interaction energy directly for use in molecular dynamics; and it is compatible with the Born–Oppenheimer approximation. However, it does not allow for the separation of individual physical contributions to the interaction, such as electrostatic, induction, dispersion, and exchange contributions.

In the perturbative approach, on the other hand, the system's Hamiltonian is written as follows:

$$\widehat{H} = \widehat{H}_A + \widehat{H}_B + \widehat{V}_{AB}$$

$\widehat{H}_A$ and $\widehat{H}_B$ are Hamiltonian of the isolated monomers, while $\widehat{V}_{AB}$ is the interaction term, which refers to the interaction.

The first-order energy correction is as follows:

$$E^{(1)} = \langle \psi_A \psi_B \mid \widehat{V}_{AB} \mid \psi_A \psi_B \rangle$$

where ψ_A and ψ_B refer to the wave function of the unperturbed A and B monomers in their ground state. This represents the electrostatic interaction between the electron densities of the two molecules. This contribution is predominant in long-range interactions.

The second-order correction, on the other hand, is given by the following formula:

$$E^{(2)} = \sum_{n \neq 0} \frac{\mid \langle \psi_n \mid \widehat{V}_{AB} \mid \psi_0 \rangle \mid^2}{E_0 - E_n}$$

The symbol ψ_0 denotes the unperturbed ground state of the combined system, A + B. It is the product of the ground-state wave functions of the two isolated molecules, ψ_A and ψ_B, and represents the reference configuration before any interaction between the molecules is introduced. By contrast, ψ_n refers to the excited states of the unperturbed system. These states involve one or both molecules being promoted to higher electronic configurations while still being considered non-interacting entities. In perturbation theory, such excited states contribute to the second-order correction to the energy because the interaction operator can couple the ground state, denoted by ψ_n, to these virtual excited states. From a physical point of view, this coupling accounts for dynamic effects beyond simple electrostatics. It describes how molecules can polarize each other (induction) and how correlated electron density fluctuations give rise to dispersion forces. Thus, while ψ_0 provides the baseline description of the system, ψ_n embodies the possible virtual excitations that mediate higher-order contributions to the intermolecular potential.

The main advantage of this approach is the ability to attribute physical meaning to the various terms, which makes analyzing the contributions easier. Furthermore, for weakly bound systems, high computational costs are not required, especially when development can be truncated at second-order perturbation. However, the method has limitations. If the interactions are strong, the method may fail because it requires the initial states to be well-defined and calculable. Furthermore, the series could diverge in the presence of degenerate states or if the perturbation is too intense. If perturbation orders higher than second are required, the method becomes complex and computationally expensive.

9.2.1 The Basis Set Superposition Error

A fundamental problem mainly concerning the supramolecular approach is the basis set superposition error (BSSE) [31, 32]. This error occurs when calculating the energy of the supramolecular complex AB. The basic functions of one monomer can improve the electronic description of the other. This results in a basis set that is larger than the one used for the isolated monomers. Consequently, the energy of the complex is artificially lower, and the interaction energy is overestimated. This effect is amplified by the variational principle since the complex benefits from a richer basis than the isolated fragments. BSSE is a consequence of finite basis sets: theoretically, an infinite basis set would eliminate the problem, but this is not computationally feasible. Several methods have been developed to overcome this error; the most widely used is the counterpoise correction method (see for example [33, 34]).

Although BSSE primarily affects the supramolecular approach, it can indirectly influence the perturbative approach [35]. In the perturbative approach, monomers and interactions are treated as perturbations to avoid directly calculating the complex, which reduces the risk of BSSE since energies obtained with different bases are not compared. However, effects similar to BSSE may arise if the bases of the monomers are not well balanced. For example, this could occur if different bases are used for A and B, or if one base is much larger than the other. Another possible cause is failing to use diffuse functions when necessary to correctly describe weak interactions.

The Counterpoise Correction

The counterpoise correction (CP) method, introduced by Boys and Bernardi in 1970, is the most common approach to compensating for basis-set superposition error (BSSE) in molecular interaction calculations [33, 34]. This approach requires calculating the energy of the AB system in a way that determines the energies of fragments A and B using the entire set of bases employed for the complex.

When calculating the energy of the monomers, the electrons of the other fragment are excluded and its nuclear charge is set to zero; however, its basis functions are retained as ghost atoms. This fictitious atoms are represented by a basis function that is positioned in space as if it belonged to a real atom but has no electrons or nuclear charge. It contributes to the electronic description of the system but does not physically interact with it. The related orbitals simulate the presence of the complete basis set of the complex when calculating the energy of a single fragment. This balances the comparison between the energy of the AB complex, which uses all of the bases of A and B, and the energies of monomers A and B, which would otherwise be calculated using less extensive bases.

The CP correction is obtained by comparing the monomer energies calculated normally with those obtained in the basis set of the complex. The corrected interaction energy is calculated by adding the energy of the complex and the CP correction. Corrected energy curves are generally higher than uncorrected curves and have energy minima at greater distances. However, the effect on the position of the minimum is less pronounced than the effect on the energy.

The CP method is advantageous because it is conceptually simple and easy to implement. It is compatible with various ab initio and DFT methods, and it does not require modifications to the Hamiltonian. However, the CP method has some limitations. In weakly bound systems, it can overestimate the correction. It does not consider geometric relaxation of fragments. In polyatomic complexes, ghost atoms can introduce numerical errors. Additionally, the CP correction lacks rigorous physical justification since BSSE is a numerical artifact, not a direct quantum effect.

Complete Basis Set Extrapolation

A common strategy to mitigate the basis set superposition error (BSSE) is to use larger and more flexible basis sets because each molecular fragment can be described more accurately when not relying on basis functions borrowed from the other fragment. Consequently, the artificial improvement in the electronic description that occurs when forming the complex is less pronounced, and the BSSE is reduced [36].

To minimize this error further, many researchers adopt the complete basis set (CBS) extrapolation approach. The goal of CBS is to estimate the theoretical energy that would be obtained with an infinite basis set—an ideal scenario that is, of course, computationally unattainable. Instead, one calculates the electronic energy using a sequence of increasingly large basis sets, such as cc-pVDZ, cc-pVTZ, and cc-pVQZ. Then, one extrapolates to the CBS limit using mathematical models. Common extrapolation formulas include exponential decay, $E(X) = E_\infty + \alpha \cdot e^{-\beta X}$, and inverse power laws like $E(X) = E_\infty + \frac{\alpha}{X^3}$, where E_∞ is the CBS limit energy for any system (e.g., for the AB system, E_∞ is referred to A, B or to the AB complex, respectively, indicated as E_A^∞, E_B^∞, and E_{AB}^∞), X is the basis set cardinal number (e.g., 2 for DZ, 3 for TZ, 4 for QZ), and α, β are fitting parameters.

After extrapolating the energies of the monomers, A and B ($E_A^\infty$and E_{AB}^∞), and the complex AB (E_{AB}^∞) to the CBS limit, one can calculate the interaction energy at the

CBS limit, denoted as $E_{\text{int}}^{\text{CBS}}$. This quantity expresses stabilization (usually negative) due to complex formation:

$$E_{\text{int}}^{\text{CBS}} = E_{AB}^{\infty} - (E_{A}^{\infty} + E_{B}^{\infty})$$

This procedure significantly reduces the BSSE since all energies are treated as if they were computed with an infinite basis set, where the BSSE theoretically vanishes. However, CBS extrapolation does not eliminate BSSE entirely; it only minimizes it. The method requires multiple high-level calculations, making it computationally demanding. Its accuracy also depends on the choice of extrapolation formula and the quality of the basis sets used. Unlike counterpoise correction, which compensates for BSSE by recalculating monomer energies with ghost orbitals, CBS extrapolation approaches an ideal limit where error is negligible.

Perturbation Theory

Calculating intermolecular interactions using perturbation theory differs significantly from the supramolecular approach. While the latter is based on a formalism that has already been derived, perturbation theory directly calculates the energy difference between the molecular complex and its monomers. It breaks down energy contributions into the following categories: electrostatic, inductive, dispersive, and short-range exchange (see for example [37]).

These methods, based on Rayleigh-Schrödinger theory, aim to determine intermolecular coefficients that appear in the long-range (LR) energy expansion:

$$E_{\text{LR}} = -\sum_{n} C_n R^{-n}$$

Here, the index n indicates the order of the distance dependence. Each value of n corresponds to a different physical mechanism of interaction. For example, terms with $n = 3$ describes dipole–dipole electrostatic interactions, terms with $n = 4$ represent induction effects where one molecule polarizes the other, and terms with $n = 6$ capture dispersion forces. As n increases, the corresponding contribution decays more rapidly with distance, meaning higher-order terms become less important at long range. The coefficients C_n quantify the strength of each contribution. They are determined by the intrinsic electrical properties of the monomers, such as permanent dipole moments, polarizabilities, and dynamic response functions. In practice, perturbation theory allows these coefficients to be calculated directly from the properties of a single molecule, so they not only describe the interaction energy but also provide valuable information about the monomers themselves.

Electrostatic Contribution

The electrostatic contribution falls within the first-order perturbation and is represented by multipolar operators. The general multipolar moments of monomers A and B are indicated by QM^{A}_{lq} and $QM^{B}_{l'q'}$, and the coefficient is proportional to the product of the two monomers' multipolar moments:

$$C_n \propto QM^{A}_{lq} \cdot QM^{B}_{l'q'}$$

where the indices l and q (as well as l' and q') are used to label the type and orientation of the multipole moments of a molecule. The index l specifies the order of the multipole: for example, $l = 0$ corresponds to the monopole or total charge, $l = 1$ to the dipole moment, $l = 2$ to the quadrupole moment, and higher values of l describe increasingly complex charge distributions. The index q identifies the component of the multipole for a given order. In the case of a dipole, q distinguishes the direction of the dipole along the x, y, or z axis; for higher multipoles, it labels the different independent components of the tensor that describe the spatial orientation of the moment.

These moments are calculated by analytically solving zero-order perturbation theory, which only requires the unperturbed wave function. We use the definition of the multipolar moment as the first negative derivative of the energy with respect to the electric field (F_{lq}):

$$QM_{lq} = -\frac{\partial \langle \psi_0 \mid H \mid \psi_0 \rangle}{\partial F_{lq}}$$

Here, the symbol ψ_0 represents the unperturbed ground-state wavefunction of the monomer. It allows one to calculate the expectation value of the Hamiltonian and then determine the multipole moments by differentiating the energy with respect to an external electric field. In this way, the multipole moments are obtained directly from the ground-state properties of the isolated molecule.

Inductive Contribution

The inductive contribution emerges in the second-order perturbation and involves the static polarizability, which describes how easily a molecule's electronic distribution can be deformed by an external electric field. Let us consider a monomer A, whose multipole moment is QM^{A}_{lq}, that interacts with a monomer B, whose multipole moment is $QM^{B}_{l'q'}$. The monomer A induces a dipole moment in B, by producing a deformation in the electronic distribution of B, represented by the polarizability $\alpha^{B}_{lq,l'q'}$. The coefficient is proportional to the product of the multipole moment of A, the polarizability of B: $\alpha^{B}_{lq,l'q'}$, and the multipole moment of B:

$$C_n \propto QM^A_{lq} \cdot \alpha^B_{lq,l'q'} \cdot QM^B_{l'q'}$$

The polarizability can be calculated analytically by solving the first-order perturbation equation or numerically as the second negative derivative of the energy with respect to the electric field.

Dispersion Contribution

Several formulas exist for calculating dispersion coefficients. The Casimir-Polder formula is provided below for informational purposes.

The dispersion coefficient is defined as follows:

$$D_n = \frac{1}{2\pi} \int_0^\infty \alpha_A(i\omega) \cdot \alpha_B(i\omega)\, d\omega$$

In this expressions, D_n the dispersion coefficient of order n. It determines how strongly the dispersion interaction contributes to the long-range potential energy between monomers A and B and shows how these forces decay with distance between A and B. The terms $\alpha_A(i\omega)$ and $\alpha_B(i\omega)$ are the dynamic polarizabilities of the two molecules, evaluated at imaginary frequency $i\omega$ that ensures the integral converges smoothly. The variable ω is the frequency over which integration is carried out, ranging from zero to infinity. Physically, this represents the spectrum of possible electronic fluctuations that contribute to dispersion forces. The factor $1/2\pi$ is a normalization factor. Altogether, the integral collects the contributions of correlated electronic fluctuations in both molecules across all frequencies. The product of the polarizabilities, $\alpha_A(i\omega) \cdot \alpha_B(i\omega)$, expresses how the fluctuations in one monomer are linked to those in the other.

The imaginary frequency polarizabilities are calculated for each molecule by solving the first-order perturbation equation. Numerical integration is often performed using Gauss–Legendre quadrature with variable substitutions to adjust the integration limits.

9.3 Case Study: H_2O–HF by Spherical Harmonics Approach

The spherical (and hyperspherical) harmonic expansion is a widely used technique for representing potential energy surfaces in intermolecular interactions, particularly in van der Waals clusters [38]. This approach has been applied to systems composed of four or five bodies, such as rigid diatomic molecules (O_2–O_2, N_2–N_2, H_2– H_2, N_2–O_2, N_2–NO [39], and CO–CO and CO–HF [40]), and has been extended to systems containing water (H_2O–H_2 [41], H_2O–N_2, and H_2O–O_2 [42]). It has also been adapted to systems with non-rigid molecules, such as noble gas aggregates with hydrogen peroxide [43] or hydrogen sulfide peroxide [44], in which torsional motion is unconstrained.

The method involves calculating a set of selected single energy points corresponding to significant system configurations, chosen according to symmetry and physical criteria. These points provide the necessary data to construct the potential energy surface, which is then expanded into (hyper)spherical harmonics. The goal is to obtain a compact and accurate potential whose precision depends on the quality and quantity of the calculated points. As an example, we consider the H_2O–HF system, in which both molecules are rigid, and an interaction model based on this technique is constructed [29] (for the H_2O–HCl system, see [45]).

9.3.1 *Construction of the Potential Energy Surface*

The Coordinates Reference Frame System

The H_2O–HF system is used as a case study to analyze the interaction between a nonlinear triatomic molecule belonging to the C_2v point group (H_2O) and a heteronuclear diatomic molecule (HF). Assuming both molecules behave as rigid rotors, the system can be described using five coordinates.

As illustrated in Fig. 9.1, the system is embedded in a Cartesian xyz coordinate frame, with the origin placed at the center of mass of the entire complex. Two additional axes, z' and z'', are introduced to represent the Jacobi vectors associated with H_2O and HF, respectively. The radial coordinate R is defined as the distance between the centers of mass of the two molecules. Four angular variables are used to describe the relative orientations: θ_a is the angle between the vector r and the axis $z\prime$, representing the orientation of H_2O, while θ_b is the angle between r and $z\prime\prime$, corresponding to the orientation of HF. Both angles are essential for characterizing the geometry of the system. The azimuthal angle α, ranging from 0 to 2π, describes the rotation of the water molecule around the $z\prime-$ axis. When H_2O lies in the yz-plane, $\alpha = 0$, which corresponds to the configuration where the oxygen atom is located at the maximum distance from HF along the rotation around the z'-axis. Configurations with $\alpha = \pi/2$ and $3\pi/2$ represent orientations where H_2O is perpendicular to the xz-plane. Finally, the dihedral angle ϕ, also ranging from 0 to 2π, defines the torsional relationship between the two Jacobi vectors $z\prime$ and z'', and is crucial for describing the relative twist between the two molecules.

Significant Configurations for the H_2O–HF System

To ensure a uniform representation of the potential energy surface across the relevant distance range, it is essential to select a set of appropriately selected configurations. In general, the number of significant configurations required is inversely related to the molecular symmetry: the higher the symmetry, the fewer configurations are needed.

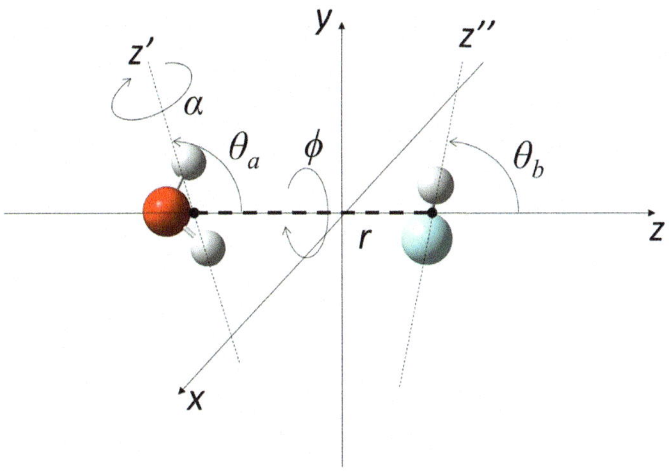

Fig. 9.1 The H_2O–HF system embedded in a Cartesian coordinate reference frame. The oxygen atom is in red, the fluorine atom in blue, and the hydrogens in white. The Jacobi axes of H_2O and HF are indicated with z' and z'', respectively. The angular variables α, θ_a,θ_b, and ϕ are reported. The radial coordinate, corresponding to the distance between the centers-of-mass of the molecules, is indicated with a bold dashed line (Figure adapted from Ref. [45])

In this study, we introduce a systematic methodology for identifying a minimal yet representative set of configurations. This approach aims to achieve an accurate description of the potential energy surface while maintaining a relatively low computational cost. For the H_2O–HF system, the selected configurations are labeled with capital letters—L, H, X, Z, and T—as summarized in Table 9.1 and shown in Fig. 9.2 (configurations with $\alpha = 0$), in Fig. 9.3 (configurations with $\alpha = \pi/2$ and $\alpha = 3\pi/2$), and in Fig. 9.4 (configurations with $\alpha = \pi$).

Ab initio calculations were performed using the Gaussian09 software program [46]. Molecular geometries were optimized using the MP2/aug-cc-pVTZ method

Table 9.1 Summary table of significant configurations for the H_2O–HF system

Conf	θ_a	θ_b	ϕ	A	Geometric description
L1	0	0	Und	Undetermined	Linear orientation, HF close to H_2O
L2	0	π	Und	Undetermined	Linear orientation, HF opposite to H_2O
H	π/2	π/2	0	0, π/2, π	Parallel axes, α is symmetric
X	π/2	π/2	π/2	0, π/2, π, 3π/2	Crossed orientation
Z1	π/4	π/4	0	0, π/2, π	Intermediate orientation
Z2	π/4	3π/4	0	0, π/2, π	Intermediate orientation
T1	π/2	0	0	0, π/2, π	Perpendicular orientation
T2	π/2	π	0	0, π/2, π	Perpendicular orientation
T3	0	0	π/2	0, π/2, π	Perpendicular orientation

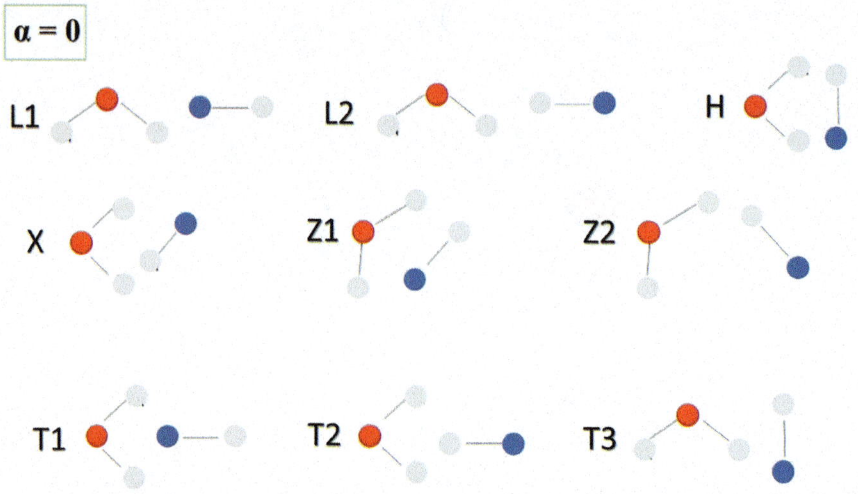

Fig. 9.2 Significant configurations for the H_2O–HF system for $\alpha = 0$. Oxygen atoms are represented by red circles, hydrogen atoms by light gray circles, and fluorine atoms by blue circles

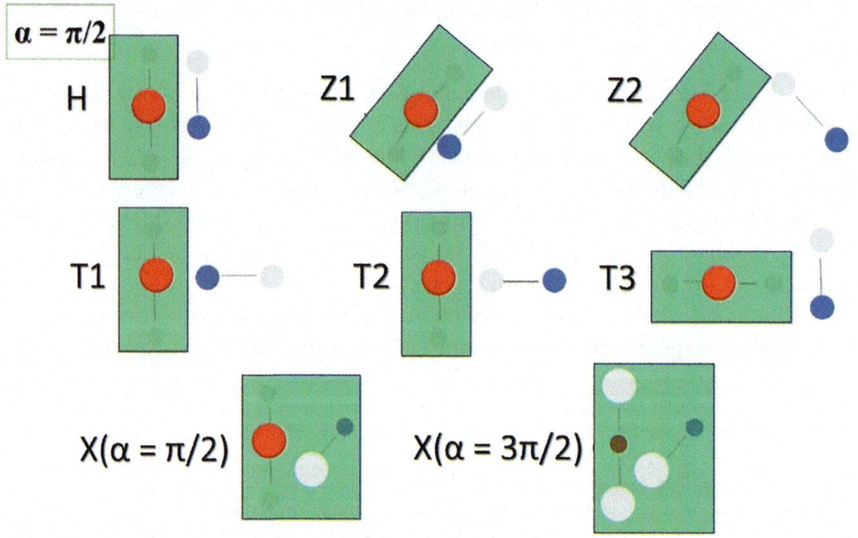

Fig. 9.3 Significant configurations of the H_2O–HF system for $\alpha = \pi/2$ and $\alpha = 3\pi/2$ (the latter applying only to the H configuration). Oxygen atoms are shown in red, hydrogen atoms in light-gray, and fluorine atoms in blue. For these values of α, the H_2O and HF molecules are oriented perpendicular to the xy-plane, which is highlighted by the transparent green rectangle. Atoms drawn larger appear closer to the viewer, whereas smaller atoms indicate positions behind the plane

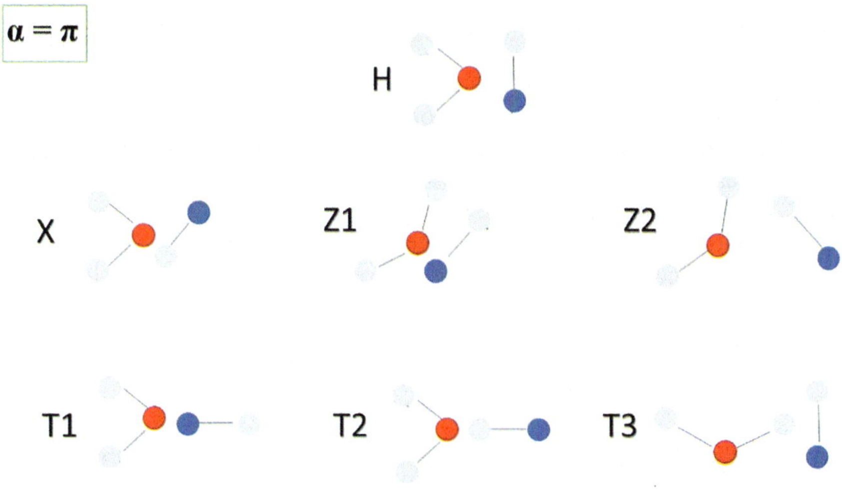

Fig. 9.4 Significant configurations for the H_2O–HF system for $\alpha = \pi$. Oxygen atoms are represented by red circles, hydrogen atoms by light gray circles, and fluorine atoms by blue circles

and interaction energies were calculated using the CCSD(T) method with the CBS (complete basis set) approximation to obtain the extrapolated complete basis set limit. MP2 stands for Møller–Plesset second-order perturbation theory, which improves upon Hartree–Fock by introducing electron correlation at second order; aug-cc-pVTZ is the *augmented correlation-consistent polarized valence triple-zeta basis set,* which uses triple-zeta functions for valence orbitals, adds polarization functions to describe electron distortion, and diffuse functions to capture long-range effects. Other basis sets are VDZ, using double-zeta functions, VQZ (quadruple zeta) and V5Z (quintuple zeta).

A total of 1,400 energy points were obtained, corresponding to 61 radial distances for each of the 24 significant configurations. The aug-cc-PVDZ basis set was used for the MP2/CBS extrapolation and the aug-cc-pVTZ and aug-cc-pVQZ basis sets were corrected by adding the energy calculated with the CCSD(T)/aug-cc-PVDZ method. The radial points are important because the range of distances considered (2.0–8.0 Å) is relevant for describing molecular collision processes.

The interaction energy, expressed as a function of five coordinates $(R, \alpha, \theta_a, \theta_b, \phi)$, is expanded into a series of bipolar spherical harmonics $Y^{L0}_{L_1L_2}(\theta_a, \theta_b, \phi)$, multiplied by the radial expansion moments $v^{L}_{L_1L_2}(R)$ and by an angular function $w_i(\alpha)$, which represents the weight of the significant configurations as a function of the parameter α. The spherical harmonic $Y^{L0}_{L_1L_2}(\theta_a, \theta_b, \phi)$ is given by the product of two spherical harmonics $Y^{m}_{L_1}(\theta_a, \phi_a)$ and $Y^{-m}_{L_1}(\theta_b, \phi_b)$, which are referred to the Jacobi vectors z' and z'', respectively, as shown in Fig. 9.1. The variables ϕ_a and ϕ_b are dihedral angles defined by the intersection of the x-axis with z' and z'', respectively. The indices L_1 and L_2 are non-negative integers that define the order of the harmonics, while L, which represents the total angular momentum quantum number and describes the shape of the angular function, results from coupling two separate angular and is constrained by the relation $| L_1 - L_2 | \leq L \leq L_1 + L_2$. The index m is also an integer number with -min $|L_1, L_2| \leq m \leq$ min $|L_1, L_2|$, and represents the magnetic quantum number describing the orientation of the angular component. In the bipolar spherical harmonics formula, $Y_{L_1L_2}^{L0}(\theta_a, \theta_b, \phi)$, the superscript 0 indicates that the total magnetic quantum number is zero. This arises because the two spherical harmonics, $Y_{L_1}^{m}(\theta_a, \varphi_a)$ and $Y_{L_2}^{-m}(\theta_b, \varphi_b)$, depend on the same integer m, but with opposite signs. When combined, their contributions cancel, yielding $m + (-m) = 0$. The bipolar spherical harmonics $Y_{L_1L_2}^{L0}$ are decomposed into products of the univariate harmonics associated with the individual angles θ_a and θ_b, according to the relation:

$$Y_{L_1L_2}^{L0}(\theta_a, \theta_b, \phi) = \sum_m \begin{pmatrix} L_1 & L_2 & L \\ m & -m & 0 \end{pmatrix} Y_{L_1}^{m}(\theta_a, 0) \cdot Y_{L_2}^{-m}(\theta_b, \phi)$$

The term $\begin{pmatrix} L_1 & L_2 & L \\ m & -m & 0 \end{pmatrix}$ is a 3-j symbol, which imposes the condition $M = m + (-m) = 0$, consistent with the notation *L0* used for composite spherical harmonics. The Wigner 3j symbol is a compact way of expressing the rules for combining angular momenta in quantum mechanics [47].

Finally, the angular function $w_i(\alpha)$, associated with configurations where $\alpha = 0, \pi/2, \pi, 3\pi/2$, is expanded into a cosine series:

$$w(\alpha) = a + b\cos\alpha + c\cos(2\alpha) + d\cos(3\alpha)$$

It is possible to derive four equations by evaluating the expression at $\alpha = 0, \pi/2, \pi$, and $3\pi/2$. In this way, we obtain a system of four equations with four unknown (the coefficients *a*, *b*, *c* and *d*), which can be solved using standard algebraic techniques, such as Cramer's rule.

Interaction Energy for H_2O–HF

The selection of spherical harmonics to be included in the expansion depends on the symmetry properties of the molecular system. In highly symmetric systems, such as H_2O–H_2, where both molecules are symmetric with respect to the exchange of hydrogen atoms, the expansion includes only spherical harmonics with even indices L_1 and L_2.

In the case of H_2O–HF, where the HF molecule exhibits lower symmetry, the harmonic expansion includes even values for L_1 and a combination of even and odd values for L_2.

The interaction energy for a system like H_2O–HF is expressed as a weighted sum over the significant configurations, according to the formula:

$$V_{H_2O-HF}(r;\alpha,\theta_1,\theta_2,\phi)=\sum_i w_i(\alpha)\Bigg[v_{000}(r)+\sqrt{3}v_{011}(r)\cos\theta_2+\frac{\sqrt{5}}{4}v_{022}(r)(1+3\cos(2\theta_2))+$$
$$\frac{\sqrt{5}}{4}v_{202}(r)(1+3\cos(2\theta_1))-\frac{1}{2}\frac{\sqrt{3}}{2}v_{211}(r)\big[(1+3\cos(2\theta_1)\cos\theta_2)+3\cos\phi\sin(2\theta_1)\sin\theta_2\big]+$$
$$\frac{3}{2}v_{213}(r)\left[\frac{1}{2}(1+3\cos(2\theta_1))\cos\theta_2-\cos\phi\sin(2\theta_1)\sin\theta_2\right]+\frac{\sqrt{5}}{4}v_{220}(r)\Bigg[\frac{1}{4}(1+3\cos(2\theta_1))(1+$$
$$\cos(2\theta_2))+3\cos\phi\sin(2\theta_1)\sin(2\theta_2)+\frac{3}{4}\cos(2\phi)(\cos(2\theta_1)-1)(\cos(2\theta_2)-1)\Bigg]-$$
$$\frac{5}{4\sqrt{14}}v_{222}(R)\Bigg[\frac{1}{2}(1+3\cos(2\theta_1))(1+3\cos(2\theta_2))+3\cos\phi\sin(2\theta_1)\sin(2\theta_2)-$$
$$\frac{3}{2}\cos(2\phi)(\cos(2\theta_1)-1)(\cos(2\theta_2)-1)\Bigg]+3\sqrt{\frac{5}{14}}v_{224}(R)\Bigg[\frac{1}{8}(1+3\cos(2\theta_1))(1+3\cos(2\theta_2))-$$
$$\cos\phi\sin(2\theta_1)\sin(2\theta_2)+\frac{1}{16}\cos(2\phi)(\cos(2\theta_1)-1)(\cos(2\theta_2)-1)\Bigg]\Bigg].$$

The expansion moments $v^L_{L_1L_2}(r)$ are constructed as linear combinations of the single-point interaction energies calculated for the significant configurations, as a function of the intermolecular distance r, and derived similarly to $w(\alpha)$ using standard algebraic methods [29]. The term $v_{000}(r)$ represents the isotropic component of the potential, obtained by averaging over all configurations. The remaining terms are referred to as anisotropic components, each carrying specific physical meaning related to angular dependence. The maximum value of L used in the expansion—set to 4 in this case—is determined by the number of significant configurations required to describe the system. With this choice, the spherical harmonic expansion includes nine terms, corresponding to nine distinct configurations. Considering four values of the angular parameter α (0, $\pi/2$, π, and $3\pi/2$), and accounting for the fact that α is undefined for the L-type configurations, and that $\pi/2$ is equivalent to $3\pi/2$ for all configurations except X, the total number of significant configurations amounts to 24.

9.3.2 *Potential Energy Surface Feature of H_2O–HF*

Here, we discuss the potential energy profiles as a function of distance for the significant configurations of the H_2O–HF system [29]. The minimum energy is considered as the absolute reference value. It is important to note that the potential expansion based on significant configurations provides an exact interpolation of the corresponding ab initio energy points. For other applications of spherical harmonics expansion to molecular systems see for example [48]. For an accurate study of the potential energy surface of H_2O–HF and its main features, in which the energy points were fitted by multipolar expansion over 3000 molecular orientations, see Ref. [49].

In Fig. 9.5, the significant configurations L and H are reported. The collinear configurations L1 and L2 exhibit opposite behaviors: L1 is attractive due to the hydrogen bond between the hydrogen atom of water and the fluorine atom of HF, whereas L2 is strongly repulsive because of the unfavorable orientation between H and F. In the parallel configuration H, with $\alpha = 0$, the potential is weakly attractive, with a minimum of approximately 1.5 kJ/mol at 3.1 Å. The mutual interaction between the hydrogen atoms of both molecules may reduce the stabilizing effect of the hydrogen bond. The potentials of the H configurations with $\alpha = \pi/2$ and $\alpha = \pi$ are exclusively repulsive due to the position of oxygen, which prevents a hydrogen bond from forming between the fluorine atom and one of the hydrogen atoms in the water molecule. This effect is slightly more evident in the configuration with $\alpha = 3\pi/2$.

Figure 9.6 shows the interaction energy as a function of the distance for the perpendicular configurations T1. The configuration with $\alpha = 0$ is stabilized by the hydrogen bond between fluorine atom and hydrogens of the water, and presents a minimum of *ca.* 5 kJ/mol at 3.0 Å. The configurations with $\alpha = \pi/2$ and $\alpha = \pi$ are exclusively repulsive due to the strong interaction between oxygen and fluorine at short-range distances.

In Fig. 9.7, the interaction energy for T2 configurations as a function of the distance is reported. The configuration with $\alpha = 0$ is repulsive, the system is arguably destabilized by the proximity of hydrogens. On the contrary, configurations with $\alpha = \pi/2$ and $\alpha = \pi$ present a strong attraction due to the increased contribution of the hydrogen bond determined by the hydrogen–oxygen interaction. For both these configurations the equilibrium distance is at 2.7 Å; the well depth is 26.3 kJ/mol for $\alpha = \pi/2$ and 34.2 kJ/mol for $\alpha = \pi$. This latter configuration is the global minimum of the potential energy surface.

In Fig. 9.8, we report the potential energy curves for T3 configurations. For $\alpha = 0$, the interaction energy presents a minimum of 12.0 kJ/mol at 2.9 Å, due to the

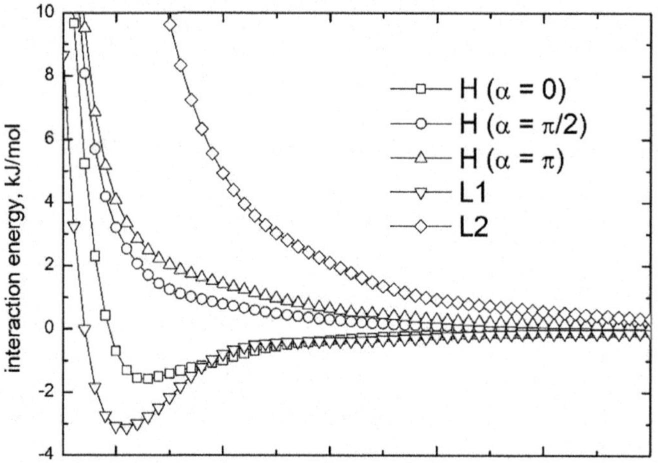

Fig. 9.5 The potential energy profiles of the parallel configurations H for $\alpha = 0$, $\pi/2$, and π, and for the collinear configurations L1 and L2

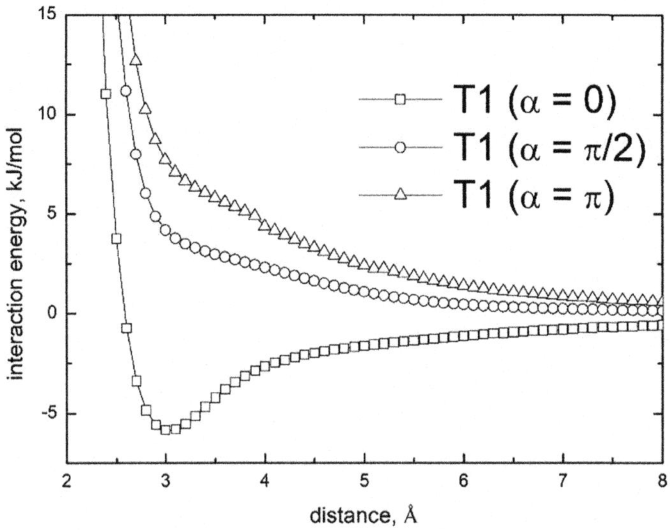

Fig. 9.6 The potential energy profiles of the perpendicular configurations T1 for $\alpha = 0$, $\pi/2$, and π

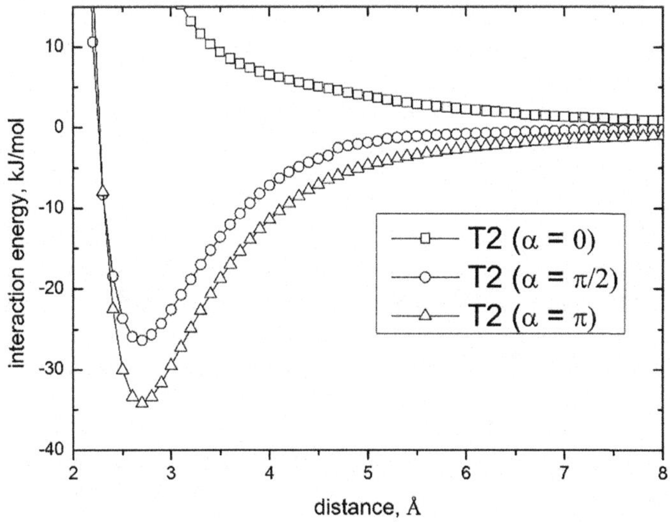

Fig. 9.7 The potential energy profiles of the perpendicular configurations T2 for $\alpha = 0$, $\pi/2$, and π

interaction between fluorine and the hydrogen of water. For $\alpha = \pi/2$, the attraction is weaker with a minimum of 5.2 kJ/mol at 3.0 Å, arguably due to the oxygen–fluorine interaction the reduces the stabilization of the system. This interaction becomes stronger for $\alpha = \pi$, making the potential exclusively repulsive.

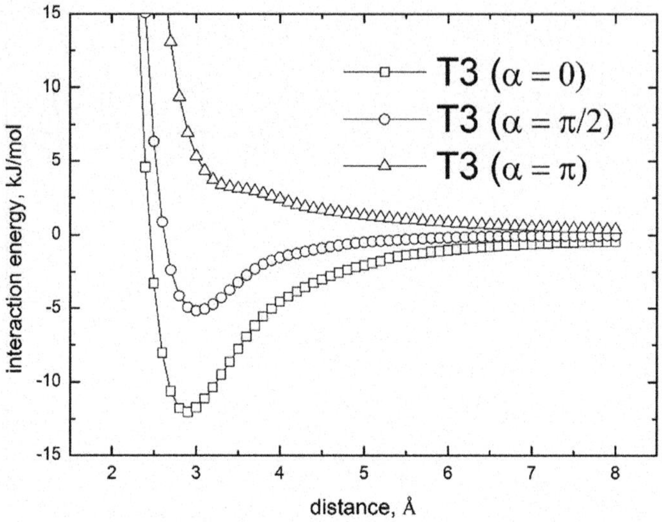

Fig. 9.8 The potential energy profiles of the perpendicular configurations T3 for $\alpha = 0$, $\pi/2$, and π

In Fig. 9.9, the case of crossed configuration X is reported. Configurations with $\alpha = \pi$ and $\alpha = 3\ \pi/2$ are repulsive, especially the latter, which is strongly repulsive due to a marked interaction between fluorine and oxygen on one side and between the three hydrogen atoms on the other. For $\alpha = 0$, the potential shows a minimum around 3.0 Å with a well depth of 5.2 kJ/mol, while for $\alpha = \pi/2$ the minimum is 3.3 at 3.1 Å.

The potential energy curves of Z configurations are shown in Fig. 9.10. The most repulsive configuration is Z2 with $\alpha = 0$; here the hydrogen of HF and one of the hydrogens of water strongly interact and destabilize the system. The other exclusively repulsive potential is that of Z1 with $\alpha = \pi$; in this case the system is destabilized by the strong interaction between oxygen and fluorine. The Z1 configuration with $\alpha = \pi/2$ has a minimum of 0.1 kJ/mol at 3.2 Å. The other intermediate configurations have an attractive character: Z1 with $\alpha = 0$, Z2 with $\alpha = \pi/2$, and Z2 with $\alpha = \pi$, the minima are, respectively, 11.3 kJ/mol at 3.0 Å, 13.3 kJ/mol at 2.8 Å, and 19.5 kJ/mol at 2.7 Å. The latter is also the most stable configuration among the Z ones; in this configuration the hydrogen–oxygen interaction determines the stabilizing effect of the system.

Interpolation tests were conducted to assess the accuracy of the expansion for non-significant configurations, for which interpolation with ab initio points is not exact. In these tests, the spherical harmonic expansions of three non-significant configurations were compared with their corresponding CCSD(T)/CBS energy points (Fig. 9.11). The first test configuration, Test1 that corresponds to ($\alpha = 116°$, $\theta_a = 105°$, $\theta_b = 77°$, $\phi = 108°$) is purely repulsive, with a deviation of about 8 kJ/mol at 2.7 Å, decreasing to 1 kJ/mol beyond 3.0 Å. The other two test configurations, Test2 with ($\alpha = 172°$, $\theta_a = 82°$, $\theta_b = 193°$, $\phi = 165°$) and Test3 corresponding to ($\alpha = 332°$, $\theta_a = 12°$, $\theta_b =$

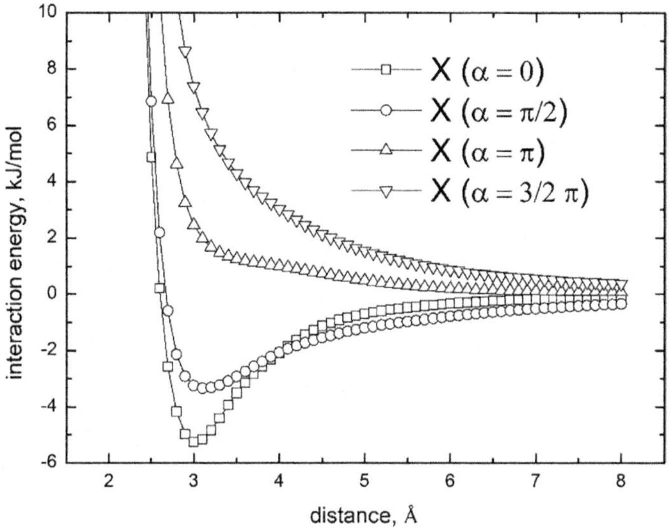

Fig. 9.9 The potential energy profiles of the crossed configurations X for $\alpha = 0$, $\pi/2$, π, and $3/2\ \pi$

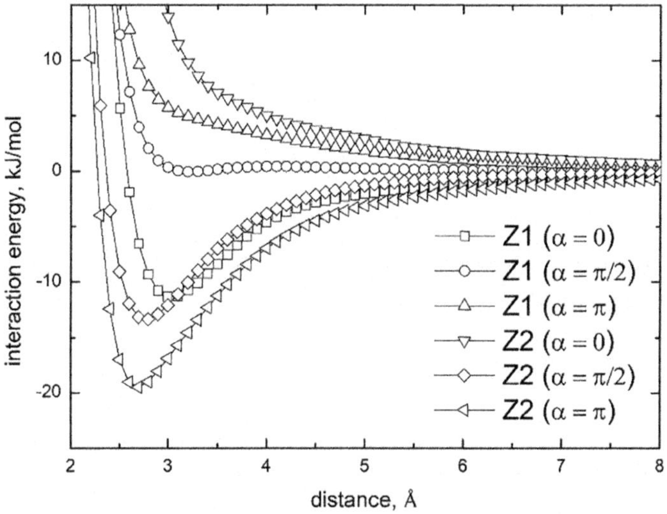

Fig. 9.10 The potential energy profiles of the intermediate configurations X for $\alpha = 0$, $\pi/2$, and π

25°, $\phi = 46°$) have attractive components and show deviations below 1 kJ/mol in the region of the potential well. The interpolation test shows that the spherical expansion is an accurate method. Such accuracy is excellent to reproduce the interaction energy in the equilibrium distance range.

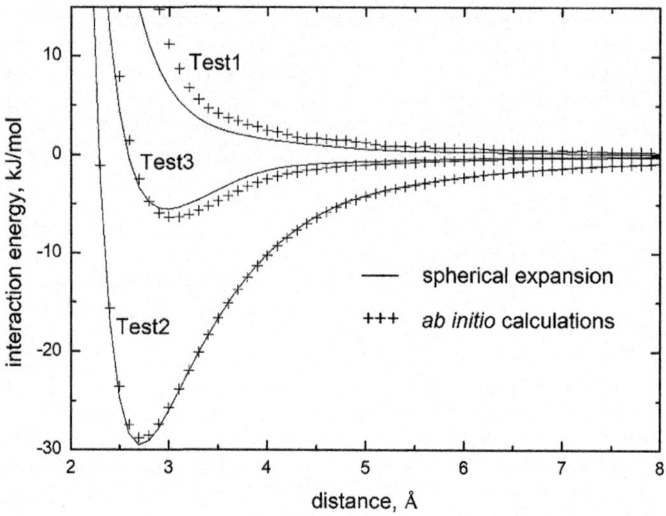

Fig. 9.11 Interaction energy as a function of the distance for three non-significant configurations: Test1, Test2, and Test3. For each configuration the potential energy curve obtained by spherical expansion is compared to that obtained by ab initio calculation

Finally, Fig. 9.12 shows that the isotropic component of the potential is obtained by averaging over the significant configurations. This term is very significant, since it can be experimentally measured through integral cross section measurements as a function of velocity. The isotropic minimum is approximately 1.1 kJ/mol at 3.3 Å.

9.4 Case Study: Propylene Oxide–He by Molecular Beam Experiment

The study of intermolecular interactions between polyatomic molecules, such as propylene oxide, and simple atoms, such as helium, is fundamental in many research areas because it enables the evaluation of the spectroscopic and dynamic properties of complex systems formed through collisions in gaseous environments.

Here, we report a study of the interaction of propylene oxide with rare gas atoms and focus on the case example of He. The multidimensional potential energy surface (PES) was analyzed. The PES is determined by anisotropic non-covalent interactions that define the structure and energy of the weakly bound adducts between propylene oxide and helium. Propylene oxide is one of the simplest chiral molecules and has been detected in the interstellar medium [49].

Experimental measurements were carried out using the molecular beam technique to evaluate the total cross section (elastic plus inelastic) of helium scattering on propylene oxide as a function of collision velocity. These measurements, with high

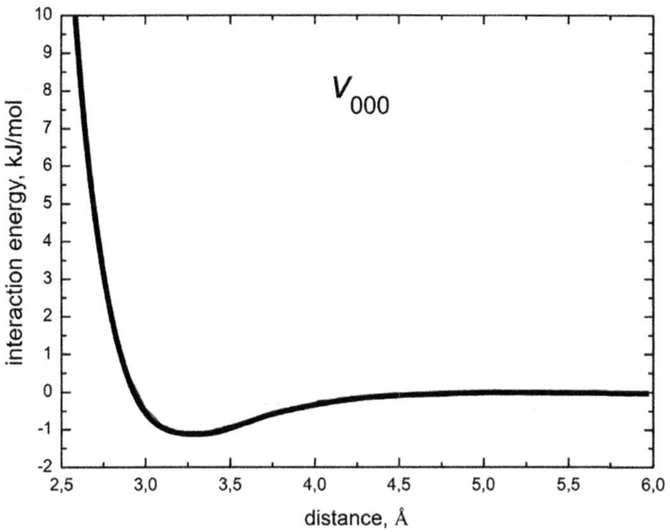

Fig. 9.12 The isotropic component of the interaction potential, v_{000} (Figure adapted from Ref. [45])

angular and velocity resolution, allowed us to observe quantum interference effects ("glory"), which are visible as oscillations in the velocity dependence of the cross section. This work is described in detail in Ref. [30].

Meanwhile, an integrated theoretical-phenomenological method was developed to characterize the strength and distance of the interaction as a function of separation and relative orientation between helium and propylene oxide. This approach enabled the construction of an analytical potential energy surface (PES) defined by the fundamental physical parameters of the interacting particles. The PES was tested against both experimental data and ab initio calculations at the highest levels of theory (CCSD(T)/CBS).

This procedure makes it possible to formulate PESs suitable for molecular dynamics simulations under various conditions of interest, providing a detailed description of the relative configurations between the partners involved.

9.4.1 Experimental Background

The experiments were conducted in the gas phase using the molecular beam (MB) technique. This technique creates high-resolution, single-collision conditions, which are useful for measuring the dependence of the total cross section (Q), including both elastic and inelastic processes, on the velocity (v) of the particles.

The molecular beam, composed of helium (He) atoms, is generated by expanding the gas through a nozzle. The temperature is maintained between 77 and 600 Kelvin

(K) and the pressure between 7 and 20 millibar (mbar) to avoid cluster formation and cover a wide range of collision velocities. Then, the beam is selected by velocity using a mechanical selector and made to collide with the target gas (propylene oxide) at the desired velocity in a scattering chamber. The chamber is maintained at room temperature and low pressure to ensure single-collision events.

The rotational temperature of the target molecules is kept high enough (T = 300 K) so that the average rotation time is comparable to the collision time. This is an important condition for limiting the effects of anisotropy in scattering and for better resolving the frequency and amplitude of quantum interference oscillations ("glory"), at least in the case of He–propylene oxide. A quadrupole mass spectrometer coupled to an ion counting system detects the beam downstream. For each selected velocity, the beam attenuation (I/I_0) is measured, where I is the intensity when the target is present, and I_0 is the intensity when the chamber is empty. Using the Lambert–Beer law, the cross section $Q(v)$, where v is the velocity, is obtained from this ratio.

In order to obtain detailed measurements, it is essential to have projectiles (i.e., He atomic beam) available over a wide range of velocities. Using propylene oxide as a target allows for experiments with high angular and velocity resolution because it is heavier and can be confined at a defined pressure and temperature. In particular, combining light atoms in the beam with heavy molecules in the target improves the resolution of quantum interference effects ("glory"), which are observable as oscillations superimposed on a regular component in $Q(v)$.

The collected experimental results allow for a detailed and consistent investigation of the absolute scale of the interaction at long and medium distances. The measured cross sections are represented as $Q(v)v2/5$ to highlight the quantum interference "glory," which exhibits reduced amplitude in both cases.

Analyzing $Q(v)$ quantitatively characterized the average strength of the intermolecular interaction at long range (based on the dependence of $Q(v)$ on velocity) and in the well region, as probed by the resolved "glory" structure. Center-of-mass cross section values were calculated using the Jeffreys–Wentzel–Kramers–Brillouin semi-classical approximation and were convolved in the laboratory system for direct comparison with experimental data. The potential parameters were optimized using an iterative procedure to best match the experimental and calculated data.

9.4.2 Derivation of Potential Energy Surface

The Improved Lennard–Jones

The pair additivity approach of the improved Lennard–Jones (ILJ) potential was adopted to analytically formulate the intermolecular potential [50]. The total interaction potential is defined as the sum of ILJ contributions between the noble gas helium (He) atom and the seven molecular centers distributed throughout the propylene oxide (C_3H_6O) molecular structure. The selected interaction centers include the asymmetric carbon atom, the carbon atom in the CH_2 group, three hydrogen atoms in the CH and CH_2 groups, the oxygen atom, and the CH_3 methyl group. The methyl group is treated as a structureless partner, i.e., as a sphere with a mass equivalent to that of the methyl group, centered on the carbon atom.

For each pair of interacting atoms, the ILJ potential contribution V depends on the distance between the He atom and the molecular center (i.e., CH_3, the central carbon C, carbon of CH_2, oxygen O, and hydrogens H) the depth of the potential well ε, and the equilibrium distance r_m:

$$V(r_x) = \left(\frac{m}{n(r_x) - m}\left(\frac{r_m}{r_x}\right)^{n(r_x)} - \frac{n(r_x)}{n(r_x) - m}\left(\frac{r_m}{r_x}\right)^{m}\right)$$

where r_x is the variable distance between two centers, i.e., the distance between He and CH_3, C, O, or H. The parameter m takes a value of six for neutral pairs, and the function $n(r_x)$ is defined as $\beta + 4\left(\frac{r_x}{r_m}\right)^2$, where β is related to the hardness of the interacting species and varies between 7.5 and 8.5. The interaction potential between propylene oxide and He is then the sum of the following interacting pairs: (He–CH_3) + (He–O) + (He–central carbon) + (He–C from CH_2 group) + 3 (He–H).

The intermolecular potential is of the van der Waals type and is determined by a balance between steric repulsion and dispersive attraction. Initial values of van der Waals interaction parameters were estimated based on polarizabilities of interacting partners identified as He–CH_3, He–C, He–O, and He–H. The polarizability of He is 0.20 Å^3, while the polarizability of propylene oxide is 6.2 Å^3, divided into partial contributions: 2.10 Å^3 (CH_3), 1.10 Å^3 (C), 0.80 Å^3 (O), and 0.35 Å^3 (H). The C, H, and O atomic centers are considered "effective" because their polarizabilities differ from those of isolated atoms in the gas phase.

The initial values of the potential parameters, which indirectly include small induction effects, were optimized to reproduce the experimental data. These values were then compared with ab initio calculations for the He–propylene oxide system. The parameters obtained for the potential energy surface are reported in Table 9.2.

Table 9.2 Parameters, equilibrium distance r_m, and well depth (i.e., the equilibrium energy), ε, used in the Improved Lennard–Jones model to describe the interaction pair individual contributions in He–propylene oxide system

	Well depth (ε), meV	Equilibrium distance (r_m), Å	*B*
CH_3–He	1.75	3.81	7.5
C–He	1.76	3.45	7.5
O–He	2.83	3.14	7.5
H–He	1.32	2.93	8.5

Cross Sectional Analysis and Potential Parametrization

In dynamic data analysis, energy barriers associated with noble gas atom orbital motion around propylene oxide molecules are compared to the molecule's average rotational energy, expressed as *RT* (where *T* is temperature and *R* is the gas constant). For the helium-propylene oxide system, the predicted energy barriers are lower than *RT* at an experimental temperature of 300 K, even within the potential well region. This implies that the barriers are low enough to prevent the system from becoming confined in rigid configurations. Under these conditions, long interaction times due to slow collisions favor dynamic behavior dominated by partial or total averaging of interaction orientations. For this reason, the measured cross sections were analyzed under the assumption that diffusion is governed by an effective central potential at low relative velocities$v$. Conversely, at high speeds, collision times are reduced, and the anisotropy of the potential energy surface (PES) becomes more relevant. This allows for direct control of collision dynamics as the molecules are "seen" by the projectile atoms in sudden conditions.

Therefore, the cross sections at lower speeds were analyzed using an effective isotropic interaction obtained as the spherical average of the PES. At higher speeds, however, the infinite order sudden approximation (IOSA) was adopted [50]. The transition between the two regimes, which occurs at intermediate velocities, is managed by a weighted combination of the two approaches depending on the speed. This method provides excellent agreement with the experimental data over the entire range of v analyzed. Furthermore, in less anisotropic cases and with rotationally hot molecules (i.e., at higher temperatures), a simple isotropic potential is sufficient to describe collisions over the entire velocity range.

Details on Ab Initio Calculations for the Propylene Oxide–He System

Calculations were performed at the MP2 and CCSD(T) levels of theory using the Gaussian09 software to determine the minimum geometries and binding energies of helium–propylene oxide complexes. The initial structures submitted for minimization were derived from the minima obtained through analytical ILJ potential energy surfaces (PESs). A preliminary optimization, in which the positions of the hydrogen atoms in the methyl group of propylene oxide were considered flexible, was carried out at the MP2/aug-cc-pVDZ level of theory to define the orientation of the methyl group, treated as a pseudo-atom in the analytical PES.

Subsequently, for the He–propylene oxide complex (with the propylene oxide molecule considered rigid), optimizations were performed at the MP2/aug-cc-pVnZ levels ($n = 3, 4, 5$) to extrapolate the complete basis set (CBS) limit using a mixed exponential/Gaussian function.

The CCSD(T) complete basis set limit energy, $E^{CBS}_{CCSD(T)}$, was estimated starting from the MP2 CBS energy, E^{CBS}_{MP2}, and adding a correction $\Delta_{CCSD(T)}$, calculated as the difference between CCSD(T) and MP2 energies using the aug-cc-pVTZ basis set.

It was observed that, for many weakly bound complexes, higher-order correlation effects are only slightly sensitive to the basis set size, provided that diffuse functions are included and the basis is not too small.

9.4.3 Potential Energy Surface Features of Propylene Oxide–He

The potential energy surface (PES) of the helium–propylene oxide system features five minima, whose geometry is described in an *xyz* Cartesian reference frame centered at the molecule's center of mass. The positions of the helium atom relative to propylene oxide are expressed in spherical coordinates (R_{CM}, θ, ϕ), where R_{CM} is the distance from the center of mass, and θ and ϕ are the polar angles (Fig. 9.13).

The five minima, obtained from the ILJ potential, are reported in Table 9.3 with their corresponding coordinates and interaction energies, and compared with results from ab initio calculations at the CCSD(T)/CBS level and with those from a previous study [51]. The differences between the current values and those from the reference are minimal (less than 0.5 meV in energy and 0.09 Å in distance), likely due to a different orientation of the methyl group.

The interaction energies of the five minima range from −9.79 to −7.62 meV. The global minimum shows geometry and energy very similar to the ab initio data, with helium interacting mainly with the oxygen and the chiral carbon atom of the ring in a perpendicular fashion. The other four minima also exhibit similar geometries, although the energy ordering is not always the same, with the ILJ potential generally yielding slightly deeper minima.

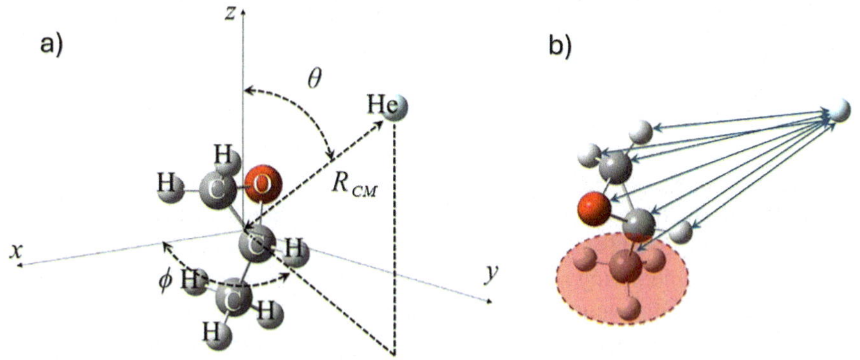

Fig. 9.13 **a** Cartesian coordinate reference frame for the He–propylene oxide system, where R_{CM} is the distance between He and the center-of-mass of the molecule, while θ andϕ are spherical angles. **b** The blue lines with double arrows indicate the binary interaction terms in the improved Lennard–Jones model, the variable distance between He and the centers in propylene oxide is indicated by r_x (Figure adapted from Ref. [30])

Table 9.3 Minimal geometries and related potential energies for helium–propylene oxide interactions obtained using the ILJ potential energy surface (bold), through ab initio calculations described here (Times) and ab initio calculations described in [51] (parentheses)

	Minimum 1	Minimum 2	Minimum 3	Minimum 4	Minimum 5
θ	**49.4** 49.3 (50.0)	**90.4** 104.1 (102.0)	**91.6** 100.9 (100.0)	**62.9** 77.2	**32.6** 25.1 (26.3)
ϕ	**110.4** 104.1 (104.0)	**36.7** 26.5 (1.7)	**148.6** 153.1 (154.0)	**256.1** 252.4	**256.5** 254.0 (245.9)
R_{CM} (Å)	**3.591** 3.528 (3.493)	**3.614** 3.675 (3.605)	**3.503** 3.357 (3.356)	**3.550** 3.564	**3.894** 3.844 (3.935)
V (meV)	**–9.79** –9.67 (–9.72)	**–8.54** –6.47 (–6.38)	**–8.49** –8.91 (–8.45)	**–8.28** –6.79	**–7.62** –7.10 (–6.86)

Such discrepancies, below 2 meV (*ca.* 0.02 kJ/mol), are negligible compared to the accuracy of ab initio calculations. Since the binding energies of the minima are very close, the ranking of the well depths may vary depending on the method used. The ILJ potential and the current ab initio calculations identify five minima on the PES, whereas the previous reference reports only four. The fifth minimum (here labeled as minimum 4) is associated with helium interacting with the CH_2 group and, partially, with the oxygen, in a manner similar to minimum 5.

Contour maps of the PES at $R_{CM} = 3.6$ and 4.0 Å (Figs. 9.14 and 9.15) show that the angle ϕ can vary from 0 to 360°, while θ has a more restricted range, with repulsive contributions predominant beyond 140°. The minima are generally found around the oxirane ring, with θ values between 50 and 100°, as also shown in the energy profiles.

The minimum energy path (MEP) is plotted as a red dashed line on the contour maps. The MEP allows one to estimate the lowest barriers between the minimum configurations of the collision complex, providing insight into the possibility of transitions between these configurations at the relevant temperature.

In the case of $R_{CM} = 3.6$ Å (Fig. 9.14), three barriers of approximately 4–5 meV are observed, which are much lower than the rotational energy RT available at 300 K (equal to 25.8 meV). This allows an almost free rotation of helium around the propylene oxide molecule during slow collisions at room temperature. Conversely, under low or ultralow temperature conditions, such as those in the interstellar medium (ISM), where temperatures can drop below 40 K (e.g., in the Sgr B2(N) cloud), the barriers become higher than RT. At 10 K, RT is less than 1 meV, suggesting effective trapping of the complex in the potential wells.

At a greater interaction distance, $R_{CM} = 4.0$ Å (Fig. 9.15), the MEP becomes smoother, with lower barriers between minima, facilitating transitions between configurations. This is confirmed by the energy profiles (see Supplementary Material in [30]), which show how energy varies with θ and ϕ as R_{CM} increases. Additionally,

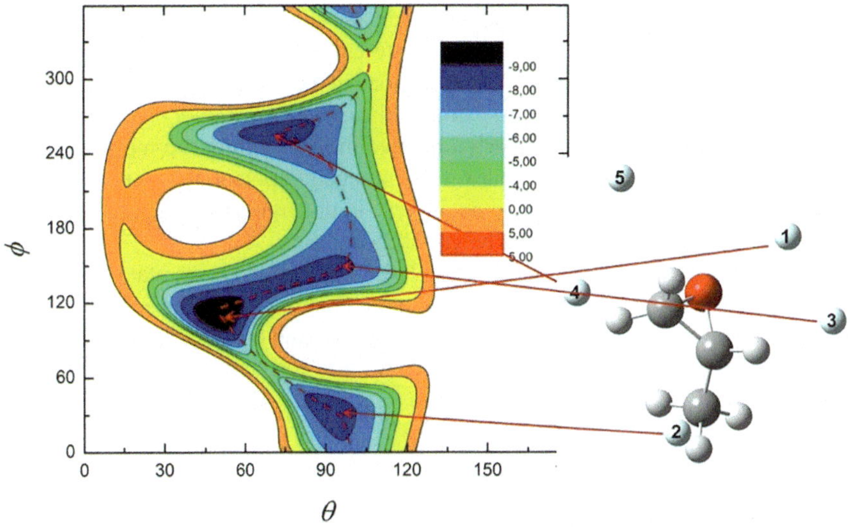

Fig. 9.14 The figure shows the color contour map for the helium–propylene oxide system at $R_{CM} = 3.6$ Å. The axes report the angles θ and ϕ in degrees, and the color scale reports the potential energy in meV. The red dashed line indicates the minimum energy path. In the bottom right are the minimum geometries for the He-propylene oxide system identified on the ILJ PES. The He atom numbering corresponds to the minima reported in Table 10.3. Figure adapted from Ref. [30]

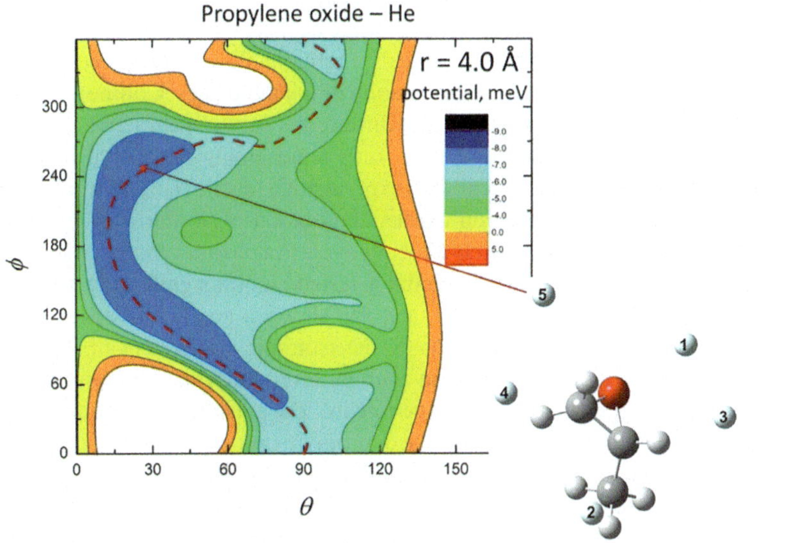

Fig. 9.15 The figure shows the color contour map for the helium–propylene oxide system at $R_{CM} = 4.0$ Å. The angles θ and ϕ are reported on the axes in degrees, and a color scale indicates the potential energy in meV. The red dashed line shows the minimum energy path. The He atom numbering corresponds to the minima reported in Table 9.3. Figure adapted from Ref. [30]

the relative depth of the wells changes with distance, and the global minimum may correspond to a different geometry (i.e., different θ and ϕ values) as R_{CM} varies. These observations have interesting implications in the low-temperature regime: the anisotropic (orientation-dependent) nature of the interaction, rather than the absolute depth of the well, can be particularly effective in determining the most populated configuration of the collision complex. Once trapped in potential minima at long or intermediate range, the system may not be able to undergo a transition to the global minimum at short range due to the height of the barriers.

The molecular beam technique has also been employed to investigate the interaction potential of the propylene oxide–Ne system. Phenomenological analysismade it possible to evaluate the parameters of the ILJ model and construct potential energy surface of the whole series of rare gases (Ar, Kr, Xe, and Rn) interacting with propylene oxide [52].

9.5 Final Remarks

9.5.1 *Spherical Harmonic Expansion of the H_2O–HF Interaction Potential*

The H_2O–HF potential energy surface has been investigated using quantum-chemical tools to compute single energy points that were expanded in spherical harmonics. The analysis of the configurations has reveals a remarkable variety of energetic behaviors, closely linked to the mutual orientation of the molecules. Collinear and perpendicular configurations show significant differences between attractive and repulsive interactions, with configuration T2 ($\alpha = \pi$) representing the global minimum, in agreement with both theoretical and experimental data. The potential expansion in spherical harmonics relies on exact interpolation of ab initio energy points of significant configurations, which are employed to construct the interaction potential expression. The expansion potential is tested on non-significant configurations, for which the interpolation is not exact, and shows a good consistency within the potential well region. The inclusion of the isotropic component of the potential, obtained as an average over the main configurations, provides a measurable parameter through integral cross section experiments. The anisotropic components, related to multipolar interactions, complete the descriptive framework of the physical contributions to the interaction potential. Overall, this study offers a detailed and coherent representation of the energy landscape of the H_2O–HF system. The spherical representation provides a compact formulation and a complete account of the symmetry properties of the interaction potential, useful for understanding molecular interaction mechanisms and guiding future theoretical and experimental investigations.

9.5.2 Mapping the Propylene Oxide–Helium Potential Energy Surface from Molecular Beam and Quantum Chemical Approaches

The propylene oxide–He interaction potential has been studied using molecular beams, measuring the total integral cross section, which allows one to derive the potential energy surface based on an improved version of the Lennard–Jones model, refined through ab initio calculations. The detailed characterization of the potential energy surface (PES) for the helium–propylene oxide system reveals five distinct minima, whose geometries and interaction energies have been thoroughly analyzed using both ILJ analytical potentials and high-level ab initio methods. The close agreement between these approaches, with discrepancies well below the threshold of computational accuracy, confirms the reliability of the ILJ model in describing weak intermolecular interactions. The contour maps and minimum energy paths (MEPs) provide valuable insight into the dynamic behavior of the collision complex, particularly under varying temperature regimes. At room temperature, low-energy barriers allow for nearly unrestricted rotational motion of helium around the propylene oxide molecule. In contrast, under cold and ultracold conditions—such as those found in the interstellar medium—these barriers become significant, leading to effective trapping within local potential wells. Furthermore, the dependence of well depth and geometry on the interaction distance highlights the importance of anisotropic effects in determining the most probable configurations of the complex. These findings underscore the role of orientation-dependent interactions in low-temperature environments, where barrier heights may prevent transitions to the global minimum, thereby influencing the distribution and stability of molecular complexes. This study offers a robust framework for understanding van der Waals interactions in weakly bound systems and provides a foundation for future investigations into molecular dynamics under astrophysical and cryogenic conditions. Finally, phenomenological analysis allows the extension of the study to other systems interacting with propylene oxide, for example the entire series of rare gases, or diatomic molecules, or even the dimer propylene oxide—propylene oxide.

References

1. Kumar S, Nussinov R. Chem Rev. 2022;122:4921.
2. Matienko LI, Mil EM, Albantova AA, Goloshchapov AN. Int J Mol Sci. 2023;24:16874.
3. Marques JMC, Prudente FV, Pirani F. Molecules. 2022;27:3072.
4. Salentin S, Schreiber S, Haupt VJ, Adasme MF, Schroeder M. Front Pharmacol. 2022;13: 879731.
5. Zhang T, Li M. Molecules. 2023;28:3072.
6. Zhao Q, Xu Y. Chem Rev. 2024;124:1456.
7. Magnasco V, McWeeny R. In: Theoretical treatment of large molecules and their interactions. Springer;1990. p. 145.

8. Griffiths DJ. Introduction to electrodynamics. 4th ed. Cambridge: Cambridge University Press; 2017.
9. Parigi G, Luchinat C. In: Paramagnetism in experimental biomolecular NMR. Cambridge: Royal Society of Chemistry;2018. p. 45.
10. Buckingham AD. Theory of intermolecular forces. 2nd ed. Oxford: Pergamon Press; 1971.
11. Chen L, Wang J. J Chem Phys. 2024;160: 024501.
12. Huang T, Petek ES, Katsumata R. JACS Au. 2025;5:774.
13. Kim S, Park J. Nat Commun. 2024;15:1123.
14. Zhang Y, Liu B. IEEE Trans Comput Biol Bioinform. 2024;21:845.
15. Wang X, Zhou H. Mater Chem Front. 2023;7:188.
16. Pirani F, Cappelletti D, Liuti G. Chem Phys Lett. 2001;350:286.
17. Pirani F, Brizi S, Ronchetti F, Casavecchia P, Cappelletti D, Vecchiocattivi F. Phys Chem Chem Phys. 2008;10:5489.
18. Grimme S, Schreiner PR. Angew Chem Int Ed. 2023;62: e202316364.
19. Cavallo G, Politzer P. Chem Sci. 2023;14:2348.
20. Szalewicz K, Patkowski K, Jeziorski B. Intermolecular forces and clusters II. 2005. p. 43.
21. Müller C, Reuter K. J Am Chem Soc. 2025;147:18150.
22. Chen CJ, Wiesendanger R. Phys Rev B. 2006;74: 113102.
23. Vargas R, Garza J, Martínez A, Ibarra IA. Chem Commun. 2024;60:3471.
24. Hermann J, DiStasio RA Jr, Tkatchenko A. Chem Rev. 2017;117:4714.
25. Ambrosetti A, Tkatchenko A. Nat Rev Phys. 2021;3:861.
26. Al-Hamdani YS, Tkatchenko A. J Chem Phys. 2019;150: 010901.
27. Montgomery JM, Herbert JM. Annual reports in computational chemistry, vol. 20. Elsevier;2024. p. 1.
28. Li W, Liu J, Li L, Hu L, Su ZM, Chen G. In: Computational materials, chemistry, and biochemistry: from bold initiatives to the last mile, Springer series in materials science, vol. 284. Cham: Springer;2021. p. 187.
29. Caglioti C, Palazzetti F. Chem Phys Lett. 2021;776: 138692.
30. Palazzetti F, Cappelletti D, Coletti C, Falcinelli S, Pirani F. J Chem Phys. 2021;155: 234301.
31. Stone AJ. The theory of intermolecular forces. Oxford: Clarendon Press; 1996.
32. Hobza P, Müller-Dethlefs K. Non-covalent interactions: theory and experiment. Cambridge: Royal Society of Chemistry; 2010.
33. Boys SF, Bernardi F. Mol Phys. 1970;19:553.
34. van Duijneveldt FB, van Duijneveldt-van de Rijdt JGCM, van Lenthe JH. Chem Rev. 1994; 94:1873.
35. Jurečka P, Hobza P. J Chem Phys. 2002;116:156.
36. Jurečka P, Šponer J, Černý J, Hobza P. Phys Chem Chem Phys. 2006;8:1985.
37. Stone AJ. In: Dobson JF, Rubio A, Cohen AJ editors. London dispersion forces in molecules, solids and nanostructures. Cambridge: Royal Society of Chemistry;2020. p. 98
38. Lombardi A, Palazzetti F, Peroncelli L, Grossi G, Aquilanti V, Sevryuk MB. Theor Chem Acc. 2007;117:709.
39. Palazzetti F, Munusamy E, Lombardi A, Grossi G, Aquilanti V. Int J Quantum Chem. 2011;111:318.
40. Barreto PRP, Cruz ACPS, Barreto RLP, Palazzetti F, Albernaz AF, Lombardi A, Maciel GS, Aquilanti V. J Mol Spectrosc. 2017;337:163.
41. Barreto PRP, Ribas VW, Palazzetti F. J Phys Chem A. 2009;113:15047.
42. Barreto PRP, Albernaz AF, Capobianco A, Palazzetti F, Lombardi A, Aquilanti V. Comput. Theor Chem. 2012;990:53.
43. Barreto PRP, Vilela AFA, Lombardi A, Maciel GS, Palazzetti F, Aquilanti V. J Phys Chem A. 2007;111:12754.
44. Maciel GS, Barreto PRP, Palazzetti F, Lombardi A, Aquilanti V. J Chem Phys. 2008;129: 164302.
45. Caglioti C, Palazzetti F. J Chin Chem Soc. 2023;70:1435.

46. Frisch MJ, Trucks GW, Schlegel HB, Scuseria GE, Robb MA, Cheeseman JR, Scalmani G, Barone V, Mennucci B, Petersson GA, Nakatsuji H, Caricato M, Li X, Hratchian HP, Izmaylov AF, Bloino J, Zheng G, Sonnenberg JL, Hada M, et al. Gaussian 09, Revision D.01. Gaussian, Inc.:Wallingford CT;2009.
47. Wigner EP. Group theory and its application to the quantum mechanics of atomic spectra. New York: Academic; 1959.
48. van der Avoird A, Wormer PES, Moszynski R. Chem Rev. 1994;94:1931.
49. McGuire BA, Carroll PB, Loomis RA, Finneran IA, Jewell PR, Remijan AJ, Blake GA. Science. 2016;352:1449.
50. Pirani F, Brizi S, Roncaratti LF, Casavecchia P, Cappelletti D, Vecchiocattivi F. Phys Chem Chem Phys. 2008;10:5489.
51. Faure A, Dagdigian PJ, Rist C, Dawes C, Qintas-Sànchez E, Lique F, Hochlaf M. ACS Earth Space Chem. 2019;3:964.
52. Palazzetti F, Coletti C, Marrone A, Pirani F. Symmetry. 2022;14:249.

Part IV
Computational Biology

Chapter 10
Encouragement of Canonical Molecular Orbital Calculations for Proteins

Toshiyuki Hirano and Fumitoshi Sato

Abstract Protein electronic structure calculations, particularly Canonical Molecular Orbital (CMO) computations, are essential for understanding protein function, interactions, and reaction mechanisms. However, such calculations are computationally demanding due to the large size and complexity of proteins. This chapter reviews and promotes the use of CMO calculations for proteins by introducing advanced methods including Density Functional Theory (DFT) with hybrid functionals, the third-generation DFT method employing Cholesky decomposition, and the Quasi-Canonical Localized Orbital (QCLO) technique for efficient initial guesses. The ProteinDF program designed for large-scale protein DFT calculations and the QCLObot orchestration tool that automates initial guess generation are described. Strategies for handling metallo-proteins, memory management, and computational optimization for scalability are presented. Applications such as molecular orbital analysis, electrostatic potential mapping, energy density analysis, and molecular orbital overlap methods are demonstrated using a SARS-CoV-2 spike protein model. This work aims to guide researchers in tackling CMO calculations for large biomolecules and to facilitate advancements in theoretical understanding and drug design.

Keywords Computer chemistry · Quantum chemistry · Canonical molecular orbital · Density functional theory · High-performance computing · Parallel computing · Quasi-canonical localized orbital (QCLO) · ProteinDF · QCLObot · Protein electronic structure

T. Hirano (✉)
Tokyo Metropolitan College of Industrial Technology, Tokyo, Japan
e-mail: hirano-t@metro-cit.ac.jp

T. Hirano · F. Sato
Institute of Industrial Science, The University of Tokyo, Tokyo, Japan
e-mail: satofumi@iis.u-tokyo.ac.jp

T. Onishi (ed.), *Advances in Quantum Science II*,
https://doi.org/10.1007/978-981-95-9721-5_10

10.1 Introduction

Most proteins are composed of hydrogen, carbon, nitrogen, oxygen, sulfur, and other light elements. These biomolecules display a wide variety of functions and physical properties, including forming muscles that generate significant power with minimal energy, acting as enzymes that metabolize chemicals, serving as sensors for light and chemicals, transmitting information as hormones, and transporting and storing compounds and ions. Proteins are indispensable to living organisms. Despite being made from common light elements, unlike the rare metals found in sophisticated electronic devices, proteins exhibit diverse physical properties in living systems. This suggests that the factors contributing to their varied physical traits stem not only from the types of atoms they contain, but also from their intricate three-dimensional structures and sizes. The steric structure of a protein is a crucial determinant of its function. The Protein Data Bank [1], an international repository of three-dimensional structures of biomolecules, including proteins, plays a vital role in both basic and applied research in the life sciences.

However, quantum chemical calculations that simulate the electronic structure of materials have become essential and practical tools for enhancing theoretical understanding in chemical research. These calculations support theoretical chemistry in tackling complex problems that are difficult to resolve through experimental methods alone, such as elucidating reaction mechanisms and interpreting various spectra. In particular, frontier molecular orbital theory has demonstrated that molecular orbitals such as the Highest Occupied Molecular Orbital (HOMO) and the Lowest Unoccupied Molecular Orbital (LUMO) significantly contribute to our understanding of seemingly inexplicable reaction mechanisms. This theory has helped clarify numerous subsequent chemical reactions. With advancements in quantum chemical calculation techniques and the development of software such as Gaussian [2] and GAMESS [3], as well as the increasing accessibility of larger, faster, and more affordable computers, quantum chemical calculations can now be executed on personal computers for common small molecules. The effects of these calculations on chemical research have been significant.

Why is the three-dimensional (3D) structure of a protein crucial for its physical properties? The answer lies in the relationship between a protein 3D atomic arrangement (i.e., molecular coordinates) and its electronic structure (i.e., the distribution and configuration of electrons), both of which are essential for understanding its physical and chemical properties. Accurate electronic structure calculations of proteins hold great promise for elucidating reaction mechanisms, analyzing interactions, correlating structure with function, designing new molecules, and developing drugs. Although it seems straightforward to calculate protein electronic structure, the reality is more complex. As detailed in a later section, the computational effort required for electronic structure calculations increases with the size of the system to the fourth power. In other words, if the system size increases by a factor of 10, the computational load increases by a factor of 10,000. This exponential growth in computational complexity poses a significant challenge when calculating the electronic structure of proteins.

Many efforts have been made to determine the electronic structure of proteins. A common approach is to divide the computational domain so that the rapid growth of quantum chemical calculations can be mitigated. The Quantum Mechanics/Molecular Mechanics (QM/MM) method [4], which earned a Nobel Prize in 2013, along with its derivative, our Own N-layered Integrated molecular Orbital and Molecular mechanics (ONIOM) method [5], is widely used to assess protein energies. Moreover, the fragment molecular orbital method [6] and divide-and-conquer approach [7] have gained traction due to advancements in computational technology and software. However, understanding the electronic structure of proteins requires obtaining molecular orbitals (or wave functions). The Canonical Molecular Orbital (CMO) calculation is the most common method for acquiring these molecular orbitals. The CMOs ϕ_i are defined as the orthonormal molecular orbitals obtained as eigenfunctions of the Fock operator $\widehat{F}$. In other words, they are the orbitals obtained as solutions of the eigenvalue problem in the canonical form of the Hartree–Fock equation, where the corresponding eigenvalues ϵ_i represent the orbital energies. Mathematically, they can be expressed by the following eigenvalue Eq. (10.1).

$$\widehat{F}\phi_i = \epsilon_i \phi_i. \tag{10.1}$$

These CMOs are mutually orthogonal and normalized and are characterized as delocalized orbitals extending over the entire molecule.

To derive the electronic structure of a protein, one must perform CMO calculations directly while navigating the complexities of increasing computational demands. Currently, many quantum chemical calculations using quantum computers require the CMO as an initial reference wave function. This study outlines our current progress in CMO calculations for proteins. The aim is to provide guidance and insights for those interested in conducting these types of calculations rather than merely presenting a comprehensive collection of results.

The remainder of the paper is organized as follows. First, we introduce a method for calculating the electronic structure of proteins. Next, we present the process for predicting the physical properties of proteins based on the results of these electronic structure calculations. Finally, we provide examples of applications utilizing protein electronic structure calculations and property predictions. We insist this study will assist many brave individuals embarking on the journey of protein CMO calculations.

10.2 Calculation Method for the Electronic Structure of Proteins

10.2.1 Schematic of the Electronic Structure Calculation Based on Density Functional Theory (DFT)

Formulation and the Necessity of the Self-consistent Field (SCF) Calculation

Here, we outline the fundamental equations of the Kohn–Sham–Roothaan method [8, 9], which is the standard approach for determining electronic states using Density Functional Theory (DFT). This method is analogous to the Hartree–Fock–Roothaan method [9–11] used in molecular orbital theory. The Roothaan method represents molecular orbitals as a linear combination of basis functions. Commonly, the ith molecular orbital $\phi_i(\boldsymbol{r})$ is represented by a linear combination of the expansion coefficients C_{pi} of the basis functions $g_i(\boldsymbol{r})$.

$$\phi_i(\boldsymbol{r}) = \sum_{p}^{N_{\mathrm{AO}}} C_{pi} g_i(\boldsymbol{r}). \tag{10.2}$$

Slater-type, Gaussian-type, and plane-wave functions are often used as basis functions. The expansion coefficient C_{pi} is also called the Linear Combination of Atomic Orbitals (LCAO) coefficient. The problem of finding the LCAO coefficient C_{pi} and the corresponding molecular orbital energy ε_i is given by the Kohn–Sham–Roothaan equation. For closed-shell systems, the Kohn–Sham–Roothaan equation can be expressed as

$$\boldsymbol{FC} = \boldsymbol{SC\varepsilon} \tag{10.3}$$

where $\boldsymbol{\varepsilon}$ and $\boldsymbol{C}$ are the eigenvalues and eigenvectors obtained by diagonalizing the Kohn–Sham matrix $\boldsymbol{F}$, respectively, and $\boldsymbol{S}$ is the matrix of overlap integrals.

$$S_{pq} = \langle pq \rangle = \int g_p^*(\boldsymbol{r}) g_q(\boldsymbol{r}) \mathrm{d}\boldsymbol{r}. \tag{10.4}$$

The Kohn–Sham matrix elements of matrix $\boldsymbol{F}$ are represented as

$$F_{pq} = h_{pq} + J_{pq} + \alpha K_{pq} + \beta \mu_{pq}^{XC} \tag{10.5}$$

$$h_{pq} = \int g_p^*(\boldsymbol{r}) \left[-\frac{1}{2}\Delta - \sum_{A}^{N_{\mathrm{atom}}} \frac{Z_A}{|\boldsymbol{r} - \boldsymbol{R}_A|} \right] g_q(\boldsymbol{r}) \mathrm{d}\boldsymbol{r} \tag{10.6}$$

$$J_{pq} = \sum_{rs} P_{rs} \langle pq|rs \rangle \tag{10.7}$$

$$K_{pq} = -\frac{1}{2} \sum_{rs} P_{rs} \langle pr|qs \rangle \tag{10.8}$$

$$\mu_{pq} = \int g_p^*(\boldsymbol{r}) \mu^{XC} g_q(\boldsymbol{r}) \mathrm{d}\boldsymbol{r} \tag{10.9}$$

where $\boldsymbol{h}, \boldsymbol{J}, \boldsymbol{K}$, and $\boldsymbol{\mu}$ are the one-electron, Coulomb, Fock exchange, and exchange–correlation terms, respectively. The four-center two-Electron Repulsive Integral (ERI), $\langle pq|rs \rangle$, and the density matrix are expressed by

$$\langle pq|rs \rangle = \iint g_p^*(\boldsymbol{r}_1) g_q(\boldsymbol{r}_1) \frac{1}{|\boldsymbol{r}_1 - \boldsymbol{r}_2|} g_r^*(\boldsymbol{r}_2) g_s(\boldsymbol{r}_2) \mathrm{d}\boldsymbol{r}_1 \mathrm{d}\boldsymbol{r}_2. \tag{10.10}$$

$$P_{pq} = 2 \sum_{i} C_{pi}^* C_{qi}. \tag{10.11}$$

Equation (10.5), in which α and β are coefficients, represents the Hartree–Fock method when $\alpha = 1$ and $\beta = 0$, whereas it represents the pure density functional method when $\alpha = 0$ and $\beta = 1$. When $\alpha \neq 0$ and $\beta \neq 0$, it is the hybrid density functional method and the current mainstream approach.

Equation (10.5) shows that the Kohn–Sham matrix $\boldsymbol{F}$ is obtained from the coefficient matrix $\boldsymbol{C}$. Therefore, $\boldsymbol{F}$ must be solved iteratively until the coefficient matrix $\boldsymbol{C}$ and the eigenvalue problem match. This procedure is known as the Self-Consistent Field (SCF) calculation. The specific calculation flow is as follows (Fig. 10.1). First, the Kohn–Sham matrix $\boldsymbol{F}$ is derived from the initial wave function $\boldsymbol{C}^{\mathrm{in}}$ (or density matrix $\boldsymbol{P}^{\mathrm{in}}$) as the initial guess. The eigenvalue problem is then solved to obtain the coefficient matrix $\boldsymbol{C}^{\mathrm{out}}$ as its solution. If $\boldsymbol{C}^{\mathrm{in}}$ and $\boldsymbol{C}^{\mathrm{out}}$ differ, the Kohn–Sham matrix $\boldsymbol{F}$ is created by updating the input $\boldsymbol{C}^{\mathrm{in}}$ again, diagonalizing, and repeating until the coefficient matrices $\boldsymbol{C}$ used for the input and output match. This SCF calculation continues until convergence is achieved, as indicated by (2.2). This iterative process accounts for a significant portion of the time required in electronic structure calculations.

Computational Cost and Efficiency Strategy for the Electronic Structure Calculation

To reduce the SCF computation time, we can reduce the computation time per SCF (i.e., the time to create the Kohn–Sham matrix $\boldsymbol{F}$). Another approach is to reduce the number of SCF iterations. Table 10.1 summarizes the contents of this section.

The most computationally expensive aspect of generating the Kohn–Sham matrix $\boldsymbol{F}$ from a computer science perspective is the ERI of (2.9). Various efforts have been

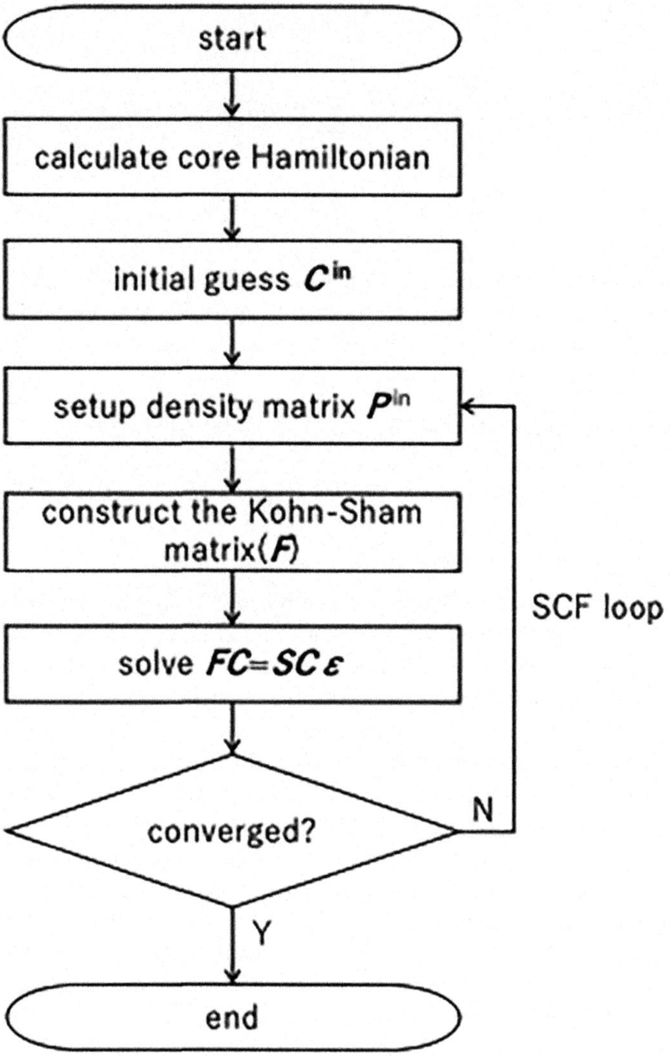

Fig. 10.1 Schematic flow of the Self-Consistent Field (SCF) calculation procedure

made to enhance computational efficiency in popular electronic structure calculation programs such as Gaussian [2] and GAMESS [3]. These improvements include using Gaussian-type functions as basis functions to simplify computations, and employing recursive algorithms, such as the Obara–Saika method [12, 13], to increase the efficiency of integral calculations. However, the task of processing quadruple-loop integral calculations is more complex than one might expect. The duration of the loop is determined by the total number of basis functions N_{AO}. As a result, the computational complexity of the four-center two-electron repulsive integral, which is a

Table 10.1 Two main strategies to reduce SCF time

Approach	Method	Advantages/strengths	Limitations/ drawbacks
Reducing the cost per SCF iteration	ERI computation improvement (DISK method, Direct method, Update method, recursive algorithms such as Obara–Saika)	Faster evaluation of two-electron integrals, reduced memory, or disk requirements	Poor scalability for large systems; difficulties in parallel load balancing
	Reducing formal scaling (symmetry consideration, cutoff techniques, linear-scaling methods)	Lower computational scaling (from $\mathcal{O}(N_{AO}{}^4)$ to $\mathcal{O}(N_{AO}{}^{2\sim3})$, or even $\mathcal{O}(N_{AO})$ in special cases)	Applicability often restricted to specific system types or approximations
Reducing the number of SCF iterations	Extrapolation techniques	Effective in the near-convergent region, helps avoid unnecessary iterations	Limited effectiveness far from convergence
	Improved initial guess for the density matrix	Reduces total iteration count; helps avoid divergence or non-convergence	The quality of the guess becomes critical for large-scale systems

quadruple loop, is formally $\mathcal{O}(N_{AO}{}^4)$. In this context, the "$\mathcal{O}$" in $\mathcal{O}(N_{AO}{}^4)$ refers to the Landau symbol (Big O notation), which mathematically describes the upper bound of algorithmic growth rate. Thus, increasing the computational scale by a factor of 10 results in a computational complexity that is 10,000 times greater. Even when accounting for the symmetry of the integral and further reducing complexity using mathematical cutoff methods, the computational complexity is still approximately $\mathcal{O}(N_{AO}{}^{2\sim3})$. Linear scale methods with computational complexity of approximately $\mathcal{O}(N_{AO})$ have also been studied, but their applicability is often limited. Consequently, the computational complexity of the ERIs continues to pose challenges for large systems.

If SCF iterations require numerous computationally intensive ERI calculations, reusing the results of these calculations must be considered. Historically, when Central Processing Unit (CPU) processing speeds were slow, the results of molecular integrals were stored on external disks for later reuse. This approach, known as the DISK method, is recognized as the first generation of ERI calculation methods. Although the DISK method worked well for small molecules, it posed challenges as the scale of calculations grew. The results of ERIs are denoted by four subscripts, and the storage area on the disk is scaled up by the $\mathcal{O}(N_{AO}{}^4)$ law. Even under symmetry considerations, the reduction in storage needs was only approximately 1/8, which is inadequate for large systems. With advancements in CPU processing speed, the direct method has become the dominant approach. This second-generation method calculates ERIs as needed, eliminating the need to store results. It also supports the Update method, which leverages the sparsity of the difference electron matrix ($\Delta \boldsymbol{P} = \boldsymbol{P}^{(n)} - \boldsymbol{P}^{(n-1)}$; $\boldsymbol{P}^{(n)}$ represents the density matrix at the n-th iteration of the SCF procedure) as the convergent solution is approached. The direct method is effective for electronic structure calculations of small- to medium-sized molecules. However, the direct/Update method, which computes ERIs on demand based on the size of the matrix elements of the difference density matrix, is not well suited to large-scale parallel processing. This is due to the difficulty in estimating the computational workload in advance, which is crucial for achieving ultra-high parallelization efficiency. In addition, the integration functions of the basis functions vary depending on the azimuthal quantum number, leading to inconsistent computational complexity based on the combinations of four basis functions. As a result, task equalization of ERIs in large-scale systems becomes complex, complicating large-scale parallel computation.

Reducing the number of SCF iterations directly decreases the overall time required for electronic structure calculations. Although achieving trouble-free and safe convergence of SCF calculations is the goal of every computational chemist, the reality is not so simple. For users of commercial and open-source software, the interruption of SCF calculations due to non-convergence or divergence is one of the most frustrating issues encountered in electronic structure work. This problem becomes increasingly pronounced with larger calculation targets. As the system size increases, as indicated by a greater number of basis functions (N_{AO}), the number of elements in the LCAO matrix $\boldsymbol{C}$ (=$N_{\mathrm{AO}} \times N_{\mathrm{MO}}$; N_{MO} represents the total number of molecular orbitals) also increases. The large number of elements in the LCAO matrix $\boldsymbol{C}$, which solves the Kohn–Sham–Roothaan equation, means that the matrix $\boldsymbol{C}$ has that many degrees of freedom, thus complicating the search for a solution and increasing the number of trials. Although many studies have attempted to accelerate SCF convergence, most focus on extrapolation and are effective primarily when the search space is near the convergent solution. Thus, success in SCF calculations for large systems often hinges on providing an initial guess that is close to the convergent solution.

10.2.2 Design and Development of the DFT Calculation Program for Proteins: ProteinDF

Design of ProteinDF

ProteinDF is an electronic structure calculation program based on DFT, primarily developed by Sato et al. [14, 15]. The name "ProteinDF" derives from its purpose: performing electronic structure calculations of proteins using DFT. It combines "Protein" and "Density Functional," indicating a DFT program capable of computing large molecules such as proteins. C++, chosen for its performance, collaborative development capabilities, and maintainability, was the selected programming language for ProteinDF at that time. The CMO calculation for cytochrome *c* was successfully completed in 2001 [16, 17]. This was followed by the CMO calculation of insulin hexamer in 2007 [18].

As previously mentioned, CMO calculations for large molecules demand significant computational resources. In this section, we outline the computational methods employed by ProteinDF to facilitate the calculation of protein electronic structures.

Handling of Metallo-proteins in ProteinDF

Trace metals in living organisms play crucial roles as cofactors in essential enzymatic reactions, including those involving enzymes such as superoxide dismutase. Estimations indicate that approximately one-third of all known proteins are metal-bearing proteins [19]. Because the d- and f-orbitals of transition metals and rare earth elements are closely aligned and their orbital energies are highly localized, electron correlation effects must be carefully considered. To accurately address electron correlation using molecular orbital methods, post-Hartree–Fock techniques such as the Møller–Plesset perturbation theory second order (MP2), Configuration Interaction (CI), and Coupled Cluster (CC) methods should be employed, even though these methods are computationally more demanding than the Hartree–Fock method. Obviously, applying post-Hartree–Fock calculations to large molecule such as a protein would be particularly resource-intensive. Alternatively, the density functional method can derive the exchange–correlation term with a computational complexity comparable to that of Hartree–Fock by using empirical parameters. ProteinDF utilizes this density functional method, specifically designed to facilitate canonical Kohn–Sham molecular orbital calculations for large molecules. For metallo-proteins that exhibit intriguing enzymatic reactions, ProteinDF, grounded in DFT, enables us to obtain the electronic structure, including electronic correlations.

Memory Management of ProteinDF

In ProteinDF, all computational data, including matrices and vectors generated from calculations, are stored on disk. The program only loads into memory the matrices essential for specific calculations, such as the one-electron term, Coulomb term, exchange–correlation term, Kohn–Sham matrix, standard orthogonalization, and matrix diagonalization, depending on the current process. By offloading unnecessary matrices to disk, ProteinDF optimizes CMO calculations to make the best use of limited physical memory. However, this approach has a drawback: the overall processing time for electronic structure calculations often increase due to the need for disk Input/Output (I/O) as well as computational processing. Nonetheless, the benefit of this system is that the matrices generated during the electronic structure calculation can be saved to disk, allowing for subsequent analysis, mid-calculation reassessment, and reproducibility of results.

Even when large matrices are moved from memory to disk, there is still the case that electronic structure calculations of sizable molecules exceed physical memory limits. For example, 1000-residue protein has approximately 100,000 basis functions, necessitating approximately 80 GB of Random-Access Memory (RAM) for a $100{,}000 \times 100{,}000$ dense matrix with double-precision elements. A simple matrix product then requires memory equivalent to three such matrices, totaling approximately 240 GB of RAM. Although RAM has become more affordable in recent years and motherboard capacities have increased, using a computer with 240 GB of RAM presents challenges. In distributed-memory parallel systems, such as PC clusters, each node often has limited RAM capacity. Accordingly, ProteinDF employs two strategies to facilitate electronic state calculations that demand significant RAM.

The first strategy is to distribute large matrices across nodes. Employing a 2D block-cyclic partitioning method enables linear operations such as matrix products and diagonalization to be executed in parallel using libraries such as Basic Linear Algebra Communication Subprograms/Scalable Linear Algebra PACKage (BLACS/ScaLAPACK) [20].

The second strategy involves utilizing the C programming function, `mmap()`. This system call, available on Portable Operating System Interface (POSIX) compliant operating systems, maps files and other objects into a virtual address space. ProteinDF employs `mmap()` as a virtual storage mechanism that treats files on hard disks or other storage devices as memory space (Fig. 10.2). This capability allows it to handle matrices larger than the physical memory limits, making it essential for large-scale calculations with limited computer resources. The advantage of `mmap()` lies in its direct access to file data within the program, eliminating the need for the extensive reading (loading) and writing (saving) process associated with large matrix files. However, because disk I/O is inherently slower than memory I/O, `mmap()` is not suitable for operations requiring frequent random access. Therefore, we recommended using a system with substantial physical memory for large-scale calculations whenever possible.

Recently, the advent of high-speed Solid-State Drives (SSDs) has alleviated concerns regarding disk I/O time. ProteinDF has determined that the benefits of leveraging disk I/O for maximizing computational scale and monitoring the computation process outweigh its disadvantages.

Improved Calculation Efficiency by the Third-Generation DF Method

When pure DFT is applied to large molecules such as proteins, the HOMO–LUMO gap becomes quite narrow, making it difficult to obtain SCF convergent solutions [21]. The hybrid exchange–correlation functional theory, represented by B3LYP, is widely used because it not only broadens the HOMO–LUMO gap by including a Fock exchange term, it also provides accurate predictions for a variety of physical properties. This hybrid approach is essential for DFT calculations involving proteins.

Fig. 10.2 Conceptual diagram of the virtual storage mechanism in ProteinDF using the mmap() function. This system treats large matrix files on storage devices as virtual memory space, enabling the program to handle matrices that exceed physical memory limits

The calculation of the Fock exchange term necessary for the hybrid exchange–correlation functional involves computationally expensive ERIs. The rapid growth in computational complexity associated with ERIs poses challenges for analyzing large systems. Various approaches to reduce the computational burden of ERIs have been developed, but most are approximate methods. Even the resolution of identity method [22], which has successfully calculated Coulomb terms, suffers from the inability to control the degree of approximation error based on the selection of auxiliary basis functions, which are truncated after a finite number. In addition, the challenges of parallel processing ERI calculations using the direct/Update method present further complications.

The first generation can be identified with the DISK method, which was commonly used before the 1990s when limited CPU performance required storing two-electron integrals on disk and reusing them in the construction of the Fock (Kohn–Sham) matrix. The second generation corresponds to the DIRECT method, which has become standard with modern CPUs by computing the necessary two-electron integrals on the fly according to the Fock (Kohn–Sham) matrix elements to be updated. Building on these developments, we have developed a third-generation DFT (3G

DF) method [23] utilizing Cholesky decomposition as a more efficient strategy for computing DFT on distributed-memory parallel computers for large-scale systems.

Cholesky decomposition is a method for decomposing a positive definite symmetric matrix $\boldsymbol{A}$ into the product of a lower triangular matrix $\boldsymbol{L}$ and its transpose $\boldsymbol{L}^T$. This can be expressed mathematically as

$$\boldsymbol{A} = \boldsymbol{L}\boldsymbol{L}^T \tag{10.12}$$

where, $\boldsymbol{A} \in \mathbb{R}^{n \times n}$ is a positive definite symmetric matrix, and $\boldsymbol{L} \in \mathbb{R}^{n \times n}$ is a lower triangular matrix with positive real numbers on the diagonal. The Cholesky decomposition exists uniquely for any positive definite symmetric matrix $\boldsymbol{A}$, and $\boldsymbol{L}$ can be computed. Given the elements of matrix $\boldsymbol{A} = (a_{ij})$ and $\boldsymbol{L} = (l_{ij})$, the relationships are

$$l_{ij} = \begin{cases} \sqrt{a_{ii} - \sum_{k=1}^{i-1} l_{ik}{}^2} & (i = j) \\ \frac{1}{l_{jj}}\left(a_{ij} - \sum_{k=1}^{j-1} l_{ik} l_{jk}\right) & (i > j) \\ 0 & (i < j) \end{cases}. \tag{10.13}$$

Thus, the diagonal elements of $\boldsymbol{L}$ are computed from the diagonal elements of $\boldsymbol{A}$ minus the sum of squares of previously computed elements, and the off-diagonal elements of $\boldsymbol{L}$ are computed sequentially using previously computed elements.

In the 3G DF method, a Cholesky matrix is formed by decomposing a super matrix comprised of ERIs.

$$V = \langle pq|rs\rangle \approx \sum_{k=1}^{M} L_{k,pq}^{\text{ERI}} L_{k,rs}^{\text{ERI}}. \tag{10.14}$$

Using this matrix $\boldsymbol{L}$, we can write the Coulomb term $\boldsymbol{J}$ and Fock exchange term $\boldsymbol{K}$ as

$$J_{pq} = \sum_{rs} P_{rs}\langle pq|rs\rangle \approx \sum_{rs}\sum_{I} L_{I,pq} L_{I,rs} P_{rs}. \tag{10.15}$$

$$K_{rs} = -\frac{1}{2}\sum_{rs} P_{rs}\langle pr|qs\rangle \approx -\frac{1}{2}\sum_{I}\sum_{rs} L_{I,pr} P_{rs} L_{I,qs}. \tag{10.16}$$

The computational accuracy δ produced by the Cholesky decomposition is mathematically rigorous, expressed as

$$\left| \langle pq|rs\rangle - \sum_{i=1}^{I} L_{i,pq} L_{i,rs} \right| \le \delta. \tag{10.17}$$

Calculating the Cholesky matrix, $\boldsymbol{L}$ can be accomplished independent of the density matrix $\boldsymbol{P}$ and prior to the SCF iterations. In addition, the Coulomb term $\boldsymbol{J}$ and Fock exchange term $\boldsymbol{K}$ can be efficiently computed using a linear computation library optimized for operations on computers such as Basic Linear Algebra Subprograms (BLAS), as the process involves only matrix operations. This means that, during SCF iterations, ERI calculations are not required, as they are inherently difficult to parallelize effectively. Although the size of the Cholesky matrix $\boldsymbol{L}$ sometimes poses challenges for large systems, the on-the-fly Cholesky decomposition using the Cholesky Decomposition with Adaptive Metric (CDAM) method [24] and pivot [25] allows for size reduction while maintaining control over computational accuracy; in this context, the "pivot operation" refers to the process in which a particular element of interest at a given step—often a diagonal element—is used as a reference value, such as the denominator, and row operations are performed centered around this element as the pivot.

In the 3G DF method, the exchange–correlation term can be calculated using a grid-free approach [26–30] instead of the standard grid method, allowing the electronic structure calculation to be conducted entirely through matrix operations. Figure 10.3 illustrates the conceptual differences between the conventional grid-based method and the grid-free approach. In the grid-free method, we first analyze the matrix representation of density $\boldsymbol{M}[\rho]$ of the density obtained from the four-center superposition integral.

$$M_{\mu\nu}[\rho] = P_{pq}\langle\mu\nu pq\rangle \tag{10.18}$$

where $\langle\mu\nu pq\rangle$ is a four-center overlap integral as follows.

$$\langle\mu\nu pq\rangle = \int g_\mu^*(\boldsymbol{r})g_\nu^*(\boldsymbol{r})g_p(\boldsymbol{r})g_q(\boldsymbol{r})\mathrm{d}\boldsymbol{r}. \tag{10.19}$$

The matrix representation of density $\boldsymbol{M}[\rho]$ can be transformed into an orthogonalized basis by the orthogonal transformation matrix $\boldsymbol{T}$.

$$\tilde{\boldsymbol{M}}[\rho] = \boldsymbol{T}^\dagger\boldsymbol{M}[\rho]\boldsymbol{T}. \tag{10.20}$$

Because it is a Hermitian matrix, the matrix representation of density in the orthogonal basis $\tilde{\boldsymbol{M}}[\rho]$ is diagonalizable.

$$\tilde{\boldsymbol{M}}[\rho] = \boldsymbol{U}\boldsymbol{\lambda}\boldsymbol{U}^\dagger. \tag{10.21}$$

When the function $f(\rho)$ is applied to this eigenvalue $\boldsymbol{\lambda}$, the exchange–correlation term in the orthogonal basis $\tilde{\boldsymbol{M}}[f(\rho)]$ can be obtained.

$$\tilde{\boldsymbol{M}}[f(\rho)] = \boldsymbol{U}f(\boldsymbol{\lambda})\boldsymbol{U}^\dagger. \tag{10.22}$$

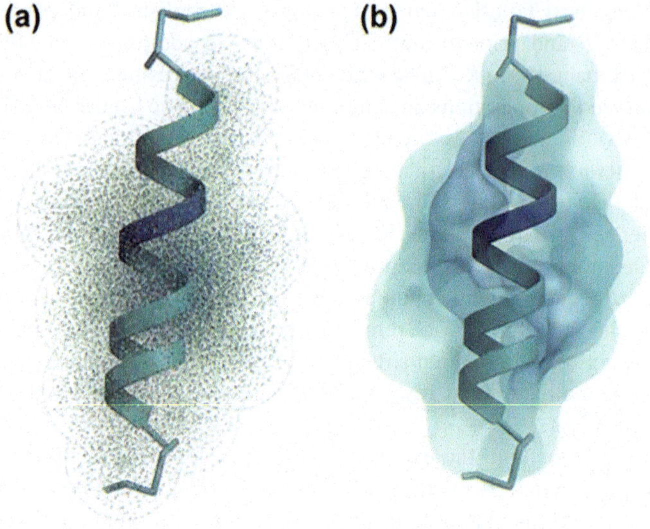

Fig. 10.3 Comparison of exchange–correlation term calculation approaches: **a** standard multi-center Lebedev grid method using discrete spherical quadrature points and **b** grid-free approach in ProteinDF based on continuous matrix representations of density

This can be converted to an exchange–correlation term in the basis function set of atoms by the following transformation:

$$\boldsymbol{M}[f(\rho)] = \boldsymbol{S'TU}f(\boldsymbol{\lambda})(\boldsymbol{S'TU})^{\dagger} \tag{10.23}$$

where $\boldsymbol{S}'$ is the overlap matrix in the basis sets for the atomic orbital and matrix representation of density.

$$S_{\mu p'} = \int g_{\mu}^{*}(\boldsymbol{r})g_{p}(\boldsymbol{r})\mathrm{d}\boldsymbol{r}. \tag{10.24}$$

In the four-center overlap integral, which is the rate-limiting step of computation, the resulting matrix is a positive definite symmetric matrix, making Cholesky decomposition possible.

$$V_{\mu p,\nu q}^{\mathrm{ovp}} = \langle \mu\nu pq \rangle \approx \sum_{I} L_{\mu p,I}^{\mathrm{ovp}} L_{\nu q,I}^{\mathrm{ovp}}. \tag{10.25}$$

Through the Cholesky matrix, $\boldsymbol{L}^{\mathrm{ovp}}$, the matrix representation of the density $\boldsymbol{M}[\rho]$ is expressed as

$$\boldsymbol{M}_{\mu\nu}[\rho] = \boldsymbol{P}_{pq}\langle \mu\nu pq \rangle \approx \sum_{I}\sum_{pq} L_{\mu p,I}^{\mathrm{ovp}} P_{pq} L_{\nu q,I}^{\mathrm{ovp}}. \tag{10.26}$$

Similar to how the Coulomb term $\boldsymbol{J}$ and Fock exchange term $\boldsymbol{K}$ are calculated, if the Cholesky vector $\boldsymbol{L}^{\mathrm{ovp}}$ is generated prior to the SCF iterations, the matrix representation of the density $\boldsymbol{M}[\rho]$ for the grid-free method can be derived solely through matrix operations during the SCF iterations.

10.2.3 Creation of the Initial Guess by the Quasi-canonical Localized Orbital Method

Importance of the Initial Guess in Large Molecules

As previously mentioned, CMO calculations require SCF iterations, necessitating a wave function or density matrix as the initial guess. For large molecules, the computational volume and time required for a single SCF cycle increase significantly. Therefore, the number of SCF iterations should be reduced as much as possible. In addition, due to the large number of matrix elements and high degrees of freedom in large molecules, an inaccurate initial guess often leads to oscillations or divergences, preventing stable convergent solutions. The challenges of achieving convergence in SCF calculations for large molecules cannot be addressed solely through techniques, such as the level shift method [31], or convergence acceleration methods, such as the Direct Inversion of the Iterative Subspace (DIIS) method [32]. To improve computational efficiency and ensure stable convergent solutions, an initial guess that is already close to the convergent solution is essential.

Quasi-canonical Localized Orbital (QCLO)

We have developed the Quasi-Canonical Localized Orbital (QCLO) method to provide high-precision initial wavefunctions for large molecules. This method features localized orbitals that possess the characteristics of orthogonalized CMOs in a specific region while also exhibiting properties of localized orbitals [33].

QCLOs are derived through the following procedure. In this context, the unit used for the CMO calculation is the frame molecule, where the substructure of this frame molecule is referred to as the fragment. The process begins with obtaining the Localized Orbital (LO) of the frame molecule using various established methods, such as the Pipek–Mezy method [34]. The orbitals belonging to the fragments used in the subsequent calculations are selected to generate the LO coefficient matrix $\boldsymbol{C}_{\mathrm{AO}}^{\mathrm{LO}}$ of the AO basis. Through this coefficient matrix $\boldsymbol{C}_{\mathrm{AO}}^{\mathrm{LO}}$, the Kohn–Sham matrix $\boldsymbol{F}_{\mathrm{LO}}$ of the LO basis is obtained by transforming the Kohn–Sham matrix $\boldsymbol{F}_{\mathrm{AO}}$ of the frame molecules, which are obtained in the AO basis.

$$\boldsymbol{F}_{\mathrm{LO}} = \boldsymbol{C}_{\mathrm{AO}}^{\mathrm{LO}\dagger} \boldsymbol{F}_{\mathrm{AO}} \boldsymbol{C}_{\mathrm{AO}}^{\mathrm{LO}}. \tag{10.27}$$

Here, the transformed Kohn–Sham matrix $\boldsymbol{F}_{\mathrm{LO}}$ is the number of LOs belonging to the fragment. The eigenvector $\boldsymbol{C}_{\mathrm{LO}}^{\mathrm{QCLO}}$, obtained by diagonalizing this matrix, is the QCLO.

$$\boldsymbol{F}_{\mathrm{LO}}\boldsymbol{C}_{\mathrm{LO}}^{\mathrm{QCLO}} = \boldsymbol{C}_{\mathrm{LO}}^{\mathrm{QCLO}}\boldsymbol{\varepsilon}^{\mathrm{QCLO}}. \tag{10.28}$$

As (2.25) shows, $\boldsymbol{C}_{\mathrm{LO}}^{\mathrm{QCLO}}$ is composed of an LO basis localized in fragments and characterized as a canonical orbital within those fragments. This characteristic makes the QCLO particularly suitable for combining fragments to generate an initial guess that closely resembles the canonical orbital. To revert this QCLO to its original AO basis, a subsequent transformation can be applied.

$$\boldsymbol{C}_{\mathrm{AO}}^{\mathrm{LO}}\boldsymbol{C}_{\mathrm{LO}}^{\mathrm{QCLO}} = \boldsymbol{C}_{\mathrm{AO}}^{\mathrm{QCLO}}. \tag{10.29}$$

Initial Guess by the QCLO Method

The QCLO method adopts a hierarchical approach for initializing large systems by recursively combining these QCLOs and expanding them gradually. Although the QCLO itself is not restricted to proteins, it can be applied to any molecular system. Here, we illustrate the process of creating initial wavefunctions using the QCLO method, with a polymer of amino acid residues (a protein) used as an example (Fig. 10.4).

First, in Step 1, the CMO calculation is conducted for each amino acid residue unit. To accommodate the dangling bonds at the N- and C-terminal ends of the main chain, acetyl (ACE) and N-methyl (NME) groups are added, as needed, to form peptide bonds. In Step 2, CMO calculations are performed for peptides containing

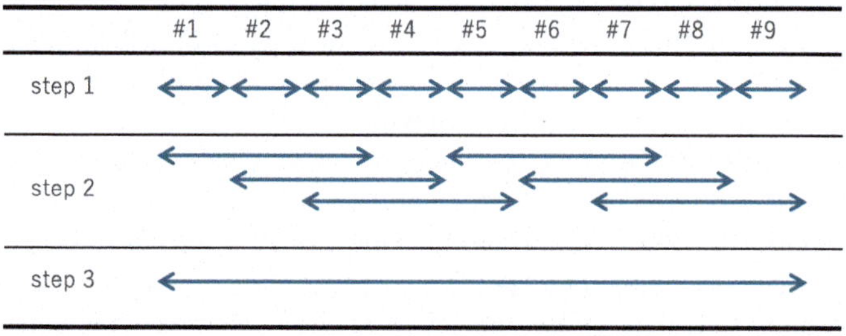

Fig. 10.4 Schematic of protein Canonical Molecular Orbital (CMO) calculation by QCLO method, where #1, #2, #3, … represent the amino acid residue numbers, with the double-headed arrow representing the frame molecule on which the CMO calculation is performed. In Step 1, the CMO calculation is conducted one residue at a time. Step 2 involves creating a frame molecule composed of three residues, each overlapping with two others. In Step 3, the initial guess for the frame molecule of nine residues is formulated based on the QCLO results of fragments #1, #2, #3, #4, …, #8, #9 obtained in Step 2. This is followed by performing the CMO calculation

three or more consecutive residues, treated as frame molecules. During this phase, adjacent frame molecules are designed to have an "overlap region" of two residues, such as 1–3, 2–4, 3–5, and so on, as the QCLO for the middle amino acid residue of these three residues will be utilized in the subsequent Step 3. In Step 3, longer frame molecules are generated, again maintaining "overlapped regions" of two residues, such as 1–7, 6–14, 13–21, etc., and the frame molecules are subsequently extended to encompass the entire protein.

In the QCLO method, the sequence of computational steps that transitions from a small frame molecule to a CMO calculation for a larger computational frame molecule is referred to as a computational scenario. A crucial aspect of the QCLO method is the establishment of an "overlap region," where the frame molecules of the subunits consistently intersect. Although the QCLO method can be implemented without this "overlap region," doing so limits the benefits gained from CMO calculations of the frame molecules. The significance of the "overlap region" is further illustrated by the interaction energy equation [35]. In the aforementioned example, the calculation scenario was structured to incorporate an "overlap region" encompassing two residues. However, a larger "overlap region" is necessary depending on the specific molecular system and its intended purpose. Previous studies have shown that a well-informed initial guess for proteins can be achieved by carefully constructing calculation scenarios that consider steric structures, such as secondary structures, salt bridges, and disulfide bonds [35]. Salt bridges are electrostatic interactions formed between oppositely charged side chains of amino acids, most commonly between acidic residues (such as aspartate or glutamate) and basic residues (such as lysine or arginine), which contribute to the stabilization of tertiary and quaternary protein structures. Disulfide bonds, on the other hand, are covalent linkages established between the thiol groups of cysteine residues, providing strong structural reinforcement by connecting distant parts of a polypeptide chain or different chains in multimeric proteins.

An Orchestration Program Based on the QCLO Method: QCLObot

The QCLO method is an essential technique for calculating CMOs for large molecules. However, the procedure of extracting subunits from a large molecule, performing CMO calculations, creating QCLOs based on the CMO calculation results of the subunits, and forming an initial guess for the large molecule from the QCLOs of multiple subunits is complex and time-consuming. The larger the target system, the more recursive the creation of QCLOs becomes, leading to increased workloads and a higher probability of errors. To streamline these tasks, we developed the QCLObot program [36].

The QCLObot program automates most aspects of frame molecule and fragment creation, as well as QCLO calculations, allowing researchers to concentrate on designing QCLO calculation scenarios. It utilizes a human-readable YAML Ain't Markup Language (YAML) file for input. Although YAML files are plain text files, they can effectively describe object structures, defining a parent–child relationship of frame molecules with one or more fragments. The input file represents the "state to be calculated" rather than outlining the operational procedure. This design feature, known as "idempotency," not only prevents side effects resulting from multiple end capping processing (such as the addition of multiple acetyl (ACE) groups and/or N-methyl (NME) groups) during recalculations, it also facilitates reuse. Furthermore, the system is programmed to avoid redundant processing by conducting necessary tasks on demand and storing once-calculated QCLOs and other data on disk for future use.

The QCLObot program was created using Python, an object-oriented scripting language. Programs that utilize the QCLO method require complex processing, including combinatorial analysis of various molecular structures and basis sets, free-fragment partitioning with nested structures, and the assignment of molecular orbitals to fragments. By leveraging the object-oriented programming features of Python and its rich data structures, we were able to develop QCLObot in an organized manner. In QCLObot, computationally intensive tasks, such as LO calculations, are delegated to external programs written in C++. This design also benefits from the role of Python as a versatile glue language. The QCLObot program is available as open-source and can be freely used and modified [36].

10.2.4 Calculation Procedure for the Electronic Structure of a Protein

Creation of Calculation Structures (1) Acquisition of Protein Structures

The electronic structure of a molecule is computed based on its 3D structure. Consequently, the quality and reliability of the desired electronic structure are primarily influenced by the accuracy of the 3D model. Given that electronic structure calculations for large molecules can be resource-intensive and time-consuming, creating the computational model must be done carefully.

Proteins are well-known as polymers of amino acids. Recent advances in AI technology [37] have enabled the prediction of protein 3D structures from amino acid sequence data. Once the 3D structure of a molecular system is obtained, an electronic structure calculation can be performed. However, when an experimental structure of the molecular system is available, it is scientifically preferable to create the calculation model based on this experimental structure rather than an AI-predicted one.

Many experimental structures of proteins and other biomolecules are recorded in a public database called the Protein Data Bank (PDB) [1]. These experimental structures are freely accessible online. Data from the PDB always includes a description of the experimental method used to obtain the structure. Understanding the characteristics of the experimental procedure and its observational limitations is essential AVE for creating a computational model based on an experimentally derived structure. For example, methods such as X-ray crystallography or cryo-electron microscopy often have difficulty revealing the positions of hydrogen atoms, which is not documented in the experimental structure. Similarly, experimental structures derived from nuclear magnetic resonance spectroscopy often present multiple conformations because they are constructed based on distances between hydrogen and carbon atoms. Consequently, a reliable stereochemistry model sometimes requires supplementing missing atoms and adjusting the positions of atoms that are abnormally close together.

Creation of Calculation Structures (2) Modeling of Computational Structure

It is uncommon for an experimental structure to be directly used as a computational model. As previously mentioned, atom coordinates are often incomplete due to experimental limitations, and simulations frequently contain nonessential components such as counterions (which is occasionally necessary), crystalline water, and surfactants. These factors can complicate the intended simulation. Therefore, computational structure modeling is crucial for achieving the desired simulation, and an overview of the process is outlined as follows.

First, the amino acid sequences recorded in the PDB should be examined. Most PDB entries include amino acid sequence information (with accession IDs from the UniProt database [38]) that serves as a reference for the registered structure. Although the structure in the PDB sometimes encompasses the entire protein of interest, it is often a partial or modified version. In modeling, this amino acid sequence information should be used as a reference to identify any missing or altered residues. When multiple amino acid residues are missing, it could be advisable to introduce ACE or NME groups into the model. These groups are effective in preserving the continuity of peptide bonds. In contrast, replacing experimentally undetermined residues with arbitrary substitutes often adversely affects the electronic structure of the regions already resolved.

In some cases, the positions of hydrogen atoms are not determined by experimental data from X-ray crystallography. Instead, hydrogen atoms are added based on the known chemical structure. Notably, histidine (His) has a side chain with an imidazole ring, allowing hydrogen to be added to either of its two nitrogen atoms at the delta and epsilon positions when in a neutral state. In a basic environment, hydrogen can be added to both nitrogen atoms at these positions. Amber and others classify His with hydrogen atoms added at the delta (HID), epsilon (HIE), and delta and epsilon (HIP) positions. The structures and hydrogen positions for these three protonation states are illustrated in Fig. 10.5. When nitrogen lone pair is used to coordinate with a metal, it is typically classified as HID or HIE. By contrast, when an acidic amino acid residue is nearby, it often forms a salt bridge, referred to as HIP. If the placement of hydrogen on the His is ambiguous, it should be resolved using additional experimental data.

When the calculated structure is prepared, the presence of charged amino acids must also be considered. In addition to the N- and C-termini of the protein backbone, acidic residues such as aspartic acid and glutamic acid as well as basic residues such

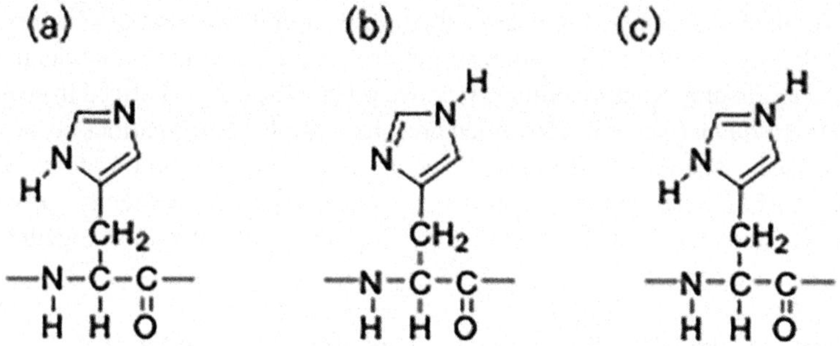

Fig. 10.5 Protonation states of histidine residues: **a** HID (hydrogen at the delta position), **b** HIE (hydrogen at the epsilon position), and **c** HIP (hydrogen at both delta and epsilon positions)

as lysine and arginine all remain. Histidine can also act as an acidic residue when both imidazole ring nitrogen atoms are protonated, classified as HIP. These charged amino acids are frequently located on the protein surface, particularly in water-soluble proteins. Generally, we want to neutralize these charged residues by placing sodium or chloride ions nearby as counter ions (Fig. 10.6). When a salt bridge forms between an acidic and a basic amino acid, the two residues can be treated as a neutralized pair. This neutralization model is adopted because proteins in neutral aqueous solutions, where various electrolytes are dissolved, are ionized; nearby carboxyl and amino groups are often considered neutralized by their interactions with water and electrolytes. When the model is constructed, sodium and chloride ions are treated as counterions, but treating them as point charges in the final electronic structure calculations is more convenient. This simplification is due to the large number of electrons associated with sodium and chloride ions, which could increase computational demands and complicate the evaluation of the electrostatic potential of proteins during analysis. QCLObot includes tools that automatically assign counterions such as sodium and chloride to the N- and C-termini of the backbone, as well as to charged amino acids.

Artificially added atoms are modeled based on limited knowledge and experience, lacking a solid scientific foundation. Ideally, optimizing the structure using

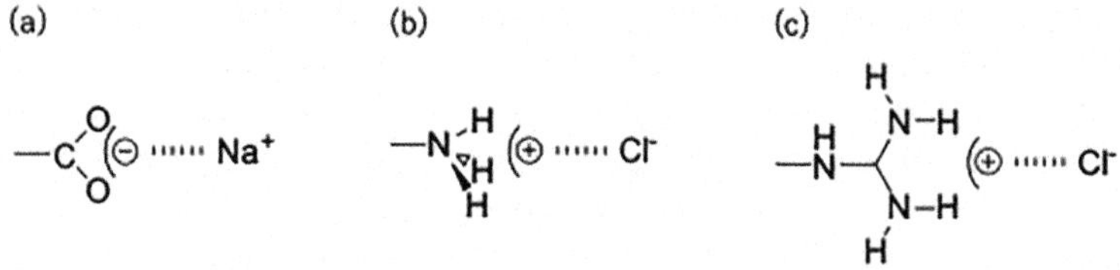

Fig. 10.6 Conceptual diagram of the charge neutralization model construction: placement of counterions (Na^+ or Cl^-) near charged amino acid residues and backbone termini to represent the ionized state of proteins in a neutral aqueous solution: **a** carboxy group, **b** amino group, **c** guanidino group

quantum chemical calculations is preferred. In fact, the common practice in computational chemistry for small molecules is to assess the electronic structure using an optimized molecular configuration. However, in today's context, calculating the electronic structure of large molecules presents significant challenges, even with a single-point calculations, making structure optimization, which requires multipoint calculations, even tougher. To reduce some of the arbitrariness associated with artificially modeled molecular structures, it is often beneficial to perform substructural optimization using molecular dynamics methods. In this approach, experimentally obtained structures (atomic coordinates) are preserved as much as possible, and only the coordinates of the atoms introduced during modeling are adjusted. This method aims to minimize the artificial and technical effects on the electronic structure.

Once the calculation model is established, the next step is to conduct CMO calculations using the QCLO method. As previously mentioned, the QCLO method begins with CMO calculations for individual residues and gradually expands the computational domain to encompass larger molecules. This method is known to produce a strong initial guess by creating a calculation scenario based on the protein 3D structure [35]. The rationale is that the greater the interaction between fragments, the larger the difference between the initial guess derived from the QCLO of the fragments and the convergent solution for the larger molecule, which accounts for fragment interactions in the CMO calculation. Therefore, we recommend integrating significant interactions that influence the electronic state early in the scenario or include them as a common overlap region in the frame molecules. The success of the scenario hinges on the ability to perform the CMO calculation for the entire protein. Currently, the development of these interaction-aware computational scenarios relies heavily on the researcher expertise. However, with advancements in AI, effective computational scenarios are expected to be generated automatically.

QCLObot includes a complete set of functions for performing CMO calculations on proteins. One key feature is its ability to add ACE and NME groups to satisfy peptide bonds. In general, the C–N bond of the peptide bond serves as a boundary for distinguishing the amino acid residues that comprise the protein. In the standard QCLO method, a frame molecule is created for each amino acid residue, followed by the CMO calculation. In this process, the peptide bond cut surface from the amino acid residue must be treated at the ends. Even if the electronic structure calculation itself succeeds without this end processing, it will not accurately represent the peptide bond formation, resulting in discrepancies with the desired electronic structure. QCLObot automates this modeling by allowing users to specify the endpoints in the input YAML file, thereby generating the input file for the electronic structure calculation program, ProteinDF. The end processing of QCLObot determines the atomic coordinates by referencing the neighboring main chain skeleton. This ensures that the QCLOs of the fragments connect smoothly, leading to well-formed initial guesses.

The fragmentation process of QCLObot permits the specification of arbitrary regions. Consequently, concerns regarding the ordering of atoms or fragments in the input file and the selection of basis functions are eliminated. QCLObot automatically extracts the matrix elements corresponding to the fragments from the LCAO and density matrices in the frame molecule. A frame molecule always consists of one or more fragments, which can be listed individually in the "task" section of the QCLObot input file. Figure 10.7 illustrates an excerpt from the YAML input file for the oxytocin (1NPO) model, showcasing how the computational tasks and the hierarchical parent–child relationships between frame molecules and fragments are organized. By defining the "state to be calculated" rather than specific procedures, the input file ensures idempotency and facilitates the reuse of computational scenarios. The electronic structure calculation is executed by calling ProteinDF in sequence. To define the fragments comprise a frame molecule, users specify one or more "fragments" in the "fragments" list within the frame molecule definition. Although fragments can be nested to create multiple components, QCLOs in QCLObot are generated in accordance with the highest-level "fragment" in the "fragments" list. In addition, users can refer previously calculated "fragments" using the "reference" element.

Depending on the protein and the simulation, computational scenarios can be rewritten flexibly, and QCLObot is designed to accommodate this flexibility. However, based on our experience with protein CMO calculations thus far, we know that some calculation scenarios are fairly rigid. One such scenario involves the formation of salt bridges. A salt bridge forms between a pair of acidic and basic amino acid residues, such as N-terminal, Lys, and Arg paired with C-terminal, Glu, and Asp. These salt bridges are clearly identifiable in the C-terminal region of the protein. When a pair of these amino acid residues is confirmed to form a salt bridge, they should be treated as a single fragment during the initial analysis. Similarly, a disulfide bond, which is a covalent bond formed between two cysteine thiols, should also be treated as a singular fragment, as with the salt bridge.

10.3 Results of Electronic Structure Calculations of Proteins

After the convergence of SCF iterations in a large-scale CMO calculation and the subsequent determination of canonical molecular orbitals and orbital energy levels, an important issue is how to analyze and interpret these results. Our prior experiences indicate that conducting a canonical molecular orbital calculation for an entire protein demands considerable time and effort. As we encountered in previous quantum chemical calculations on small molecules, modifying the model by changing the type of atoms or their coordinates can be quite challenging, and comparing these new results with previous calculations can be tedious. To address this, we present an analytical method utilizing the results of single-point calculations.

Fig. 10.7 Example of a QCLObot input file in YAML format for the oxytocin (1NPO) computational model. The file specifies the computational tasks, including the basis set (e.g., DZVP2), exchange–correlation functional (e.g., B3LYP), and the hierarchical definition of fragments within each frame molecule task

```yaml
---
tasks:
  - name: default
    brd_file: 1NPO_model.brd
    basis_set: DZVP2
    XC_functional: B3LYP

  - name: res_2-2
    sp: true
    fragments:
      - name: ACE_2
        brd_select: /model_1/_/1/
      - name: res_2-2
        brd_select: /model_1/_/2/
      - name: NME_2
        brd_select: /model_1/_/3/
  - name: res_3-3
    sp: true
    fragments:
      - name: ACE_3
        brd_select: /model_1/_/2/
      - name: res_3-3
        brd_select: /model_1/_/3/
      - name: NME_3
        brd_select: /model_1/_/4/
  - name: res_4-4
    sp: true
    fragments:
      - name: ACE_4
        brd_select: /model_1/_/3/
      - name: res_4-4
        brd_select: /model_1/_/4/
      - name: NME_4
        brd_select: /model_1/_/5/

  - name: res_2-4
    guess: QCLO
    sp: true
    fragments:
      - name: res_2-4
        fragments:
          - name: res_2-2
            reference:
              frame: res_2-2
              fragment: res_2-2
```

10.3.1 Validity of the Results of Electronic Structure Calculations of Proteins

The initial guess generated by QCLObot is employed, and it is assumed that the SCF iterations in ProteinDF have converged. Under these conditions, the obtained solution must be evaluated to determine whether it is appropriate for the intended simulation results. The computational complexity of the large-scale CMO calculation is significant, as is the time and effort required to analyze the results. Before delving into various analyses, we should first evaluate the validity of the calculation results from multiple perspectives. Table 10.2 summarizes the validation criteria discussed in this section to determine whether the CMO calculation has been successful and provides physically appropriate results. By checking these indicators, researchers can ensure the reliability of the electronic structure before proceeding to more detailed analyses. For illustration, we provide an example calculation involving the 83 amino acid residues of the Receptor-Binding Domain (RBD) of the SARS-CoV-2 spike protein. The computational model depicted in Fig. 10.8 was derived from PDB ID: 6VW1 [39]. We utilized Double Zeta Valence with Polarization (2) (DZVP2) [40] for the basis function set and B3LYP [41] for the exchange–correlation functionals. The total numbers of atoms and basis functions were 1330 and 12,692, respectively. Electronic structure calculations were performed using ProteinDF, and QCLObot was employed to generate the initial guess.

One method to verify the validity of the calculation results is to examine the evolution of the molecular orbital energy levels corresponding to the number of SCF iterations (Fig. 10.9). The horizontal axis represents the number of SCF iterations, and the vertical axis represents energy. Each black horizontal bar indicates a molecular orbital energy level, with the HOMO shown in red. The gap between the red HOMO and the black horizontal bar directly above it (i.e., the LUMO) is referred to as the HOMO–LUMO gap. The narrow spacing between the energy levels of each molecular orbital and the dense black horizontal bars is typical for large molecules. In the example illustrated in Fig. 10.9, the molecular orbital energy levels do not exhibit significant changes from the first iteration to the convergent solution, indicating stability and convergence. If the initial guess is poor due to issues with the calculation model or scenario, the HOMO rarely appears at approximately − 5 eV and tends to

Table 10.2 Validation criteria for the Canonical Molecular Orbital (CMO) calculations of proteins

Evaluation item	Normal state (reasonable result)	Signs of abnormality or failure	Probable cause and countermeasures
Transition of orbital energy levels	Levels are stable from the first iteration to convergence	Significant fluctuations or instability across iterations	Flaws in the calculation model or errors in the QCLO scenario
HOMO energy level	Positioned at approximately − 5 eV (standard for proteins)	Far from − 5 eV or varies significantly with each iteration	Extremely poor quality of the initial guess due to model/scenario issues
HOMO–LUMO gap	Remains stable in a dense state with a narrow gap	Oscillations (repeatedly narrowing and widening)	Potential problems with the model or errors in the calculation scenario
Mulliken charge (population)	Charges are within physically acceptable ranges	Anomalous values exceeding ± 1 for nonmetallic elements	Indicative of an unsuccessful CMO calculation (physically inappropriate solution)

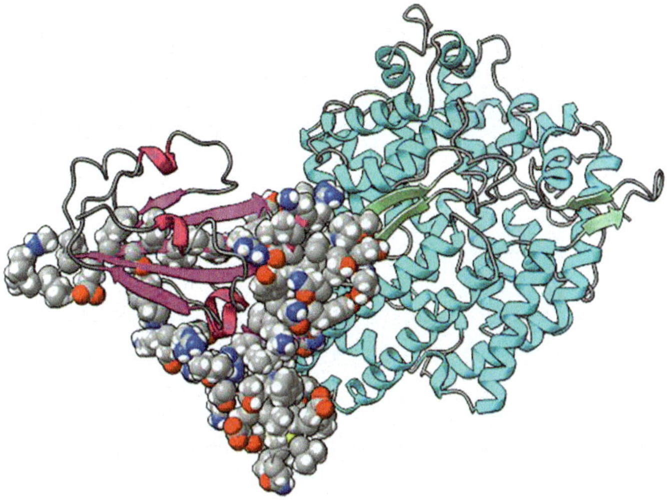

Fig. 10.8 Computational model of the SARS-CoV-2 spike protein based on PDB ID 6VW1. The spike region and the ACE2 region are illustrated with pink and light blue ribbons, respectively. The RBD region comprising 83 residues for which CMO calculations were performed is depicted using a ball model

vary significantly with each SCF iteration. This thus prevents a stable result such as that shown in Fig. 10.9. Another potential failure mode is the oscillation phenomenon, where the HOMO–LUMO gap gradually narrows and then widens with each SCF iteration. In these cases, investigating whether problems exist with the calculation model or errors are present in the calculation scenario is necessary.

The second step is to check the Mulliken population. The Mulliken charge is calculated using the matrix product of the density matrix and the overlap integral. In ProteinDF, both the density matrix and the overlap integral are stored on disk, eliminating the need for recalculation. An anomalous Mulliken charge, such as exceeding $\pm$ 1 for a nonmetallic element, is indicative of an unsuccessful CMO calculation. Because the Mulliken charge can be easily computed, it serves not only to analyze calculation results, but also to validate CMO calculations.

10.3.2 Analytical Methods for Calculating the Electronic State of Proteins

Molecular Orbitals and Orbital Energy Distribution

One of the key outcomes of CMO calculations for proteins is the determination of molecular orbitals. Figure 10.10 illustrates the frontier orbitals, specifically the HOMO and LUMO. Both orbitals extend over the protein, highlighting the necessity

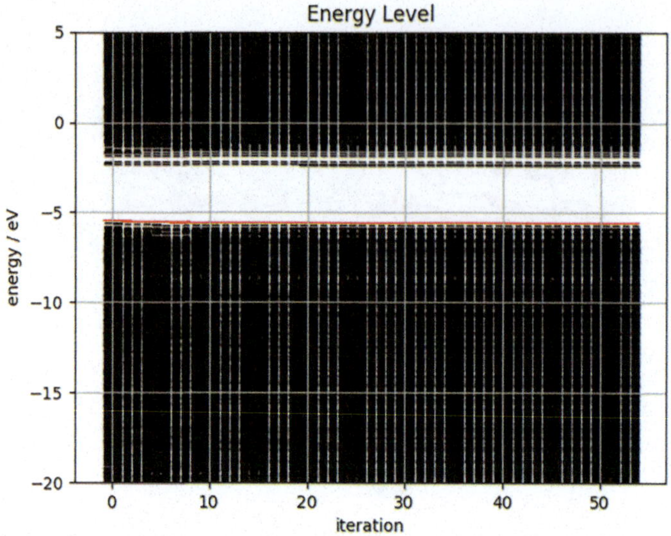

Fig. 10.9 Transition of orbital energies for each SCF iteration. The horizontal and vertical axes represent the number of SCF iterations and energy, respectively. The black horizontal lines indicate energy levels of molecular orbitals, and the red line represents the HOMO

of conducting CMO calculations for the entire protein to understand its electronic structure. Although the same isosurface values are plotted, the LUMOs tend to be more spatially extended than the HOMOs. However, discussing frontier orbital theory based solely on HOMO and LUMO is premature, as the molecular orbital energy levels of proteins are dense, and proteins are generally flexible, with active movement expected at normal temperature and pressure in aqueous solutions. Therefore, checking the orbital energy distribution and density of states is advisable to accurately identify molecular orbitals that are close in energy.

When molecular orbitals are visualized, attention should be paid to the isosurface value. If the absolute value is substantial, the principal components of the molecular orbital can be distinguished clearly. By contrast, if the absolute value is small, we can observe how the molecular orbitals are distributed across the entire molecule. Consequently, specifying the values of the isosurfaces is necessary, particularly when drawing the molecular orbitals of proteins. An effective observation method involves plotting and animating multiple isosurface values.

Electron Density and Electrostatic Potential

In addition to molecular orbitals, ProteinDF can generate electron density and electrostatic potential data in Gaussian CUBE file format, described as follows. This enables the use of general visualization software to visualize the molecular orbitals, electron density, and electrostatic potential of proteins.

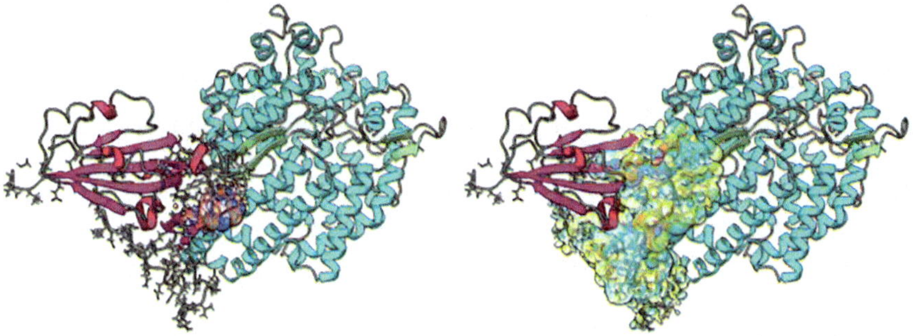

Fig. 10.10 Molecular orbital representation of the SARS-CoV-2 spike protein RBD83 residue computational model. The RBD and ACE2 regions are shown with pink and light blue ribbon, respectively, whereas the computational model is depicted as sticks. The isosurface value is ± 0.001. On the left, the HOMO is illustrated in blue and red; on the right, the LUMO is illustrated in yellow and light blue

Figure 10.11 displays the electrostatic potentials mapped onto an isoelectron density surface. The color scale in the map ranges from − 0.3 to + 0.3 in atomic units (Hartree/e). Red regions represent negative electrostatic potential, indicating electrophilic areas where electrons are abundant. Conversely, blue regions represent positive potential, indicating nucleophilic or electron-deficient areas. Traditionally, the electrostatic potential of proteins has been calculated using Restrained Electrostatic Potential (RESP) charges [42], which are constant atomic charges assigned to each amino acid residue. These charges do not vary based on the surrounding environment or structural distortions of the amino acid residues. By contrast, the electrostatic potentials derived from CMO calculations, which consider how electrons are spread throughout the protein, tend to be more accurate than those obtained from RESP charges. Our previous calculations indicate that the electrostatic potentials from CMO are generally lower in magnitude compared to those from RESP charges.

Fig. 10.11 Electro-static potential (ESP) map of the SARS-CoV-2 spike protein RBD83 residue computational model. The ESP is represented on the isoelectronic density surface of the computational model

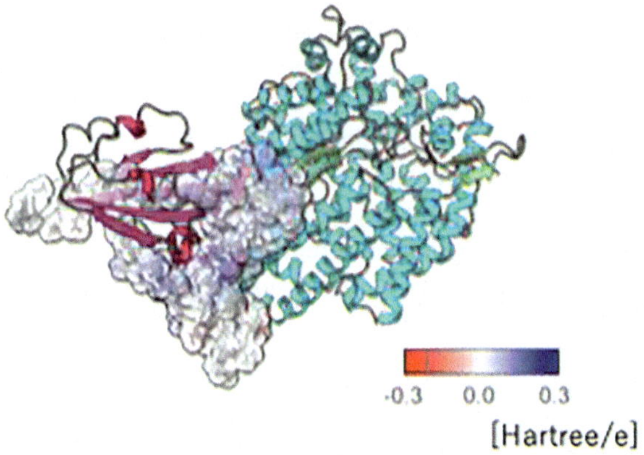

Energy Density Analysis (EDA)

Experimentally obtaining interaction energies, such as binding energies, is challenging, making them a crucial focus for simulation studies. Numerous analyses of interaction energies using CMO calculations have been documented [43]. Notably, the Morokuma analysis [44] is a fundamental method for interaction analysis and is widely utilized. However, this method requires at least three CMO calculations, which can be a significant drawback in large-scale CMO analyses, due to the high-computational cost associated with each calculation. This complexity makes estimating multiple interaction energies impractical.

The Energy Density Analysis (EDA) method [45] offers a means of estimating interaction energy from a single-point calculation. This method facilitates the determination of the interaction energy as part of the total energy. The EDA method can calculate interactions as matrix elements using basis functions, similar to the Mulliken charge calculation method, as illustrated in Eq. (10.30). Here, E_{AB}^{elec} represents the electronic interaction energy between atomic groups A and B, and $\boldsymbol{P}$, $\boldsymbol{h}$, and $\boldsymbol{F}$ denote the density matrix, one-electron term, and Kohn–Sham matrix, respectively. The indices p, q refer to the atomic orbitals.

$$E_{AB}^{\mathrm{elec}} = \frac{1}{2} \sum_{p \in A, q \in B} \left\{ P_{pq} \left(h_{pq} + F_{pq} \right) \right\}. \tag{10.30}$$

Figure 10.12 presents the energy distribution of each amino acid residue as derived from the EDA method. Both the vertical and horizontal axes represent amino acid residues, whereas red and blue shading indicates areas of high and low energy, respectively. Amino acid residues shown in blue indicate negative interaction energies, which correspond to stabilizing interactions, whereas those shown in red indicate positive interaction energies, corresponding to repulsive interactions. Because the EDA method enables calculations for all basis functions, as shown in Fig. 10.12, it enables an initial overview of each amino acid residue, followed by a detailed atomic and basis function analysis. In addition, it provides the capacity to evaluate each energy component, such as the one-electron, Coulomb, and exchange–correlation terms. The EDA method allows for an objective assessment of not only the presence of salt bridges, secondary structures, and hydrogen bonds, but also their magnitudes.

Exchange–correlation terms in density functional methods are typically calculated using quadrature methods with grid approaches. These are incompatible with EDA, which estimates energy density based on basis functions. However, in the 3G DF method, the exchange–correlation term can be calculated using the grid-free method, allowing EDA to be applied without modification.

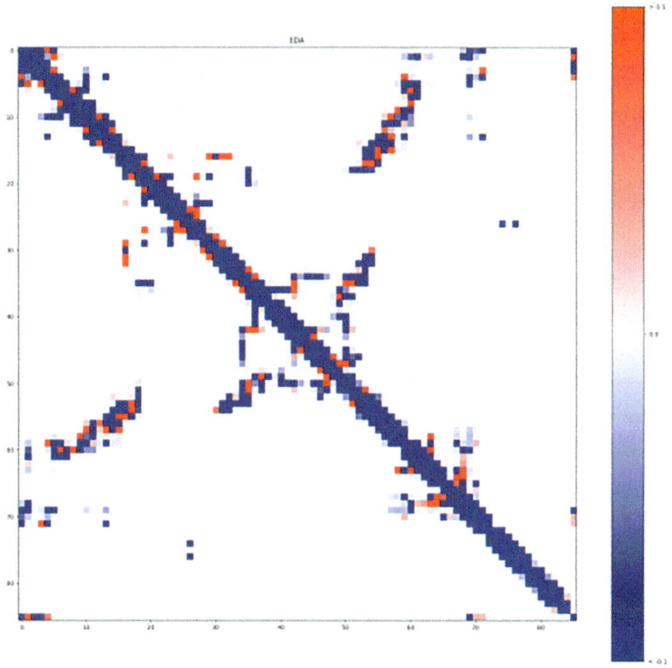

Fig. 10.12 Heat map of the EDA of the SARS-CoV-2 spike protein RBD83 residue computational model. The horizontal and vertical axes are arranged in ascending order of amino acid residue numbers. Negative interaction energies are shown in blue, and positive interaction energies are represented in red

Molecular Orbital Overlap Method

The Molecular Orbital Overlap (MOO) method evaluates the overlap of molecular orbitals [46]. The MOO method is a method for evaluating whether the molecular orbitals in two molecular systems are spatially identical and how similar they are. In the MOO method, a "system" refers to a specific molecular structure and its associated set of molecular orbitals calculated with a particular basis set. Concrete examples of these systems include a comparison between a small subunit (fragment) and the entire protein or between different mutants of the same protein to identify variations in their electronic states. The evaluation equation is given as Eq. (10.31). Here, $\boldsymbol{C}$, $\boldsymbol{C}'$ are the LCAO coefficient matrices for each system, and $\boldsymbol{S}'$ is the overlap matrix between the two systems. Additionally, p and q denote the indices of atomic orbitals, while i and j represent the indices of molecular orbitals. This method calculates the integral of the overlap between molecular orbitals in two systems. It is important to emphasize that the MOO method can be applied even if the coordinates, atomic types, and basis functions differ between the two systems, as it directly evaluates the spatial overlap of the wave functions. When two molecular orbitals overlap perfectly, the evaluated value O_{ij} is 1, whereas no overlap results in a value of 0. The MOO calculations can be performed even if the coordinates and basis functions of the molecular systems being compared differ. Because the primary process in the MOO calculation is the overlap integral of the two systems, the computational cost remains relatively low, even when all combinations of molecular orbitals are evaluated exhaustively.

$$O_{ij} = \left| \sum_{pq} C_{pi} C'_{qj} g_p g'_q \right| = \left| \left(\mathbf{C}^\dagger \mathbf{S}' \mathbf{C}' \right)_{ij} \right|. \tag{10.31}$$

One application of the MOO method is to illustrate how much the molecular orbitals in the subunit region contribute to the larger system. We can recall that the QCLO method generates molecular orbitals for large systems by combining smaller subunits. Figure 10.13 displays the results of the MOO analysis. This figure resembles the previously discussed energy level distribution diagram of molecular orbitals. On the left side, we see the subunit composed of residues 489–508 (frame molecule), whereas the right side comprises the larger unit consisting of residues 439–522 (frame molecule). The vertical axis represents energy levels, with black and gray horizontal lines indicating the energy level of a single molecular orbital. Specifically, the "energy levels" on the vertical axis refer to the orbital energies ε_i, which are obtained as eigenvalues by solving the Kohn–Sham–Roothaan equation (Eq. (10.3)). Each horizontal line (black or gray) corresponds to an individual Canonical Molecular Orbital (CMO), representing its specific energy state within the protein system. The horizontal length of each bar represents the MOO value, which signifies the ratio of molecular orbital overlap, normalized up to 1.0. More specifically, the proportion of the black line segment relative to the total gray bar length illustrates the degree of contribution from the molecular orbital of the subunit to that of the larger system. Specifically, the length of each black segment relative to the total gray bar indicates the degree of contribution from the original subunit's orbital to the molecular orbital of the larger system. The relative lengths of the black and gray lines in the large unit on the right illustrate the ratio of the evaluated values from the MOO calculation. The red dotted line in the center marks combinations of MOO evaluation values that exceed a specific threshold. The red dotted lines serve as a "tracking guide" to visually map the relationship between the Molecular Orbitals (MOs) of the subunit and those of the larger system. These lines connect orbital pairs with a MOO value (O_{ij}) exceeding a specific threshold (e.g., 0.1), thereby facilitating the identification of how an original subunit orbital is preserved or altered in the larger context. For example, if a red dotted line is tilted (not horizontal), it indicates that the orbital energy has rearranged due to interactions with other subunits in the larger system. Furthermore, if a single line from the subunit branches into multiple lines in the larger unit, it provides a concrete visual representation of orbital hybridization. Conversely, if the black lines at the ends of the dotted lines are very short, it suggests that the MO is no longer localized in the subunit but has become delocalized across the larger protein structure. In the RBD case, the molecular coordinates of the two common subunits remain unchanged, suggesting that the molecular orbitals of the large subunit closely resemble those of the small subunit. However, they are in fact quite different. As molecular size increases, the electronic structure of the subunits (molecular orbitals) becomes more unpredictable due to various factors, such as hybridization and interactions. This situation highlights the need for CMO calculations for proteins. In the QCLO method, an initial guess for large units is created stepwise from the results of subunit CMO calculations. If the calculation scenario is designed so that the CMO calculation for the subunit under MOO analysis is completed prior to the CMO calculation for the large-scale system, it becomes efficient to conduct MOO analysis without separately calculating the subunits.

Because the MOO method simply assesses the overlap of molecular orbitals, it can be applied even when the constituent atoms and coordinates in the two systems differ. For instance, MOO analysis can be beneficial in identifying differences in

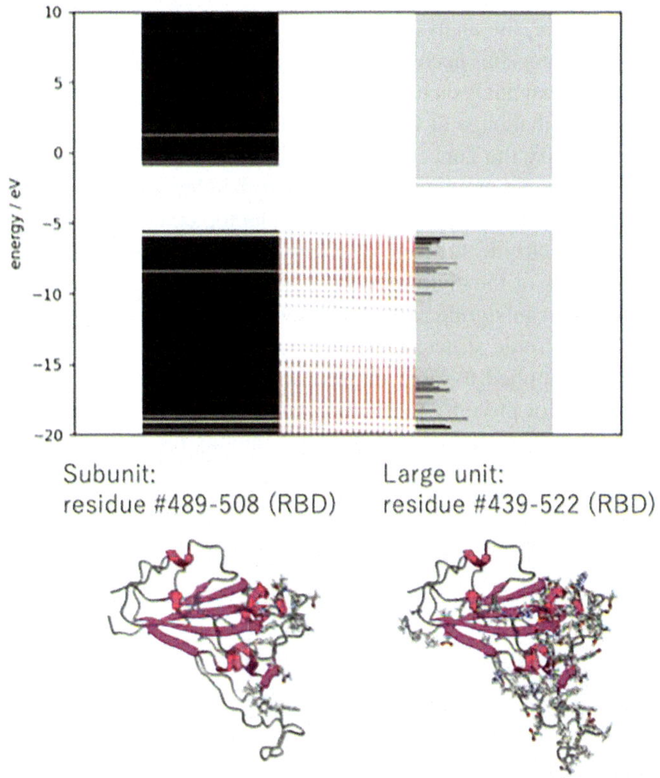

Fig. 10.13 Results of the MOO in the SARS-CoV-2 spike protein RBD83 residue computational model. The vertical axis represents energy levels. The left side shows the results of a small computational model consisting of residue numbers #489–508 of the RBD alongside a larger model that includes residues #439–522. The black and gray horizontal lines indicate the energy levels of the molecular orbitals. The red dotted line indicates the combined energy levels of the larger orbitals within the overlap region. The proportion of the black line segment relative to the gray length on the right represents the MOO value. Below, the RDB is illustrated with a pink ribbon, whereas the CMO component is represented using a ball-and-stick model

molecular orbitals between mutants with various amino acid residues. As the MOO analysis is applicable in many molecular systems, we look forward to seeing these new uses from our readers.

10.4 Summary

Even with the rapid advancement of computer speeds today, CMO calculations for proteins remain a challenging topic. However, as demonstrated so far, the collaboration between ProteinDF, a massively parallel CMO calculation engine based on the 3G DF method, and QCLObot, an orchestration program based on the QCLO method, has made it possible to perform CMO calculations for proteins safely and reliably. Furthermore, the analysis of the vast amount of data obtained from CMO calculations of proteins has become easily achievable thanks to the enhancement of various computational analysis methods. It is no exaggeration to say that now is the time to take on the challenge of CMO calculations for proteins.

Proteins exemplify the concept of “more is different” [47]. They are polymers composed of 20 distinct amino acids. Although the electronic structure of a single amino acid residue can now be calculated on a laptop computer, a basic model that extrapolates this electronic structure does not fully capture the diversity of proteins. Numerous instances of the “butterfly effect” illustrate that a single mutation in an amino acid residue can significantly alter physical properties of proteins. Remembering that the electronic structures we observe in simulations are merely models, particularly when applied to large molecular systems such as proteins, is crucial. CMO calculations for proteins continue to be a leading area of research. We hope this study aids in understanding of CMO calculations for proteins and other large molecular systems.

References

1. Berman H, Henrick K, Nakamura H. Nat Struct Mol Biol. 2003;10:980. https://doi.org/10.1038/nsb1203-980.
2. Frisch MJ, Trucks GW, Schlegel HB, Scuseria GE, Robb MA, Cheeseman JR, Scalmani G, Barone V, Petersson GA, Nakatsuji H, Li X, Caricato M, Marenich AV, Bloino J, Janesko BG, Gomperts R, Mennucci B, Hratchian HP, Ortiz JV, Izmaylov AF, Sonnenberg JL, Williams-Young D, Ding F, Lipparini F, Egidi F, Goings J, Peng B, Petrone A, Henderson T, Ranasinghe D, Zakrzewski VG, Gao J, Rega N, Zheng G, Liang W, Hada M, Ehara M, Toyota K, Fukuda R, Hasegawa J, Ishida M, Nakajima T, Honda Y, Kitao O, Nakai H, Vreven T, Throssell K, Montgomery Jr JA, Peralta JE, Ogliaro F, Bearpark MJ, Heyd JJ, Brothers EN, Kudin KN, Staroverov VN, Keith TA, Kobayashi R, Normand J, Raghavachari K, Rendell AP, Burant JC, Iyengar SS, Tomasi J, Cossi M, Millam JM, Klene M, Adamo C, Cammi R, Ochterski JW, Martin RL, Morokuma K, Farkas O, Foresman JB, Fox DJ. Gaussian 16, Revision B.01. Gaussian, Inc., Wallingford, CT; 2016.

3. Schmidt MW, Baldridge KK, Boatz JA, Elbert ST, Gordon MS, Jensen JH, Koseki S, Matsunaga N, Nguyen KA, Su S, Windus TL, Dupuis M, Montgomery JA. J Chem Phys. 2020;143: 234105. https://doi.org/10.1063/5.0005188.
4. Warshel A, Levitt M. J Mol Biol. 1976;103:227. https://doi.org/10.1016/0022-2836(76)90311-9.
5. Dapprich S, Komáromi I, Byun KS, Morokuma K, Frisch MJ. J Mol Struct (Theochem). 1999;462:1. https://doi.org/10.1016/S0166-1280(98)00475-8.
6. Kitaura K, Ikeo E, Asada T, Nakano T, Uebayasi M. Chem Phys Lett. 1999;313:701. https://doi.org/10.1016/S0009-2614(99)00874-X.
7. Yang W, Lee S. J Chem Phys. 1995;103:5674. https://doi.org/10.1063/1.470549.
8. Hohenberg P, Kohn W. Phys Rev. 1964;136:B864. https://doi.org/10.1103/PhysRev.136.B864.
9. Roothaan CCJ. Rev Mod Phys. 1951;23:69. https://doi.org/10.1103/RevModPhys.23.69.
10. Hartree DR. Proc Camb Phil Soc. 1928;24:89. https://doi.org/10.1017/S0305004100011919.
11. Fock V. Z Phys. 1930;62:795. https://doi.org/10.1007/BF01330439.
12. Obara S, Saika A. J Chem Phys. 1986;84:3963. https://doi.org/10.1063/1.450106.
13. Obara S, Saika A. J Chem Phys. 1988;89:1540. https://doi.org/10.1063/1.455717.
14. Sato F, Shigemitsu Y, Okazaki I, Yahiro S, Fukue M, Kozuru S, Kashiwagi H. Int J Quant Chem. 1997;63:245. https://doi.org/10.1002/(SICI)1097-461X(1997)63:1%3c245::AID-QUA26%3e3.0.CO;2-9.
15. ProteinDF. https://github.com/ProteinDF/ProteinDF. Accessed 30 May 2025.
16. Sato F, Yoshihiro T, Era M, Kashiwagi H. Chem Phys Lett. 2001;341:645. https://doi.org/10.1016/S0009-2614(01)00386-4.
17. Yoshihiro T, Sato F, Kashiwagi H. Chem Phys Lett. 2001;346:313. https://doi.org/10.1016/S0009-2614(01)00956-3.
18. Inaba T, Tsunekawa N, Hirano T, Yoshihiro T, Kashiwagi H, Sato F. Chem Phys Lett. 2007;434:331. https://doi.org/10.1016/j.cplett.2006.12.024.
19. Prabhulkar S, Tian H, Wang X, Zhu J-J, Li C-Z. Antioxid Redox Signal. 2012;17:1796. https://doi.org/10.1089/ars.2011.4001.
20. Scalapack. http://www.netlib.org/scalapack/. Accessed 30 May 2025.
21. Rudberg E. J Phys Condens Matter. 2012;24: 072202. https://doi.org/10.1088/0953-8984/24/7/072202.
22. Dunlap BI, Connolly JWD, Sabin JR. J Chem Phys. 1979;71:3396. https://doi.org/10.1063/1.438728.
23. Hirano T, Sato F. Phys Chem Chem Phys. 2014;16:14496. https://doi.org/10.1039/C3CP55514C.
24. Okiyama Y, Nakano T, Yamashita K, Mochizuki Y, Taguchi N, Tanaka S. Chem Phys Lett. 2010;490:84. https://doi.org/10.1016/j.cplett.2010.03.001.
25. Harbrecht H, Peters M, Schneider R. Appl Numer Math. 2012;62:428. https://doi.org/10.1016/j.apnum.2011.10.001.
26. Zheng YC, Almlöf J. Chem Phys Lett. 1993;214:397. https://doi.org/10.1016/0009-2614(93)85656-9.
27. Zheng YC, Almlöf J. J Mol Struct. 1996;388:277. https://doi.org/10.1016/S0166-1280(96)80040-6.
28. Glaesemann KR, Gordon MS. J Chem Phys. 1998;108:9959. https://doi.org/10.1063/1.476494.
29. Glaesemann KR, Gordon MS. J Chem Phys. 1999;110:6580. https://doi.org/10.1063/1.478559.
30. Glaesemann KR, Gordon MS. J Chem Phys. 2000;112:10738. https://doi.org/10.1063/1.481763.
31. Carbó R, Riera R, Arnau J. Chem Phys Lett. 1977;47:581. https://doi.org/10.1016/0009-2614(77)85046-X.
32. Pulay P. Chem Phys Lett. 1980;73:393. https://doi.org/10.1016/0009-2614(80)80396-4.
33. Kashiwagi H, Iwai H, Tokieda K, Era M, Sumita T, Yoshihiro T, Sato F. J Mol Phys. 2003;101:81. https://doi.org/10.1080/00268970210000026854.
34. Pipek J, Mezey PG. J Chem Phys. 1989;90:4916. https://doi.org/10.1063/1.456588.

35. Nishino-Uemura N, Hirano T, Sato F. J Chem Phys. 2007;127: 184106. https://doi.org/10.1063/1.2786998.
36. QCLObot. https://github.com/ProteinDF/QCLObot. Accessed 30 May 2025.
37. Jumper J, Evans R, Pritzel A, Green T, Figurnov M, Ronneberger O, Tunyasuvunakool K, Bates R, Žídek A, Potapenko A, Bridgland A, Meyer C, Kohl SAA, Ballard AJ, Cowie A, Romera-Paredes B, Nikolov S, Jain R, Adler J, Back T, Petersen S, Reiman D, Clancy E, Zielinski M, Steinegger M, Pacholska M, Berghammer T, Bodenstein S, Silver D, Vinyals O, Senior AW, Kavukcuoglu K, Kohli P, Hassabis D. Nature. 2021;596:583. https://doi.org/10.1038/s41586-021-03819-2.
38. UniProt. https://www.uniprot.org/. Accessed 30 May 2025
39. Shang J, Ye G, Shi K, Wan Y, Luo C, Aihara H, Geng Q, Auerbach A, Li F. Nature. 2020;581:221. https://doi.org/10.1038/s41586-020-2179-y.
40. Godbout N, Salahub DR, Andzelm J, Wimmer E. Can J Chem. 1992;70:560. https://doi.org/10.1139/v92-079.
41. Becke AD. J Chem Phys. 1993;98:5648. https://doi.org/10.1063/1.464913.
42. Bayly CI, Cieplak P, Cornell W, Kollman PA. J Phys Chem. 1993;97:10269. https://doi.org/10.1021/j100142a004.
43. Phipps MJS, Fox T, Tautermann CS, Skylaris C-K. Chem Soc Rev. 2015;44:3177. https://doi.org/10.1039/C4CS00375F.
44. Morokuma K. J Chem Phys. 1971;55:1236. https://doi.org/10.1063/1.1676210.
45. Nakai H. Chem Phys Lett. 2002;36:73. https://doi.org/10.1016/S0009-2614(02)01151-X.
46. Hirano T, Sato F. AIP Conf Proc. 2024;3030: 020008. https://doi.org/10.1063/5.0192921.
47. Anderson PW. Science. 1972;177:393. https://doi.org/10.1126/science.177.4047.393.

Chapter 11
Advances in Force Field Development and Molecular Dynamics Simulations Driven by Quantum Chemistry

Takefumi Yamashita

Abstract In this chapter, we provide an overview of advances in high-precision force field parameterization based on quantum chemical calculations, as well as their application to molecular simulations. We discuss how methods such as Hartree–Fock (HF), post-HF approaches, and density functional theory (DFT) have contributed to the development of force fields like FUJI and DREIDING-UT, and introduce reactive force field models (e.g., MS-EVB) for phenomena involving bond formation, cleavage, and proton transfer. Recently, the integration of artificial intelligence (AI) has accelerated, with cutting-edge methods such as AlphaFold expanding from static structure prediction toward dynamic and functional insights. By combining the "molecular-level precision" of quantum chemistry with the "large-scale, long-timescale" capabilities of molecular dynamics (MD) simulations, researchers can increasingly design complex biomolecules and advanced materials. Looking ahead, the further incorporation of machine learning is expected to boost the development of high-accuracy force fields to broaden their impact in interdisciplinary research and industrial applications.

Keywords Molecular dynamics simulation · Quantum chemical calculation · Force field development · Antigen–antibody interaction · Drug design · Biological macromolecule · Polymer material · Reactive force field · Protein structure prediction · AlphaFold

---

T. Yamashita (✉)
Department of Physical Chemistry, School of Pharmacy and Pharmaceutical Sciences, Hoshi University, Shinagawa-Ku, Tokyo, Japan
e-mail: yamashita.takefumi@hoshi.ac.jp

Research Center for Advanced Science and Technology, The University of Tokyo, Meguro-Ku, Tokyo, Japan

T. Onishi (ed.), *Advances in Quantum Science II*,
https://doi.org/10.1007/978-981-95-9721-5_11

11.1 Introduction

11.1.1 Fundamentals of Quantum Chemistry and Its Historical Background

Quantum chemistry is a field that uses the framework of quantum mechanics to describe the behavior of electrons in molecules and solids, aiming to determine their energies, structures, and properties [1]. Although attempts to analyze molecular systems using the Schrödinger equation date back to the 1920s—following the establishment of quantum mechanics—early progress was limited by the computational difficulty of many-body problems, restricting studies to very small systems. Over time, however, advances in semiconductor technology, algorithm development, and supercomputer performance have propelled quantum chemical calculations forward, enabling practical analyses of large-scale molecular systems.

The most fundamental method in quantum chemistry is the Hartree–Fock (HF) method, which uses a mean-field approximation for part of the electron–electron interaction and approximates the wavefunction of the entire system with a single Slater determinant. Because HF cannot fully capture electron correlation (i.e., the interactions among electrons) essential for accurately determining bond energies and reaction barriers, more advanced approaches—collectively called post-HF methods—are used when high-precision data are required. Key examples include MP2 (second-order Møller–Plesset perturbation theory), configuration interaction (CI), and coupled cluster methods such as Coupled Cluster with Single, Double, and perturbative Triple excitations (CCSD(T)). Among these, CCSD(T) is often called the "gold standard" of quantum chemistry because it can achieve so-called "chemical accuracy" (errors on the order of 1 kcal mol^{-1}) [2].

In parallel, density functional theory (DFT) has also developed considerably. DFT treats electron density as the fundamental variable and incorporates electron correlation via functionals [3]. Because it offers a relatively good balance between accuracy and computational cost, DFT has found wide-ranging applications—from molecular science to materials science—and is now among the most widely used methods in quantum chemistry. In recent years, improvements in density functionals (e.g., hybrid DFT, meta-GGA functionals) have enhanced accuracy, while algorithmic refinements such as linear-scaling methods and real-space grid approaches have simplified the analysis of large-scale systems.

11.1.2 High-Performance Computing and Large-Scale Quantum Chemical Calculations

Along with advances in linear-scaling algorithms and fragment-based embedding schemes, a major factor driving breakthroughs that now enable large-scale quantum chemical calculations has been the remarkable development of high-performance computing (HPC) infrastructure, including supercomputers and Graphics Processing Unit (GPU) clusters [4]. In the past, even post-HF calculations for systems with only a few dozen atoms required enormous amounts of time. Recently, however, techniques such as fragment-based partitioning combined with CCSD(T) hybrids have made it increasingly feasible to apply high-accuracy calculations to systems of hundreds or even thousands of atoms. Moreover, GPU-optimized electronic integral algorithms and linear algebra libraries have dramatically improved computational efficiency [5].

Efforts to reproduce the outputs of high-level quantum chemistry with machine-learned models have been gaining momentum. In practice, one first generates a large training set of ab initio energies and then fits a machine-learning model that instantly returns those quantities for new structures, thereby enabling larger-scale and faster computations (see Fig. 11.1) [6]. This approach preserves the high accuracy of ab initio calculations while significantly reducing computational costs. One notable approach involves subdividing CCSD(T)-level potential energy surfaces into multiple subregions for machine learning [7]. This technique enables near-quantum-accurate predictions of forces and energies at computational speeds approaching those of classical molecular dynamics (MD) simulations.

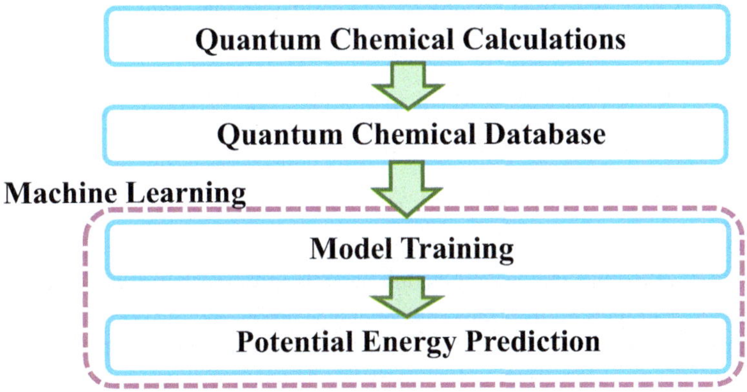

Fig. 11.1 Workflow for building a predictive machine-learning potential-energy model from quantum-chemical data. High-accuracy quantum-chemical calculations provide reference energies that are curated into a training database. A machine-learning model is then trained on this dataset to learn the underlying potential energy surface, yielding a fast, predictive surrogate that reproduces ab initio accuracy while reducing computational cost by several orders of magnitude

11.1.3 Force Field in Molecular Dynamics Simulation

Molecular Dynamics Simulation

Molecular dynamics (MD) simulation is a computational microscope that tracks the time evolution of atoms by solving Newton's equation of motion:

$$m_i \frac{\mathrm{d}^2 \mathbf{r}_i}{\mathrm{d}t^2} = \mathbf{F}_i, \tag{11.1}$$

where m_i, $\mathbf{r}_i$, and $\mathbf{F}_i$ denote the mass, position, and total force of atom i, respectively. A trajectory is generated by advancing the positions and velocities in discrete time steps, $\Delta t \approx 0.5 - 2$ fs. Thermostatting and barostatting allow one to reproduce experimental ensembles (NVE, NVT, NPT), after which structural, dynamical, and thermodynamic observables are extracted from the coordinates and velocities.

Force Field

A force field is an analytical potential-energy function that approximates the Born–Oppenheimer surface for a given chemical space:

$$U_{\mathrm{tot}}(r) = \sum_{\mathrm{bonds}} k_b (r - r_0)^2 + \sum_{\mathrm{angles}} k_\theta (\theta - \theta_0)^2 + \sum_{\mathrm{dihedrals}} \frac{1}{2} V_n \big[1 + \cos(n\phi - \gamma)\big]$$
$$+ \sum_{i<j} \Big[A_{ij}/r_{ij}^{12} - B_{ij}/r_{ij}^{6} + q_i q_j / \big(4\pi \epsilon_0 r_{ij}\big) \Big]. \tag{11.2}$$

Here, r and θ denote the instantaneous bond length and bond angle, whereas r_0 and θ_0 represent their equilibrium values; and k_b and k_θ are the corresponding force constants. The dihedral term depends on the torsion angle ϕ with amplitude V_n, periodicity n, and phase offset γ. For a non-bonded pair of atoms i and j separated by a distance r_{ij}, the Lennard–Jones parameters A_{ij} and B_{ij} reproduce short-range Pauli repulsion and London dispersion, while the Coulomb term describes the electrostatic interaction between partial charges q_i and q_j through the vacuum permittivity ϵ_0. Parameter sets are typically derived either from experimental thermodynamic data or from high-level quantum chemical calculations. Because this analytical potential can be evaluated rapidly, MD simulations routinely reach nanosecond-to-microsecond time scales and treat systems containing up to millions of atoms. Several commonly used force fields (e.g., AMBER, CHARMM, OPLS, and DREIDING) are distributed with major MD packages and can be applied directly.

11.1.4 Relationship Between Quantum Chemical Calculations and Force Field Development

As shown in Fig. 11.2, when considering the balance between computational cost and accuracy, high-level quantum calculations (e.g., MP2, CCSD(T), B3LYP) are ideal for detailed analysis of small systems and for deriving force field parameters, whereas classical force fields are better suited for large-scale, long-timescale simulations. Although advancements in supercomputers and algorithms have steadily expanded the range of system sizes amenable to quantum chemical calculations, once the number of atoms exceeds a few thousand, the computational load becomes prohibitive—making molecular mechanics methods indispensable for large-scale analyses.

Ongoing advances in quantum chemical calculations have strongly influenced MD simulations [8, 9], because they enable force fields to be reconstructed directly from ab initio data. This approach has attracted considerable attention for its ability to capture the underlying energy landscape characteristic to a system without relying solely on experimental data. By leveraging high-level quantum chemical information, researchers can refine classical force field parameters, thereby enhancing the predictive power and reliability of molecular simulations across diverse applications—from biomolecules to advanced materials. Such synergy between quantum chemistry and molecular mechanics represents a key step toward bridging small-scale precision with large-scale modeling.

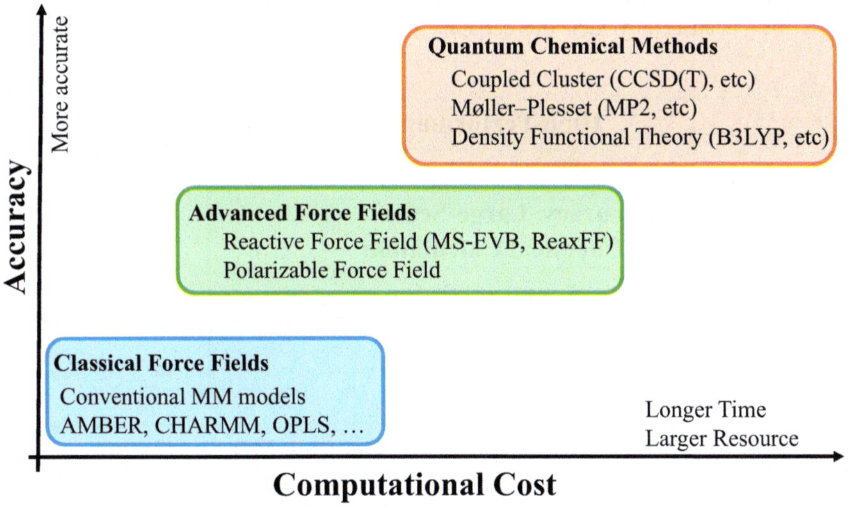

Fig. 11.2 An overview of representative computational methods. High-level quantum calculations are suited for small-scale, high-accuracy analysis, whereas classical force fields excel in large-scale, long-timescale simulations. Reactive and polarizable force fields occupy an intermediate position by incorporating effects not fully captured by classical force fields

11.1.5 *Scope and Organization of This Chapter*

In recent years, improvements in the accuracy of quantum chemical calculations and the rapid development of high-performance computing resources have made it increasingly feasible to analyze large and complex molecular systems. Figure 11.3 summarizes the influence of HPC, showing how it (i) enables high-accuracy, large-scale quantum-chemical calculations that feed realistic force-field construction and accurate MD, and (ii) simultaneously extends MD to longer time-scales and larger system sizes. Against this backdrop, this chapter provides an overview of the progress in "force field development based on quantum chemistry" and its application to molecular simulations, highlighting the cutting edge of this field.

First, we clarify the importance of reflecting accurate energy data derived from quantum chemical calculations in the parameters of classical force fields and show how refining key components—such as torsion angles and non-bonded interactions—can significantly enhance the reproducibility and predictive power of molecular simulations. We then examine specific force fields, including FUJI and DREIDING-UT (improved versions of the widely used AMBER and DREIDING force fields, respectively), through case studies involving biomolecules and polymer materials, illustrating their utility and range of applicability.

Next, we introduce a reactive force field named the multi-state empirical valence bond (MS-EVB) model, which can handle chemical reaction dynamics. We discuss its theoretical framework and application examples, emphasizing its significance as a method for large-scale analysis of phenomena like proton transfer and reaction dynamics—processes that standard classical MD methods cannot adequately describe because they lack the capacity to form or break bonds. We also explore

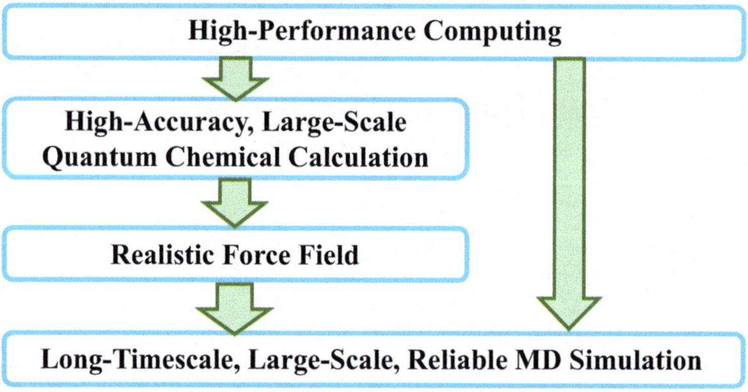

Fig. 11.3 Influence of high-performance computing on quantum chemistry and molecular dynamics (MD) simulation. High-performance computing (HPC) enables high-accuracy, large-scale quantum chemical calculations; these calculations supply data for constructing a realistic force field; the new force field delivers accurate MD simulations. Also, the HPC resources extend MD to longer time scales and larger system sizes

how the rapid convergence of machine learning and AI technologies (exemplified by AlphaFold, a deep-learning model for high-accuracy protein structure prediction) with quantum chemistry and molecular simulations is extending the scope from static structure prediction to more dynamic and functional insights, leading to a new wave of innovation.

By integrating the high precision of quantum chemical calculations with the capacity of molecular simulations to handle large-scale, long-timescale processes, it is becoming increasingly possible to predict the behavior of biomolecules and materials once considered too complex to analyze. Through this chapter, our aim is to offer a comprehensive perspective on the interdisciplinary and pioneering research that is poised to shape future developments in this field.

11.2 Quantum Chemistry and Force Field Parameter Development

11.2.1 Background of Force Field Development

In molecular mechanics calculations, the interactions between atoms are typically approximated using classical potential functions, force fields, which are typically written as Eq. (11.2). These functions generally consist of bonding terms (bond stretching, bond angles, torsions) and non-bonding terms (electrostatic and van der Waals interactions) [8, 9]. Among the bonding terms, dihedral (torsional) parameters are especially crucial because they directly determine the three-dimensional conformations of polymer chains in proteins and polymer materials. Using inaccurate torsion parameters can skew intramolecular conformational freedom, compromise assessments of secondary structure formation and flexibility, and ultimately undermine the reproducibility of simulation results. Refining torsional terms has thus been a key issue for accurately evaluating important reactions and interactions in polymers and biomolecules.

Even so, standard (non-reactive) classical force fields, in which the chemical bonds are approximated by simple harmonic potentials as in Eq. (11.2), cannot handle chemical reactions that form or break bonds. By contrast, reactive force fields—such as MS-EVB [10, 11] and ReaxFF [12]—introduce bond-order or multi-state terms so that multiple energy wells arise naturally in reactive systems. By employing these force fields in standard MD simulations, researchers can now handle chemical reactions involving bond formation and dissociation. These models incorporate changes in bond order and electronic states effectively into the force field, enabling chemical reactions to be represented within a classical simulation framework.

11.2.2 Refinement Using Quantum Chemical Calculations

Traditionally, force field parameters were often estimated by referencing experimental data or low-level quantum calculations (e.g., HF/6-31G, where 6-31G denotes a split-valence Pople double-ζ basis set) [8]. In recent years, however, high-accuracy quantum chemical calculations—such as CCSD(T)/aug-cc-pVQZ (aug-cc-pVQZ is an augmented, correlation-consistent quadruple-ζ basis set designed for high-accuracy correlated calculations)—have been employed to construct detailed energy profiles. These profiles are then incorporated into the force field (Eq. (11.2)), as shown in Fig. 11.4. This approach has gained wide recognition because it can provide both high generality and high accuracy without relying solely on experimental data.

Such efforts to capture electron correlation and reaction energy surfaces with quantum chemical precision, and then transfer this information into classical MD simulations, are critically important not only for nonreactive force fields but also for reactive force fields. For instance, the MS-EVB method approximates reaction pathways by linearly combining multiple valence-bond states (valence-bond structures). In doing so, the Hamiltonian matrix elements must be set by referencing the energies from quantum chemical calculations, enabling both a more realistic simulation of reaction phenomena and a deeper understanding of their mechanisms [10, 11].

As a result, when force fields are reparametrized using quantum chemical calculations, researchers can now conduct high-accuracy (i.e., "quantum-chemistry-level") simulations across a wide range of molecular systems without modifying the existing MD programs (Fig. 11.4). In the case of nonreactive force fields, refining internal

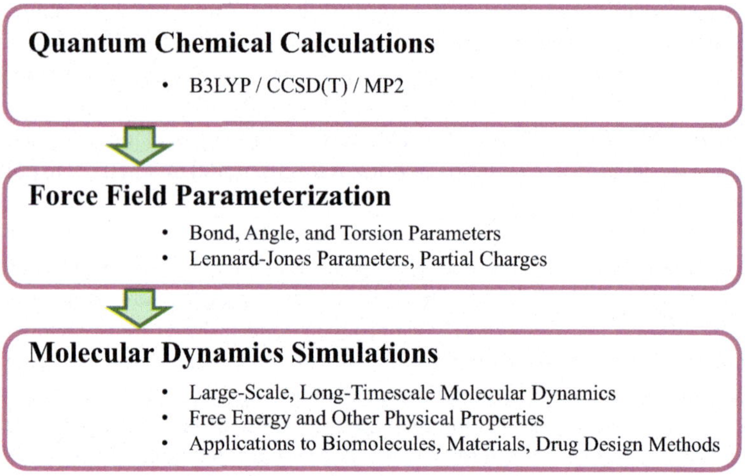

Fig. 11.4 Conceptual illustration of the workflow "quantum chemical calculations → force field parameterization → molecular dynamics simulations." First, quantum data are used to fit force field parameters. Next, large-scale MD simulations are performed with the parameterized force field, enabling applications in biomolecules, materials, and drug design

coordinates—particularly torsional angles—significantly boosts the accuracy of structural predictions and the reproducibility of intermolecular interactions. For reactive force fields, it becomes possible to investigate phenomena such as bond formation, bond breaking, and proton transfer. This approach is poised to be a key factor in advancing computational science for biomolecular and materials research.

11.3 Development and Applications of FUJI Force Field

11.3.1 Development of FUJI Force Field

Fujitani and colleagues [13] showed that for glycine-based dipeptide analogs (GDA) and alanine-based dipeptides (AD), the relative energies of the C5 and C7 conformers can differ significantly depending on the quantum chemical method and basis set. For GDA, HF/6-31G(d) indicates that C5 is 0.45 kcal mol^{-1} lower than C7, whereas MP2/6-31G(d) places C5 1.23 kcal mol^{-1} higher. In contrast, the local correlation approach DF-LCCSD(T0)/aug-cc-pVTZ yields a more moderate difference of 0.44 kcal mol^{-1} between the two conformers. Such discrepancies primarily arise from inadequate correction of basis set superposition error (BSSE) and incomplete treatment of electron correlation (including dispersion) in lower-level or canonical methods. By using a local correlation scheme at a larger Dunning-type basis set, DF-LCCSD(T0)/aug-cc-pVTZ not only reduces BSSE but also more accurately captures electron correlation effects. Even upon extrapolation to the complete basis set (CBS) limit, the remaining error stays within about 0.3 kcal mol^{-1}. As a result, these higher-level calculations predict lower conformational energies and barriers for states like C7, C5, C7_ax, and β2 of AD, compared to estimates from lower-level methods or Pople-type basis sets.

When older versions of force fields (e.g., AMBER 94, 99, 03, or 99SB) were used to model GDA and AD, discrepancies of about 1–2 kcal mol^{-1} appeared in their dihedral angle profiles compared to high-level computations, leading to significant deviations in local energy barriers or minima [13]. These inconsistencies suggest that real-world energy landscapes are not adequately reproduced by conventional parameter sets. Thus, Fujitani and colleagues [13] stressed the need for newly derived dihedral parameters to reduce the gap between conventional parameters and high-level quantum chemical calculations—thus improving the reliability of molecular mechanics simulations for protein backbones. The force field that incorporates these improved torsion parameters is named "FUJI." Indeed, it has been reported that the FUJI force field significantly improves conformational distributions in simple systems like alanine dipeptides [14]. Overestimated populations of certain conformers, commonly observed for older force fields, have been corrected, improving consistency with experimental observations.

In the context of membrane protein MD simulations, the choice of lipid force field can have a significant impact on protein structure [15]. In fact, following the

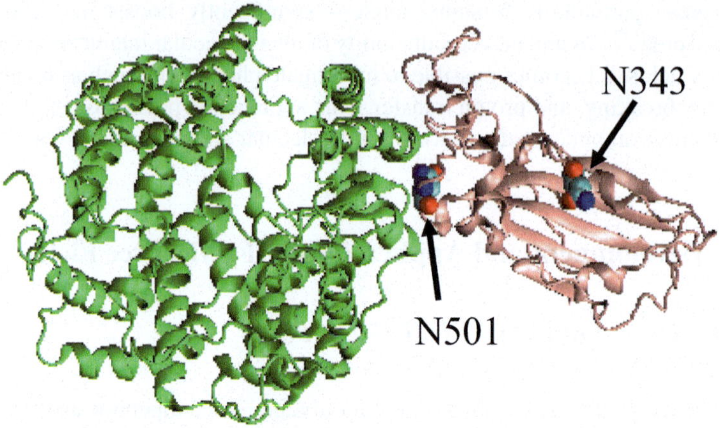

Fig. 11.5 SARS-CoV-2 spike RBD–ACE2 complex structure (PDB ID: 6LZG). Green and pink ribbons represent the human ACE2 and the SARS-CoV-2 spike receptor binding domain (RBD), respectively. Atoms of RBD residues N343 and N501 are shown as spheres

same approach used to develop the FUJI force field for protein, Kamiya et al. [16] developed a new parameter set for the lipid force field. This new parameter set reproduces membrane properties such as the area per lipid molecule and lateral diffusion rates in the lipid bilayer quite accurately.

Furthermore, the FUJI force field was extended by creating a new Restrained Electrostatic Potential (RESP) charge model for Zn^{2+}-coordinated histidine, based on HF/6-31G(d) calculations [17]. This extension enabled high-accuracy simulations of the SARS-CoV-2 spike protein receptor binding domain (RBD)–ACE2 complex (Fig. 11.5), making it possible to investigate the effects of N343 glycosylation and the N501Y mutation. By constructing four systems (with or without glycosylation, and with or without the N501Y mutation) and comparing the interaction energies and flexibility of the RBD–ACE2 interface, it was shown that glycosylation lowers the interaction energy, while the N501Y mutation itself lowers the interaction energy but significantly increases fluctuations in certain regions. This suggests that the N501Y mutation enhances the entropic contribution and improves the binding affinity between the spike protein and ACE2.

11.3.2 Applications to Antigen–Antibody Interactions

Highly specific molecular recognition between antibodies and antigens is one of the most critical issues in life sciences and drug discovery research. Experimentally, techniques such as X-ray crystallography, nuclear magnetic resonance (NMR), and isothermal titration calorimetry (ITC) have provided valuable insights into the structure and thermodynamics of antigen–antibody complexes (Fig. 11.6), yet it remains difficult to observe their dynamic binding process or dissect its intricate energetic components in detail. On the other hand, molecular dynamics (MD) simulations are a powerful tool that statistically incorporate the dynamic fluctuations of proteins and solvent molecules, providing a more realistic assessment of both structure and energy than analyses based on a single static conformation. In this subsection, we provide an overview of how MD simulations using the FUJI force field have been applied to antibody research.

Refinement of Potential of Mean Force Calculation Methods for the Hen Egg-White Lysozyme–HyHEL-10 System

Hen egg-white lysozyme (HEL) and the antibody HyHEL-10 (Fig. 11.6a) have served as a prototypical antigen–antibody pair for decades, yet it remains difficult to accurately determine their binding free energies and potential of mean force (PMF). Conventional umbrella sampling method with steered MD (SMD) simulations often applies artificially large external forces, leading to non-physical structural disruption. To address this, Yamashita and Fujitani proposed a new approach that combines multistep targeted MD (TMD) simulation method with conventional umbrella sampling, allowing the unbinding pathway to be explored incrementally while maintaining the native shape [18]. This method resulted in more reasonable free energy barriers and significantly improved agreement with experimentally determined dissociation constants. The study provides valuable guidance on how to handle structural constraints and path dependencies in free energy evaluations of antibody–antigen interactions, which is a longstanding challenge.

Prolyl *Cis–trans* Isomerization in the Heavy-Chain CDR-H3 Loop of an Anti-epiregulin Antibody

A highly unusual phenomenon has been reported in the anti-epiregulin (EPR) antibody 9E5, in which the proline residue (Pro103) in the heavy-chain CDR-H3 loop changes from the *cis* form to the *trans* form upon binding to EPR (Fig. 11.6b) [19]. The *cis–trans* isomerization of proline generally requires an activation energy of around 20 kcal mol^{-1}, making it an extremely slow reaction without enzymatic assistance. However, in this system, local structural rearrangements triggered by EPR binding appear to accelerate the proline isomerization. Based on crystallographic analysis and MD simulations, it is suggested that the interaction network near the CDR-H3 loop dynamically reorganizes after EPR binding, ultimately stabilizing the entire complex when Pro103 adopts the *trans* form.

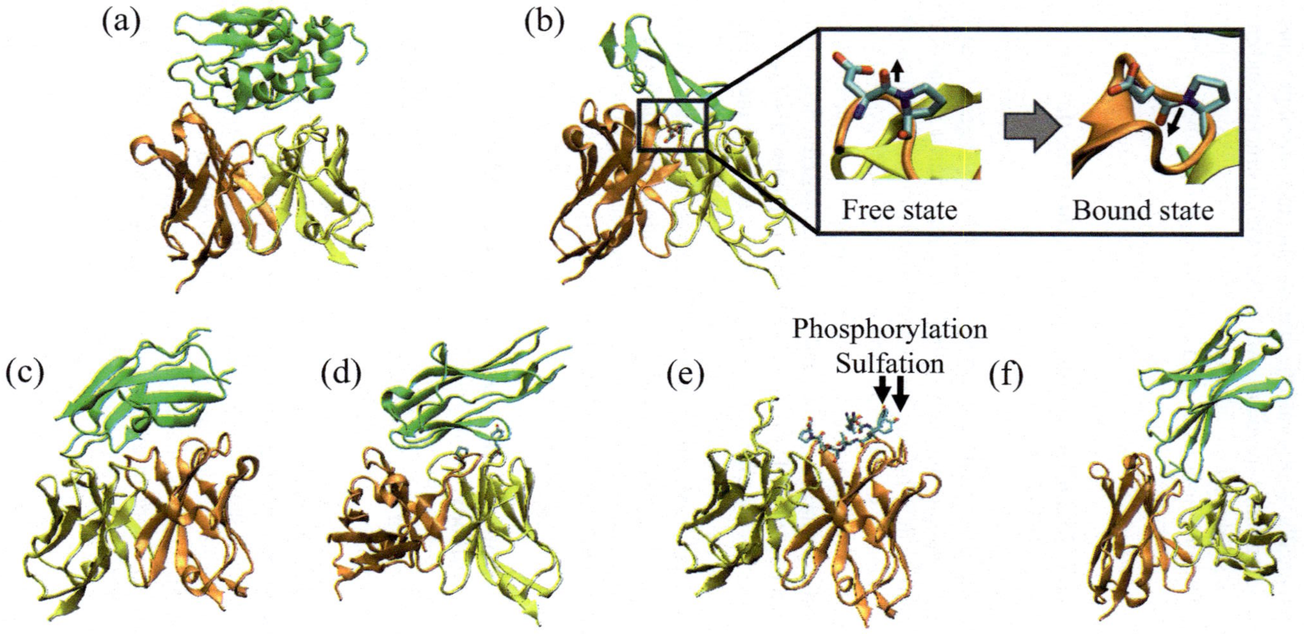

Fig. 11.6 Antigen–antibody complex structures (**a**) hen egg-white lysozyme (HEL) bound by HyHEL-10 (PDB ID: 2DQJ). (**b**) Epiregulin bound by 9E5 (PDB ID: 5A8D); inset highlights *cis–trans* isomerization of heavy-chain Pro103. (**c**) ROBO1 Fn3 domain bound by B2212A (PDB ID: 3WIH). (**d**) ROBO1 Ig5 domain bound by B5209B (PDB ID: 6A77). (**e**) CCR5-derived peptide bound by 4B08 (PDB ID: 5YD3); the antigenic peptide is shown as sticks. (**f**) PD-1 bound by nivolumab (PDB ID: 5WT9). In all panels, ribbon representations are colored green for antigen, yellow for the antibody light chain, and orange for the antibody heavy chain

Analysis of ROBO1 Antibody Targeting Hepatocellular Carcinoma

Nakayama and colleagues [20] elucidated the complex structure of the Fn3 domain of ROBO1—identified as a novel immunotherapy target in hepatocellular carcinoma—and a specific antibody, B2212A, using MD simulations along with ITC and X-ray crystallography (Fig. 11.6c). It was found that the light-chain $Tyr50^{L}$ of B2212A plays a critical role at the binding interface, forming CH/π interactions and hydrogen bonds with $Phe68^{F}$ to yield the principal binding energy. Additionally, substituting $Tyr50^{L}$ with Ala lowered the binding enthalpy but increased the binding entropy, leading to a minimal net change in free energy. These findings emphasize the importance of considering both enthalpy and entropy when developing therapeutic antibodies.

Yamashita and colleagues [21] further reported efforts to optimize the binding affinity of another ROBO1-specific antibody, B5209B (Fig. 11.6d). They performed alanine-scanning mutations and measured affinity and thermodynamic parameters by ITC, combining X-ray crystallographic analysis with MD simulations to elucidate the detailed mechanism. While conventional antibody design for increasing affinity (affinity maturation) often introduces larger side chains to create new interactions, this study replaced bulky side chains with alanine to reduce the steric hindrance and entrapment of water molecules. This approach yielded entropy gains and stabilized peripheral salt bridges, achieving up to a 30-fold increase in binding affinity. It thus highlights a new possibility in antibody optimization.

MD simulations were used to investigate the subtle interfacial structural changes triggered by alanine substitutions, making it possible to quantify how salt bridges, π interactions, and water dynamics influence binding free energy [21]. For example, substituting $P103^{H}$ with alanine slightly altered the ROBO1 interaction network, promoting local salt-bridge formation and improving binding enthalpy. In contrast, substituting $Y30^{L}$ released interfacial water molecules, resulting in the entropic gain [21]. One notable aspect of this study is how it combines computational methods and experimental analysis to show how single-site mutations reorganize interaction networks at the molecular level. These insights suggest that, beyond the conventional strategy of introducing large side chains to enhance interactions, fine-tuning the interface through alanine substitutions can also be an effective design concept.

Analysis of the Interactions Between Modified Tyrosine Peptide and Its Antibody

Extension of the FUJI force field enabled a detailed MD simulation analysis, clarifying how post-translational modifications such as tyrosine phosphorylation (pTyr) and sulfation (sTyr) influence peptide–antibody interactions. Focusing on a tyrosine-modified peptide (pep1) derived from CCR5 in complex with antibody 4B08 (Fig. 11.6e), it was observed that pTyr exerts a strong influence on the peptide's conformation and flexibility, whereas sTyr produces somewhat milder changes. It was also confirmed that the peptide's N-terminus can detach from the antibody during simulation, regardless of the presence or absence of modification, aligning with experimental observations. These results reveal that even two chemically similar modifications can lead to distinct structural and interaction profiles, offering crucial insights into how post-translational modifications shape the molecular recognition [22].

Stability of Salt Bridges in the Antigen–Antibody Interface

A salt bridge is a non-covalent electrostatic interaction between oppositely charged ionizable residues—typically Asp/Glu and Lys/Arg. Because salt bridge stability can significantly impact the binding affinity of antigen–antibody complexes, Okajima and colleagues [23] examined two salt bridges ($D32^H$–$K97^Y$ and $D99^H$–$K97^Y$) at the interface of hen egg-white lysozyme (HEL) and antibody HyHEL-10 (Fig. 11.6a), using MD simulations. They found that the stability of each salt bridge depends sensitively on surrounding residue arrangements, the orientations of aromatic side chains, and even slight distortions in the main chain. The $D32^H$–$K97^Y$ salt bridge is strongly stabilized by the spatial configuration of aromatic residues ($Y33^H$ and $W98^H$), whereas $D99^H$–$K97^Y$ is unstable in the wild-type interface but can become stabilized through the main-chain deformational effect induced by alanine substitutions. These findings underscore that not only electrostatic interactions but also the entire surrounding environment (such as aromatic side-chain orientations and main-chain conformation) must be considered when introducing or optimizing salt bridges in antibody engineering.

Mechanisms of Hen Egg-White Lysozyme–HyHEL-10 Complex Formation

In a recent study by Takamatsu and colleagues [24], more than 300 parallel sub-microsecond MD simulations—totaling 57 μs of simulation data—were used to reproduce the formation of the HEL–HyHEL-10 complex. Since the conventional MD analysis rarely exceeds 5 μs of cumulative sampling per system, the present 57 μs data set constitutes an order-of-magnitude larger, truly large-scale analysis. This large-scale analysis revealed that numerous encounter complexes—transient, loosely oriented pre-binding assemblies that arise upon diffusive collision—form in solution, and some of them can proceed through multistep stabilization pathways leading to a correctly oriented epitope–paratope interface. Notably, the side-chain conformation of residue $N32^L$ was shown to mediate a "structural locking" in the late stage of binding. The $N32^L$D mutation accelerates binding through electrostatic interactions but compromises this locking mechanism, resulting in a looser complex with a substantial entropic gain. These insights offer a detailed view of the dynamic binding mechanisms and energy landscapes—information difficult to obtain solely from static crystal structures—and provide valuable guidance for strategies to enhance antibody affinity.

Role of the N-Terminal Loop in PD-1–Nivolumab Interactions

MD simulations have also been employed to investigate interactions between PD-1 (Programmed Cell Death-1) and the anti-PD-1 antibody nivolumab (Fig. 11.6f) [25]. In a model lacking the PD-1 N-terminal loop, the interaction energy between PD-1 and nivolumab decreased markedly, significantly reducing the binding affinity. Furthermore, the computational alanine scan analysis of the N-terminal loop revealed that replacing any single residue did not diminish binding to the same extent as the complete loop deletion, implying that the N-terminal loop structure collectively enhances the PD-1–nivolumab interaction. This work contributes quantitative knowledge about the mechanism of the clinically used immune checkpoint inhibitor nivolumab and offers guidance for examining PD-1 variants and designing antibodies.

Concluding Remarks on Antibody–Antigen Interactions

In this subsection, we have introduced various research examples that combine MD simulations with experimental techniques to gain a deeper understanding of antigen–antibody interactions. These studies demonstrate how we can elucidate, at the molecular level, (1) the specific pathways of complex formation and dissociation, and (2) the subtle factors—such as salt bridges, aromatic residues, and hydration dynamics—that govern antigen–antibody recognition. Moreover, in the pursuit of enhancing binding affinity, it was shown that considering entropic aspects—such as releasing interfacial water—will offer advantages beyond the conventional strategy of simply increasing interactions. This underscores the complex nature of antibody engineering and highlights new avenues for refining antibody design.

11.3.3 Applications in Drug Discovery

The MP-CAFEE method (Massively Parallel Computation of Absolute Free Energy with well-equilibrated states), which estimates absolute protein–ligand binding free energies by launching hundreds of short alchemical trajectories in parallel from thoroughly equilibrated ensembles, enables the direct calculation of the binding free energy between proteins and ligands [26, 27]. Yamashita and colleagues [26] demonstrated that MP-CAFEE calculations employing the FUJI force field produce results closer to actual affinities than simple scoring approaches, highlighting the importance of incorporating the dynamic effects of proteins and solvents. Moreover, it was shown that by using supercomputers, it is possible to accelerate the evaluation of numerous compounds, indicating the potential for compound screening [27].

Negami et al. [28] applied MP-CAFEE calculations to an ion channel hERG (Fig. 11.7) to quantify binding affinities of various drugs that directly relate to cardiotoxicity. It was confirmed that the calculated data are in good agreement with experimental data. Note that they used the CHARMM force field in this study. Because the risk of lethal arrhythmias resulting from hERG channel inhibition is a major cause of discontinued drug development, it is extremely valuable to have a computational method for preliminary evaluation. Furthermore, Negami and colleagues [29] analyzed the binding energies of 25 CiPA reference drugs against both the hERG channel and the hNaV1.5 channel. By revisiting initial assumptions such as docking poses and drug protonation states, they improved accuracy and observed that the calculated binding energies for both the hERG channel and the hNaV1.5 channel correlated well with experimental data, pointing toward a possible drug discovery approach that simultaneously considers two types of ion channels.

Mitsui et al. [30] used the FUJI force field to precisely model the SARS-CoV-2 spike protein and performed MD simulations to analyze its interactions with cepharanthine (CEP). First, they explored putative binding candidates via docking, focusing on the receptor binding domain (RBD). Next, they performed broad conformational sampling of the binding structure using the Replica-Exchange with Solute Tempering (REST2) method. Finally, they combined accurate energy evaluations with the FUJI force field and long-timescale MD to verify the stability of the binding poses. As a result, they discovered that the most stable binding mode exists at a site distant from the ACE2-binding region, suggesting that CEP inhibits structural changes in the spike protein rather than serving as a competitive inhibitor.

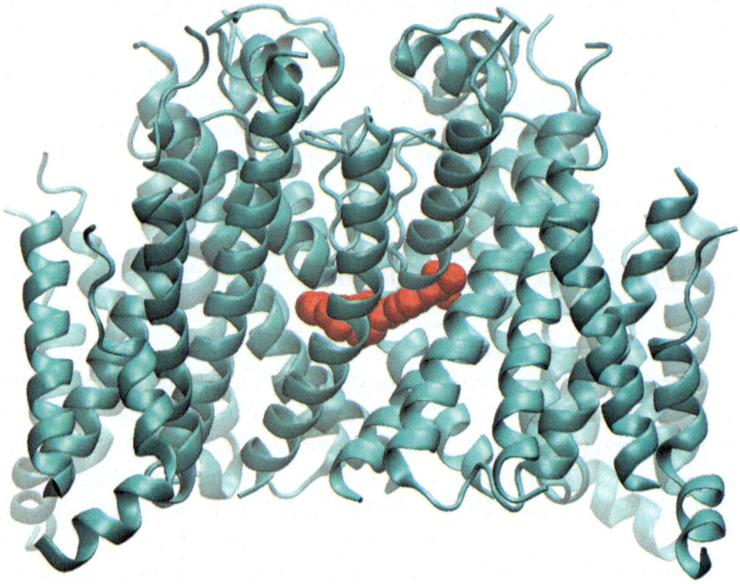

Fig. 11.7 Structure of hERG (PDB ID: 8ZYP). Cyan ribbons represent the transmembrane helices of the human hERG channel. The inhibitor bound in the central cavity is shown as red spheres

This series of studies strongly supports the utility of MD simulation–based drug discovery for wide-ranging applications, from novel lead discovery to safety assessments, thereby reinforcing the value of computational drug discovery.

11.3.4 Additional Applications and Future Perspectives

The FUJI force field is among the most reliable force fields for protein–ligand simulations, and its applications extend beyond the examples mentioned above. For instance, Nasrin et al. [31] combined high-speed atomic force microscopy (AFM) experiments with FUJI-based MD simulations to investigate how slight bending of microtubules (protein filaments that make up the cell's internal scaffold) affects the walking speed of kinesin (a motor protein that uses ATP to carry cargo along microtubules). They found that even minor curvatures increase the interaction energy between microtubules and kinesin, thereby hindering kinesin's progress. In a subsequent study, the same group discovered that low-level tensile deformation boosts dynein-driven cargo transport (dynein is another motor protein that moves in the opposite direction on microtubules using ATP), whereas high-level tension results in microtubule breakage or a complete halt in movement [32]. Taken together, these findings suggest that microtubules are not merely passive "rails" but can function as "mechanosensors," detecting geometric stress (e.g., bending or stretching) and consequently altering motor protein dynamics.

Turning to an example of enzyme activity analysis, a case in point is the functional evaluation of a missense single-nucleotide polymorphism (SNP)—a genomic position where alternative alleles exist in the population and, when one allele replaces another, the resulting codon encodes a different amino acid—within the 17β-hydroxysteroid dehydrogenase 1 (Hsd17b1) gene, which is involved in the ZZ/ZW sex determination of *Seriola* fish [33]. In this specific SNP, the allele carried on the W chromosome converts the codon for glutamic acid (Glu) to one for glycine (Gly). MD simulations based on the FUJI force field found that this SNP substitution disrupts the hydrogen-bond network near the enzyme's active site, leading to reduced steroid-conversion activity. The W-type allele (Gly) therefore predisposes individuals to develop a female phenotype (ovarian differentiation), providing genetic evidence that supports the longstanding hypothesis that steroid hormones contribute to primary sex determination in fish.

The FUJI force field contributes to elucidating molecular mechanisms in a wide range of biological macromolecules, spanning from the mechanical interactions between microtubules and motor proteins to activity modulation resulting from structural changes in hydrogen-bond networks within enzymes. Looking ahead, it is expected to see extensive use not only in drug development but also in the analysis of diverse protein functions.

11.4 Development and Applications of DREIDING-UT Force Field

11.4.1 Development of DREIDING-UT Force Field

The general-purpose DREIDING force field can cover a wide variety of molecular species, but it has been criticized for insufficient accuracy in certain aspects, such as predicting liquid densities. Sasaki and Yamashita [34] revised Lennard–Jones (LJ) parameters (see Eq. 11.2) and partial charges, creating a new version named "DREIDING-UT." To derive the LJ parameters from Buckingham-type potentials:

$$V_{\mathrm{Buckingham}}(r) = A\exp(-Br) - \frac{C}{r^6},$$

they introduced a numerical fitting method. In parallel, partial charges were determined with the RESP method, including solvation effects via the polarizable continuum model (PCM), an implicit-solvent approach that treats the surrounding solvent as a homogeneous dielectric medium, and incorporating dynamic electron correlation (e.g., at the MP2 level). As a result, the calculation accuracy for liquid densities and self-diffusion coefficients has improved dramatically.

11.4.2 Application to Polymer Material and Multiscale Simulation

Although DREIDING-UT was originally developed and refined to accurately reproduce molecular liquids [34], its enhanced accuracy and broader utility also benefit polymer material simulations. For example, the accuracy of force field parameters can significantly affect predictions of the cross-linked network and resulting material properties (such as density and mechanical strength) in epoxy resins. In comparison with MD simulations using the original DREIDING force field [35], those using DREIDING-UT achieved markedly better agreement with experimental densities, demonstrating that the revised parameters effectively enhance predictive accuracy. In the same study [35], a highly accurate coarse-grained (CG) model—where groups of roughly four to six covalently connected heavy atoms are merged into a single interaction "bead," greatly lowering computational cost while preserving key stereochemical features—was successfully developed based on the DREIDING-UT force field and was used to investigate the effect of angular flexibility on density.

In polymer materials like epoxy resins, the degree of cross-linking strongly affects mechanical properties. Shoji and colleagues [36] constructed epoxy resin models with varying cross-link structures and evaluated stress–strain behavior. By combining

CG-MD and finite element method (FEM) in a multiscale approach, they quantitatively demonstrated how microscopic heterogeneities (variations in cross-link structures) impact macroscopic mechanical properties (strain localization, stress reduction). This series of findings indicates that the improved accuracy of DREIDING-UT, in combination with multiscale analyses, can substantially aid structural design and functional prediction of polymer materials.

11.4.3 Incorporating Electronic Polarization Effects

DREIDING-UT refines its Lennard–Jones parameters and partial charges so that the liquid densities and diffusion coefficients of many organic solvents are reproduced accurately. Because it still employs a fixed-charge scheme, however, it cannot capture subtle dielectric behavior. To remedy this limitation, Sasaki and Yamashita [37] introduced explicit electronic polarization by adding a Drude-oscillator model. In this model, every heavy (non-hydrogen) atom carries a negatively charged "Drude particle," whose displacement in the local electric field allows the molecular dipole to fluctuate. The resulting, extended force field—DREIDING-UT(D)—reproduces experimental dielectric constants for hydrocarbon liquids far better than conventional fixed-charge force fields.

A Drude-polarizable force field is computationally more demanding than its fixed-charge counterpart, yet continuing improvements in hardware are steadily closing the cost gap. Applying polarizable force fields such as DREIDING-UT(D) to broader classes of organic molecules, composite materials, and biomolecules should provide more realistic descriptions of solvation effects and of polarization responses in complex systems [38].

11.5 Development and Applications of Multi-state Empirical Valence Bond (MS-EVB) Model

11.5.1 Overview of the MS-EVB Method

The Empirical Valence Bond (EVB) approach, originally proposed by Warshel and Weiss on the basis of the valence bond concept [39], offers an efficient way to model bond formation and cleavage in chemical reactions. A two-state EVB framework was later introduced for proton transfer in solution and enzymes.

First, consider the simple 2-state EVB for the reaction, $AH^+ + B \rightarrow A + BH^+$. We define two diabatic states, $|1\rangle$ (the state with the proton on A) and $|2\rangle$ (the state with the proton on B). They have corresponding potential energies $V_1(R)$ and $V_2(R)$, where R denotes the coordinates. The 2×2 EVB Hamiltonian $H(R)$ is defined as follows:

$$H(R) = \begin{pmatrix} V_1(R) & \varepsilon(R) \\ \varepsilon(R) & V_2(R) \end{pmatrix}.$$

Here, $\varepsilon(R)$ represents the coupling between the diabatic states. Diagonalizing this Hamiltonian yields the eigenvalues $E_{\pm}(R)$:

$$E_{\pm}(R) = \frac{V_1 + V_2}{2} \pm \sqrt{\left(\frac{V_1 - V_2}{2}\right)^2 + \varepsilon^2}.$$

The EVB ground-state energy is given by $E_{\mathrm{EVB}}(R) = E_{-}(R)$. Denoting the corresponding eigenvector by $\boldsymbol{c} = (c_1, c_2)^T$, the energy E_{EVB} and force $F(R)$ are calculated as follows.

$$E_{\mathrm{EVB}} = \boldsymbol{c}^T H \boldsymbol{c}$$

$$F(R) = -\nabla_R E_{\mathrm{EVB}} = -\sum\nolimits_{i,j=1}^{2} c_i c_j, \nabla_R H_{ij}(R)$$

The model can be generalized to N diabatic states (MS-EVB model) [10, 11]. An $N \times N$ Hamiltonian matrix $H(R) = \left[H_{ij}(R)\right]$ $(i, j = 1, \ldots, N)$ is constructed, where its diagonal elements, $H_{ii} = V_i(R)$, represent the molecular mechanics potential for diabatic state i. The off-diagonal elements $H_{ij} = H_{ji}$, $(i \neq j)$ are coordinate-dependent coupling terms. The eigenproblem for this Hamiltonian is solved as follows.

$$H\boldsymbol{c}^{(k)} = E_k \boldsymbol{c}^{(k)}$$

The lowest eigenvalue $E_0(R)$ is the MS-EVB ground-state energy. With its eigenvector $\boldsymbol{c} = \boldsymbol{c}^{(0)}$, the MS-EVB energy $E_{\text{MS-EVB}}(R)$ and force $F(R)$ are calculated as follows.

$$E_{\text{MS-EVB}}(R) = \sum\nolimits_{i,j=1}^{N} c_i c_j H_{ij}(R)$$

$$F(R) = -\nabla_R E_{\text{MS-EVB}} = -\sum\nolimits_{i,j=1}^{N} c_i c_j \nabla_R H_{ij}(R)$$

At each MS-EVB MD simulation step, the relevant states are selected on-the-fly, the MS-EVB Hamiltonian matrix is constructed and diagonalized, and forces are obtained, as explained above.

Standard force fields cannot describe reactive events; ab initio MD simulations can, but they require enormous computational resources. The MS-EVB simulation occupies the middle ground—dramatically cheaper than ab initio MD simulation, while still retaining the essential quantum character of proton transfer—making it particularly suitable for studying proton-transport processes in protein channels, membranes, and other biomolecular environments.

11.5.2 Application to Lipid Bilayer Interfaces

At interfaces of lipid bilayers, where hydrophobic and hydrophilic regions coexist, excess protons often accumulate preferentially on the interfacial side (Fig. 11.8). This phenomenon is commonly called "proton antenna effect." Using the third-generation MS-EVB (MS-EVB3) method [40], Yamashita and Voth [41] performed MD simulations that explicitly treated proton hopping in hydrated bilayers composed of DOPC, DOPE, and DOPG. The calculations revealed strong solvation of the excess proton around the phosphatidyl headgroups: the free energy barrier for migration from bulk water to the membrane surface is essentially negligible, supporting an efficient proton-collection antenna at the interface. Such behavior is difficult for conventional force fields to reproduce, because they cannot describe the breaking and formation of chemical bonds. The MS-EVB framework captures these interfacial solvation mechanisms at molecular resolution by explicitly allowing proton transfer to evolve on-the-fly, and thus provides a powerful tool for analyzing proton dynamics in complex membrane environments.

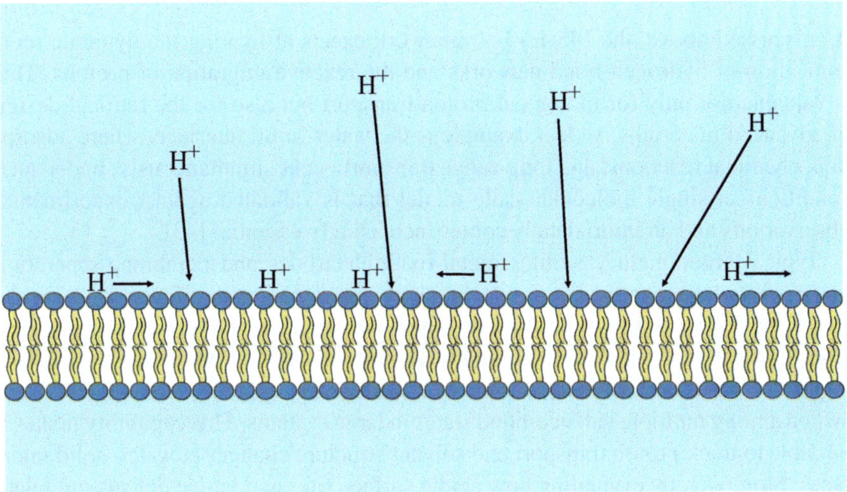

Fig. 11.8 Schematic of the proton antenna at a lipid-bilayer interface. Excess protons (H^+) accumulate at the interfacial zone, where they diffuse laterally

11.5.3 Application to Cytochrome C Oxidase

Cytochrome *c* oxidase (CcO), the terminal enzyme of the mitochondrial electron-transport chain, pumps protons across the membrane while reducing O_2 to H_2O. To dissect the underlying proton-pump mechanism, Yamashita and Voth [42] performed large-scale molecular-dynamics simulations with the third-generation MS-EVB (MS-EVB3) model [40], considering three redox combinations of heme *a* and heme a_3. The MS-EVB calculations reveal a marked redox dependence: when heme a is reduced and heme a_3 is oxidized, the free-energy barrier for a proton to cross the hydrophobic cavity falls to ≈ 5 kcal mol^{-1}, whereas in the other redox states it remains ≥ 15 kcal mol^{-1}. In this low-barrier state, the hydrogen-bond network reorganizes, the incoming proton is stabilized by the D-propionate of heme a_3 (PRD_{a3}), and it is subsequently expelled toward the periplasmic side—supporting a unidirectional pumping step. These results provide a molecular-scale free-energy picture that corroborates experimental evidence for tight coupling between heme redox chemistry and proton translocation, and they demonstrate that the MS-EVB approach is a powerful quantitative tool for investigating proton transport in large membrane proteins.

11.5.4 Application to Advanced Materials

As discussed above, the MS-EVB framework excels at treating the dynamic reorganization of hydrogen-bond networks and the reactive migration of protons. This is valuable not only for biological proton transport but also for the rational design of advanced materials. A clear example is the water–solid interface, where adsorption, chemical reactions, and long-range transport occur simultaneously; under such conditions, a single molecular-scale model that is validated against experimental observations and ab initio data becomes increasingly essential [43].

Typical targets include zeolites, metal-oxide electrodes, and membrane-separation materials, where central questions are how interfacial water forms and breaks hydrogen bonds and how protons or other ions migrate via surface charge sites. Conventional non-reactive force fields cannot model the bond-breaking events required for proton conduction, whereas a reactive approach such as MS-EVB can switch among multiple valence-bond states in large systems. This capability makes it possible to track proton transport and solvent structure changes at water–solid interfaces. Moreover, by evaluating how acidic surface sites and lattice defects modulate water structure and ion mobility on the MS-EVB energy surface, one can directly relate microscopic behavior to catalytic activity or separation performance.

In this way, MS-EVB MD simulations are expected to have a significant impact in materials science. As surface chemistry, electrochemical devices, and separation technologies demand finer control of interfacial phenomena, continued expansion and refinement of MS-EVB models will be crucial for connecting the microscopic dynamics of water and ions to macroscopic materials design and performance.

11.5.5 Prospect of the MS-EVB Method

This section (11.5) has surveyed how the MS-EVB framework enables large-scale molecular-dynamics simulations of bond formation and cleavage, in both biological and materials systems. By optimizing MS-EVB parameters against high-level quantum-chemical data for Zundel- and Eigen-type proton complexes, the method reproduces key experimental observables. Consequently, phenomena that were difficult to capture with conventional force fields, including the "proton-antenna" effect at membrane interfaces and Grötthuss-type shuttling within enzymes, can now be examined accurately in large systems. Looking ahead, still tighter integration with quantum-chemical calculations will allow the MS-EVB parameters to be refined further, paving the way for molecular-level design guidelines in both biological and high-performance materials contexts.

11.6 The Role of Force Fields in Molecular Dynamics Simulations During the AlphaFold Era

Recent advances in AI technology, typified by AlphaFold, have revolutionized the accuracy of three-dimensional (3D) protein-structure prediction and have profoundly influenced life science research and drug discovery (Fig. 11.9). At the same time, MD simulations, whose predictive power has increased in step with continual improvements in force fields, remain indispensable. This section first reviews the historical and theoretical background of protein-structure prediction, then traces the parallel evolution of MD simulations and AlphaFold, and finally outlines the remaining challenges and prospects.

11.6.1 Background: The Protein-Structure Prediction Problem

A key starting point is Anfinsen's dogma, proposed by Anfinsen [44]. It states that the final, native 3D structure of a protein is determined solely by its amino-acid sequence. In contrast, Levinthal highlighted a seeming contradiction, now known as Levinthal's paradox [45]. If a protein were to sample all possible conformations randomly, astronomical timescales would be required, yet folding is observed experimentally on the millisecond-to-second scale. Resolving this paradox led to the idea

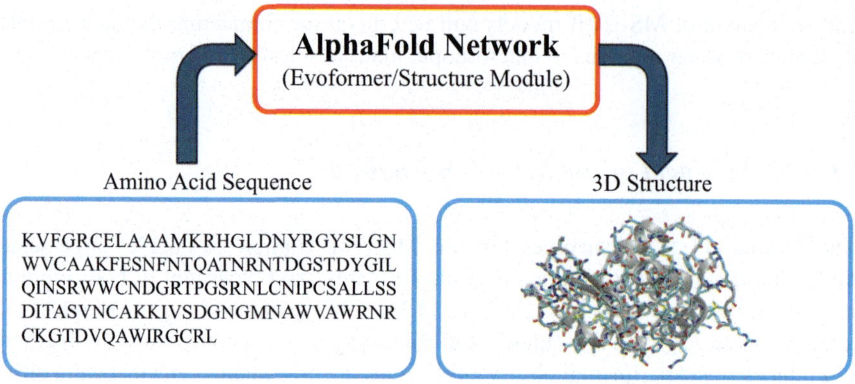

Fig. 11.9 Schematic workflow of AlphaFold. An input amino-acid sequence is processed by the AlphaFold network. The network outputs an all-atom three-dimensional (3D) model of the protein

that proteins follow energetically favorable folding pathways on a funnel-shaped energy landscape. Taken together, Anfinsen's dogma and Levinthal's paradox shaped a theoretical basis of structure prediction.

Since the late 1980s, a variety of computational strategies, including homology modeling and ab initio modeling, have been developed, yet no systematic framework initially existed for comparing their performance. To fill this gap, the Critical Assessment of Protein Structure Prediction (CASP) was launched in 1994 [46]. By challenging participants to predict newly solved but still unpublished experimental structures, CASP has become the premier biennial competition for objectively gauging progress in protein-structure-prediction methodology.

11.6.2 Impact of Anton on Molecular Dynamics Simulations

The special-purpose MD supercomputer, Anton, developed by D. E. Shaw Research (DESRES), attracted worldwide attention, making it possible for simulations to span hundreds of microseconds and even reach the millisecond timescale. These timescales had been virtually impossible on general-purpose hardware [47, 48]. This breakthrough rests on application-specific integrated circuits (ASICs) designed exclusively for MD simulations. These ASICs implement the heavy kernels, including many-body force evaluations and boundary-condition handling, directly in hardware, which enable massive parallelization. In addition, Anton employs a proprietary high-bandwidth interconnect that enables neighboring ASICs to exchange neighbor-list and domain-decomposition data with minimal latency, further boosting strong scaling [49]. Together, these hardware and algorithmic optimizations created the first platform on which millisecond-scale protein simulations became tractable in real wall-clock time.

Beginning around 2010, Anton ran very long MD simulations, allowing researchers to observe complex protein motions at atomic resolution. For example, a 1 ms simulation of bovine pancreatic trypsin inhibitor (BPTI) was reported, revealing repeated transitions among multiple metastable states [50]. Lindorff-Larsen and co-workers [51] employed Anton to simulate several fast-folding proteins and showed that they reached their experimentally determined native states spontaneously. By reproducing, at atomic detail and over realistic timescales, how amino-acid sequences alone encode a protein's higher-order structure, these studies marked a major breakthrough that strengthened the link between structural biology and computational biophysics.

11.6.3 Evolution of Artificial Intelligence (AI) and Impact of AlphaFold

Historical Background

Since the late 2000s, deep learning has driven an extraordinary acceleration in AI research—an advance underscored by the 2024 Nobel Prize in Physics, awarded to Hopfield and Hinton. Hopfield pioneered the Hopfield network, importing ideas from statistical physics to explain how the brain can recall complete patterns from partial cues [52]. Hinton, for his part, laid the theoretical foundations of Boltzmann machines and deep neural networks, demonstrating how multi-layer architectures can be trained efficiently [53].

In 2012, the success of AlexNet, a deep convolutional neural network (CNN) architecture, on the ImageNet challenge popularized large-scale GPU-based training [54], and in 2016 AlphaGo defeated Go champion Lee Sedol, showcasing the power of combining deep neural networks with reinforcement learning [55]. Building on this foundation, DeepMind soon turned its attention to the protein-folding problem.

The Emergence of AlphaFold

By the late 2010s, DeepMind's AlphaFold system had become the focus of intense interest as a method for predicting protein 3D structures directly from amino-acid sequences. AlphaFold has progressed through three generations: AlphaFold1, which won the thirteenth round of the Critical Assessment of Techniques for Protein Structure Prediction (CASP13, 2018); AlphaFold2, which outperformed all other entrants at CASP14 in 2020—the fourteenth biennial installment of CASP—and was widely hailed as a breakthrough [56]; and AlphaFold3 (2024), which extends the approach to protein–protein complexes, nucleic acids, and ligand binding.

AlphaFold2 attained a median Cα RMSD_95 of ≈ 0.96 Å—far superior to the 2.8–4.0 Å typical of earlier approaches—and many targets exhibited all-atom RMSDs below 2 Å, indicating high side-chain accuracy.

How AlphaFold Works

AlphaFold's core advances lie in its deep-learning architecture, which integrates multiple-sequence alignment (MSA) information with predictions of inter-residue distances and contacts:

1. **MSA generation**: Homologous sequences are retrieved and aligned to build an MSA for the query protein.
2. **Neural inference**: The network processes MSA-based co-evolution signals (AlphaFold1 [57] first predicts a residue-distance distogram and then minimizes it to obtain coordinates, whereas AlphaFold2 predicts full-atom coordinates directly, treating the distogram merely as an auxiliary training target.)
3. **Iterative refinement**: Predicted structures are recycled into the network, progressively improving the model. (This recycling step is an enhancement introduced in AlphaFold2 and is not present in AlphaFold1.)

This synergy of large-scale data and sophisticated deep-learning algorithms delivers accuracy well beyond traditional homology or ab initio methods [56, 57].

Developments Since the Critical Assessment of Protein Structure Prediction 14 (CASP14)

DeepMind and EMBL-EBI have released the AlphaFold Protein Structure Database (AlphaFold DB), which now provides more than 200 million high-confidence structure predictions—roughly covering every known protein sequence—and this freely accessible resource has already transformed many areas of biology [58]. Moreover, the 2024 Nobel Prize in Chemistry was awarded to Hassabis and Jumper—the principal architects of AlphaFold—and to Baker, a pioneer of computational protein design [59]. These recognitions highlight the growing centrality of in silico approaches to modern life-science and drug-discovery research.

11.6.4 Future Challenges and Prospects in Protein-Structure Prediction

Anfinsen's dogma—that a protein's amino-acid sequence uniquely determines its fold—together with Levinthal's paradox—the question of how a protein can find its native state so rapidly within an astronomically large conformational space—has long underpinned research in structure prediction. Today, physics-based MD simulations and data-driven, AI-based approaches typified by AlphaFold have continued to develop in a complementary fashion, each leveraging its own strengths.

AlphaFold attains remarkable accuracy for static monomeric structures, yet challenges remain for complexes, large conformational changes, and ligand binding. These limitations make time-resolved MD simulations indispensable; ongoing improvements in force-field accuracy, together with special-purpose hardware such as Anton, now make microsecond-scale simulations routine for such challenging systems and extend to millisecond timescales for small- to medium-sized proteins [50]. Meanwhile, a landmark paper published recently presents AlphaFold3, which extends AI-based prediction to protein–protein complexes, nucleic-acid assemblies, and even protein–ligand interactions, thereby narrowing the gap that once required purely physics-based methods [60]. Even so, MD remains essential for exploring long-timescale dynamics and entropy-driven processes.

Hybrid strategies that integrate AI with MD are also attracting attention. Examples include refining high-accuracy AlphaFold models by MD, or conversely, analyzing the rich dynamical ensembles generated by MD with machine-learning techniques [61]. As force fields are further enriched with quantum-chemical data and machine-learning corrections, and as ever-larger experimental datasets become available, the field is poised to move from structure prediction to comprehensive function prediction. Harnessing both data-driven and physics-based modeling in tandem should usher in a new era in which increasingly complex biomolecular phenomena can be elucidated—thereby accelerating progress in life-science research and drug discovery.

11.7 Conclusions

This chapter has provided an integrated overview of recent progress in force-field development grounded in quantum chemistry and the molecular dynamics simulation techniques that exploit those force fields. Case studies such as the FUJI and DREIDING-UT force fields demonstrate that refining torsional and non-bonded parameters greatly improves the fidelity with which molecular structures and thermodynamic properties are reproduced, thereby boosting predictive power in both biological and materials applications. Reactive models like MS-EVB, which can handle proton transfer and other chemistry that lies beyond the reach of traditional fixed-bond MD, are likewise emerging as indispensable tools for large-scale, atomistic investigations of complex reactive phenomena.

These advances build on a succession of pioneering studies that have been recognized by the Nobel Prize in Chemistry (Table 11.1). In 2013, Karplus, Levitt, and Warshel were honored for multiscale models that seamlessly couple quantum mechanics with molecular mechanics, enabling tractable simulations of sizeable biomolecular systems; Karplus helped drive the development of the CHARMM force field, while Warshel laid the foundations of EVB theory. Earlier, Pople and Kohn (1998 laureates) revolutionized quantum chemistry methodologies, the GAUSSIAN program package, and density-functional theory, thereby delivering practical electronic-structure calculations that now underpin modern force-field parameterization.

In parallel, deep-learning breakthroughs, typified by AlphaFold, have taken static protein-structure prediction to unprecedented levels of accuracy. The field is now poised to merge data-driven AI approaches with high-accuracy, quantum-informed force fields. Such integration promises three main advances: (i) AI and force fields will increasingly play complementary roles; (ii) predictive scope will expand from static structures to functions and reactions; and (iii) multiscale simulations will be

Table 11.1 Nobel Prize laureates related to theoretical or computational chemistry

Year	Award	Laureate(s)	Reason
1922	Physics	Niels Bohr	For the development of the theory of atomic structure
1932	Physics	Werner Heisenberg	For the establishment of quantum mechanics (matrix mechanics) and the uncertainty principle
1933	Physics	Erwin Schrödinger Paul Dirac	For pioneering new forms of quantum theory (wave mechanics)
1954	Chemistry	Linus Pauling	For his work on the nature of the chemical bond
1966	Chemistry	Robert Mulliken	For the theoretical development of the molecular orbital (MO) method
1981	Chemistry	Kenichi Fukui Roald Hoffmann	For developing theories that explain chemical reactions via molecular orbitals
1992	Chemistry	Rudolph Marcus	For developing the theory of electron transfer reactions
1998	Chemistry	Walter Kohn John Pople	For the development of density functional theory (DFT) and the establishment of quantum chemical computational methods
2013	Chemistry	Martin Karplus Michael Levitt Arieh Warshel	For the development of multiscale models that enable the study of complex chemical systems
2024	Physics	John Hopfield Geoffrey Hinton	For pioneering work in machine learning using artificial neural networks
2024	Chemistry	David Baker Demis Hassabis John Jumper	For the development of computational protein design methods and AI-based protein structure prediction

refined and accelerated. Progress in "AI force fields," though not covered here in detail [6, 7, 62–68], will be central to this evolution.

11.8 Outlook

As these trends converge, the "molecular-level precision" afforded by quantum chemistry and the "long-time, large-system reach" of classical MD will be unified through machine-learning pipelines. The resulting platforms should enable comprehensive, predictive insight into systems ranging from biomacromolecules to advanced materials. Beyond drug discovery and materials design, such integrated frameworks are expected to spur innovation across condensed-matter physics, chemical engineering, environmental science, and other disciplines. Deeper synergy among quantum chemistry, machine learning, and multiscale simulation will, in turn, allow us to unravel the behavior of ever more complex systems and to translate that knowledge into transformative technologies.

Acknowledgements The preparation of this book was made possible by the generous support of my colleagues, mentors, and collaborators. I am especially grateful to those who offered incisive advice and detailed feedback on the manuscript, as well as to the many conference and workshop participants who enriched my thinking through lively discussions over the years. Their guidance has enabled me to engage deeply with pioneering work in quantum chemistry and molecular simulation and has been instrumental in shaping the vision presented here.

Glossary

1. Basis Set Superposition Error (BSSE)

Basis Set Superposition Error (BSSE) arises in quantum-chemical calculations of intermolecular interaction energies. When two monomers A and B are brought together, each monomer can "borrow" basis functions from the other, effectively using a larger basis set than in the isolated calculations. This results in the overestimation of the interaction energy.

2. Complete Basis Set (CBS) Limit

Complete Basis Set (CBS) limit is the theoretical result you would obtain if your quantum-chemical calculation used a mathematically perfect, infinitely large basis set. Because only finite basis sets can be used in practice, the CBS limit is approximated by performing calculations with systematically larger basis sets and then extrapolating their results.

3. Steered Molecular Dynamics (SMD)

Steered Molecular Dynamics (SMD) applies an external force to one or more atoms or collective coordinates, dragging the system along a predefined direction or distance.

4. Targeted Molecular Dynamics (TMD)

Targeted Molecular Dynamics (TMD) applies a restraint that gradually lowers the deviation between the instantaneous structure and a chosen target conformation, effectively changing the system toward the endpoint.

5. Computational Alanine Scan Analysis

Computational alanine scan analysis systematically replaces each residue at a protein–protein or protein–ligand interface with alanine in silico and calculates the resulting change in binding free energy to identify energetic hot spots critical for affinity and specificity.

6. Replica Exchange with Solute Tempering (REST2)

Replica Exchange with Solute Tempering (REST2) accelerates conformational sampling by running multiple replicas in which only the solute and solute–solvent interactions are scaled—effectively heating the molecule while keeping the bulk solvent at physiological temperature—and exchanging configurations between replicas.

7. Finite Element Method (FEM)

Finite Element Method (FEM) solves partial differential equations on complex domains by dividing the continuum into discrete, element-wise polynomial sub-regions, assembling their local stiffness matrices into a global system, and numerically computing field variables such as displacement, stress, or temperature.

References

1. Pople JA. Nobel lecture: quantum chemical models. Rev Mod Phys. 1999;71:1267–74. https://doi.org/10.1103/RevModPhys.71.1267.
2. Bartlett RJ, Musiał M. Coupled-cluster theory in quantum chemistry. Rev Mod Phys. 2007;79:291–352. https://doi.org/10.1103/RevModPhys.79.291.
3. Verma P, Truhlar DG. Status and challenges of density functional theory. Trends Chem. 2020;2:302–18. https://doi.org/10.1016/j.trechm.2020.02.005.
4. de Jong WA, Bylaska EJ, Govind N, et al. High-performance computing in quantum chemistry. Phys Chem Chem Phys. 2010;12:6896–920. https://doi.org/10.1039/C002859B.
5. Ufimtsev IS, Martinez TJ. Quantum chemistry on graphical processing units. 1. Strategies for two-electron integral evaluation. J Chem Theory Comput. 2008;4:222–231. https://doi.org/10.1021/ct700268q
6. Behler J, Parrinello M. Generalized neural-network representation of high-dimensional potential-energy surfaces. Phys Rev Lett. 2007;98: 146401. https://doi.org/10.1103/PhysRevLett.98.146401.

7. Houston PL, Qu C, Yu Q, Pandey P, Conte R, Nandi A, Bowman JM, Kukolich SG. Formic acid-ammonia heterodimer: A new Δ-machine learning CCSD(T)-level potential energy surface allows investigation of the double proton transfer. J Chem Theory Comput. 2024;20:1821–8. https://doi.org/10.1021/acs.jctc.3c01273.
8. Jorgensen WL, Tirado-Rives J. The OPLS potential functions for proteins: energy minimizations for crystals of cyclic peptides and crambin. J Am Chem Soc. 1988;110:1657–66. https://doi.org/10.1021/ja00214a001.
9. Cornell WD, Cieplak P, Bayly CI, et al. A second generation force field for the simulation of proteins, nucleic acids, and organic molecules. J Am Chem Soc. 1995;117:5179–97. https://doi.org/10.1021/ja00124a002.
10. Schmitt UW, Voth GA. A multistate empirical valence bond model for proton transport in water. J Phys Chem B. 1998;102:5547–51. https://doi.org/10.1021/jp9818131.
11. Voth GA. Computer simulation of proton solvation and transport in aqueous and biomolecular systems. Acc Chem Res. 2006;39:143–50. https://doi.org/10.1021/ar0402098.
12. Senftle TP, Hong S, Islam MM, et al. The ReaxFF reactive force-field: development, applications and future directions. npj Comput Mater. 2016;2:15011. https://doi.org/10.1038/npjcompumats.2015.11.
13. Fujitani H, Matsuura A, Sakai S, Sato H, Tanida Y. High-level ab initio calculations to improve protein backbone dihedral parameters. J Chem Theory Comput. 2009;5:1155–65. https://doi.org/10.1021/ct8005437.
14. Yamashita T. Improvement in empirical potential functions for increasing the utility of molecular dynamics simulations. JPS Conf Proc. 2015;5: 010003. https://doi.org/10.7566/JPSCP.5.010003.
15. Mahmood MI, Yamashita T. Influence of lipid bilayer on the GPCR structure: comparison of all-atom lipid force fields. Bull Chem Soc Jpn. 2021;94:2569–74. https://doi.org/10.1246/bcsj.20210244.
16. Kamiya N, Kayanuma M, Fujitani H, Shinoda K. A new lipid force field (FUJI). J Chem Theory Comput. 2020;16:3664–76. https://doi.org/10.1021/acs.jctc.9b01195.
17. Yamashita T, Mitsui T, Sasaki K, et al. Effect of N343 glycosylation and N501Y mutation on the SARS-CoV-2 spike protein: modeling and MD simulations. AIP Conf Proc. 2022;2611: 020008. https://doi.org/10.1063/5.0119713.
18. Yamashita T, Fujitani H. On accurate calculation of the potential of mean force between antigen and antibody: a case of the HyHEL-10–hen egg white lysozyme system. Chem Phys Lett. 2014;609:50–3. https://doi.org/10.1016/j.cplett.2014.06.028.
19. Shinoda K, Fujitani H. Initiation of prolyl cis–trans isomerisation in the CDR-H3 loop of an antibody in response to antigen binding. Sci Rep. 2017;7:16964. https://doi.org/10.1038/s41598-017-16766-8.
20. Nakayama T, Mizohata E, Yamashita T, et al. Structural features of interfacial tyrosine residue in ROBO1 fibronectin domain–antibody complex: crystallographic, thermodynamic, and molecular dynamic analyses. Protein Sci. 2015;24:328–40. https://doi.org/10.1002/pro.2619.
21. Yamashita T, Mizohata E, Nagatoishi S, et al. Affinity improvement of a cancer-targeted antibody through alanine-induced adjustment of antigen–antibody interface. Structure. 2019;27:519-27.e5. https://doi.org/10.1016/j.str.2018.11.002.
22. Yamashita T, Okajima R, Miyanabe K, Tsumoto K. Modified AMBER force-field (FUJI) parameters for sulfated and phosphorylated tyrosine residues: development and application to CCR5-derived peptide systems. AIP Conf Proc. 2019;2186: 030013. https://doi.org/10.1063/1.5137924.
23. Okajima R, Hiraoka S, Yamashita T. Environmental effects on salt bridge stability in the protein–protein interface: the case of hen egg-white lysozyme and its antibody, HyHEL-10. J Phys Chem B. 2021;125:1542–9. https://doi.org/10.1021/acs.jpcb.0c09248.
24. Takamatsu Y, Hamakubo T, Yamashita T. Molecular dynamics simulation of the antigen–antibody complex formation process between hen egg-white lysozyme and HyHEL-10. Bull Chem Soc Jpn. 2022;95:1611–9. https://doi.org/10.1246/bcsj.20220239.

25. Yamashita T. A molecular dynamics study on the PD-1–Nivolumab interaction: roles of the N-terminal loop and N58-glycosylation of PD-1. AIP Conf Proc. 2024;3030: 020009. https://doi.org/10.1063/5.0192852.
26. Yamashita T, Ueda A, Mitsui T, et al. Molecular dynamics simulation-based evaluation of the binding free energies of computationally designed drug candidates: importance of the dynamical effects. Chem Pharm Bull. 2014;62:661–7. https://doi.org/10.1248/cpb.c14-00132.
27. Yamashita T, Ueda A, Mitsui T, et al. The feasibility of an efficient drug design method with high-performance computers. Chem Pharm Bull. 2015;63:147–55. https://doi.org/10.1248/cpb.c14-00596.
28. Negami T, Araki M, Okuno Y, Terada T. Calculation of absolute binding free energies between the hERG channel and structurally diverse drugs. Sci Rep. 2019;9:16586. https://doi.org/10.1038/s41598-019-53120-6.
29. Negami T, Terada T. Calculations of the binding free energies of the Comprehensive in vitro Proarrhythmia Assay (CiPA) reference drugs to cardiac ion channels. Biophys Physicobiol. 2023;20: e200016. https://doi.org/10.2142/biophysico.bppb-v20.0016.
30. Mitsui T, Wada M, Kamiya N, et al. Binding pose prediction of a drug candidate, cepharanthine, targeting the SARS-CoV-2 spike protein using large-scale MD simulations. AIP Conf Proc. 2022;2611: 020009. https://doi.org/10.1063/5.0119741.
31. Nasrin SR, Ganser C, Nishikawa S, et al. Deformation of microtubules regulates translocation dynamics of kinesin. Sci Adv. 2021;7:eabf2211. https://doi.org/10.1126/sciadv.abf2211.
32. Nasrin SR, Yamashita T, Ikeguchi M, et al. Tensile stress on microtubules facilitates dynein-driven cargo transport. Nano Lett. 2024;24:9129–36. https://doi.org/10.1021/acs.nanolett.4c00209.
33. Koyama T, Nakamoto M, Morishima K, et al. A SNP in a steroidogenic enzyme is associated with phenotypic sex in Seriola fishes. Curr Biol. 2019;29:1901-9.e8. https://doi.org/10.1016/j.cub.2019.04.069.
34. Sasaki K, Yamashita T. Modification and validation of the DREIDING force field for molecular liquid simulations (DREIDING-UT). J Chem Inf Model. 2021;61:1172–9. https://doi.org/10.1021/acs.jcim.0c01169.
35. Shoji N, Sasaki K, Uedono A, et al. Effect of conversion on epoxy resin properties: combined molecular dynamics simulation and experimental study. Polymer. 2022;254: 125041. https://doi.org/10.1016/j.polymer.2022.125041.
36. Shoji N, Yamashita T. Tensile stress reduction and strain concentration of epoxy resin caused by heterogeneity: a multiscale approach combining molecular dynamics simulation and finite element method. J Phys Chem B. 2023;127:9066–73. https://doi.org/10.1021/acs.jpcb.3c04391.
37. Sasaki K, Yamashita T. Structural effect on the dielectric constant of hydrocarbon liquids: a molecular dynamics study using the Drude polarizable force field. Bull Chem Soc Jpn. 2023;96:1258–61. https://doi.org/10.1246/bcsj.20230100.
38. Lemkul JA, Huang J, Roux B, MacKerell AD Jr. An empirical polarizable force field based on the classical Drude oscillator model: development history and recent applications. Chem Rev. 2016;116:4983–5513. https://doi.org/10.1021/acs.chemrev.5b00505.
39. Shaik S, Danovich D, Hiberty PC. Valence bond theory—its birth, struggles with molecular orbital theory, its present state and future prospects. Molecules. 2021;26:1624. https://doi.org/10.3390/molecules26061624.
40. Wu Y, Chen H, Wang F, Paesani F, Voth GA. An improved multistate empirical valence bond model for aqueous proton solvation and transport. J Phys Chem B. 2008;112:467–82. https://doi.org/10.1021/jp076658h.
41. Yamashita T, Voth GA. Properties of hydrated excess protons near phospholipid bilayers. J Phys Chem B. 2010;114:592–603. https://doi.org/10.1021/jp908768c.
42. Yamashita T, Voth GA. Insights into the mechanism of proton transport in cytochrome c oxidase. J Am Chem Soc. 2012;134:1147–52. https://doi.org/10.1021/ja208638w.
43. Barry E, Burns R, Chen W, et al. Advanced materials for energy-water systems: the central role of water/solid interfaces in adsorption, reactivity, and transport. Chem Rev. 2021;121:9450–501. https://doi.org/10.1021/acs.chemrev.1c00069.

44. Anfinsen CB. Principles that govern the folding of protein chains. Science. 1973;181:223–30. https://doi.org/10.1126/science.181.4096.223.
45. Levinthal C. How to fold graciously. In: Mössbauer spectroscopy in biological systems (Proc. Allerton House, Monticello, IL, 1969). p. 22–6.
46. Moult J, Pedersen JT, Judson R, Fidelis K. A large-scale experiment to assess protein structure prediction methods. Proteins. 1995;23(S2):ii–iv. https://doi.org/10.1002/prot.340230303.
47. Shaw DE, Deneroff MM, Dror RO, et al. Anton: a special-purpose machine for molecular dynamics simulation. ACM SIGARCH Comput Archit News. 2007;35:1–12. https://doi.org/10.1145/1273440.1250664.
48. Shaw DE, Deneroff MM, Dror RO, et al. Anton: a special-purpose machine for molecular dynamics simulation. Commun ACM. 2008;51:91–7. https://doi.org/10.1145/1364782.1364802.
49. Dror RO, Dirks RM, Grossman JP, Xu H, Shaw DE. Biomolecular simulation: a computational microscope for molecular biology. Annu Rev Biophys. 2012;41:429–52. https://doi.org/10.1146/annurev-biophys-042910-155245.
50. Shaw DE, Maragakis P, Lindorff-Larsen S, et al. Atomic-level characterization of the structural dynamics of proteins. Science. 2010;330:341–6. https://doi.org/10.1126/science.1187409.
51. Lindorff-Larsen S, Piana S, Dror RO, Shaw DE. How fast-folding proteins fold. Science. 2011;334:517–20. https://doi.org/10.1126/science.1208351.
52. Hopfield JJ. Neural networks and physical systems with emergent collective computational abilities. Proc Natl Acad Sci U S A. 1982;79:2554–8. https://doi.org/10.1073/pnas.79.8.2554.
53. Hinton GE. Connectionist learning procedures. Artif Intell. 1989;40:185–234. https://doi.org/10.1016/0004-3702(89)90049-0.
54. Krizhevsky A, Sutskever I, Hinton GE. ImageNet classification with deep convolutional neural networks. Adv Neural Inf Process Syst. 2012;25:1097–105.
55. Silver D, Huang A, Maddison CJ, et al. Mastering the game of Go with deep neural networks and tree search. Nature. 2016;529:484–9. https://doi.org/10.1038/nature16961.
56. Jumper J, Evans R, Pritzel A, et al. Highly accurate protein structure prediction with AlphaFold. Nature. 2021;596:583–9. https://doi.org/10.1038/s41586-021-03819-2.
57. Senior AW, Evans R, Jumper J, et al. Improved protein structure prediction using potentials from deep learning. Nature. 2020;577:706–10. https://doi.org/10.1038/s41586-019-1923-7.
58. Varadi M, Anyango S, Deshpande M, et al. AlphaFold protein structure database: expansions and open access to protein structure predictions. Nucleic Acids Res. 2022;50:D439–44. https://doi.org/10.1093/nar/gkab1061.
59. Huang P-S, Boyken SE, Baker D. The coming of age of de novo protein design. Nature. 2016;537:320–7. https://doi.org/10.1038/nature19946.
60. Abramson J, Adler J, Dunger J, et al. Accurate structure prediction of biomolecular interactions with AlphaFold 3. Nature. 2024;630:493–500. https://doi.org/10.1038/s41586-024-07487-w.
61. Yamashita T, Miyamura N, Kawai S. Classification of the HCN isomerization reaction dynamics in Ar buffer gas via machine learning. J Chem Phys. 2023;159: 124116. https://doi.org/10.1063/5.0156313.
62. Schran C, Michaelides A. Machine learning potentials for complex aqueous systems made simple. Proc Natl Acad Sci U S A. 2021;118: e2110077118. https://doi.org/10.1073/pnas.2110077118.
63. Takamoto S, Ibuka T. Towards universal neural network potential for material discovery (PFP). Nat Commun. 2022;13:2991. https://doi.org/10.1038/s41467-022-30687-9.
64. Musaelian A, Kozinsky B. Learning local equivariant representations for large-scale atomistic dynamics (Allegro). Nat Commun. 2023;14:579. https://doi.org/10.1038/s41467-023-36329-y.
65. Deng B, Ceder G. CHGNet as a pretrained universal neural network potential for charge-informed atomistic modelling. Nat Mach Intell. 2023;5:1031–41. https://doi.org/10.1038/s42256-023-00716-3.
66. Wang T, He X, Li M, et al. Ab initio characterization of protein molecular dynamics with AI^2BMD. Nature. 2024;635:1019–27. https://doi.org/10.1038/s41586-024-08127-z.

67. Unke OT, Müller KR. Biomolecular dynamics with machine-learned quantum-mechanical force fields (GEMS). Sci Adv. 2024;10:eadn4397. https://doi.org/10.1126/sciadv.adn4397.
68. Seute L, Gräter F. Grappa—a machine learned molecular mechanics force field. Chem Sci. 2025;16:2907. https://doi.org/10.1039/D4SC05465B.

Part V
Applied Physics and Physical Chemistry

Chapter 12
Efficient Spin-Charge Conversion in Two-Dimensional Electron-Gas Systems in Oxides for Future Energy-Efficient Data Processing Technology

Shinobu Ohya

Abstract This chapter discusses the interplay between spin and charge currents, which is a key phenomenon for next-generation low-power spin orbitronics. As the demand for energy-efficient devices surges, the reduction of power consumption has become a critical challenge. One promising solution lies in the development of non-volatile memory technologies. The spin current can be converted to electrical currents (and vice versa) and holds immense potential for energy-efficient applications. They could be particularly useful in devices like spin-orbit torque magnetic random-access memory, which is a promising candidate for significantly reducing power consumption in modern memory technologies. Materials with strong spin-orbit interactions such as Pt, Ta, and W are generally used for spin-charge conversions through the spin Hall effect. Recently, however, two-dimensional electron gases (2DEGs) that are formed at the interfaces of materials like $LaAlO_3/SrTiO_3$ have demonstrated exceptional efficiency in spin-current conversion. This is driven by the strong spin-orbit coupling arising from intense electric fields at the interfaces. This chapter also highlights a promising heterostructure of $(La,Sr)MnO_3/LaTiO_3/SrTiO_3$, which enhances the spin-charge conversion efficiency. Currently, outstanding device performance has been achieved at only low temperatures, but future research will be done to push the boundaries and achieve high efficiency at higher temperatures, which could pave the way for applications such as magnetization reversal.

Keywords Spin-charge conversion · Spin-orbitronics · Spin Hall effect · Two-dimensional electron gas · Non-volatile memory

---

S. Ohya (✉)
Department of Electrical Engineering and Information Systems, The University of Tokyo, Tokyo, Japan
e-mail: ohya@cryst.t.u-tokyo.ac.jp

Center for Spintronics Research Network, Graduate School of Engineering, The University of Tokyo, Tokyo, Japan

Institute for Nano Quantum Information Electronics, The University of Tokyo, Tokyo, Japan

T. Onishi (ed.), *Advances in Quantum Science II*,
https://doi.org/10.1007/978-981-95-9721-5_12

12.1 Introduction

12.1.1 *Challenges in Magnetization Switching for Non-volatile Spintronic Devices*

The energy consumed for data processing is rapidly increasing with the growing volume of data, which is especially being driven by the recent extraordinary advancements in artificial intelligence and the Internet of Things. A significant portion of this power consumption stems from transistors. Due to the volatility of stored data, a constant supply of power is required to maintain it. Spintronics is a field of research that aims to reduce the power consumption of electronic devices by utilizing the spin degree of freedom of electrons. Magnetic random-access memory (MRAM) has emerged as a promising candidate for future non-volatile spintronics memory technologies. This memory relies on a tri-layer structure known as a magnetic tunnel junction (MTJ), which consists of a thin insulating tunnel barrier sandwiched between two ferromagnetic layers (Fig. 12.1a). Data are stored based on the parallel or antiparallel magnetization configurations of these ferromagnetic layers. However, one of the key challenges in MRAM development is reducing the power consumption required for writing data. Previously, the spin transfer torque method was widely used. In this approach, a large current is applied to transfer the spin of carriers to the magnetization. However, this necessitates a high bias voltage across the thin tunnel barrier, which reduces device durability.

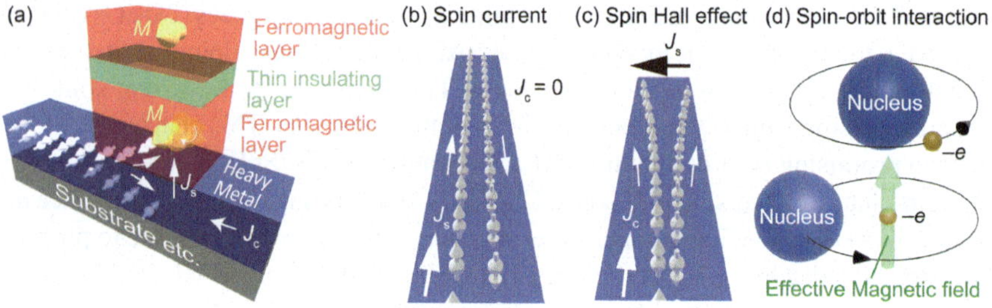

Fig. 12.1 **a** Device structure of spin-orbit torque—magnetic random-access memory (SOT-MRAM). Due to the spin hall effect, the charge current flowing in the heavy metal layer is converted to the spin current, which is injected into the magnetic tunnel junction (MTJ) composed of the ferromagnetic layer/insulating layer/ferromagnetic layer. Then, the magnetization of the bottom ferromagnetic layer is reversed. **b, c** Schematic illustration of the spin current **b** and the spin Hall effect **c**. Here, J_S, J_C, and M represent spin current, charge current, and magnetization. **d** Schematic illustration of the origin of spin–orbit interaction. In the classical picture, while electrons orbit the nucleus, viewing the system from the electron's rest frame makes it appear as though the positively charged nucleus revolves around the electron. This apparent motion generates an effective magnetic field at the electron's position. The interaction between this field and the electron's spin gives rise to the spin–orbit interaction

12.1.2 Spin-Orbit Torque Magnetization Reversal

More recently, spin-orbit torque (SOT) magnetization reversal has been developed as an alternative (Fig. 12.1a). A (pure) spin current—where up-spin and down-spin electrons flow in opposite directions, resulting in zero net charge current, as shown in Fig. 12.1b—is generated by the spin Hall effect induced by a charge current flowing through a non-magnetic layer, which is typically a heavy metal such as Pt, Ta, or W, as illustrated in Fig. 12.1c. The spin current is injected into the MTJ and reverses the magnetization of its bottom ferromagnetic layer. This highlights the importance of developing efficient methods to convert a charge current into a spin current. Conversely, materials can also exhibit the reverse process where a spin current is converted into a charge current. In bulk materials, this phenomenon is referred to as the inverse spin Hall effect.

12.1.3 Spin-Charge Interconversion

Such spin-to-charge conversion processes have inspired innovative devices, including Intel's magneto-electric spin-orbit (MESO) device, which leverages this principle for advanced computing applications. Both the charge-to-spin and spin-to-charge conversion effects generally originate from carrier scattering facilitated by spin-orbit interactions. Consequently, materials containing heavy atoms are often employed for spin-charge interconversion, as they exhibit strong spin-orbit coupling—a relativistic interaction between an electron's spin and its orbital motion around the nucleus (Fig. 12.1d)—that facilitates the conversion between spin and charge currents.

Spin-charge interconversion can be induced at the interface between different materials by using the perpendicular electric field acting at the interface to generate strong spin-orbit interaction. In this case, the materials themselves are not necessarily required to exhibit large spin-orbit interaction. As a result, a wide variety of material combinations can serve as candidates for such systems.

12.1.4 Spin-Charge Interconversions in Two-Dimensional Electron Gas Systems

Inverse Edelstein Effect

A two-dimensional electron gas (2DEG), which is a confined layer of electrons free to move only in two dimensions, can sometimes form at the interface between materials. In general, 2DEG is often created due to the strong electric field present at the interface. The magnitude of this electric field depends on the combination of materials forming the interface, and it allows significant spin-orbit interaction to be achieved without relying on heavy atoms. The phenomenon of converting a spin current into a charge current within the 2DEG region is referred to as the inverse Edelstein effect (IEE), which has garnered attention as a promising mechanism for spin-charge interconversion in spintronic devices.

Rashba-Type Spin-Orbit Interaction

Figure 12.2 illustrates the principle of spin-charge interconversion in a 2DEG that forms at the interface of materials with simple parabolic bands. As mentioned, the 2DEG region experiences an electric field in the direction perpendicular to the film plane due to the confinement potential of the electrons. This electric field gives rise to Rashba-type spin-orbit interaction described by the Hamiltonian:

$$H_R = \alpha_R\left(k_y\sigma_x - k_x\sigma_y\right)/\hbar,$$

where α_R is the Rashba coefficient, k_x and k_y are electron wave numbers, and σ_x and σ_y are Pauli's operators in the x and y directions, respectively. As a result of this effect, if the horizontal axis is defined as k_x, the bands split along this axis direction depending on the spin orientation, as shown in Fig. 12.2a. This splitting occurs isotropically within the two-dimensional k_x–k_y plane, resulting in the formation of outer and inner circles at the Fermi level (E_F), each of which has opposite spin winding directions (dotted lines in Fig. 12.2b). When a spin current is injected into the system, up-spin electrons are primarily injected into the right side of the outer circle, while down-spin electrons are emitted from the left side of the same circle. This imbalance causes a slight increase in the chemical potential on the right side due to the excess of up-spin electrons and a slight decrease on the left side due to the deficit of down-spin electrons. Consequently, the outer Fermi circle shifts toward the right side. The opposite phenomenon occurs for the inner circle, but the outer circle is larger than the inner circle, which leads to a net movement of electrons to the right. The net electron motion generates a charge current from the injected spin current. The efficiency of this spin-to-charge current conversion is quantified by the inverse Edelstein length λ_{IEE}, which is defined as the ratio of the generated two-dimensional current density j_{2D} to the injected three-dimensional spin current density j_s. Thus, λ_{IEE} has units of length. In the simple parabolic band model, λ_{IEE} is expressed as $\alpha_R\tau/\hbar$, where τ is the momentum relaxation time.

$LaAlO_3/SrTiO_3$ (LAO/STO) Interface

The IEE was first discovered at an Ag/Bi interface, for which λ_{IEE} was reported to be 0.3 nm [1]. Following this discovery, the phenomenon was observed at various other interfaces [2–5]. It is well known that a 2DEG forms at the interface between the insulating perovskite oxides $LaAlO_3$ (LAO) and $SrTiO_3$ (STO) [6], both of which are complex transition metal oxides with a perovskite crystal structure as illustrated

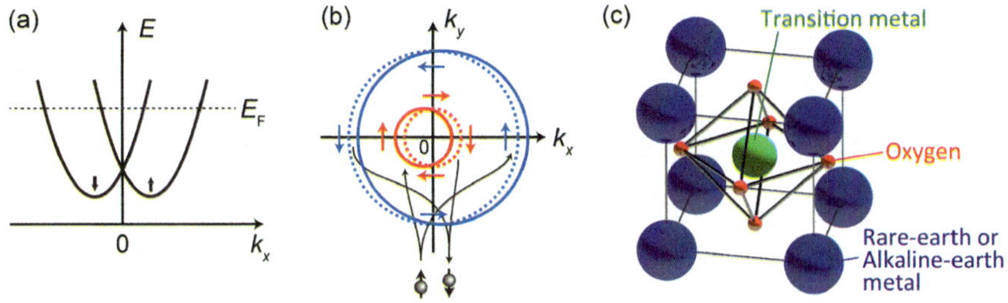

Fig. 12.2 **a** Simple parabolic band split by the Rashba spin-orbit interaction in a two-dimensional electron gas (2DEG). The dotted line represents the Fermi level (E_F). **b** Principle of spin-to-charge conversion via the inverse Edelstein effect (IEE). When the spin current is injected into the $LaAlO_3$ (LAO)/$SrTiO_3$ (STO) interface, it causes a shift of the outer and inner Fermi circles, resulting in the generation of a charge current in the x direction. Here, the dotted lines represent the original Fermi circles, while the solid lines indicate their positions after the spin current injection. **c** Schematic illustration of a perovskite-oxide lattice structure

in Fig. 12.2c. This system has demonstrated the ability to modulate spin-orbit interactions through gate voltage application and has garnered significant attention for spintronics applications [7]. Subsequently, several groups have also reported IEE in the 2DEG that forms at the $LaAlO_3$ (LAO)/$SrTiO_3$ (STO) interface [8–11] with large λ_{IEE} of up to $-$ 6.4 nm [8]. In all these studies, polycrystalline ferromagnetic materials such as permalloy were used as spin injectors into the 2DEG. However, it is well known that spin scattering can occur due to crystal irregularities or randomness. Considering this limitation, studies on single-crystalline all-epitaxial heterostructures are essential for understanding the intrinsic physics of spin-charge conversion.

12.2 Intrinsic Inverse Edelstein Effect in an all-Epitaxial Single-Crystalline ($La_{0.67}$, $Sr_{0.33}$) MnO_3/LAO/STO Heterostructure

12.2.1 Heterostructure Growth Using Molecular Beam Epitaxy

We grew an all-epitaxial single-crystalline heterostructure consisting of ($La_{0.67}$, $Sr_{0.33}$) MnO_3 (LSMO) [30 unit cell (uc) = 12 nm]/LAO (2 uc = 0.8 nm) on an STO (001) substrate using molecular beam epitaxy (Fig. 12.3a) [12, 13]. We employed a shuttered growth technique for all layers. A mixture of 20% ozone and 80% oxygen was used as the oxygen source during growth. We grew the LSMO and LAO layers at 730 °C. The thickness of the LAO layer was set to 2 uc because a 2DEG is formed at the LAO/STO interface when the LAO film thickness reaches 2 uc or more in the presence of a metallic overlayer [14, 15]. All layers were single crystalline and coherently grown on the STO substrate, as shown in the high-angle annular dark-field scanning transmission electron microscopy (HAADF-STEM) lattice image in Fig. 12.3b. Atomic force microscopy (AFM) analysis showed that the substrate surface was atomically flat with uc-height steps (~ 0.4 nm) (Fig. 12.3c). We prepared samples A and B: In sample A, we adjusted the composition ratio of La/Al in LAO to 0.83 to induce the formation of a 2DEG, which prevents potential divergence in LAO [16]. In sample B, the ratio was set to 1.01, at which no 2DEG forms. Additionally, we prepared samples ref-A and ref-B, which were composed of LAO (8 uc)/STO with the same La/Al ratios as samples A and B, respectively, but without an LSMO layer.

12.2.2 Confirmation of Two-Dimensional Electron Gases (2DEG) Formation

To confirm the presence of a 2DEG in samples A and ref-A, we compared the transport properties for samples ref-A and ref-B (Fig. 12.3d). With decreasing temperature, sample ref-B showed increasing resistance, while sample ref-A exhibited decreasing resistance, which indicates the presence of a 2DEG in only sample ref-A, as expected. As shown in Fig. 12.3e, the electron mobility in sample ref-A increased with decreasing temperature and reached 3.7×10^3 cm^{2}V^{-1} s^{-1}, which is typical for LAO/STO interfaces. As shown in Fig. 12.3f, sample B shows the same temperature dependence of the sheet resistance as that of LSMO, which is consistent with the absence of a 2DEG. In contrast, sample A exhibits much lower sheet resistance, which reflects the contribution of the 2DEG. These results confirm that the 2DEG is present in only samples A and ref-A.

12.2.3 Temperature-Dependent Ferromagnetic Resonance and Electromotive Force Generation

We cut samples A and B into 2-mm × 1-mm pieces and positioned them at the center of an electron spin resonance cavity (Fig. 12.3a). Microwaves with a frequency of 9.1 GHz were applied with the magnetic field component aligned along the x-axis (//$[\bar{1}10]$). A static magnetic field H was aligned along the y-axis (//[110]), which is the easy magnetization axis of the LSMO layer. This alignment is critical for eliminating the effects of the magnetoresistance and anomalous Hall effect on the observed electromotive force (EMF).

Figures 12.4a–c show the data obtained for sample A. As shown in Fig. 12.4a, ferromagnetic resonance (FMR) was observed at all measured temperatures. The resonance magnetic field H_r decreases as the temperature is lowered, which can be attributed to temperature-dependent changes in the anisotropy field of the LSMO

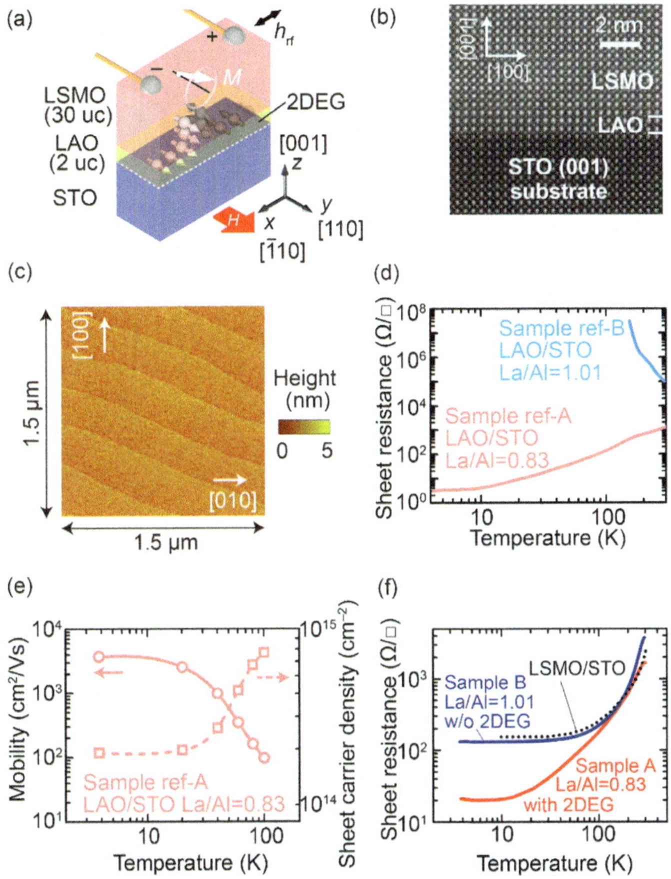

Fig. 12.3 **a** Schematic sample structure consisting of ($La_{0.67}$, $Sr_{0.33}$) MnO_3 (LSMO) (30 uc)/LAO (2 uc)/STO. For the spin pumping experiment, the sample was placed in an electron spin resonance cavity under a static magnetic field applied along the [110] direction with an ac magnetic field along the $[\bar{1}10]$ axis. **b** Scanning-transmission electron microscopy (STEM) lattice image of the LSMO (30 uc)/LAO (2 uc)/STO heterostructure (sample A) projected along the [010] axis. **c** Atomic-force-microscope (AFM) image of the surface of sample A showing clear atomic steps. **d** Temperature dependence of the sheet resistance for the reference samples of LAO (8 uc)/STO with La/Al ratios of 83% (sample ref-A) and 101% (sample ref-B). **e** Temperature dependences of the mobility and carrier density for sample ref-A. **f** Temperature dependence of the sheet resistance of the LSMO (30 uc)/LAO (2 uc)/STO samples with a 2DEG (sample A) and without a 2DEG (sample B). The dotted curve represents the sheet resistance of LSMO/STO reported in Ref. [17], which was reproduced with the assumption of the same film thickness as our LSMO layer. Copyright 2020 American Physical Society, Ref. [12]

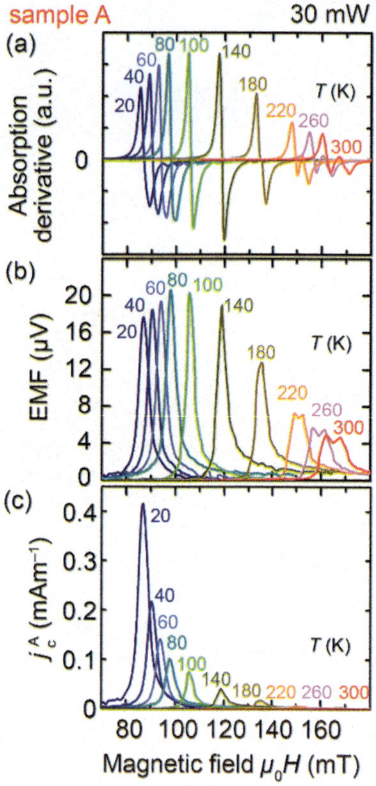

Fig. 12.4 a–c Magnetic-field $\mu_0 H$ dependences of **a** the microwave absorption derivative, **b** Observed electromotive force (EMF), and **c** j_c^A measured for sample A (LSMO/LAO/STO with a 2DEG) at various temperatures. The microwave power used was 30 mW. Copyright 2020 American Physical Society, Ref. [12]

layer. In Fig. 12.4b, the EMF peak is observed at H_r for each temperature, which indicates that the FMR induces the EMF. The sheet current density j_c^A estimated from the EMF in sample A increases as the temperature decreases (Fig. 12.4c).

12.2.4 Evaluation of the Inverse Edelstein Effect (IEE) and its Temperature Dependence

To extract the IEE signal from j_c^A, we obtained the two-dimensional current density j_c^{2D} by subtracting j_c^B, which was obtained from measurements of sample B, from j_c^A (Fig. 12.5a). We estimated the injected spin current density j_s using

$$j_s = \frac{g^{\uparrow\downarrow}\gamma^2(\mu_0 h_{\text{rf}})^2\hbar\left(4\pi M_s\gamma + \sqrt{(4\pi M_s\gamma)^2 + 4\omega^2}\right)}{8\pi\alpha^2\left[(4\pi M_s\gamma)^2 + 4\omega^2\right]}\left(\frac{2e}{\hbar}\right), \tag{12.1}$$

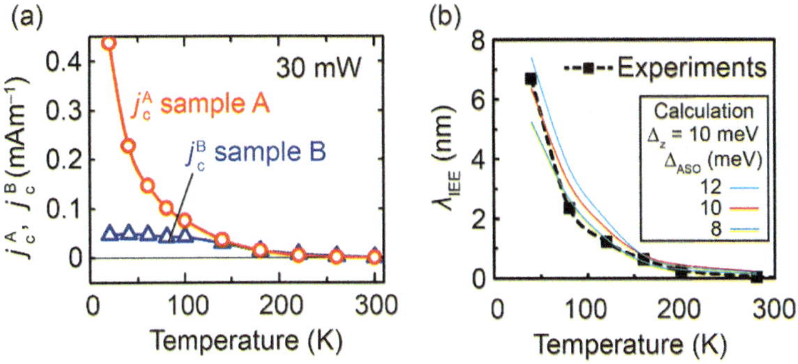

Fig. 12.5 **a** Comparison between j_c^A (with a 2DEG) and j_c^B (without a 2DEG) as a function of temperature. Copyright 2020 American Physical Society, Ref. [12]. **b** Temperature dependences of experimental λ_{IEE} and theoretical λ_{IEE} for the atomic spin-orbit interaction energy $\Delta_{\mathrm{ASO}} = 8$, 10, and 12 meV. Copyright 2022 IOP Publishing, Ref. [13]

where $\hbar$ is the Dirac constant, e is the elementary charge, γ is the gyromagnetic ratio of LSMO, α is the Gilbert damping constant, M_s is the saturation magnetization of LSMO, and $g^{\uparrow\downarrow}$ is the real part of the spin-mixing conductance, which is obtained from the experimental FMR linewidth of sample A and the intrinsic linewidth of LSMO in the absence of spin-current injection[12]. Figure 12.5b shows the obtained temperature dependence of λ_{IEE} $(= j_c^{2\mathrm{D}}/j_s)$. λ_{IEE} monotonically increases with decreasing temperature and reaches 6.7 nm at 20 K, which was one of the highest values obtained for 2DEG systems at the time.

12.2.5 *Theoretical Estimation of the Inverse Edelstein Effect Via a Tight-Binding Band Structure Calculation*

Band Structure Calculation

To verify that the observed signal was intrinsic, we theoretically estimated λ_{IEE} by calculating the band structure of LAO/STO. The 2DEG that forms at the LAO/STO interface is known to originate from the d_{xy}, d_{yz}, and d_{zx} band electrons on the STO side. Due to the strong confinement of electrons at the interface, these bands are strongly quantized. The d_{xy} band has a heavier effective mass than d_{yz} and d_{zx} due to the weak wavefunction overlap in the film normal direction [19]. Thus, whereas for the d_{xy} band the first and second subbands appear below the E_{F}, for the d_{yz} and d_{zx} bands only the first subband appears below E_{F}. We constructed an 8×8 effective tight-binding Hamiltonian based on the 6×6 Hamiltonian proposed by Kim et al. [20] and added the second d_{xy} subband ($= \mathrm{d}_{xy2}$) as an additional basis to the d_{xy} ($= \mathrm{d}_{xy1}$), d_{yz}, and d_{zx} bands. We considered the Hamiltonian H_0 to represent the

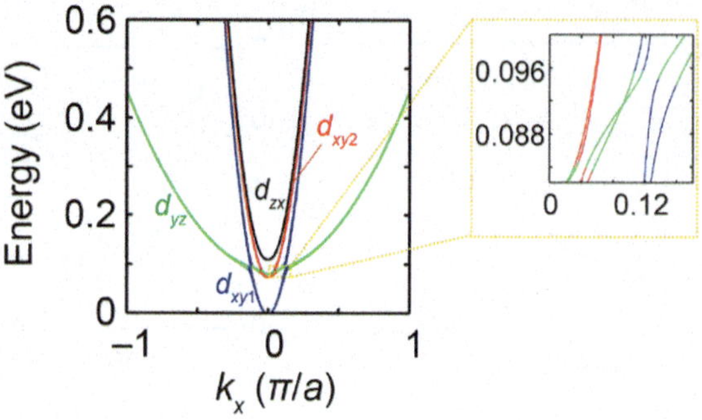

Fig. 12.6 Band structure obtained with the 8 × 8 effective tight-binding model. The black curve is the d_{zx} band, the red curve is the second d_{xy} subband (d_{xy2}), the blue curve is the first d_{xy} subband (d_{xy1}), and the green curve is the d_{yz} band. The intensity Δ_z of H_a was set to 10 meV, and the atomic spin-orbit interaction energy $\Delta_{\mathrm{ASO}} = 10$ meV. The enlarged image shows the band structure near the band crossing points. Copyright 2022 IOP Publishing, Ref. [14]

onsite energy and nearest-neighbor hopping, while H_{ASO} represents atomic spin-orbit interaction, and H_a represents the nearest-neighbor hopping induced primarily by polar lattice distortion due to the electric field originating from broken inversion symmetry.

The calculated band structures are shown in Fig. 12.6 [13]. The inset shows that significant band repulsion occurs near the band crossing points, which results in large band splitting. This increase in band splitting produces an effect equivalent to enhancing the spin-orbit interaction, which is the origin of the enhanced IEE at the LAO/STO interface. The band crossing points are called the Lifshitz points, which are key to enhancing the IEE.

Theoretical Estimation Method of the Inverse Edelstein Length

We can calculate λ_{IEE} using the eigenstate at each wave vector $\boldsymbol{k}$ from the Hamiltonian $H = H_0 + H_{\mathrm{ASO}} + H_a$. The sheet current density $j_c^{\mathrm{FS}_n}$ and spin accumulation δs^{FS_n} carried by electrons on the n-th Fermi contour (FS_n) are expressed by the following equations, which are derived from the Boltzmann equation [21]:

$$j_c^{\mathrm{FS}_n} = \frac{e^2}{4\pi^2\hbar}\int^{\mathrm{FS}_n} F_x(\boldsymbol{k})\,\mathrm{d}S_F,\ \delta s^{\mathrm{FS}_n} = \frac{e}{4\pi^2\hbar}\int^{\mathrm{FS}_n} \left|S_y(\boldsymbol{k})\right|\mathrm{d}S_F, \tag{12.2}$$

where $F_x(\boldsymbol{k})$ and $S_y(\boldsymbol{k})$ are defined as:

$$F_x(\boldsymbol{k}) = F\,\mathrm{sgn}\big(S_y(\boldsymbol{k})\big)\tau(\boldsymbol{k})v_x(\boldsymbol{k})\frac{v_x(\boldsymbol{k})}{|v(\boldsymbol{k})|}. \tag{12.3}$$

$$S_y(\boldsymbol{k}) = F\tau(\boldsymbol{k})\sigma_y(\boldsymbol{k})\frac{v_x(\boldsymbol{k})}{|v(\boldsymbol{k})|}. \tag{12.4}$$

Here, F is the magnitude of the electric field in the x-direction acting on each electron, $v(\boldsymbol{k}) = (v_x(\boldsymbol{k}), v_y(\boldsymbol{k}), v_z(\boldsymbol{k}))$ is the group velocity vector of an electron at $\boldsymbol{k}$, $\tau(\boldsymbol{k})$ is the momentum relaxation time, and $\sigma_y(\boldsymbol{k})$ is the spin component in the y-direction. These equations indicate that the magnitude of $v_x(\boldsymbol{k})$ strongly influences the magnitude of $j_c^{\mathrm{FS}_n}$. The relationships between j_c^{2D}, the total spin accumulation δs, and λ_{IEE} are expressed as follows:

$$j_c^{\mathrm{2D}} = \sum_n j_c^{\mathrm{FS}_n}, \quad \delta s = \sum_n \delta s^{\mathrm{FS}_n}, \qquad \lambda_{\mathrm{IEE}} = \frac{j_c^{\mathrm{2D}}}{j_s} = \frac{\tau}{e}\frac{j_c^{\mathrm{2D}}}{\delta s}. \tag{12.5}$$

Calculated Fermi Contours and Group Velocities

Figure 12.7a shows the Fermi contours and the group velocities along the x direction (v_x) just above the Lifshitz point at an energy of 140 meV above the bottom of the d_{xy1} band, where significant band splitting is observed. Notably, $v_x(\boldsymbol{k})$ is large in the d_{xy1} and d_{xy2} bands near $k_y = 0$, which significantly contributes to the IEE. Figure 12.7b shows the spin distribution on the Fermi surface. The spin distribution is highly complex; however, one can see that purely up-spin and down-spin components are observed in the electron states near $k_y = 0$ in the d_{xy1} band, which has a large group velocity. These states make a significant contribution to the IEE.

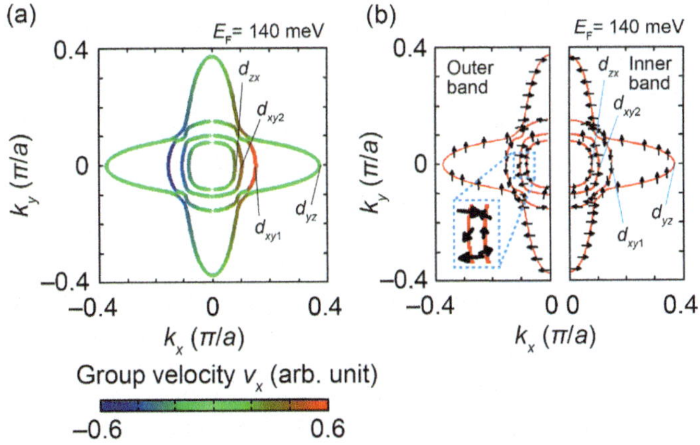

Fig. 12.7 **a** Fermi contours mapped with the group velocity v_x in the x direction at $E_F = 140$ meV. **b** Fermi contours mapped with the spin direction at each $\boldsymbol{k}$ point at $E_F = 140$ meV with the left and right sides showing only the outer and inner bands, respectively. The Fermi contours at $E_F = 140$ meV are shown as an illustrative example of the band crossing generating complicated Fermi contours near this energy. Copyright 2022 IOP Publishing, Ref. [14]

Comparison with the Experimental Data

The E_F position at each temperature was determined such that the carrier concentration obtained from the band calculations matched the experimentally measured carrier concentration [13]. Figure 12.5b shows the theoretically obtained temperature dependence of λ_{IEE} (solid lines). In the calculations, the varied parameter was the magnitude of the atomic spin-orbit interaction energy (Δ_{ASO}), which represents the magnitude of H_{ASO}. This model can reproduce the experimentally observed trend of decreasing λ_{IEE} with increasing temperature. This result supports the conclusion that the observed IEE was an intrinsic phenomenon arising from the band structure of LAO/STO.

One reason for the decrease in the IEE with increasing temperature is the reduction in mobility as the temperature increases. This reduction is attributed to the significant change in the dielectric constant of STO. The dielectric constant can reach values of up to 20,000 and mitigates Coulomb scattering in STO. Improving the mobility would further enhance λ_{IEE}, even at room temperature.

12.3 Large Inverse Edelstein Effect in a Strongly Correlated all-Epitaxial ($La_{0.67}$, $Sr_{0.33}$) MnO_3 (LSMO)/$LaTiO_{3+\delta}$/STO Heterostructure

12.3.1 Addressing Spin Injection Limitations in Oxide 2DEG Interfaces Via $LaTiO_3$ Integration

As described above, we achieved a large IEE using the 2DEG at the LAO/STO interface in an all-epitaxial single-crystal LSMO/LAO/STO heterostructure [13]. After the publication of the paper, a CNRS group in France reported an even larger λ_{IEE} of approximately 60 nm in the 2DEG at an AlO_x/STO interface [19, 22]. However, the "insulating" property of LAO and AlO_x is thought to largely hinder spin injection from ferromagnetic materials into the 2DEG region, and numerous studies have shown that paramagnetic insulating layers can greatly affect spin injection [10, 11, 23].

To address this issue, we replaced the LAO layer with $LaTiO_3$. Although $LaTiO_3$ is intrinsically an antiferromagnetic insulator, it becomes metallic when grown on STO due to the absorption of excess oxygen atoms from the STO substrate [24, 25]. Strain is also thought to contribute to this metallization process [27]. It has long been known that a 2DEG forms at a $LaTiO_3$/STO interface [28]. In 2018, Veit et al. reported that the 2DEG region at a $LaTiO_3$/STO interface exhibits large Rashba splitting [29]. Moreover, $LaTiO_3$ is a well-known correlated material where only one direction of spin can be transported at the Fermi level because the opposite spin band lies far above the Fermi level [30]. This makes it suitable for spin-current transport.

12.3.2 Heterostructure Growth Via Molecular Beam Epitaxy

We grew LSMO (30 uc)/$LaTiO_{3+\delta}$ (LTO, 4 uc) on an STO (001) substrate (Fig. 12.8a) [29]. LTO is a relatively unstable material that can easily transition to pyrochlore-type $La_2Ti_2O_7$ due to excessive oxygen absorption. To prevent LTO from becoming oxygen-rich, we reduced the oxygen pressure during growth to approximately 10^{-7} Pa, which is close to the chamber's background level. Figure 12.8b shows the HAADF-STEM lattice image of the heterostructure after growth, which indicates that the LSMO and LTO layers coherently grew on the STO substrate, and no $La_2Ti_2O_7$ phase was detected. As shown in Fig. 12.8c, the temperature dependence of the sheet resistance of the LSMO/LTO/STO sample exhibited metallic behavior.

We also fabricated the same LSMO/LTO structure on an insulating (La, Sr) (Al, Ta)O_3 (LSAT) substrate, which does not contain a 2DEG. The temperature dependence of the sheet resistance was almost identical to that of an LSMO single layer of the same thickness. In contrast, the sheet resistance of the LSMO/LTO/STO sample

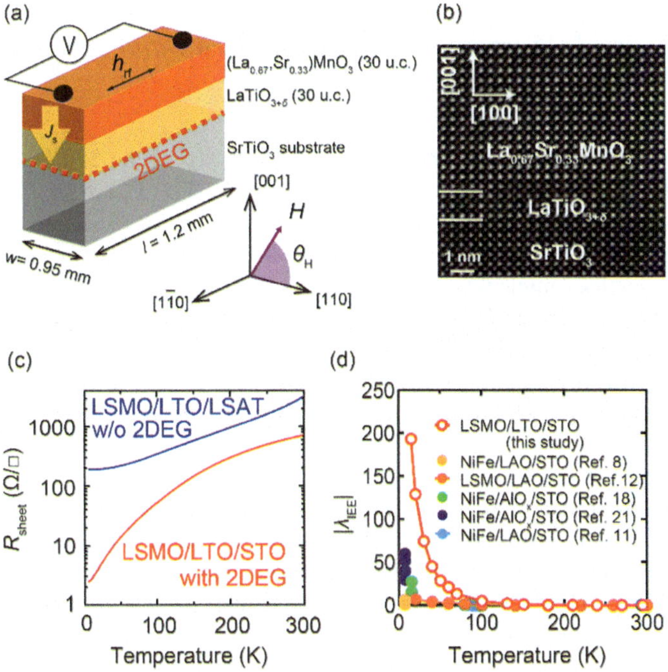

Fig. 12.8 **a** Schematic heterostructure of the LSMO/LTO/STO sample used for spin pumping measurements. **b** HAADF-STEM lattice image of LSMO (30 uc)/LTO (3 uc)/STO. **c** Temperature dependence of the sheet resistance R_{sheet} of the LSMO (30 uc)/LTO (3 uc)/STO and the LSMO (30 uc)/LTO (3 uc)/LSAT samples. **d** Summary of the temperature dependences of λ_{IEE} in various material systems. Copyright 2022 Nature Publishing Group, Ref. [31]

was lower than that of the LSMO layer. This result indicates the presence of a 2DEG in the LSMO/LTO/STO sample.

12.3.3 Giant Inverse Edelstein Effect at the $LaTiO_3$/STO Interface

Figure 12.8d shows the temperature dependence of λ_{IEE}. The λ_{IEE} monotonically increases with decreasing temperature. At 4 K, a notably large value of 193 nm was achieved. When comparing λ_{IEE} with the spin-charge current conversion efficiency of bulk materials, the product of the spin Hall angle and spin diffusion length is used as an equivalent value to λ_{IEE}. For materials such as Pt and W, this value is less than 1 nm, and for BiSb, it is approximately 2.5 nm [30]. These comparisons highlight the exceptionally large λ_{IEE} achieved in the LTO/STO system.

12.3.4 Direct Observation of the Quantized Band Structure at the LTO/STO Interface

To understand the significant increase in λ_{IEE} observed in the 2DEG at the LTO/STO interface, we measured the band structure of the LTO (3 uc)/STO sample using resonance angle-resolved photoemission spectroscopy (R-ARPES). The sample used was grown under the same conditions as the LSMO/LTO/STO sample. Since LTO is unstable in air, the sample was transferred to a nitrogen-filled container immediately after removing it from the MBE chamber and then transported to the Photon Factory synchrotron radiation facility. Previous ARPES measurements of 2DEG systems on STO substrates have primarily focused on systems that formed at the surface of STO. However, measuring "embedded interfaces," as in this study, is generally extremely challenging.

The measurements were conducted using a horizontal-incidence and horizontal-detection slit configuration. While the d_{zx} band is predominantly detected, signals from the d_{xy} and d_{yz} bands are also slightly included in the results. The measurement results are shown in Fig. 12.9a, b. The dotted horizontal line in Fig. 12.9a represents the Fermi level. One key point in this figure is that the states in the k_x direction observed at the Fermi level have a width ($\approx$ the distance between the two intersections of the dashed horizontal line and the red line, ~ $0.4\pi/a$) that is nearly the same as the horizontal width at $k_y = 0$ of the Fermi contour in Fig. 12.9b ($\approx$ the diameter of the red circle). This indicates that both of these (red lines in Fig. 12.9a, b) originate from the same band.

Another important observation is that the Fermi contour in Fig. 12.9b has an elongated elliptical shape. This reflects the Fermi contour shape, as shown in Fig. 12.7b, which arises from the quantization of Ti 3*d* electrons. This differs from the Fermi

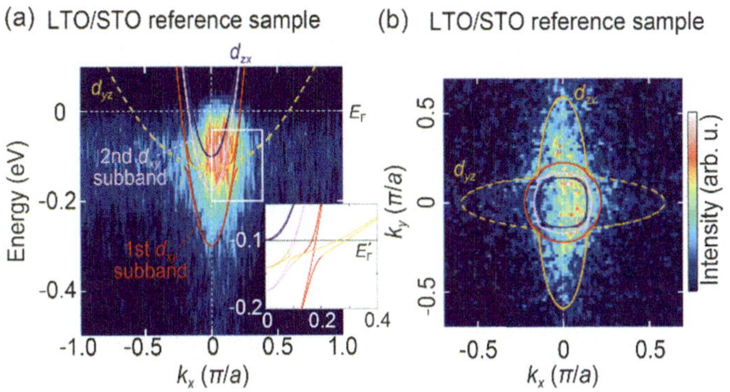

Fig. 12.9 R-ARPES measurement results for the band dispersion and Fermi surface with the theoretical curves. **a** Band dispersion of a reference sample LTO (3 uc)/STO along k_x measured by R-ARPES. The 2DEG states were observed using the resonant photon energy $h\nu$ for the Ti L_3 absorption edge of the Ti^{3+} component ($h\nu = 459.7$ eV). The solid and dashed curves represent the band dispersions calculated by the effective tight-binding model. The inset shows the enlarged band dispersion of the calculation near the crossing point of the d_{xy} and d_{yz} bands. The dotted line in the inset corresponds to the Fermi level used for the calculation of λ_{IEE} in the LSMO/LTO/STO sample ($E'_F = -102$ meV). E_F and E'_F are the Fermi levels of LTO/STO obtained by the R-ARPES measurement and the Hall effect measurement, respectively. **b** Fermi surface of the reference sample LTO (3 uc)/STO obtained by R-ARPES. In this measurement, the Fermi contour elongated in the k_y direction is visible, but the surface elongated in the k_x direction is almost not visible due to the experimental geometry of the measurements [31]. The solid and dashed curves are the Fermi contours calculated by the effective tight-binding model at $E_F = 0$ meV. Copyright 2022 Nature Publishing Group, Ref. [29]

surfaces observed in the $3d^1$ systems, as shown for $SrVO_3$, which demonstrates that the 2DEG originates from the *d*-orbitals of STO rather than LTO.

12.3.5 Quantitative Modeling of the Inverse Edelstein Effect in LTO/STO Using an Experimentally Tuned Tight-Binding Band Structure

In the model used for effective tight-binding approximation, we adjusted the band parameters to reproduce the band structure experimentally obtained by R-ARPES. The solid lines in Fig. 12.9a are the final band dispersions obtained. An adjusted Fermi level position (denoted as E'_F in the inset of Fig. 12.9a) was also used to replicate the experimentally observed λ_{IEE} values.

As shown in Fig. 12.10a, this model successfully reproduced the temperature dependence of λ_{IEE} obtained in LSMO/LTO/STO. The resulting Fermi contours are shown in Fig. 12.10b. α_R was estimated using the spin splitting Δk at $k_y = 0$ in the d_{xy1} subband and the relation $\alpha_R = \hbar^2\Delta k/(2\,m)$, and the result was 1.0×10^{-11} eVm. This value shows good agreement with the Rashba coefficient estimated from weak

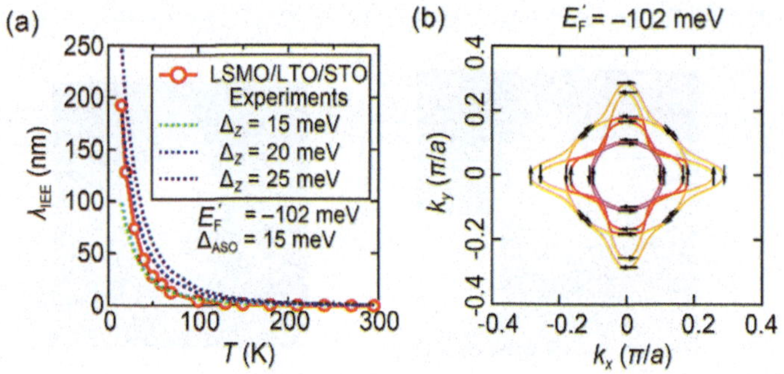

Fig. 12.10 **a** Dependence on temperature T of experimental λ_{IEE} for the LSMO/LTO/STO sample and calculated λ_{IEE} for Δz (intensity of H_a) = 15, 20, and 25 meV with Δ_{ASO} fixed at 15 meV. **b** Calculated Fermi contours and spin expectation values (arrows) at $E'_F = -102$ meV by the effective tight-binding model. Copyright 2022 Nature Publishing Group, Ref. [29]

antilocalization experiments at the LTO/STO interface (1.8×10^{-11} eVm) [29]. Our data suggest the great potential of the single-crystalline correlated heterostructures for efficient spin-to-charge conversion.

12.4 Summary

We explored spin-to-charge conversion via the IEE in oxide 2DEG systems. In epitaxial LSMO/LAO/STO heterostructures, a large λ_{IEE} (6.7 nm at 20 K) was observed. To overcome spin-injection limitations from the insulating LAO layer, we replaced LAO with LTO, which becomes metallic on STO. This LTO/STO interface exhibited an exceptionally large λ_{IEE} of 193 nm at 4 K. Resonant ARPES measurements revealed that the 2DEG originates from quantized Ti 3*d* orbitals in the STO side. The observed Fermi contour and subband structure were well reproduced using a tailored 8×8 tight-binding model. The model also replicated the temperature dependence of λ_{IEE} and yielded a Rashba coefficient α_R of 1.0×10^{-11} eVm. These results demonstrate a strong Rashba effect and efficient spin-charge conversion in LTO/STO systems. Compared to LAO/STO, LTO/STO offers improved spin injection and larger conversion efficiency. Our study highlights the promise of oxide heterostructures for next-generation spintronic devices.

12.5 Future Perspective or Future Challenging

We have demonstrated through both experimental and theoretical approaches that highly efficient spin current-to-charge current conversion can be achieved in all-epitaxial 2DEG systems based on oxides. The results can be attributed to the strong spin-orbit interaction in the 2DEG system, high crystallinity, and the high mobility enabled by the large dielectric constant of STO. Currently, this high conversion efficiency has only been achieved at low temperatures. In the future, it will be essential to realize this efficiency at higher temperatures.

Since electron mobility plays a significant role in the IEE, the development of 2DEG systems with high mobility even at room temperature will potentially address this issue. Another challenge is improving the efficiency of charge-to-spin current conversion, the inverse effect of the IEE, and applying it to practical technologies such as magnetization reversal. At present, magnetization reversal using 2DEG systems remains a challenging task. We hope that the insights gained from spin-to-charge current conversion will contribute to overcoming these challenges in the future.

Acknowledgments The research was supported by the Spintronics Research Network of Japan (Spin-RNJ), JSPS KAKENHI (Grants No. 18H03860, 20H05650, 22H04948, and 23H03802), JST CREST (Grant No. JPMJCR1777), the Nanotechnology Platform Program of the Ministry of Education, Culture, Sports, Science, and Technology (MEXT), and the Cryogenic Research Center at The University of Tokyo.

References

1. Sánchez JCR, Vila L, Desfonds G, Gambarelli S, Attané JP, De Teresa JM, Magén C, Fert A. Nat Commun. 2013;4:2944.
2. Noel P, Thomas C, Fu Y, Vila L, Haas B, Jouneau P-H, Gambarelli S, Meunier T, Ballet P, Attané JP. Phys Rev Lett. 2018;120:167201.
3. Oyarzún S, Nandy AK, Rortais F, Rojas-Sánchez J-C, Dau M-T, Noël P, Laczkowski P, Pouget S, Okuno H, Vila L, Vergnaud C, Beigné C, Marty A, Attané J-P, Gambarelli S, George J-M, Jaffrès H, Blügel S, Jamet M. Nat Commun. 2016;7:13857.
4. Rojas-Sánchez J-C, Oyarzún S, Fu Y, Marty A, Vergnaud C, Gambarelli S, Vila L, Jamet M, Ohtsubo Y, Taleb-Ibrahimi A, Le Fèvre P, Bertran F, Reyren N, George J-M, Fert A. Phys Rev Lett. 2016;116:096602.
5. Rojas-Sánchez J-C, Fert A. Phys Rev Applied. 2019;11:054049.
6. Ohtomo A, Hwang HY. Nature. 2004;427:423.
7. Caviglia D, Gabay M, Gariglio S, Reyren N, Cancellieri C, Triscone J-M. Phys Rev Lett. 2010;104:126803.
8. Lesne E, Yu F, Oyarzun S, Rojas-Sánchez JC, Vaz DC, Naganuma H, Sicoli G, Attané J-P, Jamet M, Jacquet E, George J-M, Barthélémy A, Jaffrès H, Fert A, Bibes M, Vila L. Nat Mater. 2016;15:1261.
9. Song Q, Zhang H, Su T, Yuan W, Chen Y, Xing W, Shi J, Sun J, Han W. Sci Adv. 2017;3:e1602312.
10. Wang Y, Ramaswamy R, Motapothula M, Narayanapillai K, Zhu D, Yu J, Venkatesan T, Yang H. Nano Lett. 2017;17:7659.
11. Chauleau J-Y, Boselli M, Gariglio S, Weil R, de Loubens G, Triscone J-M, Viret M. Europhys Lett. 2016;116:17006.
12. Ohya S, Araki D, Anh LD, Kaneta S, Seki M, Tabata H, Tanaka M. Phys Rev Res. 2020;2:012014. (R).
13. Arai S, Kaneta-Takada S, Anh LD, Tanaka M, Ohya S. Appl Phys Express. 2022;15:013005.
14. Nakagawa N, Hwang HY, Muller DA. Nat Mater. 2006;5:204.

15. Lesne E, Reyren N, Doennig D, Mattana R, Jaffrès H, Cros V, Petroff F, Choueikani F, Ohresser P, Pentcheva R, Barthélémy A, Bibes M. Nat Commun. 2014;5:4291.
16. Warusawithana MP, Richter C, Mundy JA, Roy P, Ludwig J, Paetel S, Heeg T, Pawlicki AA, Kourkoutis LF, Zheng M, Lee M, Mulcahy B, Zander W, Zhu Y, Schubert J, Eckstein JN, Muller DA, Hellberg CS, Mannhart J, Schlom DG. Nat Commun. 2013;4:2351.
17. Boschker H, Verbeeck J, Egoavil R, Bals S, van Tendeloo G, Huijben M, Houwman EP, Koster G, Blank DHA, Rijnders G. Adv Funct Mater. 2012;22:2235.
18. Vaz DC, Noël P, Johansson A, Göbel B, Bruno FY, Singh G, McKeown-Walker S, Trier F, Vicente-Arche LM, Sander A, Valencia S, Bruneel P, Vivek M, Gabay M, Bergeal N, Baumberger F, Okuno H, Barthélémy A, Fert A, Vila L, Mertig I, Attané J-P, Bibes M. Nat Mater. 2019;18:1187.
19. Kim Y, Lutchyn RM, Nayak C. Phys Rev B. 2013;87:245121.
20. Gambardella P, Miron IM. Phil Trans R Soc A. 2011;369:3175.
21. Noël P, Trier F, Arche LMV, Bréhin J, Vaz DC, Garcia V, Fusil S, Barthélémy A, Vila L, Bibes M, Attané J-P. Nature. 2020;580:483.
22. Du H, Wang HL, Pu Y, Meyer TL, Woodward PM, Yang FY, Hammel PC. Phys Rev Lett. 2013;111:247202.
23. Lichtenberg F, Widmer D, Bednorz JG, Williams T, Reller AZ. Phys. B. 1991;82:211.
24. Zhang TT, Gu CY, Mao ZW, Chen XF, Gu ZB, Wang P, Nie YF, Pan XQ. Appl Phys Lett. 2019;115:261604.
25. Dymkowski K, Ederer C. Phys Rev B. 2014;89:161109(R).
26. Ohtomo A, Muller DA, Grazul JL, Hwang HY. Nature. 2002;419:378.
27. Veit MJ, Arras R, Ramshaw BJ, Pentcheva R, Suzuki Y. Nat Commun. 2018;9:1458.
28. Gu M, Rondinelli JM. Sci Rep. 2016;6:25121.
29. Kaneta-Takada S, Kitamura M, Arai S, Arai T, Okano R, Anh LD, Endo T, Horiba K, Kumigashira H, Kobayashi M, Seki M, Tabata H, Tanaka M, Ohya S. Nat Commun. 2022;13:5631.
30. Goel S, Huynh N, Khang D, Osada Y, Anh LD, Hai PN, Tanaka M. Sci Rep. 2023;13:2181.
31. Cancellieri C, Reinle-Schmitt ML, Kobayashi M, Strocov VN, Willmott PR, Fontaine D, Ghosez P, Filippetti A, Delugas P, Fiorentini V. Phys Rev B. 2014;89:121412(R).

Chapter 13
Nuclear Quantum Effects of Hydrogen in Metals

Takahiro Ozawa and Katsuyuki Fukutani

Abstract Hydrogen, the lightest element, apparently exhibits quantum properties. Quantum effects such as zero-point vibration and quantum tunneling are crucial for the structures, reactions, and diffusion of hydrogen. Especially at low temperatures, where thermal processes are suppressed, these quantum effects have a significant impact on hydrogen-related phenomena. On the other hand, since it is generally difficult to detect hydrogen atoms, the lack of experimental observations has been a longstanding issue to examine the quantum effects of hydrogen. In this chapter, we present the recent progress in experimental techniques to directly probe the position and migration of hydrogen atoms in solid materials. The contribution of the quantum nature to the structure and diffusion of hydrogen is experimentally and theoretically discussed.

Keywords Nuclear quantum effects · Zero-point vibration · Quantum tunneling · Isotope effect · Hydrogen · Diffusion · Metal hydrides · Nuclear reaction analysis · Channeling ^{15}N-NRA

13.1 Introduction

Hydrogen, composed of a proton and an electron, readily migrates into various materials due to its small atomic size and moderate electron affinity. The recent technological innovations on green energy systems utilizing hydrogen as an energy carrier are founded on these intrinsic properties of hydrogen. Solid hydride materials, including metal hydrides, metal oxyhydrides, and polymers, have attracted considerable attention as efficient high-density hydrogen storage media, aiming for reversible hydrogen storage and release under moderate temperature and pressure conditions [1–3]. As

T. Ozawa · K. Fukutani (✉)
Institute of Industrial Science, The University of Tokyo, Tokyo, Japan
e-mail: fukutani@iis.u-tokyo.ac.jp

T. Ozawa
e-mail: t-ozawa@iis.u-tokyo.ac.jp

T. Onishi (ed.), *Advances in Quantum Science II*,
https://doi.org/10.1007/978-981-95-9721-5_13

absorption and desorption processes are often rate-limited by hydrogen dynamics in subsurface regions, and advanced devices are now being fabricated at the nanometer scale, understanding of the dynamics of hydrogen, particularly in nanoscale regions, is crucial for the further advancement of hydrogen carrier systems.

From the perspective of solid-state properties, hydrogen also serves as a valuable tool in materials engineering. Absorbed and/or adsorbed hydrogen atoms have the potential to drastically alter electronic, magnetic, and optical properties of host materials, as demonstrated in superconducting transitions, metal-insulator transitions, magnetic phase transitions and so on [4–11]. Since these hydrogen-induced phenomena are closely associated with electronic modifications caused by charge transfer between hydrogen and host atoms and with the lattice deformations induced by incorporated hydrogen, the lattice location and concentration of hydrogen play key roles in controlling these effects.

13.2 Quantum Nature of Hydrogen

Nuclear quantum effects (NQEs), such as zero-point vibration and quantum tunneling, have a significant impact on the properties of light particles in materials. Since the lightest element of hydrogen is a prototypical example that exhibits pronounced quantum behaviors, hydrogen in metals is one of the suitable platforms to investigate these quantum characteristics in atomic motion. Metal hydride systems have been extensively studied over the past several decades. Absorbed hydrogen atoms in metals occupy an interstitial site within the host lattice. As schematically illustrated in Fig. 13.1, hydrogen is commonly found in octahedral (O) (Fig. 13.1a) and/or tetrahedral (T) (Fig. 13.1b) sites, depending on the crystal structure and hydrogen concentration. Figure 13.2 shows a schematic potential for hydrogen as a function of the hydrogen position coordinate, where either of the O or T sites corresponds to the stable site and the other becomes a metastable site. The zero-point vibration and quantum tunneling have a significant influence on the static and dynamic behaviors of hydrogen in metals as detailed in the following subsections.

13.2.1 Zero-Point Vibration

Zero-point vibration is an intrinsic vibrational motion of molecules and atoms arising from quantum mechanical uncertainty. In a one-dimensional harmonic approximation, the zero-point energy (ZPE), the energy due to the zero-point vibration, is described as

$$\mathrm{ZPE} = \frac{1}{2}\hbar\sqrt{\frac{k}{m}},$$

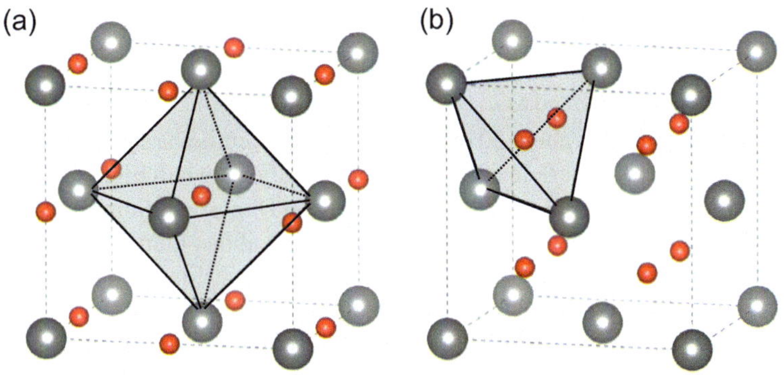

Fig. 13.1 Face-centered cubic (FCC) structures showing **a** octahedral (O) and **b** tetrahedral (T) interstitial sites. Hydrogen in the O site is surrounded by six host atoms and hydrogen in the T site is surrounded by four host atoms, respectively. Large black and small red circles represent host and hydrogen atoms, respectively

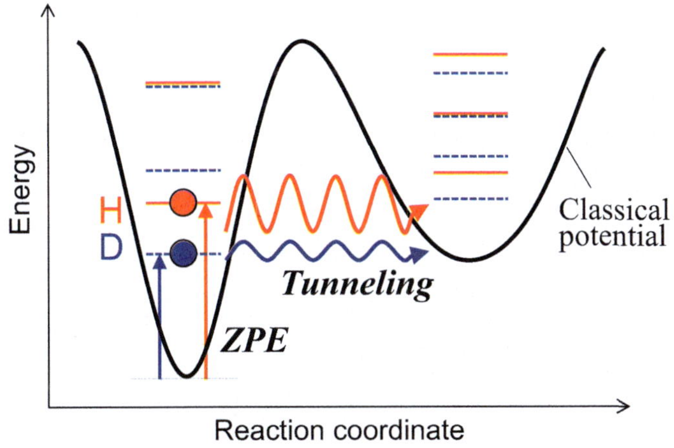

Fig. 13.2 Schematic illustration of nuclear quantum effects (NQEs) on hydrogen isotopes. The solid curve represents the classical potential energy surface, and the solid and dotted bars represent the quantum energy levels of the isotopes

where $\hbar$ is the reduced Planck constant and m and k are the particle mass and potential curvature, respectively. This expression indicates that the ZPE gets larger for lighter particles and in systems with sharper potential energy surfaces as illustrated by the solid bars in Fig. 13.2. Note that in an anharmonic potential, the effects of zero-point vibration are more complex because the mean vibrational position shifts away from the equilibrium point. NQEs arising from zero-point motion have been observed as strong isotope dependencies in phenomena such as the dielectric phase transition of KH_2PO_4, phase diagrams of metal hydrides and the structure of water [12–15].

Given that NQEs strongly depend on particle mass, studying isotope effects provides a valuable approach to understanding the quantum behaviors. The typical vibrational energy, expressed as $\hbar\sqrt{\frac{k}{m}}$, for H in metals is approximately 100 meV, and the corresponding ZPE is about 50 meV. Since these quantities are proportional to the inverse square root of a particle mass, as schematically shown in Fig. 13.2, this leads to a ZPE difference of about 15 meV between H and D in a one-dimensional system. In a three-dimensional potential, the value amounts to about 45 meV. When energetically close sites exist with an energy difference of 45 meV or less, H and D atoms potentially take different structural configurations resulting in different electronic structures. These are regarded as NQEs in the static properties of hydrogen. Although isotope-dependent site occupancy by hydrogen in crystals and corresponding isotope-dependent physical properties have rarely been discussed due to the difficulty in isotope-sensitive structure analysis, the recent advancements in experimental and theoretical techniques clarify NQEs on static properties of hydrogen.

13.2.2 Quantum Tunneling

Regarding dynamics of hydrogen in metals, an important elementary process is diffusion, which occurs as successive jumps of hydrogen from a potential minimum to neighboring minima. While such a jump process is thermally activated at high temperatures to overcome a potential barrier between neighboring sites, at low temperatures, thermal activation is suppressed and quantum tunneling is considered to become dominant. These are often called the thermal and quantum regimes, respectively. In the thermal regime, the diffusion coefficient D obeys the Arrhenius law and is expressed as

$$D = D_0 \exp\left(-\frac{E_a}{k_B T}\right),$$

where D_0, E_a, k_B, and T denote the pre-exponential factor, the activation energy, the Boltzmann constant, and temperature, respectively. The activation energy can be obtained from the slope of the Arrhenius plot. Isotope-dependent energy barriers for diffusion have been reported in many metals [1]. The difference in the energy barrier between isotopes is attributed to the mass dependence in ZPE, highlighting that the zero-point vibration is also important in the dynamics of hydrogen. In the quantum regime, hydrogen diffusion is governed by quantum tunneling and strongly dependent on mass. In a one-dimensional system, the Wentzel–Kramers–Brillouin (WKB) approximation gives an expression for the tunneling probability (W) for a particle with mass m as

$$W = \exp\left(\frac{-2}{\hbar}\int_{l_1}^{l_2}\sqrt{2m(V(x)-E)}\mathrm{d}x\right),$$

where E is the energy level of the particle, $V(x)$ is the potential along the hopping path x, and $x = l_1$ and l_2 are the positions determined by the condition of $V(x) = E$. This expression suggests that the particle penetrates the barrier with a finite probability via quantum tunneling, and its contribution becomes more significant for light particles going through a shallower and narrower potential barrier.

Hydrogen existing in metals interacts with its environment, such as phonons and conduction electrons of host metals, during hopping processes. The temperature dependence of the hopping rate offers valuable insights into the influence of the surrounding environment. When a hydrogen atom occupies an interstitial site, the surrounding lattice is deformed to stabilize the structure leading to a self-trapped state. In ionic crystals, an excess electron is coupled with lattice vibrations forming a quasiparticle known as a polaron, as described by polaron theory [16, 17]. Under thermally activated conditions, the electron mobility can be facilitated by phonon excitations. By analogy with this electron polaron, a hydrogen atom in metals is regarded as a polaron, and phonon excitation is considered to assist the tunneling of hydrogen [18]. This process is called phonon-assisted tunneling. When phonon excitation is suppressed at lower temperatures, a coherent tunneling without changing the phonon number is expected to occur, which is theoretically analyzed to reveal a negative power-law dependence of the hopping rate on temperature [19]. In addition to phonons, furthermore, electronic excitation of metals affects the tunneling of hydrogen as discussed by Kondo [20]: The tunneling rate is predicted to follow a T^{2K-1} dependence, where T and K are temperature and the proton-electron coupling constant, respectively. This behavior is attributed to the non-adiabatic effects of conduction electrons near the Fermi surface in metals, where electronic excitations with infinitesimally small energy are allowed due to the continuous density of states around the Fermi level. Yamada calculated the coupling constant K using the phase shift of electron scattering, predicting that K lies between 0 and 1/2 [21]. This suggests a negative temperature dependence of the tunneling rate of hydrogen in metals. Regelmann proposed that the asymmetry of the potential affects the magnitude of the coupling [22].

The recent advances in quantum mechanical calculations have enabled us to treat NQEs through numerical approaches. Path integral molecular dynamics (PIMD) calculations, where a quantum particle is represented as a ring polymer, allow for the inclusion of NQEs in molecular dynamics simulations. Figure 13.3 shows the calculated hydrogen hopping rates on narrow and broad potential energy surfaces found on metal surfaces such as Pd and Ni [23]. The bottom and top axes represent the inverse temperature, expressed as $1000/T$, and the corresponding absolute temperature, respectively. The dark curves in the insets illustrate the potential energy profiles used in the calculations. When the NQEs are included, upward deviations from the Arrhenius law are observed as temperature decreases, indicating the contribution of quantum tunneling. A high tunneling probability is obtained for the hopping across a narrower potential, demonstrating the influence of the potential shape and agreement with predictions from the WKB approximation. Interestingly, the gradual deviation (left panel in Fig. 13.3) suggests that quantum tunneling plays an important role in

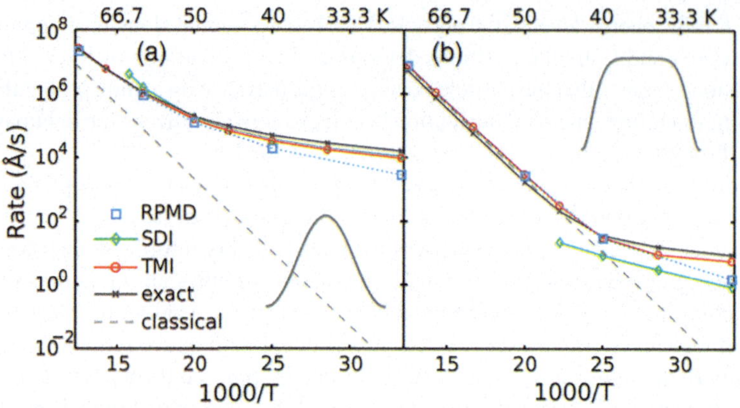

Fig. 13.3 Calculated hopping rates of hydrogen on metal surfaces by several calculations considering NQEs. PIMD, TMI, and "exact" in the left panel denote the hopping rates calculated by ring polymer molecular dynamics, thermalized microcanonical instanton, and the solution of the Schrödinger equation, respectively, where the NQEs are taken into account. SDI represents the hopping rates obtained by the steepest-descent approximation, which underestimates the NQEs. Classical denotes thermal hopping rates calculated without considering NQEs, which obeys the Arrhenius law. With the NQEs, upward deviation from the Arrhenius law is more prominent in a narrow potential (left) than in a wide potential (right). Reprinted with permission from Ref. [23]. Copyright 2017American Physical Society

hydrogen diffusion not only at low temperatures but also at intermediate temperatures around the crossover between thermal and quantum regimes. Importance of NQEs on the hydrogen diffusion around the crossover region is also indicated in studies of hydrogen in metals [24]. Figure 13.4 shows the calculated hopping rates of proton in Pd, indicating a gradual transition from thermally activated hopping to quantum tunneling attributed to NQEs. These behaviors are regarded as NQEs in the dynamic behavior of hydrogen. Although the importance of NQEs in hydrogen diffusion is highlighted in these theoretical calculations, the influence of the surrounding environment and detailed mechanism of quantum tunneling are yet to be clarified.

13.2.3 Difficulty in Detecting Hydrogen: Previous Experimental Research

Compared to advancements in theoretical investigations of NQEs, experimental NQEs data of the structure and quantum tunneling of hydrogen remain limited, hindering a comprehensive understanding of the quantum nature of hydrogen. Since the scattering cross sections of hydrogen for X-rays and electrons are substantially small compared with other elements, it is generally difficult to directly probe hydrogen atoms in conventional diffraction techniques. Although neutron scattering

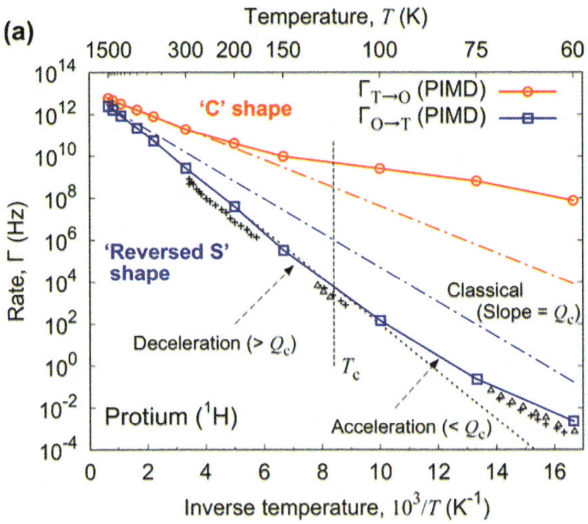

Fig. 13.4 Calculated hopping rates of H from T to O sites ($\Gamma_{T \to O}$) and from O to T sites ($\Gamma_{O \to T}$) in Pd by Path integral molecular dynamics (PIMD) calculations. The dashed-dotted curves represent the classical limits excluding NQEs. The dotted line is an Arrhenius fit of $\Gamma_{O \to T}$ between 200 and 600 K. The "C shape" and "Reversed S shape" describe the characteristic temperature dependence of the respective hydrogen hopping rates. Reprinted with permission from Ref. [24]. Copyright 2019 American Physical Society

is a powerful way for structure analysis of hydride materials, its sensitivity is insufficient for investigating nanoscale materials. Moreover, isotope-selective structure analysis is more challenging.

As for hydrogen diffusion, ordinary macroscopic techniques have been used to derive the diffusion coefficients of hydrogen in various metals, which obey the Arrhenius law at thermal regimes [1]. A slight upward deviation from the Arrhenius law with decreasing temperature was observed in the thermal regime, suggesting the contribution of NQEs. No detailed measurements on the temperature dependence, however, have been obtained at lower temperatures in the quantum regime due to the difficulty of probing slow diffusion via quantum tunneling. On solid surfaces, several experimental techniques have been developed to observe quantum tunneling of hydrogen including field emission microscopy (FEM), optical reflectance, ^{3}He echo, and atom-track Scanning Tunneling Microscopy (STM). Using FEM, Gomer and co-workers reported nearly temperature-independent diffusion of hydrogen at low temperatures [25], as shown in Fig. 13.5, which represents the hopping rates of H and D on a W surface. This nearly temperature-independent behavior is indicative of tunneling diffusion. Similar transitions from the thermal to quantum regimes were also observed by optical measurements [26–28]. Two types of transitions, abrupt and gradual transitions, have been found on different metals to date. By using atom-track STM, Ho and co-workers observed slightly positive and negative temperature dependences above and below 25 K, respectively [29, 30], consistent with the phonon-assisted tunneling and non-adiabatic effects on tunneling, as shown in Fig. 13.6, which represents the H and D hopping rates on a Cu surface. By taking advantage of the pseudo-isotope character of a muon (μ) with a mass 1/9 that of H, on the other hand, detailed analysis of the effects of phonons and electrons on the tunneling of muons was performed by μ spin rotation and relaxation (μSR) studies, where the muon hopping rate is obtained from the time evolution of spin relaxation [31–33]. Figure 13.7 shows the muon hopping rate in Cu, where a negative temperature dependence was clearly observed. At even lower temperatures, coherent tunneling of a particle is expected to occur. This phenomenon was reported by a μSR study as the presence of a Bloch state in the insulating material KCl [34]. However, the roles of NQEs on hydrogen hopping around the crossover from thermal to quantum regimes still remain unclear, and the influence of the surrounding environment on quantum tunneling is yet to be elucidated.

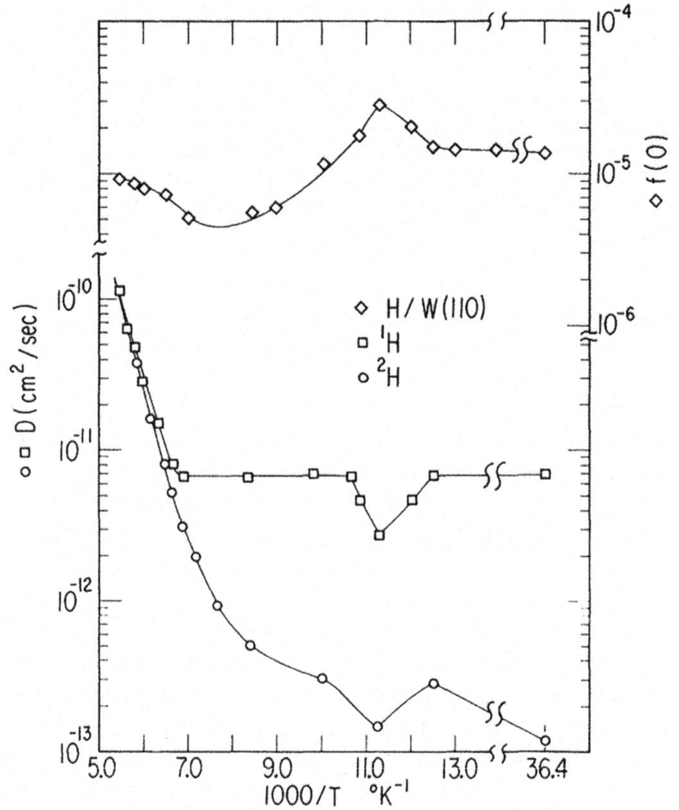

Fig. 13.5 Experimentally obtained diffusion coefficients of H and D at a coverage of $0.6\theta_{max}$ on W(110) measured by FEM measurements. The correlation function (◇) in the analysis is also shown. The bottom axis represents the inverse temperature. Reprinted with permission from Ref. [25]. Copyright 1980 American Physical Society

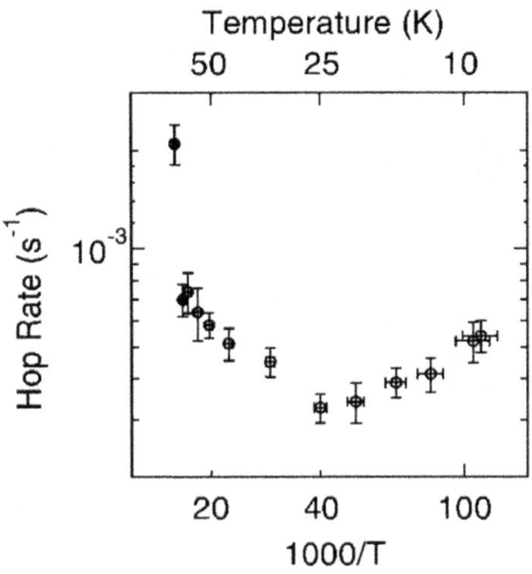

Fig. 13.6 Experimentally obtained hopping rates of H on Cu(001) measured by single-atom tracking using STM. The derived coupling constant K is 0.25 ± 0.05. Reprinted with permission from Ref. [29]. Copyright 2000 American Physical Society

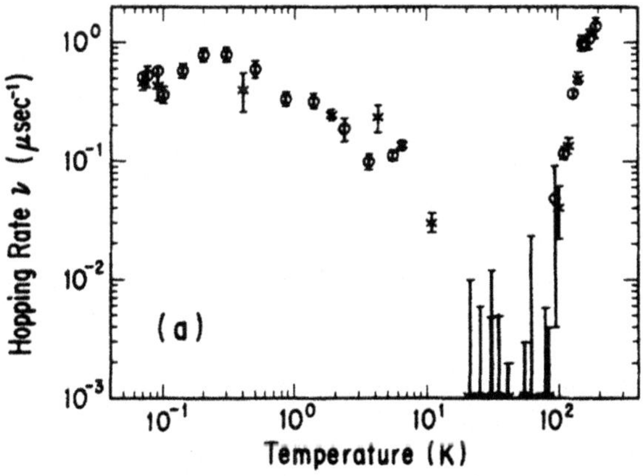

Fig. 13.7 Experimentally obtained hopping rates of muon in Cu measured by spin-relaxation measurements. The derived coupling constant K is 0.16 ± 0.01. Reprinted with permission from Ref. [33]. Copyright 2000 American Physical Society

To clarify NQEs on these static and dynamic behaviors of hydrogen, technological advancements for identifying the lattice location of hydrogen isotopes and measuring the hopping rate are recently reported, which are detailed in Sect. 13.3.

13.3 Recent Advancement in Experimental Methods

A recent experimental concept to probe hydrogen location and motion in materials is to utilize ion beam analysis combined with electrical-resistance measurement, which we have recently developed. The principles of the experimental methods are introduced in this section.

13.3.1 Quantitative Depth Profiling of Hydrogen by ^{15}N-NRA (Nuclear Reaction Analysis)

Ion beam analysis, which utilizes scattering and reactions between atomic nuclei, is widely employed for analyzing the composition and depth distribution of impurities in materials. Ions are accelerated to MeV-scale energies by an accelerator and irradiate samples. Nuclear reaction analysis (NRA) is one of the ion beam analysis techniques, where target atoms are selectively detected by measuring characteristic particles or γ-rays emitted from nuclear reactions. For H detection, the nuclear reaction between ^{15}N and ^{1}H, designated as $^{1}H(^{15}N,\alpha\gamma)^{12}C$, is used, which is called ^{15}N-NRA [35]. A monochromatized beam of ^{15}N ions accelerated to an energy of ~ 6.5 MeV is injected into a sample, and the reaction proceeds as follows: $^{15}N + H \rightarrow {}^{16}O^{*} \rightarrow {}^{12}C + \alpha + \gamma(4.43\ \text{MeV})$. Here, $^{16}O^{*}$ represents an oxygen in an excited state, which subsequently decays to ^{12}C by emitting an alpha particle and a γ-ray with a characteristic energy of 4.43 MeV. Hydrogen is quantitatively detected by counting the emitted γ-ray yield. A key feature of this reaction is that the reaction resonantly occurs at an incident energy (E_N) of 6.385 MeV (=E_R) with a resonance width of 1.8 keV, which allows for high-resolution depth profiling: While H atoms on a sample surface are detected at $E_N = E_R$, at $E_N > E_R$ the reaction occurs at the depth where E_N decreases to E_R due to the energy loss as the incident ions penetrate the sample. Since the resonance width of 1.8 keV is comparable with the typical stopping power of 2–4 keV/nm in solids, the depth profiling of H can be performed by scanning E_N with a depth resolution on the order of nanometers. Therefore, this technique is a highly suitable technique for detecting H in nanoscale materials and subsurface regions.

13.3.2 Structure Analysis by Channeling ^{15}N-NRA

The combination of ^{15}N-NRA and ion channeling techniques allows for detailed structure analysis of H atoms in crystals, which is called channeling ^{15}N-NRA [36, 37, 38]. Ion channeling is a phenomenon, where incident ions are guided by the Coulombic force from the atomic nuclei of the host lattice when the ion beam enters along a crystallographic axis of the crystal. As a result, the incident ions travel through the open spaces (channels) between rows or planes of atoms in the crystal lattice, as schematically shown in Fig. 13.8. There are two types of channelings, axial and planar channelings, as illustrated in Fig. 13.9, where incident ions travel through the gaps between atomic rows and atomic planes, respectively. Under the channeling conditions, there are crystal regions behind the host atoms where the incident ions cannot reach, referred to shadow cones, as illustrated by gray regions in Fig. 13.8. When H atoms occupy lattice sites within these shadow cones, the probability of the nuclear reaction with the incident ions decreases, leading to a reduction in the NRA yield. On the other hand, if H atoms are located in the channels, the NRA yield

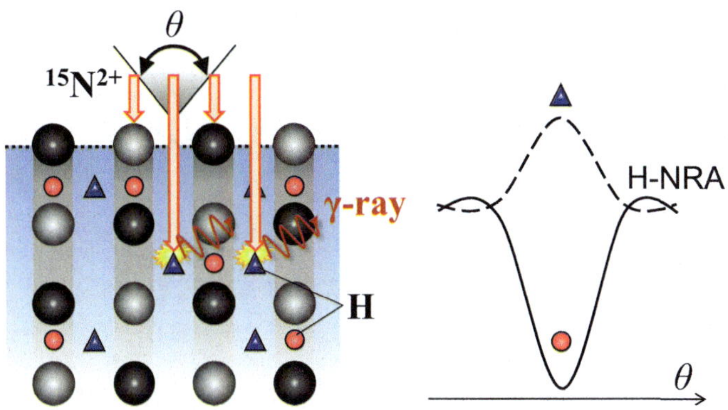

Fig. 13.8 Schematic illustration of channeling nuclear reaction analysis (^{15}N-NRA). Gray regions represent shadow cones. Black and gray circles indicate metal lattice atoms, while blue triangles and red circles represent H occupation sites located along channels and within shadow cones, respectively. The expected angular (θ) dependence of H-NRA yields for H atoms at each site is also illustrated

exhibits an increase, as shown in Fig. 13.8. Thus, variations in the NRA yield as a function of the incident beam angle allow for identification of H lattice location in crystals. In the channeling measurement, Rutherford backscattering spectrometry (RBS) is simultaneously performed, where the amount and energy of incident ions back-scattered from a sample are analyzed. Under channeling conditions, the RBS yield decreases. The analysis of the RBS yield thus provides information on the degree of channeling. ^{15}N-NRA and RBS are conducted using scintillation detectors and a solid-state detector, respectively. Although the sample has to be a single crystal, the channeling ^{15}N-NRA has capability to perform structural analysis of hydrogen even in nanoscale thin films and subsurface regions, owing to the depth-sensitive detection of H via the nuclear reaction of ^{1}H(^{15}N,$\alpha\gamma$)^{12}C.

Another advantage of channeling ^{15}N-NRA is its isotope-selective detection. The ^{15}N ions undergo nuclear reactions not only with H but also with D, though the reaction with D is not resonant and lacks depth resolution [39]. H and D atoms can be distinguished by analyzing the energy of the emitted γ-rays; while the reaction with H yields a γ-ray with an energy of 4.43 MeV, γ-rays with energies of 2.7, 3.8, 6.1 and 7.1 MeV are emitted from the reactions with D. Hereafter, NRA yields arising from reactions with H and D atoms are referred to as H-NRA and D-NRA, respectively. This technique enables isotope-specific analysis of hydrogen locations in a single sample containing both H and D using a single ^{15}N beam [38]. Consequently, channeling ^{15}N-NRA uniquely evaluates isotope effects on hydrogen site occupancy.

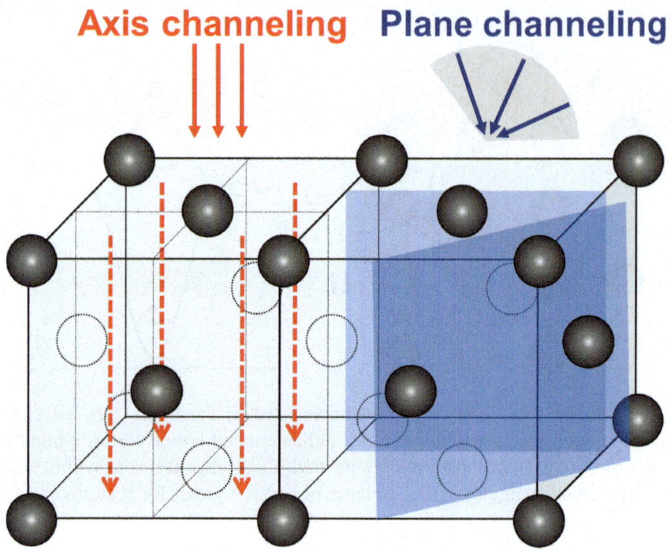

Fig. 13.9 Schematic illustration of axis channeling at normal incidence along a crystallographic axis and plane channeling at oblique incidence along a crystallographic plane. The dashed arrows and colored regions represent the beam trajectories

13.3.3 Probing Hydrogen Hopping Via Electrical Resistance

Electrical resistance is a sensitive probe for hydrogen-related dynamics such as absorption, desorption, and migration owing to the influence of hydrogen on the electronic structure. Whereas electrical resistance does not change during hydrogen hopping between equivalent states, hydrogen hopping between inequivalent states, such as from metastable to stable states, is accompanied by variations in resistance. Resistance is also sensitive to the macroscopic change in H depth distribution. These variations enable accurate determination of hydrogen hopping rates through time-evolution measurements of resistance. Metastable hydride states can be formed either by quenching a high-temperature phase or implanting energetic hydrogen ions, where H atoms possibly occupy metastable sites in the lattice or spatially localize in the near-surface region. A combination of the hopping rate measurements using resistance analysis and the structure analysis via channeling ^{15}N-NRA provides valuable insights into hopping mechanisms of hydrogen isotopes, including their migration pathways.

13.4 Recent Experimental Studies

Here, we report the recent experimental results demonstrating the importance of NQEs on static and dynamic properties of hydrogen in metals.

13.4.1 Nuclear Quantum Effects (NQEs) on Static Properties of Hydrogen in Titanium

The location of hydrogen in titanium hydride (TiH_x) nanofilms reflects NQEs, which is clarified by channeling ^{15}N-NRA [38]. According to the previous studies on bulk TiH_x, it consists of face-centered cubic (FCC) Ti lattice with H atoms occupying the T site [40]. For a 90-nm-thick $TiH_{1.47}$(110) epitaxial thin film grown on an MgO(110) substrate by reactive sputtering, RBS and ^{15}N-NRA were simultaneously performed at room temperature around the surface normal direction with the $^{15}N^{2+}$ beam at an incident energy of 6.45 MeV, which corresponds to the H probing depth of ~ 30 nm from the surface. Figure 13.10b, c shows the two-dimensional (2D) angular mappings of Ti-RBS and H-NRA yields as functions of incident angles θ and φ defined in Fig. 13.10a. Here, Ti-RBS denotes the RBS signal originating from Ti atoms in the film. In Fig. 13.10b, the dip at the center corresponds to the (110) axis channeling, while the dip lines in multiple directions represent the plane channelings. In contrast to the Ti-RBS, the horizontal plane shows a decrease and other planes reveal increases in the H-NRA yield (Fig. 13.10c). Figure 13.11a, b represents top views of TiH_x(110) crystals with H at the T and O sites, respectively. Considering that the H atoms in the O site are located between the planes of (1–11) and (1–1-1) but hidden behind the host Ti atoms at the (001) and (−110) plane channelings (Fig. 13.11b), whereas the H atoms in the T site are all located between these planes except under the (−110) plane channeling (Fig. 13.11a), the H-NRA pattern obtained for the $TiH_{1.47}$(110) is qualitatively understood in terms of the T-site occupation by H.

For the detailed analysis of the H location, the experimental results were compared with ion beam trajectory simulations using FLUX7 [41], which estimates the average nuclear close-encounter probabilities (NEP) of $^{15}N^{2+}$ ions to Ti and H atoms. The NEP value is normalized to unity under random incidence conditions. NEPs for Ti (NEP(Ti)) and for H (NEP(H)) can be quantitatively compared with the normalized Ti-RBS and H-NRA yields,. respectively. It is important to note that the H-NRA yield is influenced not only by the location of H atoms but also by the degree of ion channeling, i.e., the trajectory of the incident beam in a crystal. For instance, the variation in the H-NRA yield gets smaller under an insufficient channeling condition. Since hydrogen is much lighter than ^{15}N ions, it has little influence on the ^{15}N ion beam trajectory, and thus the degree of channeling is primarily determined by the quality of the host crystal and the incident beam conditions, which is reflected in the Ti-RBS yield. Consequently, the detailed structure analysis of H can be achieved

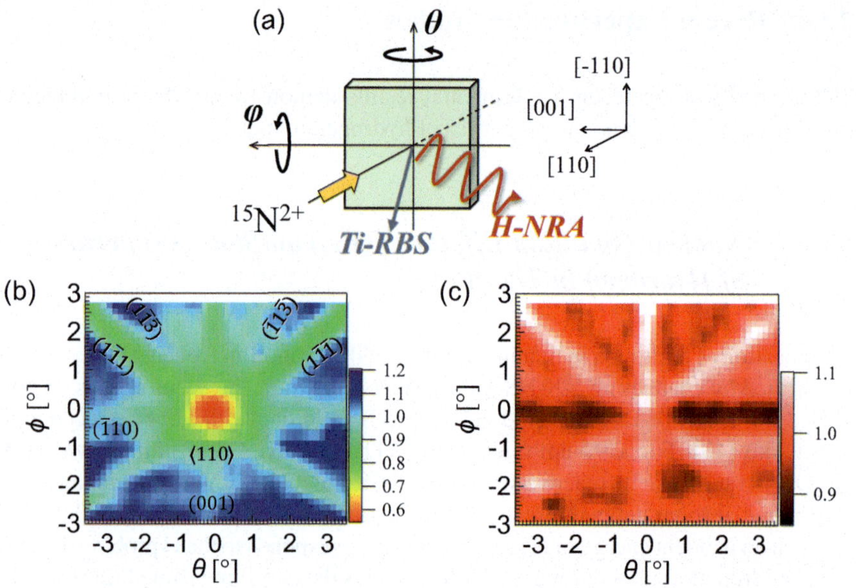

Fig. 13.10 Results of structure analysis on the 90-nm-thick $TiH_{1.47}(110)$ epitaxial film by channeling ^{15}N-NRA. **a** Schematic illustration of experimental setup. Two-dimensional (2D) mappings of **b** Ti-RBS and **c** H-NRA yields as functions of incident angles θ and ϕ defined in (*a*). Adapted with permission from Macmillan Publishers Ltd.: Ref. [38]. Copyright 2024

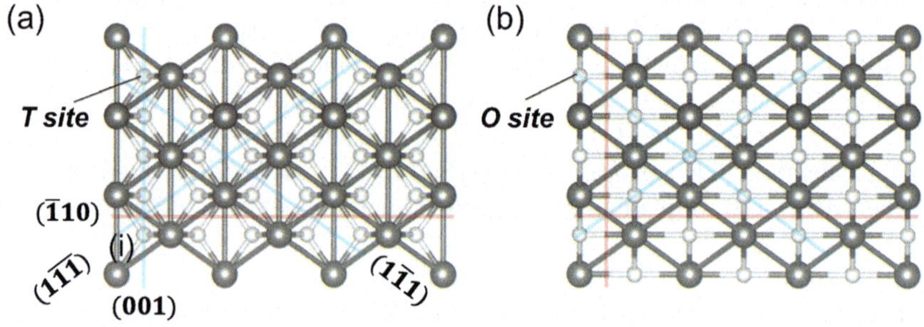

Fig. 13.11 The (110) top views of Ti crystal structures with H atoms in **a** the T site and **b** the O site, respectively. Black and white circles represent Ti and H atoms, respectively. Blue and red lines represent plane channels in which H is located or not, respectively. Adapted with permission from Macmillan Publishers Ltd.: Ref. [38]. Copyright 2024

by evaluating the degree of channeling from the angular dependence of the Ti-RBS yield and subsequently analyzing the angular dependence of the H-NRA yield.

Figure 13.12a shows the calculated 2D mapping of NEP(Ti) after applying a correction for the blocking effect of the $< 100 >$ atomic row for the back-scattered ions, which corresponds to the time-reversal process of ion channeling for incident ions. This corrected mapping (corr. NEP(Ti)) is in good agreement with the experimental result of Fig. 13.10b, showing that the beam trajectory in the nanofilm is accurately reproduced by the simulation. After confirming the validity of the beam trajectory, the location of H was subsequently analyzed by comparing the H-NRA and NEP(H). Figure 13.12b, c shows the calculated 2D mappings of NEP's for H assuming the T- and O-site occupation (NEP(H_T) and NEP(H_O)), respectively. Although the NEP(H_T) rather than the NEP(H_O) resembles the H-NRA pattern, the intensities under the channeling conditions are not quantitatively consistent between the H-NRA and NEP(H_T); while the intensity at the (001) plane channeling is clearly higher than those at the (1–11) and (1–1-1) plane channelings in the NEP(H_T), the intensities are almost the same at these three plane channelings in the H-NRA. This discrepancy reveals that H atoms do not simply occupy the T site. To investigate this, the possibility of partial occupation of the O site by H is examined. Corresponding NEP($H_T + H_O$) obtained by the linear combination of NEP(H_T) and NEP(H_O) with an O-site occupancy of 11 at.% (Fig. 13.12d) showed the best agreement with the experimental result, revealing that H atoms co-occupy the T and O sites in the $TiH_{1.47}$(110) film.

The origin of the partial O-site occupation by H atoms was theoretically investigated using density functional theory (DFT) calculations. The projector augmented wave (PAW) method was employed, along with the Perdew-Burke-Ernzerhof exchange-correlation functional for solids (PBEsol). The formation energies were calculated for a supercell composed of Ti_8H_n with nH atoms in the T site and for a supercell with (n-1) H atoms in the T site and an H atom in the O site, as shown in Fig. 13.13a. As the H concentration is decreased from $x = 2$ ($n = 16$), the partial O-site occupation structures are found to be stabilized. Figure 13.13b shows the changes in the electronic density of states (DOS) for $x = 1.5$ ($n = 12$) with partial O-site occupation. The DOS for the structure with exclusive T site occupation reveals a sharp peak at the Fermi level, originating from Ti 3d electrons, and this sharp peak is split into two peaks forming a deep valley (so-called pseudogap) with partial O-site occupation. This pseudogap formation is caused by lowering of the structural

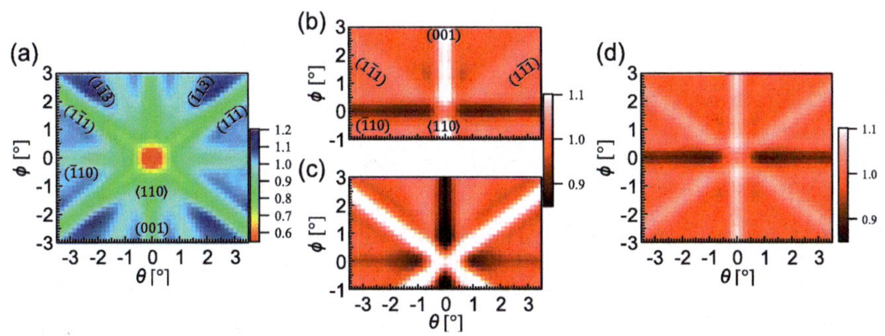

Fig. 13.12 Calculated nuclear close-encounter probabilities (NEP) mappings for **a** Ti, **b** H in T site, **c** H in O site, and **d** H in both T and O sites. Adapted with permission from Macmillan Publishers Ltd.: Ref. [38]. Copyright 2024

symmetry due to the co-occupation of T and O sites by H and the Fermi level shift owing to the H deficiency from $x = 2$. Figure 13.13c shows the schematic illustration of the local structure around a Ti atom, where the local symmetry around a Ti atom is apparently lowered due to the partial transfer of H atoms from T to O sites. This lowering of the symmetry splits the near-degenerate states at the Fermi level, which lowers the electronic energy. This stabilization mechanism can be recognized as the hydrogen-induced Jahn–Teller effect, which is in keen contrast with the conventional Jahn–Teller effect: While atoms are marginally displaced (< 0.1 Å) in ordinary Jahn–Teller systems, H atoms are located in substantially different sites of T and O (> 1 Å) in the present case. The hydrogen-induced Jahn–Teller effect identified in this study will serve as a general mechanism for determining hydrogen site occupancy in hydride materials.

Interestingly, this titanium hydride system reveals static NQEs. Isotope effects on the lattice location of hydrogen were explored taking advantage of the isotope-selective detection of H and D atoms using channeling ^{15}N-NRA. The H-NRA and

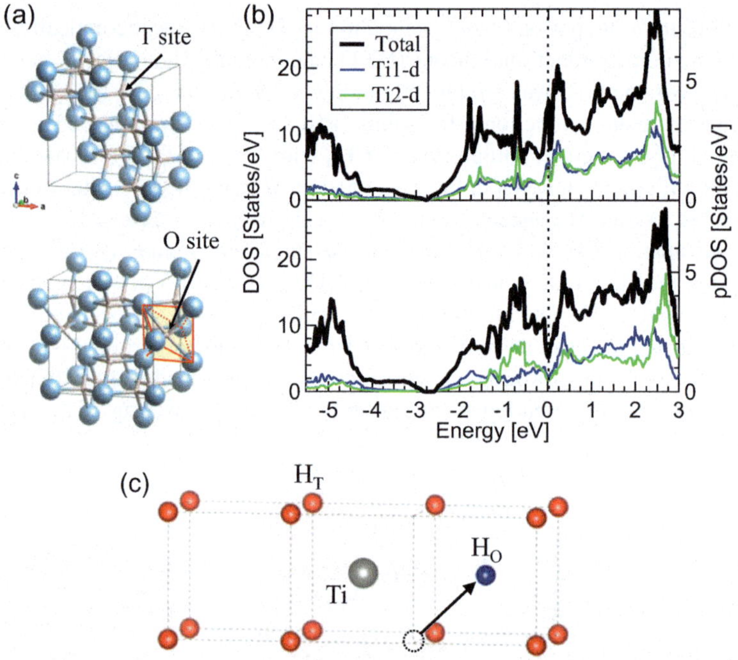

Fig. 13.13 Results of density functional theory (DFT) calculations for TiH_x. **a** Supercells used in the calculations with H atoms at T sites (top) and at both T and O sites (bottom). **b** Corresponding DOS for these structures at H concentration $x = 1.5$. The dotted line represents the Fermi level. **c** Local structure around a Ti atom. Black, red and blue balls represent Ti, H in T site, and H in O site, respectively. Each Ti atom is surrounded by eight H atoms in the T site, and the local symmetry around the Ti atom is lowered by transfer of an H atom from the T to O sites. Adapted with permission from Macmillan Publishers Ltd.: Ref. [38]. Copyright 2024

D-NRA yields as functions of incident angles were simultaneously obtained for a 100-nm-thick $TiH_{0.60}D_{1.0}$(110) epitaxial film. It was found that whereas part of H is located at the O site with the remaining H at the T site as the $TiH_{1.47}$ film, D atoms solely occupy the T site. This is explained by the difference in ZPE, which depends on both the atomic mass and the shape of the potential energy surface. While ZPE for H tends to be larger than D due to the mass difference, ZPE for isotopes in the T site is larger than in the O site because the T site is spatially narrower. As a result, D atoms in the T site are relatively more stabilized compared to H atoms, as also confirmed by ZPE calculations. This isotope-dependent site occupation is a consequence of NQEs on the static properties of hydrogen. Furthermore, it enables tuning of the hydride structure by varying the isotopic composition, opening new avenues for exploring novel phenomena in hydrides through controlled manipulation of hydrogen site occupancy and crystal symmetry.

13.4.2 NQEs on Dynamic Properties of Hydrogen in Palladium at Intermediate Temperatures

Contribution of NQEs to hydrogen diffusion at intermediate temperatures was investigated in palladium hydrides (PdH_x) using electrical-resistance measurements [42]. In the FCC lattice of Pd, the O site is the most stable and the T site is metastable for hydrogen atoms, as shown in the schematic energy diagram of Fig. 13.14. The diffusion of hydrogen in Pd occurs between the O sites via the T site. Also shown in the figure are the vibrational energy levels of H in the O and T sites. A characteristic feature of H in Pd is that the energy level of the second excited state at the O site is close to that of the ground state at the T site. PdH_x exhibits a particular order-disorder phase transition related to the hydrogen configuration around 50 K, which is recognized by anomalous behavior in the resistance called the 50 K anomaly. The high-temperature disordered phase can be quenched to low temperatures, which slowly relaxes to the ordered phase accompanied by hydrogen diffusion as probed by resistance measurements. Therefore, the hydrogen diffusion between the O sites can be investigated by the resistance change associated with this structure change at low temperatures.

A 50-nm-thick polycrystalline Pd film was hydrogenated with H_2 gas at 200 K and subsequently quenched to 3 K in a vacuum, resulting in the formation of a metastable quenched state. Figure 13.15 shows the time evolutions of temperature and resistance of $PdH_{0.56}$ kept at a constant temperature after the quenching. The resistance relaxation reflects the transition from a metastable disordered state to the stable ordered state, driven by H hopping. Figure 13.16 shows the temperature dependence of the H and D hopping rates (ν) derived from the relaxation curves by applying exponential fitting to the resistance relaxation data in Fig. 13.15. These rates are described as the inverse of the time constant ($1/\tau$). The hopping rate reveals a thermally activated behavior at high temperatures above ~ 55 K and a nearly temperature-independent

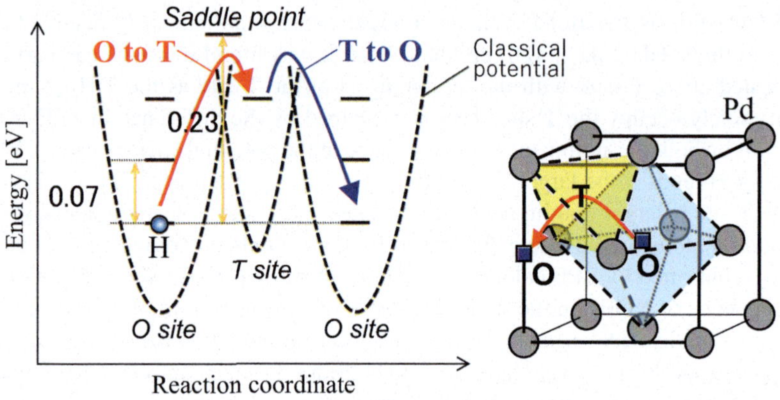

Fig. 13.14 Schematic illustration of potential and vibrational energy states of H in Pd. Hopping from the O to T sites and the T to O sites are represented by red and blue arrows, respectively, and the former is the rate-limiting step for the diffusion. The crystal structure is also shown [42]

feature at lower temperatures, which corresponds to a crossover from thermal to quantum regimes. Whereas a gradual crossover was observed for H in PdH_x with small x, an abrupt transition was found in PdH_x with large x. Furthermore, an abrupt transition was observed for D regardless of its concentration. The hopping rate of H in PdH_x with small x is found to be expressed by the sum of three components: classical thermal diffusion with an activation energy of 220 meV, thermally activated tunneling with an activation energy of 150 meV, and quantum tunneling with a power-law dependence on temperature. The gradual crossover observed for H at low concentrations is attributed to the thermally activated tunneling, and its activation energy of 150 meV corresponds to the thermal excitation of H from the ground state to the second excited vibrational state at the O site. This behavior agrees well with the PIMD calculation results for H in Pd [24].

Fig. 13.15 Time evolutions of (A) resistance and (B) temperature after the quenching in $PdH_{0.56}$. Adapted from Ref. [42], Copyright 2024, with permission from Elsevier

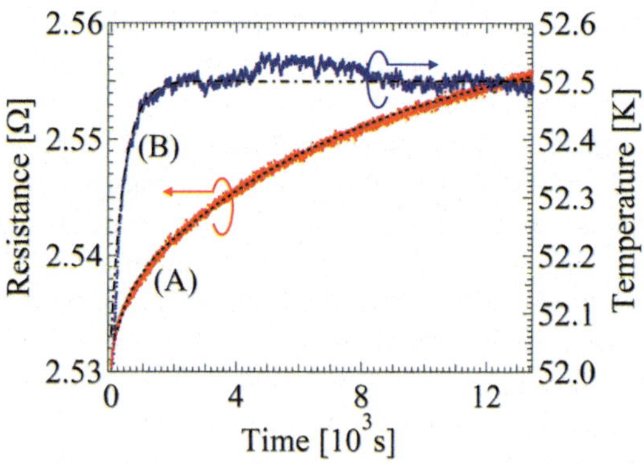

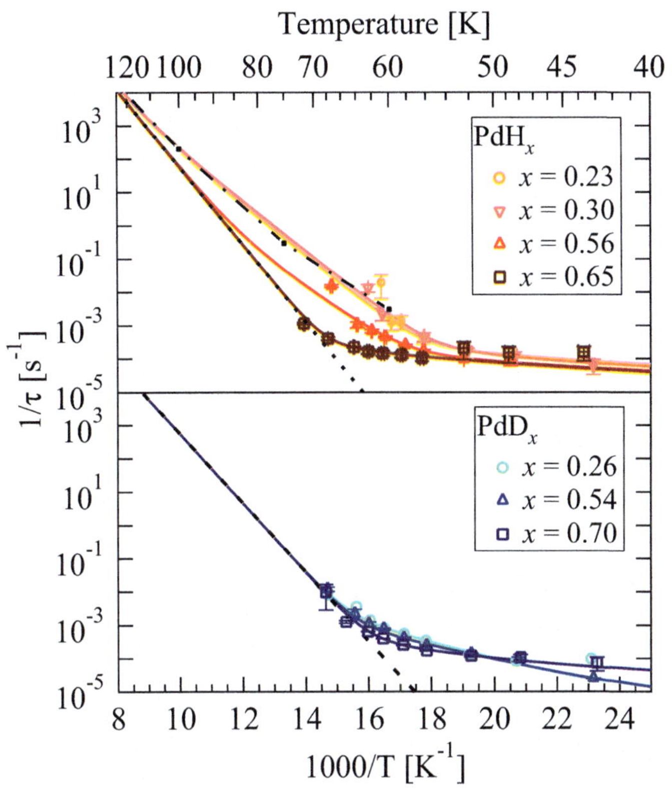

Fig. 13.16 Temperature dependences of H and D hopping rates at various hydrogen concentrations derived from the resistance relaxation measurements. The bottom and top axes represent the inverse temperature and the corresponding temperature, respectively. Adapted from Ref. [42], Copyright 2024, with permission from Elsevier

The hopping mechanism was theoretically investigated by quantum simulations. The quantum states of hydrogen in Pd were calculated by solving the Schrödinger equation for the hydrogen nuclear motion on the three-dimensional potential energy surface derived from the DFT calculations for a $Pd_{32}H_1$ supercell. A large supercell with one H atom was employed to examine the quantum state of an isolated H atom. The DFT calculations were performed using the PAW method with the Perdew-Burke-Ernzerhof (PBE) functional. Figure 13.17 shows the calculated wave functions of H. While the ground states at the O and T site are localized (Fig. 13.17a, b), an excited state corresponding to the second vibrational state at the O site is extended over both O and T sites (Fig. 13.17c). This extended state is attributed to the coupled state via quantum tunneling between the second excited state at the O site and the ground state at the T site. From the analysis of the quantum state splitting, the tunneling matrix element is derived to be 0.501 meV. It is noted that the degree of the wavefunction extension between the O and T sites depends on the energy-level matching. Figure 13.18 shows the relation between the experimentally obtained hopping rate by tunneling (ν) and the calculated energy-level difference between the second excited state at the O site $\left(E_{\mathrm{O}}^{2}\right)$ and the ground state at the T site $\left(E_{\mathrm{T}}^{0}\right)$ for different H concentrations. As the concentration increases, the energy-level difference becomes larger. As seen in the figure, the hopping rate due to tunneling reveals a resonance feature with respect to the energy-level matching between the O and T sites. It should be noted that no extended state was observed for D because of the mismatch of the energy levels as well as the mass effect on the tunneling probability.

While both abrupt and gradual transitions from the thermal to the quantum regimes have been reported depending on the system, the influence of NQEs on the crossover region has been unclear. The above results demonstrate that the hydrogen hopping behavior in the crossover region at intermediate temperatures is governed by quantum tunneling via an excited vibrational state, where energy-level matching plays a crucial role.

13.4.3 NQEs on Dynamic Properties of Hydrogen in Palladium at Low Temperatures

As discussed above, the tunneling of H in Pd occurs between the second excited state of the O site and the ground state of the T site. However, because the hopping by the tunneling from the O to T sites in the previous section is induced by thermal excitation of the hydrogen vibration at the O site, the intrinsic temperature dependence of the tunneling rate, which reflects the effects of the environment on the tunneling process, is obscured by the vibrational excitation. This is because the exponential temperature dependence of vibrational excitation is substantially stronger than the weak power-law dependence of quantum tunneling. To overcome this problem, H atoms were placed into the T site as an initial metastable state by adopting hydrogen ion implantation at low temperatures, and the tunneling of hydrogen from the T to O sites was investigated through structure analysis and resistance measurements [43, 44].

To confirm the T-site occupation by H, channeling ^{15}N-NRA was performed on a Pd(100) single crystal following H implantation at 50 K. RBS for Pd (Pd-RBS) was simultaneously obtained. Figure 13.19 shows the angular profiles of the H-NRA yields obtained at 50 K before and after the H-implanted sample was annealed at ~ 90 K. The Pd-RBS yield is also shown in the figure. A decrease in the H-NRA was observed at the < 100 > axis channeling after the annealing. The crystal top views shown in Fig. 13.20a, b suggest that this decrease is attributed to the O-site occupation. From the analysis based on beam trajectory simulations, the H atoms implanted at 50 K were found to be located equally at the T and O sites, and migrate from the T to O sites at ~ 90 K. Since the hydrogen migration from the T to O sites accompanies a change in the resistance, the hopping rate of hydrogen from the T to

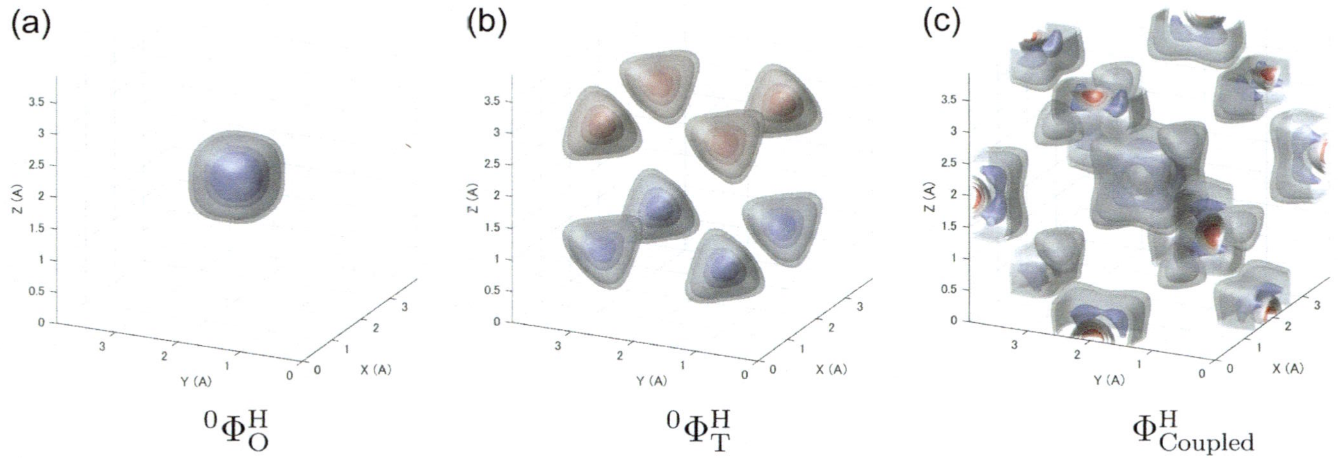

Fig. 13.17 Spatial wave functions ($\Phi(x, y, z)$) of **a** the ground state of H localized at the O site, **b** the ground state of H localized at the T site, and **c** the coupled state of H extended between the O and T sites. Contour lines of the amplitude of $\Phi(x, y, z)$ are shown, and the color scale represents the polarity. Adapted from Ref. [42], Copyright 2024, with permission from Elsevier

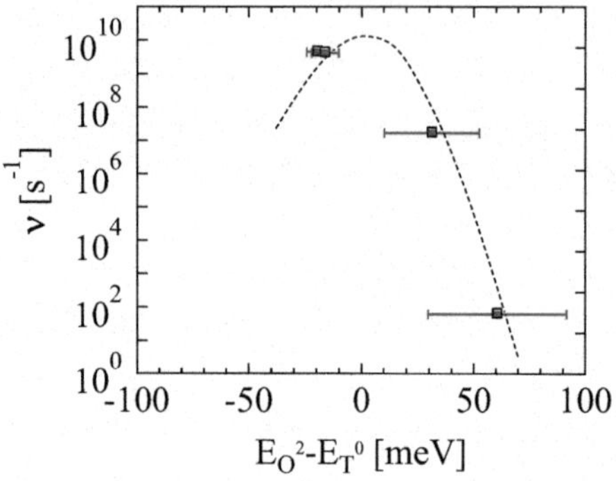

Fig. 13.18 The hopping rate of H by tunneling (ν) from the second excited state at the O site to the ground state at the T site as a function of the energy-level difference between the two states. E_O^2 and E_T^0 denote the energies of H in the second excited state at the O site and the ground state at the T site, respectively. The hopping rates were experimentally obtained in Fig. 13.16, and the energy levels were derived from the DFT calculations for various concentrations. The dashed curve is a guide for the eye [42]

O sites was successfully derived from the time evolution of the resistance at various temperatures. Figure 13.21a shows the temperature dependence of the hopping rates of H and D, which clearly shows a large isotope effect. The H hopping reveals a linear relation above 70 K in the Arrhenius plot indicating a thermal hopping, and a nearly temperature-independent feature below 70 K suggesting quantum tunneling. In contrast, the D hopping is significantly suppressed below 70 K.

In the thermal regime above 70 K in Fig. 13.21a, both H and D hopping rates obey the Arrhenius law. Based on the Arrhenius equation, the activation energies were derived to be 64.1 ± 4.2 and 73.0 ± 3.9 meV for H and D, respectively. This isotope effect is ascribed to the zero-point vibration. According to the transition state theory, the activation energy corresponds to the energy-level difference between the initial and transition states, which, in the present case of hopping from the T to O sites, correspond to the T site and a saddle point located between the T and O sites, respectively. Note that there exists a confining potential in the coordinates other than the reaction coordinate at the saddle point. The smaller activation energy observed for H hopping compared to D hopping reveals that the ZPE is larger at the T site than at the saddle point, suggesting the narrower potential of the T site. On the other hand, an inverse isotope effect in the activation energy has been reported for the thermal macroscopic diffusion of hydrogen in Pd [45], where the hopping from O to T site is rate-limiting, indicating that the potential at the saddle point is narrower than that of the O site. Eventually, as schematically shown in Fig. 13.22a, these clarify the shape of the potential landscape that the potential for hydrogen is sharper in the order of the T site, the saddle point, and the O site, which is in good agreement with recent theoretical calculations with the ZPE corrections [24].

At the quantum regime below 70 K, as shown in Fig. 13.21b, c, the H hopping rate by quantum tunneling reveals slightly positive and negative temperature dependences

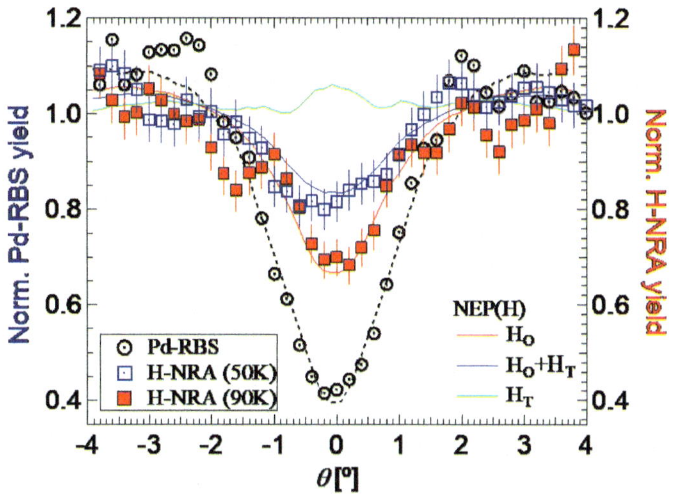

Fig. 13.19 Angular dependence of the H-NRA yields for the H implanted Pd(100) crystal around the < 100 > axis, obtained by channeling ^{15}N-NRA before and after annealing at ~90 K. The Pd-RBS is also shown. H-NRA (right axis) and Pd-RBS (left axis) yields are presented in normalized units. θ denotes the incident angle of the ^{15}N ions with respect to the surface normal. Adapted from [44], 2025 under CC-BY license

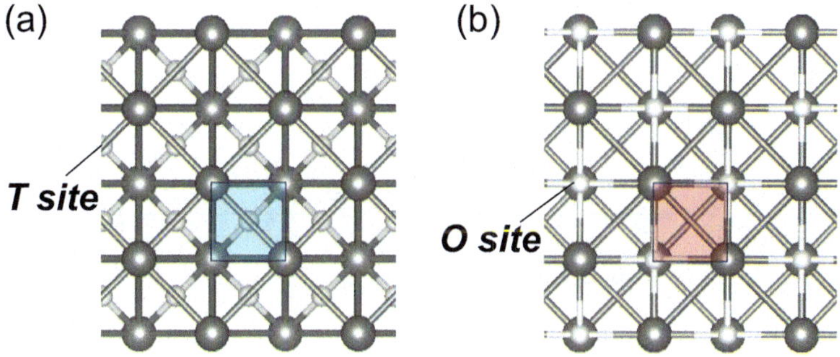

Fig. 13.20 (100) top views for Pd structures with H at **a** the T site and **b** the O site, respectively. Black and white circles represent Pd and H atoms, respectively. Blue and red regions indicate axis channels in which H is located or not, respectively. Adapted from [44], 2025 under CC-BY license

above and below 20 K, respectively, which are attributed to the phonon effects and non-adiabatic effects of conduction electrons on the proton tunneling. From the fit to the data between 20 and 50 K, the activation energy was estimated to be 3.1 ± 0.5 meV, which is consistent with the energy difference of 1.8 meV between the second excited state at the O site and the ground state at the T site, as obtained from quantum calculations. This indicates that H tunneling occurs with the assistance of phonons between 20 and 50 K. On the other hand, the latter behavior below 20 K

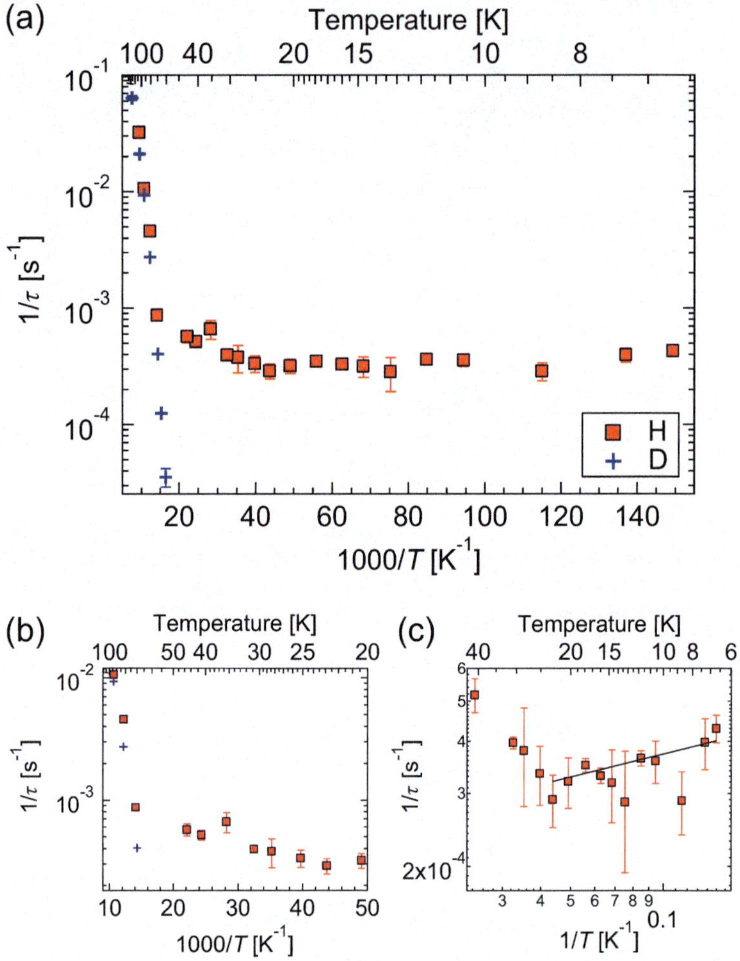

Fig. 13.21 **a** Temperature dependence of the H and D hopping rates derived from the resistance relaxation measurements. The enlarged views at **b** intermediate- and **c** low-temperature regions. The solid curve in (c) represents the fitting line described as T^{2K-1}. The bottom and top axes represent the inverse temperature and the corresponding temperature, respectively. Adapted from [44], 2025 under CC-BY license

agrees with Kondo's theory dealing with the effects of conduction electrons on the tunneling. According to the theory, the temperature dependence of the tunneling probability is described as T^{2K-1} with the proton-electron coupling constant K. Here, the tunneling matrix element is renormalized as T^{2K} due to the orthogonality of the electronic wavefunctions with an H atom located at different sites, while an effective level broadening leads to a T^{-1} term. The former term corresponds to the non-adiabatic effects of conduction electrons, where the electrons scattered by H do not perfectly follow the H tunneling and get excited, as schematically illustrated

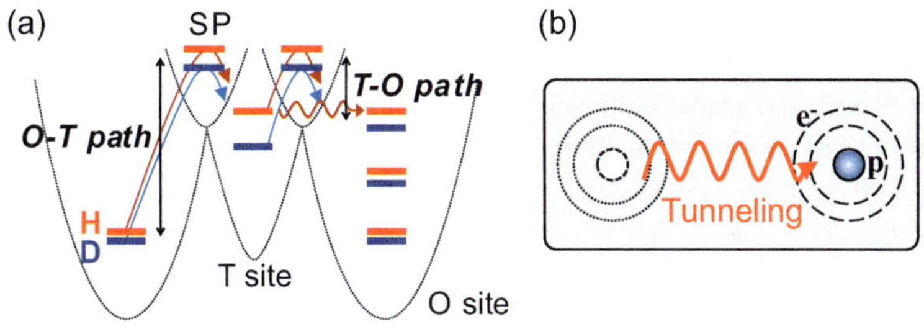

Fig. 13.22 **a** Schematic energy diagram of H and D along the O-T and T-O pathways. **b** Schematic illustration of the non-adiabatic effect of conduction electrons on proton tunneling. The electrons of Pd are scattered by proton as schematically shown by dotted and dashed circles, and the scattered electrons do not perfectly follow the instantaneous position change of the tunneling proton, which corresponds to an electronic excitation. This induces a temperature dependence of the proton tunneling rate proportional to $T^{2K\text{-}1}$. Adapted from [44], 2025 under CC-BY license

in Fig. 13.22b. The coupling constant K was derived to be 0.41 ± 0.03 from its temperature dependence in Fig. 13.21c. The derived value of H in Pd is larger than those reported for Cu in the previous studies (Figs. 13.6 and 13.7), suggesting that the asymmetric potential pathway from the T to O sites, as is discussed by Regelmann [22], and/or the involvement of d-electrons influences the proton-electron coupling.

These findings highlight the importance of measuring hopping rate and identifying hopping pathway, both of which are essential for understanding the detailed mechanism of hydrogen hopping in metals. Measuring the bidirectional hopping rates of hydrogen isotopes, such as between T and O sites, enables experimental investigation of the potential landscape. These measurements also provide a framework for examining how the surrounding environment affects hydrogen tunneling in relation to the potential shape.

13.5 Summary

Nuclear quantum effects of hydrogen are clarified for the structure and hopping dynamics of hydrogen in metals in a highly precise manner by channeling ^{15}N-NRA and resistance measurements as follows:

- The crystal symmetry due to the hydrogen configuration in crystals can be crucial for the structure stability and the electronic state. It was demonstrated that H atoms co-occupy the tetrahedral and octahedral sites in Ti crystals by lowering the crystal symmetry accompanied with the electronic energy gain. Furthermore, the lattice location of H and D atoms was found to be different, proposing the possibility to tune the lattice location of hydrogen via ZPE of NQEs.

- Quantum tunneling as a manifestation of NQEs plays an essential role in determining the crossover behavior of the hydrogen hopping at intermediate temperature. Whereas a gradual transition from thermal to quantum hoppings, attributed to the tunneling via the thermally excited vibrational state, was observed in Pd, an abrupt transition was observed for H with high concentration and for D because of the suppression of tunneling process caused by energy-level mismatching.
- The surrounding environment, such as phonons and conduction electrons, affects the quantum tunneling of hydrogen at low temperatures. Slightly positive and negative temperature dependence of H tunneling from metastable T to stable O sites were observed in Pd, which are the signatures of proton coupling with phonon and electron excitations, respectively, in the tunneling process.

The recent experimental advancements have enabled the direct observation of nuclear quantum effects of hydrogen in its static and dynamic properties. These developments are expected to drive further innovations in hydrogen-based applications, while also providing new insights into the quantum nature of atomic motion and facilitating the exploration of novel quantum phenomena.

References

1. Fukai Y. The metal-hydrogen system basic bulk properties. 2nd ed. Berlin: Springer; 2005.
2. Sang SH, Furukawa H, Yaghi OM, Goddard WA. Covalent organic frameworks as exceptional hydrogen storage materials. J Am Chem Soc. 2008;130:11580.
3. Kato R, Yoshimasa K, Egashira T, Oya T, Oyaizu K, Nishide H. A ketone/alcohol polymer for cycle of electrolytic hydrogen-fixing with water and releasing under mild conditions. Nat Commun. 2016;7:1.
4. Satterthwaite CB, Toepke IL. Superconductivity of hydrides and Deuterides of thorium. Phys Rev Lett. 1970;25:741.
5. Skoskiewicz T. Superconductivity in the palladium-hydrogen and palladium-nickel-hydrogen systems. Phys Status Solidi A. 1972;11:K123.
6. Drozdov AP, Eremets MI, Troyan IA, Ksenofontov V, Shylin SI. Conventional superconductivity at 203 kelvin at high pressures in the sulfur hydride system. Nature. 2015;525:73.
7. Drozdov AP, et al. Superconductivity at 250 K in lanthanum hydride under high pressures. Nature. 2019;569:528.
8. Shi J, Zhou Y, Ramanathan S. Colossal resistance switching and band gap modulation in a perovskite nickelate by electron doping. Nat Commun. 2014;5:4860.
9. Yoon H, Choi M, Lim TW, Kwon H, Ihm K, Kim JK, Choi SY, Son J. Reversible phase modulation and hydrogen storage in multivalent VO2 epitaxial thin films. Nat Mater. 2016;15:1113.
10. Matsuzawa I, Ozawa T, Nishiya Y, Sidik U, Hattori AN, Tanaka H, Fukutani K. Controlling dual Mott states by hydrogen doping to perovskite rare-earth nickelates. Phys Rev Mater. 2023;7:085003.
11. Wertheim GK, Buchanan DNE. Magnetic properties of hydrogen-saturated nickel. Phys Lett. 1966;21:255.
12. Koval S, Kohanoff J, Migoni RL, Tosatti E. Ferroelectricity and isotope effects in hydrogen-bonded KDP crystals. Phys Rev Lett. 2002;89:1.
13. Luo W, Clewley JD, Flanagan TB. Thermodynamics and isotope effects of the vanadium-hydrogen system using differential heat conduction calorimetry. J Chem Phys. 1990;93:6710.

14. Ben-Naim A, Wilf J, Yaacobi M. Hydrophobic interaction in light and heavy water. J Phys Chem. 1973;77:95.
15. Marcus Y, Ben-Naim A. A study of the structure of water and its dependence on solutes, based on the isotope effects on solvation thermodynamics in water. J Chem Phys. 1985;83:4744.
16. Holstein T. Studies of polaron motion I. Ann Phys (N Y). 1959;**8**:325.
17. Holstein T. Studies of polaron motion II. Ann Phys (N Y).1959;**8**:343.
18. Flynn CP, Stoneham AM. Quantum theory of diffusion with application to light interstitials in metals. Phys Rev B. 1970;1:3966.
19. Kagan Y, Klinger MI. Theory of quantum diffusion of atoms in crystals. J Phys C Solid State Phys. 1974;7:2791.
20. Kondo J. Adiabatic vs. non-adiabatic state of a heavy particle in a metal. Physica B+C. 1984;124:25.
21. Yamada K. Diffusion of positive muon and orthogonality theorem. Prog Theor Phys. 1984;72:195.
22. Regelmann T, Schimmele L. Low-temperature quantum diffusion of light particles in solids. J Alloys Compd. 1995;231:208.
23. Fang W, Richardson JO, Chen J, Li XZ, Michaelides A. Simultaneous deep Tunneling and classical hopping for hydrogen diffusion on metals. Phys Rev Lett. 2017;119:1.
24. Kimizuka H, Ogata S, Shiga M. Unraveling anomalous isotope effect on hydrogen diffusivities in fcc metals from first principles including nuclear quantum effects. Phys Rev B. 2019;100:024104.
25. DiFoggio R, Gomer R. Tunneling of hydrogen in surface diffusion on the tungsten (110) plane. Phys Rev Lett. 1980;44:1258.
26. Lee A, Zhu XD, Deng L, Linke U. Observation of a transition from over-barrier hopping to activated tunneling diffusion: H and D on Ni(100). Phys Rev B. 1992;46:15472.
27. Cao GX, Nabighian E, Zhu XD. Diffusion of hydrogen on Ni(111) over a wide range of temperature: exploring quantum diffusion on metals. Phys Rev Lett. 1997;79:3696.
28. Zheng CZ, Yeung CK, Loy MMT, Xiao X. Quantum diffusion of H on Pt(111): step effects. Phys Rev Lett. 2006;97:166101.
29. Lauhon LJ, Ho W. Direct observation of the quantum Tunneling of single hydrogen atoms with a scanning Tunneling microscope. Phys Rev Lett. 2000;85:4566.
30. Sundell PG, Wahnström G. Quantum motion of hydrogen on cu(001) using first-principles calculations. Phys Rev B. 2004;70:081403.
31. Kehr KW, Richter D, Welter JM, Hartmann O, Karlsson E, Norlin LO, Niinikoski TO, Yaouanc A. Muon diffusion and trapping in aluminum and dilute aluminum alloys: experiments and comparison with small-polaron theory. Phys Rev B. 1982;26:567.
32. Hartmann O, Karlsson E, Wäckelgrd E, Wäppling R, Richter D, Hempelmann R, Niinikoski TO. Diffusion and trapping of muons in aluminum: new experiments and comparison with Kondo theory. Phys Rev B. 1988;37:4425.
33. Kadono R, Imazato J, Matsuzaki T, Nishiyama K, Nagamine K, Yamazaki T, Richter D, Welter JM. Quantum diffusion of positive muons in copper. Phys Rev B. 1989;39:23.
34. Kadono R, Higemoto W, Nagamine K, Pratt FL. An atom in the bloch state. Phys Rev Lett. 1999;83:987.
35. Wilde M, Fukutani K. Hydrogen detection near surfaces and shallow interfaces with resonant nuclear reaction analysis. Surf Sci Rep. 2014;69:196.
36. Feldman LC, Hill M, Mayer JW. Materials analysis by ion Channeling, in materials analysis by ion Channeling. Elsevier; 1982.
37. Komander K, Tran T, Saha J, Moro MV, Pálsson GK, Wolff M, Primetzhofer D. Interstitial hydrogen in Fe/V superstructures: lattice site location and thermal vibration. Phys Rev Lett. 2021;127:136102.
38. Ozawa T, Sugisawa Y, Komatsu Y, Shimizu R, Hitosugi T, Sekiba D, Yamauchi K, Hamada I, Fukutani K. Isotope-dependent site occupation of hydrogen in epitaxial titanium hydride nanofilms. Nat Commun. 2024a;15:9558.

39. Wilde M, Matsumoto M, Gao L, Schwarz-Selinger T, Manhard A, Jacob W. Cross section of 15N-2D nuclear reactions from 3.3 to 7.0 MeV for simultaneous hydrogen and deuterium quantitation in surface layers with 15N ion beams. Nucl Instrum Methods Phys Res B. 2020;478:56.
40. Sidhu SS, Heaton L, Zauberis DD. Neutron diffraction studies of hafnium–hydrogen and titanium–hydrogen systems. Acta Crystallogr. 1956;9:607.
41. Smulders PJM, Boerma DO. Computer simulation of channeling in single crystals. Nucl Instrum Methods Phys Res B. 1987;29:471.
42. Ozawa T, Nakanishi H, Kato K, Shimizu R, Hitosugi T, Fukutani K. Observation of resonant tunneling of proton from octahedral to tetrahedral sites in Pd. J Phys Chem Solids. 2024b;185:111741.
43. Ozawa T, Shimizu R, Ogura S, Hitosugi T, Fukutani K. Fabrication and observation of metastable to stable relaxation of palladium hydride thin films. Vacuum and Surface Science. 2019;62:492.
44. Ozawa T, Shimizu R, Harayama I, Hitosugi T, Sekiba D, Fukutani K. Observation of proton tunneling correlated with phonons and electrons in Pd. Sci Adv. 2025;11:eady8495.
45. Völkl J, Wollenweber G, Klatt K-H, Alefeld G. Reversed isotope dependence for hydrogen diffusion in palladium. Zeitschrift Für Naturforschung A. 1971;26:922.

Chapter 14
Porphyrin Organization—Assembly Processes and Photophysics

Joe Otsuki

Abstract Porphyrins, which are cyclic tetrapyrrolic compounds often complexed with metal ions, possess redox and spectroscopic properties that make them well-suited for use as photosensitizers and catalysis. As a result, they play crucial roles in various physiologically important processes such as light-harvesting, electron transfer, and oxygen transport. Synthetic porphyrin analogues are also studied extensively. When porphyrin derivatives are organized into assemblies, not only their properties are altered, but new functions arise through intermolecular intermolecular interactions. This chapter reviews the basic spectroscopic characteristics of porphyrins, chlorins (which share the chlorophyll framework), and their assemblies, followed by case studies illustrating some porphyrin assemblies prepared in our laboratory and their photophysical properties. These assemblies have been investigated for their potential applications in artificial photosynthesis and catalysis.

Keywords Chlorin · Coordination · Energy transfer · Excited states · Exciton coupling · Molecular orbitals · Photosynthesis · Porphyrin · Self-assembly

14.1 Porphyrin Assemblies—Natural Versus Artificial Assemblies

Chlorophylls and porphyrins, fittingly called the "Pigments of Life," play essential roles in biology. Chlorophylls serve as the primary sensitizers for photosynthesis, while porphyrins play broader roles in various biological processes. For instance, the iron-porphyrin complex in hemoglobin transports oxygen from the lungs to oxygen-demanding cells, the heme c complex mediates electron transfer within the respiratory chain. Organized multichromophore structures are critical for fine-tuning of excited-state energy and electron transfer processes, which are beautifully exemplified by

J. Otsuki (✉)
College of Science and Technology, Nihon University, Chiyoda, Tokyo, Japan
e-mail: otsuki.joe@nihon-u.ac.jp

T. Onishi (ed.), *Advances in Quantum Science II*,
https://doi.org/10.1007/978-981-95-9721-5_14

the intricate bacteriochlorophyll arrays in the photosynthetic antenna complexes of purple bacteria [1].

Chlorophylls and porphyrins, particularly the latter, have been used in artificial systems as sensitizers and as catalysts [2, 3]. These molecules can be modified and synthesized in ways that enable them to self-assemble into higher-order structures. By examining natural as well as artificial porphyrin assemblies, researchers gain insights into how organizing porphyrin analogues can yield highly sophisticated photo- and electroactive molecular devices [4].

Two principal strategies exist for constructing porphyrin assemblies. In the covalent approach, two or more porphyrins bonded via covalent linkages, leading to the creation of some remarkable "superporphryins" [5]. This method produces well-ordered, robust multiporphyrin arrays, but the required multistep syntheses can sometimes be challenging and limit the amount of material obtained. In contrast, the supramolecular approach uses noncovalent bonds, such as reversible coordination or hydrogen bonds, to build multiporphyrin arrays. In this method, porphyrin units are specifically designed to incorporate these intermolecular interactions, allowing the assemblies to adopt the most thermodynamically stable structure. However, their reversibility can also make the resulting aggregates fragile and prone to dissociation. At the same time, this dynamic nature provides new opportunities for adaptability enabling the assembled structure to respond or switch arrangements in response to environmental changes [6].

14.2 Electronic Structure of Porphyrins

14.2.1 Excited States of Porphyrins

Figure 14.1a shows the chemical structure of the basic unit of porphyrin, known as porphine. Although porphine contains a total of 26-π electrons (excluding any metal electrons), it is generally treated as an 18-electron system by counting along the cyclic path along alternating single and double bonds. A simple Hückel molecular orbital (MO) calculation provides the frontier orbitals shown in Fig. 14.1b. In point group D_{4h} notation, the a_{1u} and a_{2u} orbitals are occupied, whereas the pair of e_g orbitals are the LUMOs. The orthogonal pair of degenerate LUMOs are commonly denoted e_{gx} and e_{gy}, reflecting their nodal planes at $x = 0$ and $y = 0$, respectively. Gouterman's four orbital model uses these four orbitals to explain the absorption spectra of porphyrins [7, 8].

x-polarized light excites an electron from the a_{1u} orbital to e_{gy} and from a_{2u} to e_{gx}, while y-polarized light excites an electron from the a_{1u} orbital to e_{gx} and from a_{2u} to e_{gy} (Fig. 14.1c), due to the symmetry of the molecular orbitals. Focusing on the x-polarized case, we write the single-configuration excited states as $|a_{1u}, e_{gy}\rangle$ and $|a_{2u}, e_{gx}\rangle$, indicating that an electron in the a_{1u} or a_{2u} orbital is promoted to the e_{gy} or e_{gx} orbital. Both excited states transform as E_u in the D_{4h} point group. Because

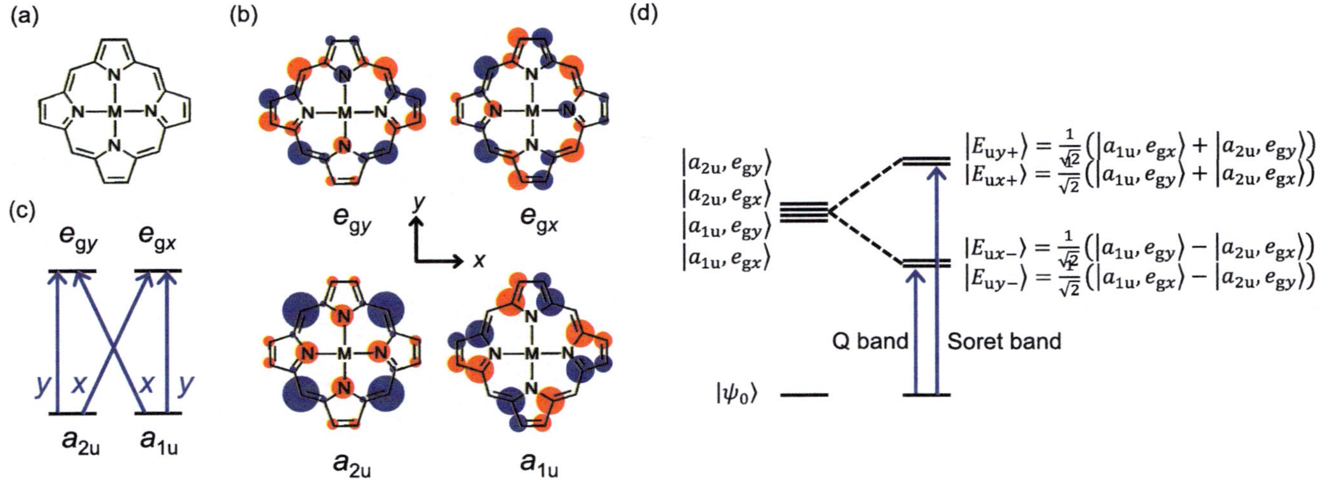

Fig. 14.1 Electronic structure of porphyrin. **a** Chemical structure. **b** Hückel molecular or bitals (MOs). Coulomb integrals C: –11.2 (inonization energy in electron volts), N: –11.9 [9]; resonance integrals C–C: –0.7 [10], C–N: –0.7 [9]. The circle radii are proportional to the molecular-orbital coefficients, and red and blue represent positive and negative values, respectively. **c** Single-configuration transitions. **d** Configuration interaction

the direct product $E_u \times E_u$ contains the totally symmetric irreducible representation A_{1g}, the integrals $\langle a_{1u}, e_{gy} | \widehat{H} | a_{2u}, e_{gx} \rangle$ and $\langle a_{2u}, e_{gx} | \widehat{H} | a_{1u}, e_{gy} \rangle$ are nonzero. Consequently, neither $|a_{2u}, e_{gx}\rangle$ nor $|a_{1u}, e_{gy}\rangle$ is an eigenstate of $\widehat{H}$. Such mixing of electronic configurations is called configuration interaction. In this situation, we search for true eigenstates as linear combination of the two configurations (Eq. (14.1)).

$$|E_{ux}\rangle = c_1 |a_{1u}, e_{gy}\rangle + c_2 |a_{2u}, e_{gx}\rangle \tag{14.1}$$

The two coefficients c_1 and c_2 have two independent solutions, which we denote (c_1^+, c_2^+) and (c_1^-, c_2^-). To obtain them, we construct the 2×2 Hamiltonian matrix (Eq. (14.2)):

$$H = \begin{pmatrix} \langle a_{1u}, e_{gy} | \widehat{H} | a_{1u}, e_{gy} \rangle & \langle a_{1u}, e_{gy} | \widehat{H} | a_{2u}, e_{gx} \rangle \\ \langle a_{2u}, e_{gx} | \widehat{H} | a_{1u}, e_{gy} \rangle & \langle a_{2u}, e_{gx} | \widehat{H} | a_{2u}, e_{gx} \rangle \end{pmatrix} \tag{14.2}$$

and diagonalize it. That is, we find a unitary matrix C such that Eq. (14.3) holds:

$$C^\dagger H C = E, \tag{14.3}$$

where

$$C = \begin{pmatrix} c_1^+ & c_1^- \\ c_2^+ & c_2^- \end{pmatrix} \textit{and } E = \begin{pmatrix} E_{ux+} & 0 \\ 0 & E_{ux-} \end{pmatrix}.$$

Here, $C^\dagger$ is the Hermitian (complex) conjugate of C. The transformation with $C^\dagger$ and C renders H diagonal, and the eigenvalues E_{ux+} and E_{ux-} are obtained. For a 2×2 matrix, the diagonalization can be done manually following a prescription found in most quantum-chemistry textbooks, while for larger matrices one generally employs a computer.

Because a_{1u} and a_{2u} have nearly the same energy for porphyrins, they can be treated as degenerate. In this limit, we set $\langle a_{1u}, e_{gy} | \widehat{H} | a_{1u}, e_{gy} \rangle = \langle a_{2u}, e_{gx} | \widehat{H} | a_{2u}, e_{gx} \rangle (= E_1)$, and the coefficients become $c_1^+ = c_2^+ = c_1^- = -c_2^-$. Choosing $c_1^+ = 1/\sqrt{2}$ for normalization, the resulting wavefunctions, now eigenstates of $\widehat{H}$, are given in Eq. (14.4), and a schematic representation is shown in Fig. 14.1d.

$$|E_{ux\pm}\rangle = \frac{1}{\sqrt{2}}\left(|a_{1u}, e_{gy}\rangle \pm |a_{2u}, e_{gx}\rangle\right) E_{ux\pm} = E_1 \pm \left|\langle a_{1u}, e_{gy} | \widehat{H} | a_{2u}, e_{gx} \rangle\right| \tag{14.4}$$

The intensity of a transition from a ground state $|\psi_0\rangle$ to an excited state $|\psi_1\rangle$ is proportional to the square of the transition dipole moment $\boldsymbol{\mu} = \langle \psi_1 | (-e\boldsymbol{r}) | \psi_0 \rangle$, where $-e$ is the elementary charge and $\boldsymbol{r}$ is its location. In the case of $|E_{ux\pm}\rangle$ being the excited state with x-polarized light, the transition dipole moment is given by

Table 14.1 Processes from the S_1 of porphyrins

	Solvent	τ/ns	k_f^{-1}/ns	k_{isc}^{-1}/ns	k_{ic}^{-1}/ns	ϕ_f	ϕ_{isc}	ϕ_{ic}
H_2TPP [a]	Toluene	12.8	140	16	120	0.090	0.80	0.11
	DMF	12.4	140	15	110	0.090	0.80	0.11
ZnTPP [b]	Toluene	2.10	70	2.4	23	0.030	0.88	0.090
	MeCN	2.00	45	2.0	2.3	0.040	0.88	0.080

[a] [12]. [b] [13]

Eq. (14.5):

$$\mu_{x\pm} = \langle E_{ux\pm}|(-ex)|\psi_0\rangle = \frac{-e}{\sqrt{2}}\left(\langle a_{1u}, e_{gy}\middle|x\middle|\psi_0\rangle \pm \langle a_{2u}, e_{gx}\middle|x\middle|\psi_0\rangle\right), \quad (14.5)$$

where e is the elementary charge. The plus combination produces a higher-energy state with a large transition moment, giving rise to the strong near-UV absorption known as the Soret band (molar absorption coefficients of 5–6 × 10^5 M^{-1} cm^{-1}). The term "band" here simply refers to a broad absorption, as opposed to an absorption line, and should not to be confused with the electronic bands in solid-state physics. The minus combination yields a lower-energy excited state with a nearly zero transition moment, observed as the weaker Q-band at 500–600 nm (molar absorption coefficients of ~ 2 × 10^4 M^{-1} cm^{-1}). Similar arguments hold for y-polarized light, leading to Eq. (14.6).

$$\mu_{y\pm} = \langle E_{uy\pm}|(-ey)|\psi_0\rangle = \frac{-e}{\sqrt{2}}\left(\langle a_{2u}, e_{gy}\middle|y\middle|\psi_0\rangle \pm \langle a_{1u}, e_{gx}\middle|y\middle|\psi_0\rangle\right) \quad (14.6)$$

From the S_1 state, a molecule can undergo three primary processes: fluorescence emission, internal conversion, and intersystem crossing to triplet states. Table 14.1 presents experimental rate constants for these pathways, showing that intersystem crossing dominates over the other processes. It may seem surprising that such efficient intersystem crossing occurs in molecules lacking heavy atoms, which facilitate spin–orbit coupling. However, recent theoretical calculations have produced rates for these processes that match well with experimental values [11].

14.2.2 Excited States of Chlorin

Chlorin is a reduced porphyrin in which one of the pyrrole double bonds is hydrogenated (Fig. 14.2a) forming the central framework of common chlorophylls. A Hückel calculation reveals the MOs shown in Fig. 14.2b, with the single bond placed on the x-axis accordance to standard conventions [12]. The two lower MOs are occupied, while the two upper MOs are unoccupied. Comparing the lobe distributions indicates a clear correspondence between the MOs of porphyrins and chlorins. However, the MOs with greater electron density on the reduced bond are shifted to higher energy, causing the a_{1u} and e_{gy} orbitals to rise in energy, as illustrated in Fig. 14.2c. Consequently, the energies of $|a_{1u}, e_{gy}\rangle$ and $|a_{2u}, e_{gx}\rangle$ remain largely unchanged, whereas $|a_{2u}, e_{gy}\rangle$ is shifted upward and $|a_{1u}, e_{gx}\rangle$ is lowered. As a result, the single-configuration states $|a_{1u}, e_{gy}\rangle$ and $|a_{2u}, e_{gx}\rangle$ exhibit strong interaction, splitting into higher- and lower-energy states (Fig. 14.2d), while the interaction between $|a_{2u}, e_{gy}\rangle$ and $|a_{1u}, e_{gx}\rangle$ is weaker. Based on these interactions, the Soret band partially splits into blue-shifted component, while the Q-band partially splits into a stronger red-shifted component compared with typical porphyrin spectra [14].

14.2.3 Exciton Coupling

Interaction of two or more porphyrin molecules in their ground state is manifested in the absorption spectra. Because the excited states of individual molecule are degenerate, it is necessary to treat the degeneracy explicitly when describing their interactions. Consider a dimer of identical porphyrin molecules, labeled A and B. We expect that the excited-state wavefunctions that incorporate the interaction are linear combinations of the excites states of individual molecules, as shown in Eq. (14.7).

$$|\Phi\rangle = c_1|A_{Eux}B_0\rangle + c_2|A_{Euy}B_0\rangle + c_3|A_0B_{Eux}\rangle + c_4|A_0B_{Euy}\rangle \tag{14.7}$$

The Hamiltonian matrix for their excited states in these basis functions can be written as in Eq. (14.8):

$$H^{(1)} = \begin{pmatrix} \langle A_{Eux}B_0|\widehat{H}^{(1)}|A_{Eux}B_0\rangle & \langle A_{Eux}B_0|\widehat{H}^{(1)}|A_{Euy}B_0\rangle & \langle A_{Eux}B_0|\widehat{H}^{(1)}|A_0B_{Eux}\rangle & \langle A_{Eux}B_0|\widehat{H}^{(1)}|A_0B_{Euy}\rangle \\ \langle A_{Euy}B_0|\widehat{H}^{(1)}|A_{Eux}B_0\rangle & \langle A_{Euy}B_0|\widehat{H}^{(1)}|A_{Euy}B_0\rangle & \langle A_{Euy}B_0|\widehat{H}^{(1)}|A_0B_{Eux}\rangle & \langle A_{Euy}B_0|\widehat{H}^{(1)}|A_0B_{Euy}\rangle \\ \langle A_0B_{Eux}|\widehat{H}^{(1)}|A_{Eux}B_0\rangle & \langle A_0B_{Eux}|\widehat{H}^{(1)}|A_{Euy}B_0\rangle & \langle A_0B_{Eux}|\widehat{H}^{(1)}|A_0B_{Eux}\rangle & \langle A_0B_{Eux}|\widehat{H}^{(1)}|A_0B_{Euy}\rangle \\ \langle A_0B_{Euy}|\widehat{H}^{(1)}|A_{Eux}B_0\rangle & \langle A_0B_{Euy}|\widehat{H}^{(1)}|A_{Euy}B_0\rangle & \langle A_0B_{Euy}|\widehat{H}^{(1)}|A_0B_{Eux}\rangle & \langle A_0B_{Euy}|\widehat{H}^{(1)}|A_0B_{Euy}\rangle \end{pmatrix}, \tag{14.8}$$

where $\hat{H}^{(1)}$ is the perturbation due to intermolecular electronic interaction. Because the excited states are degenerate, they can be represented by any linear combination of E_{ux} and E_{uy} components. As such, no definite assignment of the polarization directions of local transition moments can be made. However, an appropriate unitary transformation to align the axes enables to the calculations to proceeds [15]. The same procedure that carried us from Eqs. (14.1–14.4) in the configuration-interaction treatment can be applied here as well. Before diagonalizing the Hamiltonian matrix, however, we introduce a few approximations.

When the distance between the porphyrin centers is significantly larger than their π systems, one can approximate $\hat{H}^{(1)}$ with a dipole–dipole interaction. Under this assumption, the integrals are limited to each molecule individually, giving Eq. (14.9):

$$\hat{H}^{(1)} = \frac{1}{4\pi\varepsilon_0}\left\{\frac{\hat{\mu}_A\cdot\hat{\mu}_B}{r_{AB}^3} - \frac{3(\hat{\mu}_A\cdot\boldsymbol{r}_{AB})(\boldsymbol{r}_{AB}\cdot\hat{\mu}_B)}{r_{AB}^5}\right\}, \tag{14.9}$$

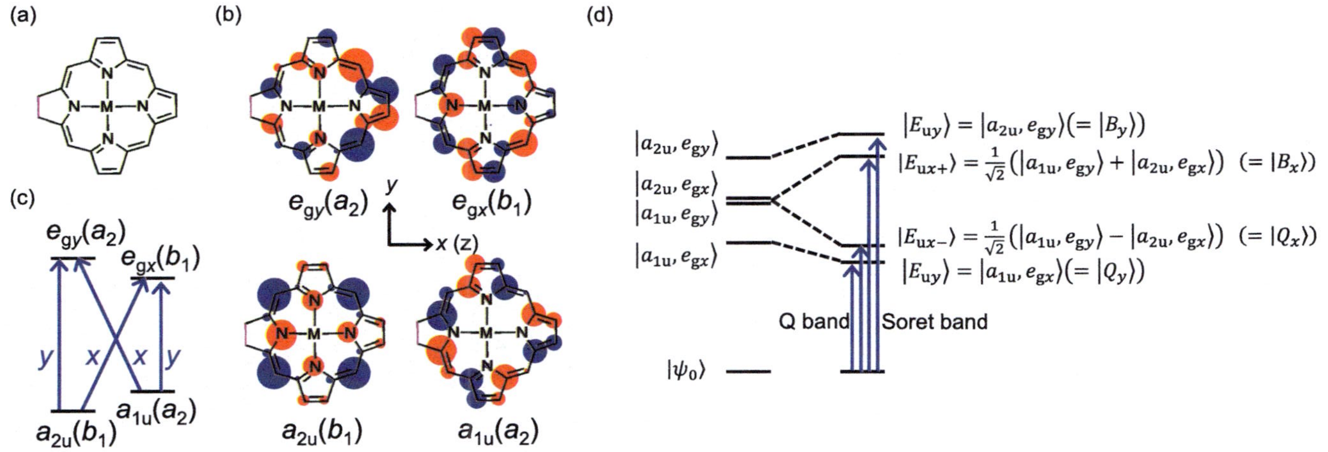

Fig. 14.2 Electronic structure of chlorin. The irreducible representation symbols are borrowed from the D_{4h} point group. The symbols of the C_{2v} point group are shown in parentheses. **a** Chemical structure. **b** Hückel MOs. Coulomb integrals C: –11.2 (inonization energy in electron volts), N: –11.9 [9]; resonance integrals C–C: –0.7 [10], C–N: –0.7 [9]. The circle radii are proportional to the molecular-orbital coefficients, and red and blue represent positive and negative values, respectively. **c** Single-configuration transitions. **d** Configuration interaction

where ε_0 is the vacuum permittivity, $\hat{\mu}_A$ and $\hat{\mu}_B$ are dipole operators for molecules A and B, respectively, $\boldsymbol{r}_{\mathrm{AB}}$ is the center-to-center vector joining A and B, and r_{AB} is its magnitude. For instance, the 1,1-element of the Hamiltonian matrix in Eq. (14.8) is zero because the ground state of porphyrin has no permanent dipole moment (Eq. (14.10)).

$$H^{(1)}{}_{11} = \frac{1}{4\pi\varepsilon_0}\left\{\frac{A_{Eux}|\hat{\mu}_A|A_{Eux}\cdot B_0|\hat{\mu}_B|B_0}{r_{\mathrm{AB}}{}^3} - \frac{3(A_{Eux}|\hat{\mu}_A|A_{Eux}\cdot r_{\mathrm{AB}})(r_{\mathrm{AB}}\cdot B_0|\hat{\mu}_B|B_0)}{r_{\mathrm{AB}}{}^5}\right\}$$
$$= 0 \tag{14.10}$$

The 1,3-element in Eq. (14.8) becomes Eq. (14.11):

$$H_{13}^{(1)}$$
$$= \frac{1}{4\pi\varepsilon_0}\left\{\frac{A_{Eux}|\hat{\mu}_A|A_0\cdot B_0|\hat{\mu}_B|B_{Eux}}{r_{\mathrm{AB}}^3} - \frac{3(A_{Eux}|\hat{\mu}_A|A_0\cdot \boldsymbol{r}_{\mathrm{AB}})(\boldsymbol{r}_{\mathrm{AB}}\cdot B_0|\hat{\mu}_B|B_{Eux})}{r_{\mathrm{AB}}^5}\right\}$$
$$= \frac{1}{4\pi\varepsilon_0}\left\{\frac{\boldsymbol{\mu}_{Ax}\cdot\boldsymbol{\mu}_{Bx}}{r_{\mathrm{AB}}^3} - \frac{3(\boldsymbol{\mu}_{Ax}\cdot\boldsymbol{r}_{\mathrm{AB}})(\boldsymbol{r}_{\mathrm{AB}}\cdot\boldsymbol{\mu}_{Bx})}{r_{\mathrm{AB}}^5}\right\} \equiv (\boldsymbol{\mu}_{Ax},\boldsymbol{\mu}_{Bx}), \tag{14.11}$$

which is denoted $(\boldsymbol{\mu}_{Ax}, \boldsymbol{\mu}_{Bx})$. Here, $\boldsymbol{\mu}_{Ax}$ and $\boldsymbol{\mu}_{Bx}$ are the transition dipole moments of x-polarization *in respective molecular frameworks*. The matrix will eventually be simplified to Eq. (14.12).

$$H^{(1)} = \begin{pmatrix} 0 & 0 & (\boldsymbol{\mu}_{Ax},\boldsymbol{\mu}_{Bx}) & (\boldsymbol{\mu}_{Ax},\boldsymbol{\mu}_{By}) \\ 0 & 0 & (\boldsymbol{\mu}_{Ay},\boldsymbol{\mu}_{Bx}) & (\boldsymbol{\mu}_{Ay},\boldsymbol{\mu}_{By}) \\ (\boldsymbol{\mu}_{Ax},\boldsymbol{\mu}_{Bx}) & (\boldsymbol{\mu}_{Ay},\boldsymbol{\mu}_{Bx}) & 0 & 0 \\ (\boldsymbol{\mu}_{Ax},\boldsymbol{\mu}_{By}) & (\boldsymbol{\mu}_{Ay},\boldsymbol{\mu}_{By}) & 0 & 0 \end{pmatrix} \tag{14.12}$$

In general, diagonalization yields four energy levels. For certain symmetrical geometries, such as those shown in Fig. 14.3, where the x-polarized and y-polarized transition dipole moments do not mix, the Hamiltonian further reduced to Eq. (14.13).

$$H^{(1)} = \begin{pmatrix} 0 & 0 & (\boldsymbol{\mu}_{Ax},\boldsymbol{\mu}_{Bx}) & 0 \\ 0 & 0 & 0 & (\boldsymbol{\mu}_{Ay},\boldsymbol{\mu}_{By}) \\ (\boldsymbol{\mu}_{Ax},\boldsymbol{\mu}_{Bx}) & 0 & 0 & 0 \\ 0 & (\boldsymbol{\mu}_{Ay},\boldsymbol{\mu}_{By}) & 0 & 0 \end{pmatrix} \tag{14.13}$$

The wavefunctions and energies are obtained as Eqs. (14.14) and (14.15) by diagonalizing matrix (14.13).

$$|\Phi_{x\pm} = \frac{1}{\sqrt{2}}(|A_0 B_{Eux} \pm |A_{Eux} B_0) \pm (\boldsymbol{\mu}_{Ax},\boldsymbol{\mu}_{Bx}) \tag{14.14}$$

$$|\Phi_{y\pm} = \frac{1}{\sqrt{2}}(|A_0 B_{Euy} \pm |A_{Euy} B_0) \pm (\boldsymbol{\mu}_{Ay},\boldsymbol{\mu}_{By}) \tag{14.15}$$

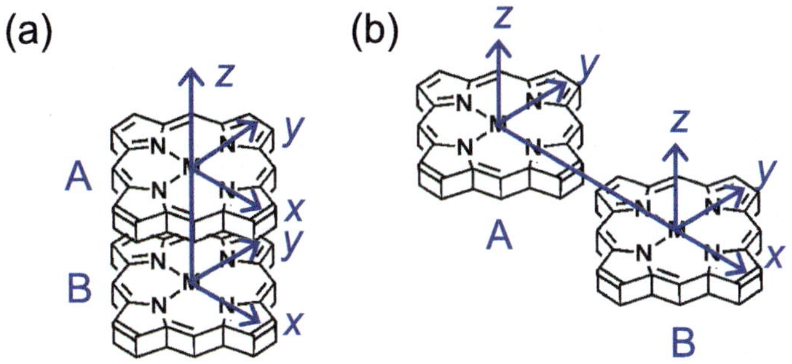

Fig. 14.3 Dimers of porphyrins. **a** Face-to-face dimer (H-dimer). **b** Side-by-Side dimer (J-dimer)

The transition dipole moments from the ground state to these excited states (Eqs. (14.16) and (14.17)):

$$\frac{1}{\sqrt{2}}(A_0B_{Eux}| \pm A_{Eux}B_0|)(\hat{\mu}_{Ax} + \hat{\mu}_{Bx})|A_0B_0 = \frac{1}{\sqrt{2}}(\boldsymbol{\mu}_{Bx} \pm \boldsymbol{\mu}_{Ax}) = \begin{cases} \sqrt{2}\mu(+) \\ 0(-) \end{cases}, \tag{14.16}$$

$$\frac{1}{\sqrt{2}}\left(A_0B_{Euy}\right| \pm A_{Euy}B_0\left|\right)\left(\hat{\mu}_{Ay} + \hat{\mu}_{By}\right)|A_0B_0 = \frac{1}{\sqrt{2}}\left(\boldsymbol{\mu}_{By} \pm \boldsymbol{\mu}_{Ay}\right) = \begin{cases} \sqrt{2}\mu(+) \\ 0(-) \end{cases}, \tag{14.17}$$

where $|\boldsymbol{\mu}_{Ax}| = |\boldsymbol{\mu}_{Bx}| = \left|\boldsymbol{\mu}_{Ay}\right| = \left|\boldsymbol{\mu}_{By}\right| = \mu$ show that the plus combination is optically allowed while the minus combination is optically forbidden. This leads to one strong and one weak (or forbidden) transition for each pair of degenerate excitations.

In dimers such as the one illustrated in Fig. 14.3a, the two porphyrin rings adopt a face-to-face arrangement (H-dimer). In this arrangement, $\boldsymbol{\mu}_{A_x}$ is aligned side-by-side with $\boldsymbol{\mu}_{Bx}$, and $\boldsymbol{\mu}_{Ay}$ with $\boldsymbol{\mu}_{By}$. The corresponding perturbation energies are therefore equal and positive, $(\boldsymbol{\mu}_{Ax}, \boldsymbol{\mu}_{Bx}) = \left(\boldsymbol{\mu}_{Ay}, \boldsymbol{\mu}_{By}\right) > 0$. Because the plus combination of states is raised in energy by this interaction, the absorption shifts to higher energy (i.e., a blue shift) upon H-dimer formation. An example of this Soret-band blue shift in porphyrins will be illustrated in a later section.

Another limiting geometry is the side-by-side arrangement of porphyrin rings, which is shown in Fig. 14.3b (J-dimer). In this arrangement, the $\boldsymbol{\mu}_{Ax}$ and $\boldsymbol{\mu}_{Bx}$ align end-to-end in line, while $\boldsymbol{\mu}_{Ay}$ and $\boldsymbol{\mu}_{By}$ align side-by-side. For x-polarization, the perturbation energy is negative, $(\boldsymbol{\mu}_{Ax}, \boldsymbol{\mu}_{Bx}) < 0$, lowering the energy of the allowed transition and producing a red shift in the absorption spectrum. Conversely, for y-polarization, the interaction is positive, raising the allowed transition energy and causing a blue shift. Consequently, the overall absorption band splits compared to that of the monomer [16].

14.2.4 Energy Transfer

Excited-state energy transfer is a critical process involving porphyrins and chlorophylls. In this process, a pair of an excited-state donor and a ground-state acceptor, D*A (D and A can be the same molecular species), evolves into a pair of a ground-state donor and an excited-state acceptor, DA*. According to Fermi's golden rule, the rate of energy transfer is proportional to the square of the matrix element coupling these initial and final states, where the perturbation Hamiltonian is the electronic interaction between D and A. Focusing on the case of singlet energy transfer, the initial and final states in terms of Slater determinant can be represented by Eqs. (14.18) and (14.19).

$$|\mathrm{D}^*\mathrm{A} = \frac{1}{\sqrt{2}}\left\{\left|\cdots a_\mathrm{D}\bar{b}_\mathrm{D}\cdots a_\mathrm{A}a_\mathrm{A}\right\rangle + |\cdots b_\mathrm{D}\bar{a}_\mathrm{D}\cdots a_\mathrm{A}\bar{a}_\mathrm{A}\rangle\right\} \tag{14.18}$$

$$|\mathrm{DA}^* = \frac{1}{\sqrt{2}}\left\{\left|\cdots a_\mathrm{D}\bar{a}_\mathrm{D}\cdots a_\mathrm{A}\bar{b}_\mathrm{A}\right\rangle + |\cdots a_\mathrm{D}\bar{a}_\mathrm{D}\cdots b_\mathrm{A}\bar{a}_\mathrm{A}\rangle\right\} \tag{14.19}$$

Here, a and b denote occupied and unoccupied orbitals, respectively, while unbarred and barred labels refer to alpha and beta spins. Within each bracket, the donor orbitals appear first, followed by the acceptor orbitals. Figure 14.4 offers a schematic representation of energy transfer from state $|\mathrm{D}^*\mathrm{A}\rangle$ to $|\mathrm{DA}^*\rangle$.

Calculating the matrix element between the initial and final states and integrating out the spin variables, one obtains Eq. (14.20):

$$\begin{aligned} H_\mathrm{if}^{(1)} &= 2\int d\boldsymbol{r}_1 d\boldsymbol{r}_2 \psi_{\mathrm{D}b}^*(\boldsymbol{r}_1)\psi_{\mathrm{D}a}(\boldsymbol{r}_1)\frac{e^2}{4\pi\varepsilon_0 r_{12}}\psi_{\mathrm{A}a}^*(\boldsymbol{r}_2)\psi_{\mathrm{A}b}(\boldsymbol{r}_2) \\ &\quad - \int d\boldsymbol{r}_1 d\boldsymbol{r}_2 \psi_{\mathrm{D}b}^*(\boldsymbol{r}_1)\psi_{\mathrm{A}b}(\boldsymbol{r}_1)\frac{e^2}{4\pi\varepsilon_0 r_{12}}\psi_{\mathrm{A}a}^*(\boldsymbol{r}_2)\psi_{\mathrm{D}a}(\boldsymbol{r}_2), \end{aligned} \tag{14.20}$$

where each symbol follows conventional notation, expressed in terms of the spatial orbitals ψ [17]. The first integral represents electron 1 remaining in D and electron 2 in A, as shown schematically by the red arrows in Fig. 14.5a. This Coulomb term (often called through-space energy transfer) does not require orbital overlap between donor and acceptor. Approximating the Coulomb interaction as a dipole–dipole interaction, combined with the fluorescence of D and absorption of A to define the density of states needed for Fermi's golden rule, yields the well-known Förster resonance energy transfer (FRET) rate formula in Eq. (14.21):

$$k_\mathrm{FRET} = \frac{9(\ln 10)\kappa^2\phi_\mathrm{f}}{128\pi^5 n^4 r_\mathrm{DA}^6 N_\mathrm{A}\tau_\mathrm{D}}\frac{\int d\lambda\lambda^4 F_\mathrm{D}(\lambda)\varepsilon_\mathrm{A}(\lambda)}{\int d\lambda F_\mathrm{D}(\lambda)}, \tag{14.21}$$

where ϕ_f is the fluorescence quantum yield of D (in the absence of A), n is the refractive index of the medium, r_DA is the center-to-center distance between D and A, N_A is Avogadro's constant, τ_D is the excited-state lifetime of D (in the absence of A), $F_\mathrm{D}(\lambda)$ is the fluorescence spectrum of D, and $\varepsilon_\mathrm{A}(\lambda)$ is the molar absorption coefficient as A as a function of wavelength λ. The orientation factor κ represents the dependence of dipole–dipole interaction on the mutual orientation, which is defined

Fig. 14.4 Schematic representation of energy transfer from D* to A in spin-adapted configurations. The initial and final states correspond to Slater determinants (Eqs. (14.18) and (14.19), respectively)

by Eq. (14.22).

$$\kappa = \overline{\mu}_{\mathrm{D}} \cdot \overline{\mu}_{\mathrm{A}} - 3(\overline{\mathbf{r}}_{\mathrm{DA}} \cdot \overline{\mu}_{\mathrm{D}})(\overline{\mathbf{r}}_{\mathrm{DA}} \cdot \overline{\mu}_{\mathrm{A}}) \quad (14.22)$$

where $\overline{\mu}_{\mathrm{D}}$, $\overline{\mu}_{\mathrm{A}}$, and $\overline{\mathbf{r}}_{\mathrm{DA}}$ are the unit vectors pointing along the transition dipole moments of D and A, and the line connecting their centers, respectively, as depicted in Fig. 14.6. In porphyrins, the transition dipole moment can be two-dimensionally degenerate; the orientation factors for such cases have been discussed by Mårtensson [18].

Figure 14.5a (blue arrows) also shows the schematic for the second integral of Eq. (14.20). In this situation, both electron 1 and electron 2 occupy orbitals in both D and A. This can be viewed as an exchange of electrons between D and A, and is therefore referred to as the exchange mechanism or the Dexter mechanism. Because substantial overlap between the orbitals of D and A is required for this integral to

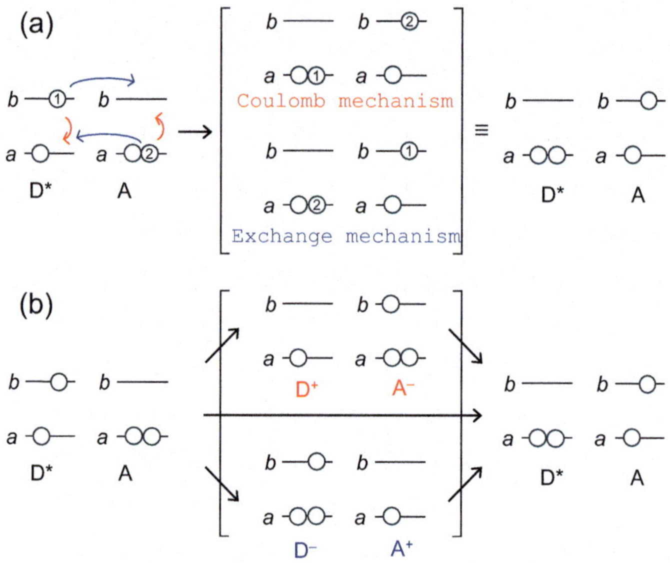

Fig. 14.5 Energy transfer mechanisms. **a** Two-configuration model. Coulomb and exchange (Dexter) mechanisms. **b** Four-configuration model. Direct- and through-configuration energy transfer mechanisms

Fig. 14.6 Relative orientations of transition dipole moments

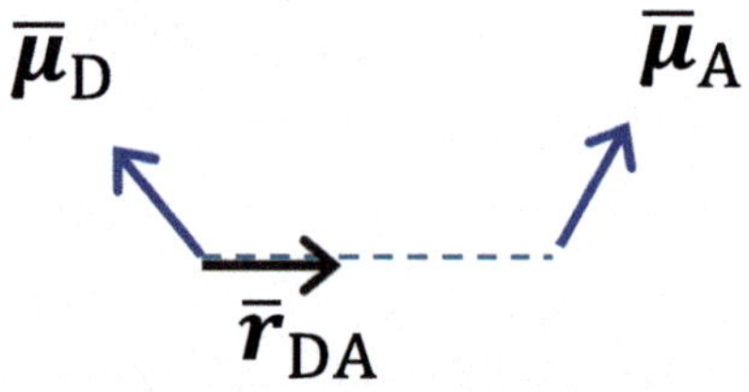

become significant, physical contact of the two molecules is typically necessary for the exchange mechanism to contribute to energy transfer.

The analysis so far relies on two basis states, D*A and DA*. However, charge-separated states D^+A^- and D^-A^+ often lie at comparable energies and can mix into these states through-configuration interactions (Fig. 14.5b). Harcourt and co-workers demonstrated that incorporating these additional states can dominate the transfer mechanism at short donor–acceptor separations [19]. The involvement of virtual states D^+A^- and D^-A^+ implies that this mechanism may also be viewed as an exchange-type mechanism, albeit one distinct from the direct Dexter mechanism. It is important to note that the Coulomb and exchange mechanisms (whether the Dexter-type or the through-configuration type) represent different terms within the same equation rather than mutually exclusive processes.

14.3 Porphyrin Organization—Case Studies of Assemblies and Photophysics

14.3.1 Cyclic Tetramer of Chlorophylls—A Photosynthesis Model

A straightforward supramolecular strategy for porphyrin organization involves exploiting the axial coordination of metal porphyrins. Zn(II) porphyrin complexes are widely studied, because they accept an axial ligand and their d^{10} configuration makes them photophysically active. Our group and others have investigated Zn–porphyrin/chlorophyll derivatives that include a peripheral pyridine unit [20, 21]. In this motif, the pyridine moiety on one porphyrin vertically coordinates to the Zn center of another. Repeated vertical coordination can yield either linearly extended, zigzag polymer-like structures or cyclic *n*-mers, as illustrated in Fig. 14.7a. In solution, distinct *n*-mers (most often tetramers) are generally favored whereas polymer-like assemblies are more frequently observed in crystal structures [21, 22].

This behavior can be rationalized by an equilibrium model illustrated in Fig. 14.7a. Each step of the linear polymerization reaction is represented by $P_n + P \rightleftharpoons P_{n+1}$ and the binding constant for this step is given by $K_L = [P_{n+1}]/([P_n][P])$, for which a common value is assumed for each successive addition. For any *n*-mer, a separate equilibrium exists between the linear species P_n and its cyclic counterpart cP_n, which is governed by $K_C = [cP_n]/[P_n]$. If the cyclic form is particularly stable for a certain *n* ($n = 4$ for our molecular design with a 4-pyridyl moiety), K_C becomes large due to an enthalpic gain. As an example, a simulation was conducted using $K_L = 1.1 \times 10^4 M^{-1}$, $K_{C4} = 1.1 \times 10^3$, and $K_{Cn} = 0$ for $n \neq 4$ values determined for our chlorophyll derivatives, which we discuss in the next paragraph. The resulting distribution is plotted in Fig. 14.7b. Under dilute conditions where the initial porphyrin concentration $[P]_0 < 10^{-5}$M, the porphyrin remains predominantly monomeric. Over the intermediate concentration range 10^{-3} M $< [P]_0 < 10^{-1}$M,

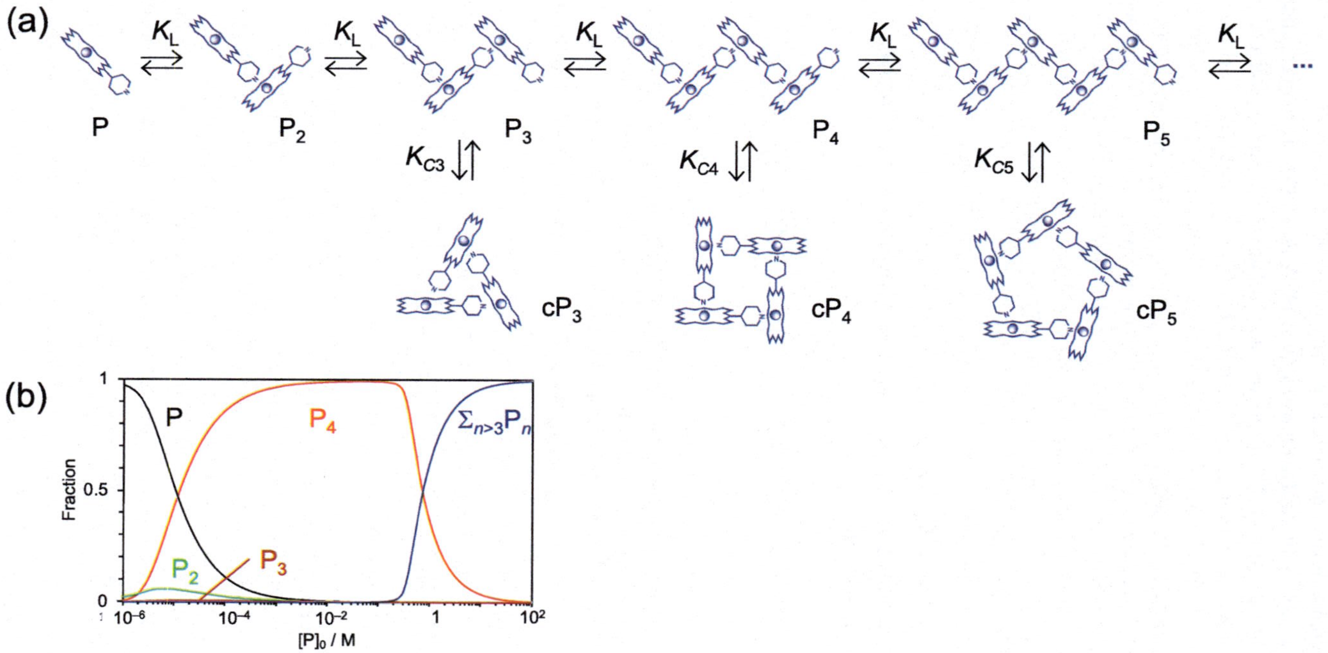

Fig. 14.7 Equilibria among linear and cyclic aggregates. **a** Elementary processes. **b** Fraction of species as a function of initial porphyrin concentration ($[P]_0$). Parameters used were $K_L = 1.1 \times 10^4 M^{-1}$, $K_{C4} = 1.1 \times 10^3$, and $K_{Cn} = 0 (n \neq 4)$

the tetrameric assembly dominates. As the concentration increases further, the fraction of the tetramer diminishes while linear polymeric species eventually prevail once $[P]_0$ exceeds about 1 M. During crystallization or deposition, the local concentration at the solid–liquid interface becomes high enough to favor formation of polymeric structures, yielding crystals that incorporate these extended assemblies.

We synthesized Zn-chlorophyll derivatives bearing a pyridyl moiety, which we denote ZnPy and $ZnPyC_{60}$ as depicted in Fig. 14.8a [23]. In nonpolar solvents, four of these molecules assemble into a cyclic tetramer, one of possible conformations of which is shown in Fig. 14.8b. Because chlorophyll chromophores lie in close proximity within this tetramer, energy transfer is anticipated. Given that their π-systems do not overlap the process is likely mediated by the Förster mechanism. The lowest excited state of chlorophyll derivatives is the Q_y state with its the transition dipole moment oriented along the y-axis (see Fig. 14.2) [24]. Depending on which face of the chlorophyll is coordinated by the pyridyl moiety, orientation of the transition dipole moments can vary within the assembly [25]. Using Eq. (14.21), we estimated the Förster energy transfer rates for various orientations to be roughly $k_{FRET} \approx 1 \times 10^{11} s^{-1}$ for both nearest-neighbor and diagonal energy-transfer pathways. We recorded the transient absorption of a ZnPy solution at varying excitation laser powers; the resulting traces at 850 nm are shown in Fig. 14.8c. A notable feature is the appearance of a short-lived component ($\tau = 1.9$ps) that becomes more pronounced with increasing laser intensity. The behavior is attributed to singlet–singlet annihilation, which indicates that singlet energy transfer does indeed occur. The observed lifetime τ relates to the energy-transfer rate k_{EnT} by $2k_{EnT} = \tau^{-1}$, from which we obtain $k_{EnT} = 2.5 \times 10^{11} s^{-1}$. This value agreed with the estimated value of $k_{Fö\,rster}$, suggesting that the energy transfer occurs is dominated by the Förster mechanism, without the need to invoke an exchange-type pathway. These findings demonstrate that the tetrameric chlorophyll assembly effectively function as a light-harvesting antenna complex.

We also prepared a fullerene appended derivative, $ZnPyC_{60}$, which similarly forms cyclic tetramers. Transient absorption spectroscopy (Fig. 14.8d) reveals the appearance of fullerene radical anion C_{60}^{-} via a distinct absorption peak at 1000 nm. This radical anion is generated by forward electron transfer at a rate $2.6 \times 10^{11} s^{-1}$ and then undergoes back electron transfer at $1.9 \times 10^{9} s^{-1}$. In nanosecond transient absorption, recorded 2 s after excitation, the chlorophyll triplet state was detected and decayed with two distinct rate constants $4.3 \times 10^{3} s^{-1}$ and $4.3 \times 10^{4} s^{-1}$. To explore mixed assemblies, we combined ZnPy and $ZnPyC_{60}$ in a 3:1 ratio. Under the assumption of random formation of tetramer from these components, the most abundant species is $(ZnPy)_3(ZnPyC_{60})1$ comprising 42% of all tetramers. Using the kinetic parameters measured for individual assemblies along with the data from the mixed system, we constructed the sequence of events within the tetramer as illustrated in Fig. 14.8e. Upon excitation, one of the four chlorophyll units becomes excited; energy transfer then occurs in a 3.8 ps timescale. If, either by direct absorption or by energy transfer, the C_{60}-functionalized chlorophyll is excited, it undergoes further energy transfer or charge separation to form C_{60}^{-} within another 3.8 ps. Given that these processes happen well within the singlet excited state lifetime of 4.3 ns, which is three orders

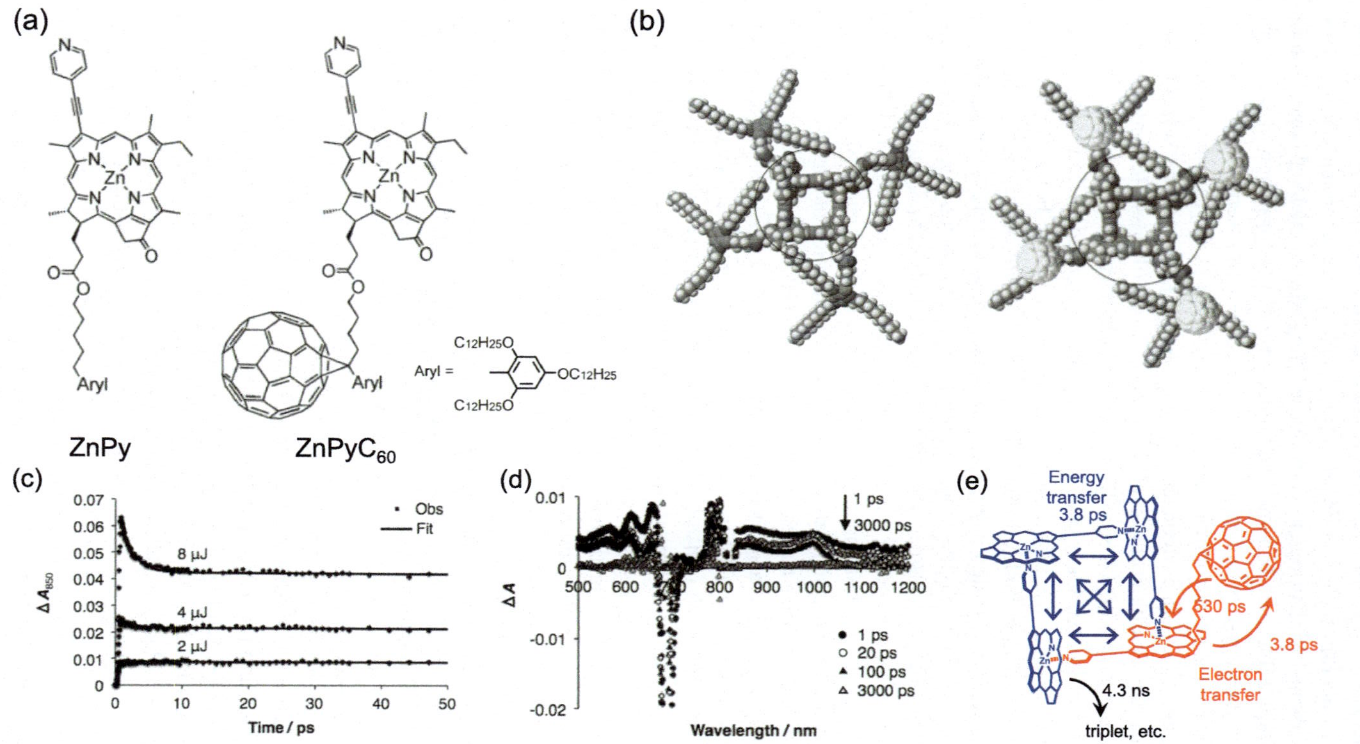

Fig. 14.8 Zn-chlorophyll derivatives bearing a pyridyl moiety. Part of the figure was reproduced with permission from Wiley [23]. **a** Chemical structures. **b** A possible conformation of the tetramers of ZnPy and ZnPyC$_{60}$. **c** Dependence of transient absorption decay for ZnPy on the excitation laser power. **d** Transient absorption of ZnPyC$_{60}$. **e** Compilation of excited-state processes occurring in the mixed assembly of ZnPy and ZnPyC$_{60}$

of magnitude longer, the resulting charge-separated state forms with an effectively 100% quantum yield.

14.3.2 Cyclic Trimer of Porphyrins—A Complementary Photosynthesis Model

We synthesized Zn–porphyrin derivatives featuring a *meta*-pyridyl moiety, termed ZnPPyC, as shown in Fig. 14.9a [26]. In their most stable conformation, the *meso*-phenyl or pyridyl substituents twist almost perpendicular to the porphyrin plane. Consequently, if the *meta*-pyridyl group coordinates upright to the zinc center of another porphyrin, the interporphyrin angle would be 30°, as illustrated in Fig. 14.9b. Repeating this coordination motif could, in principle, yield a cyclic 12-mer. However, both vapor pressure osmometry and ^{1}H NMR data revealed the formation of a trimer (Fig. 14.9c), in which three porphyrin planes aggregate much like three faces of a tetrahedron. Evidently, the entropic cost of forming a larger 12-mer ring favors the smaller trimeric assembly.

We also synthesized a ferrocene-conjugated porphyrin, ZnPFcPyC, which is structurally similar to ZnPPyC, except for the ferrocene moiety. The transient absorption measurements revealed three kinetic components. The fastest component ($<$ 1ps) likely corresponds to charge separation, forming a ferrocene radical cation and porphyrin radical anion. A subsequent process with a 2.3 ps time scale likely reflects conformation relaxation following the charge separation. Finally, the longest component, with a 26 ps lifetime, is attributed to the charge recombination step.

When ZnPPyC and ZnPFcPyC were mixed at a 2:1 ratio, statistical assembly into trimers yielded $(ZnPPyC)_2(ZnPFcPyC)_1$ species in 44% abundance. The energy transfer rates between porphyrins in the trimer were estimated via the Förster mechanism, giving transfer times of 7.0 ps from ZnPPyC to another and 8.1 ps from ZnPPyC to ZnPFcPyC. Both timescales are more than two orders of magnitude faster than the 1.2 nm singlet lifetime of ZnPPyC, ensuring that energy transfer occurs well before the excited state decays. Notably, energy transfer from ZnPFcPyC back to ZnPPyC is expected to be negligible because (1) ZnPFcPyC has extremely low fluorescence yield, and (2) its singlet excited-state energy lies below that of ZnPPyC, making the integrand $F_D(\lambda)\varepsilon_A(\lambda)$ very small, as manifested in their absorption spectra shown in Fig. 14.9d. Consequently, once energy is funneled to the ferrocene-substituted porphyrin, efficient charge separation follows without any back energy transfer. As in the tetrameric chlorophyll assembly described earlier, the energy and electron transfer steps occur orders of magnitude faster than the excited-state decay, implying an effectively 100% quantum yield for generating the charge-separated state. Here, electron transfer proceeds from ferrocene to the porphyrin in the excited state, in contrast to the fullerene-bearing chlorophyll system where the electron transfer proceeds from

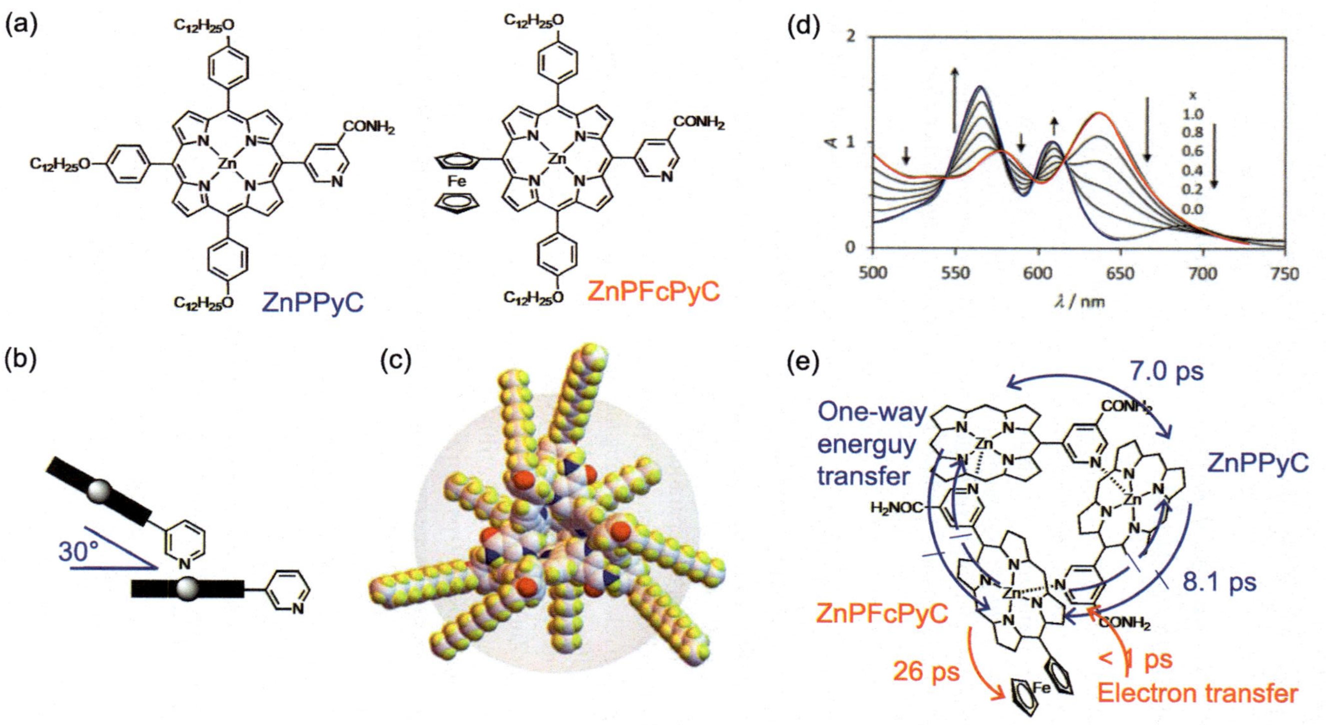

Fig. 14.9 Zn-*meta*-pyridylporphyrins. Part of the figure was reproduced with permission from Wiley [26]. **a** Chemical structures. **b** Ideal coordination geometry. **c** Experimental geometry. **d** The Q band absorption of a mixture of ZnPPyC and ZnPFcPyC in a ratio of $(1-x):x$ (total concentration of 0.1 mM) in chloroform. The blue and red traces are of pure ZnPPyC and ZnPFcPyC, respectively. **e** Compilation of excited-state processes occurring in the mixed assembly of ZnPPyC and ZnPFcPyC

the chlorophyll to the fullerene. In this regard, the ferrocene system is complementary to the fullerene-based counterpart as photosynthetic models. The excited-state processes in the trimer are depicted in Fig. 14.9e.

14.3.3 Face-to-Face Porphyrin Dimer

Face-to-face porphyrin dimers form a particularly intriguing class of assemblies. Early studied by the Collman group in the 1970s demonstrated that face-to-face Co-porphyrin dimers bridged by covalent linkages catalyze electrochemical O_2 reduction to H_2O more efficiently than the corresponding monomeric Co-porphyrins [27, 28]. More recently, face-to-face porphyrin dimes have been constructed using coordination interactions. One approach employs *meso*-tetrapyridylporphyrins and bimetallic "clips," which bridge the two porphyrins via pyridyl-metal coordination [29]. Various clipping fragments have been evaluated for their catalytic performance in Co-porphyrin dimers. Another strategy involves introducing bipyridine [30] or terpyridine [31] groups at the four *meso*-positions as coordination sites. Addition of metal ions then generates tetrahedral or octahedral metal complexes at these sites, respectively, resulting in face-to-face porphyrin dimers.

Our strategy followed the latter approach. The designed chemical structure is shown in Fig. 14.10a [32], in which a 2,2'-bipyridine moiety is tethered at its 6-position to the meta position of the *meso*-phenyl groups. Molecular modeling indicated that suitable rotations about the porphyrin-phenyl and phenyl-bipyridine bonds would orient the bipyridine for tetrahedral coordination to a metal ion when paired with another porphyrin in a face-to-face arrangement. Experimental evidence confirmed this arrangement in both solution and the solid state, with single crystal X-ray diffraction providing direct structural validation.

Figure 14.10b shows the results of a titration experiment in which aliquots of $Zn(OTf)_2$ were added to a solution $ZnPor(bpy)_2$, and absorption spectra were recorded at each step. The most prominent change is a 13-nm blue shift of the Soret band, indicating that the transition dipoles adopt a parallel side-by-side arrangement in line with previous discussions. The magnitude of the shifts (13 nm) suggests an interporphyrin distance of approximately 5 Å [33, 34], an inference that is corroborated by ^1H NMR measurements in solution and single crystal X-ray diffraction analysis in the solid state.

Another notable feature of the titration experiment is the pronounced S-shaped changes in the absorbances, which is shown in Fig. 14.10c, suggesting that cooperative binding is involved. Cooperative binding processes are widespread in nature. which is exemplified by the O_2 binding by hemoglobin undertaking efficient O_2 transport from the lungs to tissues. Artificially implementing such cooperativity is crucial for designing smart supramolecular systems with regulatory functionalities, and has thus gained considerable attention in the chemical community [35, 36].

The principle steps in the binding of porphyrins and Zn^{2+} ions are outlined schematically in Fig. 14.11. Beginning from a monomeric porphyrin (P) at the upper

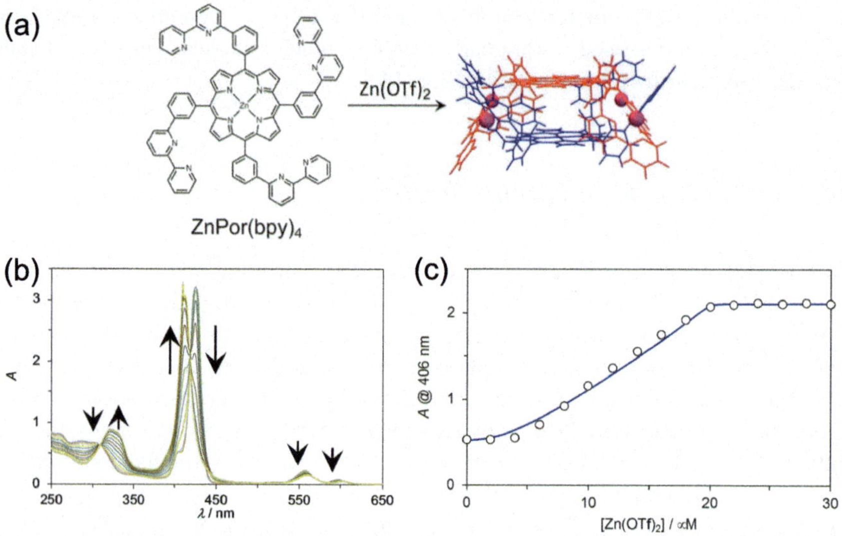

Fig. 14.10 Coordination-directed face-to-face porphyrin dimer. Part of the figure was reproduced with permission from Wiley [32]. **a** Molecular structure of ZnPor(bpy)$_4$ and its dimerization upon the addition of Zn(OTf)$_2$. **b** Absorption spectral changes upon addition of Zn(OTf)$_2$. **c** Absorbance plot at 406 nm as a function of [Zn(OTf)$_2$]

left corner, horizontal arrows represent Zn^{2+} (M) binding, whereas vertical arrows denote porphyrin dimerization. The binding constants K_1 through K_8 are defined as a set of Eqs. (14.23).

$$K_1 = \frac{[\mathrm{PM}]}{[\mathrm{P}][\mathrm{M}]}, K_2 = \frac{[\mathrm{PM_2}]}{[\mathrm{PM}][\mathrm{M}]}, K_3 = \frac{[\mathrm{PM_3}]}{[\mathrm{PM_2}][\mathrm{M}]}, K_4 = \frac{[\mathrm{PM_4}]}{[\mathrm{PM_3}][\mathrm{M}]} \quad (14.23)$$

$$K_5 = \frac{[\mathrm{P_2M}]}{[\mathrm{PM}][\mathrm{P}]}, K_6 = \frac{[\mathrm{P_2M_2}]}{[\mathrm{P_2M}][\mathrm{M}]}, K_7 = \frac{[\mathrm{P_2M_3}]}{[\mathrm{P_2M_2}][\mathrm{M}]}, K_8 = \frac{[\mathrm{P_2M_4}]}{[\mathrm{P_2M_3}][\mathrm{M}]}$$

If ion binding to each porphyrin site were independent, one would expect $K_1 = 4K_\mathrm{m}$, $K_2 = (3/2)K_\mathrm{m}$, $K_3 = (2/3)K_\mathrm{m}$, and $K_4 = (1/4)K_\mathrm{m}$ where K_m is the microscopic binding constant for the monomer. Analogously, for the dimer, $K_6 = (3/2)K_\mathrm{d}$, $K_7 = (2/3)K_\mathrm{d}$, and $K_8 = (1/4)K_\mathrm{d}$ with K_d as the microscopic binding constant for the dimer. The remaining three vertical steps are not independent and their binding constants can be expressed by combination of the other constants as indicated in the scheme. However, this straightforward parameterization could not reproduce the observed S-shaped binding curve. To resolve this, we introduced correction factors f_m for the second and subsequent ion-binding events in the monomer case, so that $K_2 = (3/2)K_\mathrm{m}f_\mathrm{m}$, $K_3 = (2/3)K_\mathrm{m}f_\mathrm{m}$, and $K_4 = (1/4)K_\mathrm{m}f_\mathrm{m}$.

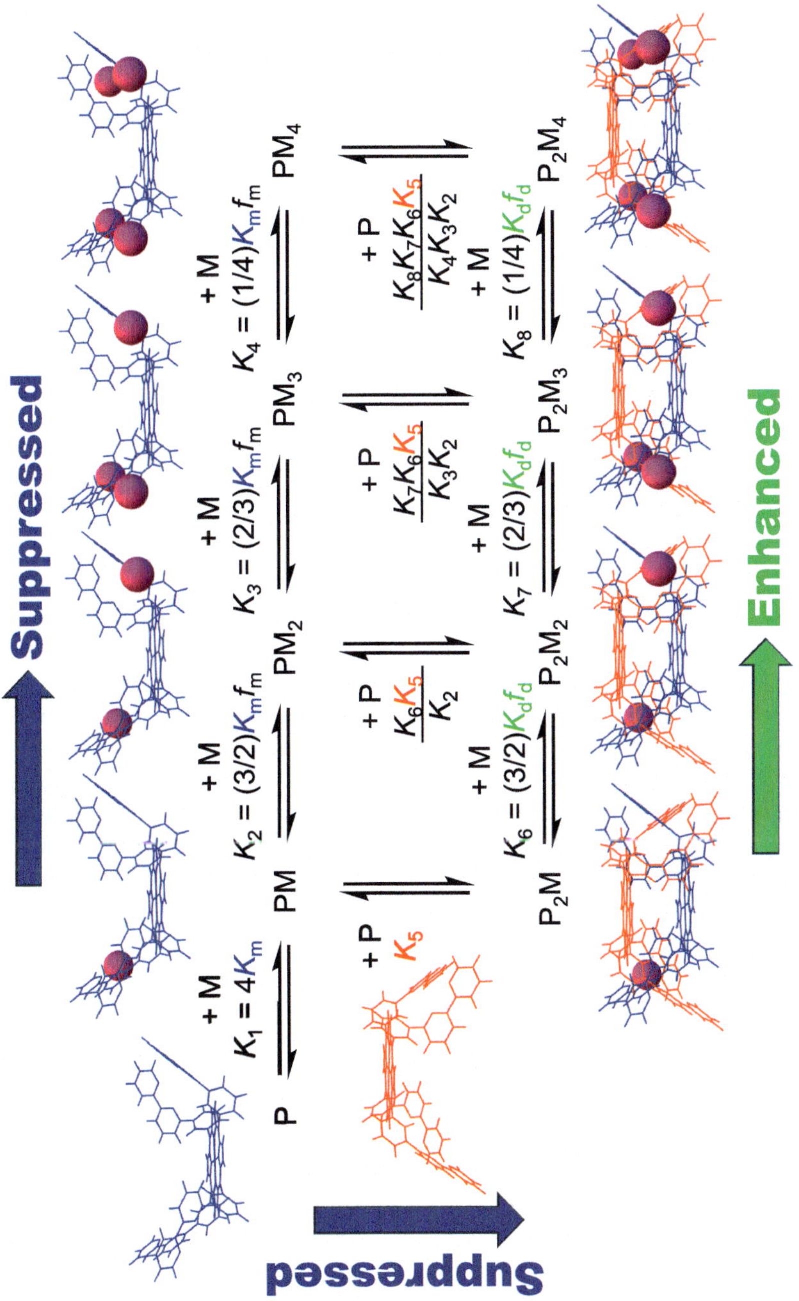

Fig. 14.11 Cooperative ion binding steps towards porphyrin dimer formation. Magenta spheres represent the Zn^{2+} ion (M)

Here, f_m may be viewed as an average correction factor for these processes. Similarly, for the dimer, $K_6 = (3/2)K_d f_d$, $K_7 = (2/3)K_d f_d$, and $K_8 = (1/4)K_d f_d$. (In some portions of the original publication, $K_d f_d$ is collectively denoted as K_d). Although the curve fitting did now permit all parameters to be determined precisely, several key observations emerged.

(i) $f_m \ll 1$ ($\log f_m \leq -2.1$). Once the first ion binds to the porphyrin, subsequent ion binding is greatly suppressed, likely because of electrostatic repulsion.
(ii) $K_5 \ll K_1 or K_m (2.7 \leq \log K_5/M^{-1} \leq -2.1$ vs. $7.4 \leq \log K_m)$. Binding the second porphyrin to an ion to form a dimer is considerably less favorable than the initial porphyrin binding step, presumably due to steric constraints.
(iii) $K_5 \ll K_d f_d$ ($9.0 \leq \log K_d f_d$). Once a dimer forms, subsequent ion binding is strongly enhanced, demonstrating the effect of preorganization.

These intricate intra-supramolecular interactions give rise to the cooperative, synergistic binding behavior evident in the pronounced S-shaped binding curve.

14.4 Concluding Remarks and Prospects

We have described three types of representative supramolecular porphyrin assemblies from our laboratory: cyclic chlorophyll tetramers, cyclic porphyrin trimers, and a face-to-face porphyrin dimer. The chlorophyll tetramer containing a fullerene unit serves as a model for the initial steps in photosynthesis. Following photoexcitation, energy transfer occurs, then charge separation proceeds with a quantum yield of effectively 100%. A porphyrin trimer incorporating a ferrocene unit also achieves 100% quantum yield for photoinduced charge separation. In this case, energy transfer is directed exclusively toward the ferrocene-bearing porphyrin, and the electron transfer proceeds from ferrocene to porphyrin, opposite to the electron-transfer direction in the fullerene modified system. The face-to-face porphyrin dimer discussed here can provide a platform for catalysis by incorporating suitable metal ions at the porphyrin centers with an interporphyrin distance of approximately 5 Å; however, the current design does not allow for guest-molecule or ion encapsulation. Ongoing work aims to modify the dimer geometry and metal ions to enable host–guest chemistry, develop light-energy conversion devices, and expand catalytic applications.

Acknowledgements We acknowledge Kakenhi (24K08386) from the JSPS for financial support.

References

1. Papiz MZ, Prince SM, Howard T, Cogdell RJ, Isaacs NW. The structure and thermal motion of the B800–850 LH2 complex from rps. acidophila at 2.0 A ° resolution and 100 k: new structural features and functionally relevant motions. J Mol Biol. 2003;326:1523–38.
2. Charisiadis A, Nikolaou V, Nikoloudakis E, Ladomenou K, Charalambidis G, Coutsolelos AG. Metalloporphyrins in bio-inspired photocatalytic conversions. Chem Commun. 2025;61:4630–46.
3. Otsuki J. Supramolecular approach towards light-harvesting materials based on porphyrins and chlorophylls. J Mater Chem A. 2018;6:6710–53.
4. Kim D (ed) Multiporphyrin Arrays, Fundamentals and Applications. Singapore: Pan Stanford.
5. Gotfredsen H, Hergenhahn J, Duarte F, Claridge TDW, Anderson HL. Bimolecular sandwich aggregates of porphyrin nanorings. J Am Chem Soc. 2024;146:25232–44.
6. Lehn J-M. Perspectives in chemistry—aspects of adaptive chemistry and materials. Angew Chem Int Ed. 2015;54:3276–89.
7. Gouterman M. Study of the effects of substitution on the absorption spectra of porphin. J Chem Phys. 1959;30:1139–61.
8. Gouterman M. Spectra of porphyrins. J Mol Spectrosc. 1961;6:138–63.
9. Streitwieser AJ. Molecular orbital theory for organic chemists. New York:Wiley;1961.
10. Cotton FA. Chemical applications of group theory. New York: Wiley; 1990.
11. Bhandari S, Sarkar S, Schubert A, Yamada A, Payne J, Ptaszek M, Geva E, Dunietz BD. Intersystem crossing in tetrapyrrolic macrocycles. A first-principles analysis. J Phys Chem C. 2021;125:13493–500.
12. Mandal AK, Taniguchi M, Diers JR, Niedzwiedzki DM, Kirmaier C, Lindsey JS, Bocian DF, Holten D. Photophysical properties and electronic structure of porphyrins bearing zero to four Meso-phenyl substituents: new insights into seemingly well understood Tetrapyrroles. J Phys Chem A. 2016;120:9719–31.
13. Magdaong NCM, Taniguchi M, Diers JR, Niedzwiedzki DM, Kirmaier C, Lindsey JS, Bocian DF, Holten D. Photophysical properties and electronic structure of Zinc(II) porphyrins Bearing 0–4 meso-phenyl substituents: zinc porphine to zinc tetraphenylporphyrin (ZnTPP). J Phys Chem A. 2020;124:7776–94.
14. Senge MO, Ryan AA, Letchford KA, MacGowan SA, Mielke T. Chlorophylls, symmetry, chirality, and photosynthesis. Symmetry. 2014;6:781–843.
15. Shingu K, Imajo S, Kuritani H, Hagishita S, Kuriyama K. Optical activity due to exciton interaction between 1B transitions of benzene chromophores. J Am Chem Soc. 1983;105:6966–7.
16. Villari V, Micali N, Nicosia A, Mineo P. Water-soluble non-ionic pegylated porphyrins: a versatile category of dyes for basic science and applications. Top Curr Chem. 2021;379:35.
17. Szabo A, Ostlund NS. Modern quantum chemistry. Introduction to Advanced Electronic Structure Theory. New York: McGraw-Hill;1989.
18. Mårtensson J. Calculation of the Förster orientation factor for donor-acceptor systems with one chromophore of threefold or higher symmetry: zinc porphyrin. Chem Phys Lett. 1994;229:449–56.
19. Harcourt RD, Scholes GD, Ghiggino KP. Rate expressions for excitation transfer. II. Electronic considerations of direct and through–configuration exciton resonance interactions. J Chem Phys. 1994;101:10521–5.
20. Gunderson VL, Smeigh AL, Kim CH, Co DT, Wasielewski MR. Electron transfer within self-assembling cyclic tetramers using chlorophyll-based donor−acceptor building blocks. J Am Chem Soc. 2012;134:4363–72.
21. Shinozaki Y, Richards G, Ogawa K, Yamano A, Ohara K, Yamaguchi K, Kawano S-i, Tanaka K, Araki Y, Wada T, Otsuki J. Double helices of a pyridine-appended zinc chlorophyll derivative. J Am Chem Soc. 2013;135:5262–5.
22. Howard IA, Zutterman F, Deroover G, Lamoen D, Van Alsenoy C. Approaches to calculation of exciton interaction energies for a molecular dimer. J Phys Chem B. 2004;108:19155–62.

23. Shinozaki Y, Ohkubo K, Fukuzumi S, Sugawa K, Otsuki J. Cyclic tetramers of zinc chlorophylls as a coupled light-harvesting antenna–charge-separation system. Chemistry—A Eur J. 2016; 22:1165–1176
24. Linke M, Lauer A, von Haimberger T, Zacarias A, Heyne K. Three-dimensional orientation of the Qy electronic transition dipole moment within the chlorophyll a molecule determined by femtosecond polarization resolved VIS pump-IR probe spectroscopy. J Am Chem Soc. 2008;130:14904–5.
25. Kelley RF, Goldsmith RH, Wasielewski MR. Ultrafast energy transfer within cyclic self-assembled chlorophyll tetramers. J Am Chem Soc. 2007;129:6384–5.
26. Otsuki J, Okumura T, Sugawa K, Kawano S-i, Tanaka K, Hirao T, Haino T, Lee YJ, Kang S, Kim D. A light-harvesting/charge-separation model with energy gradient made of assemblies of meta-pyridyl zinc porphyrins. Chem Eur J. 2021;27:4053–63.
27. Collman JP, Anson FC, Barnes CE, Bencosme CS, Geiger T, Evitt ER, Kreh RP, Meier K, Pettman RB. Further studies of the dimeric.beta.-linked "face-to-face four" porphyrin: FTF4. J Am Chem Soc. 1983;105:2694–9.
28. Collman JP, Denisevich P, Konai Y, Marrocco M, Koval C, Anson FC. Electrode catalysis of the four-electron reduction of oxygen to water by dicobalt face-to-face porphyrins. J Am Chem Soc. 1980;102:6027–36.
29. Oldacre AN, Friedman AE, Cook TR. A self-assembled cofacial cobalt porphyrin prism for oxygen reduction catalysis. J Am Chem Soc. 2017;139:1424–7.
30. Nakamura T, Ube H, Shionoya M. Silver-mediated formation of a cofacial porphyrin dimer with the ability to intercalate aromatic molecules. Angew Chem Int Ed Engl. 2013;52:12096–100.
31. Li M, Shi Y-Q, Gan X, Su L, Liang J, Wu H, You Y, Che M, Su P, Wu T, Zhang Z, Zhang W, Yao L-Y, Wang P, Xie T-Z. Coordination-driven tetragonal prismatic cage and the investigation on host-guest complexation. Inorg Chem. 2023;62:4393–8.
32. Otsuki J, Sato K, Sugawa K. A Cofacial porphyrin dimer generated by cooperative zinc ion binding. Eur J Inorg Chem. 2024;27: e202400188.
33. Durot S, Flamigni L, Taesch J, Dang TT, Heitz V, Ventura B. Synthesis and solution studies of silver(I)-assembled porphyrin coordination cages. Chem Eur J. 2014;20:9979–90.
34. Rose C, Lebrun A, Clement S, Richeter S. Cofacial porphyrin dimers assembled from N-heterocyclic carbene-metal bonds. Chem Commun. 2018;54:9603–6.
35. Blanco V, Leigh DA, Marcos V. Artificial switchable catalysts. Chem Soc Rev. 2015;44:5341–70.
36. Pan T, Wang Y, Xue X, Zhang C. Rational design of allosteric switchable catalysts. Exploration. 2022;2:20210095.

Chapter 15
Chirality at Metal Surfaces Through Scanning Tunneling Microscopy and Tip-Enhanced Raman Spectroscopy: Helicenes as a Model System

Paweł Krukowski and Yuji Kuwahara

Abstract Scanning probe techniques provide unparalleled access to structural, electronic, and vibrational properties of matter at the atomic and molecular scales. In this review, we present a comprehensive overview of scanning tunneling microscopy (STM) and tip-enhanced Raman spectroscopy (TERS) as complementary tools for investigating chiral molecular systems on solid surfaces, with particular emphasis on thiaheterohelicene derivatives. We first outline the fundamental principles of STM, including the quantum–mechanical origin of the tunneling current and its sensitivity to both surface topography and local electronic density of states. We then introduce Raman spectroscopy and its surface- and tip-enhanced variants, discussing the role of localized surface plasmon resonances in achieving dramatic signal enhancement and nanoscale spatial resolution beyond the optical diffraction limit. The core of this review focuses on molecular-scale observations of helicene-based chiral systems adsorbed on metal substrates, by critically surveying recent experimental studies. By combining high-resolution STM and tip-enhanced Raman spectroscopy, with surface-enhanced Raman spectroscopy (SERS) serving as a complementary and comparative technique, and X-ray photoelectron spectroscopy (XPS), we analyze adsorption geometries, intermolecular interactions, and self-assembly processes of racemic and enantiopure helicene derivatives on Ag(111) and Au(111) surfaces. Special attention is given to thiaheterohelicenes, whose thiophene units introduce distinct electronic characteristics and surface interactions compared to purely hydrocarbon helicenes. The results reveal how subtle variations in molecular structure, functionalization, and substrate choice govern chiral recognition, enantiomeric pairing, and the emergence of long-range ordered assemblies. By integrating

P. Krukowski (✉)
Department of Solid State Physics, Faculty of Physics and Applied Informatics, University of Lodz, Łódź, Poland
e-mail: pawel.krukowski@uni.lodz.pl

Y. Kuwahara
Department of Precision Engineering, Graduate School of Engineering, The University of Osaka, Suita, Japan
e-mail: kuwahara@prec.eng.osaka-u.ac.jp

T. Onishi (ed.), *Advances in Quantum Science II*,
https://doi.org/10.1007/978-981-95-9721-5_15

structural imaging and spectroscopic characterization, this work highlights the power of STM and TERS for unraveling chiral phenomena at surfaces. These insights are expected to contribute to the rational design of chiral interfaces and functional materials for applications in molecular electronics, enantioselective catalysis, and chiral photonics.

Keywords Scanning tunneling microscopy (STM) · Tip-enhanced Raman spectroscopy (TERS) · Chirality · Helicenes · Thiaheterohelicenes · Chiral recognition · Self-assembly · Surface-enhanced Raman spectroscopy · Metal surfaces · Single-molecule spectroscopy

15.1 Fundamental Principles

15.1.1 Scanning Tunneling Microscopy (STM)

History of STM

Scanning tunneling microscopy (STM), invented in 1982 by Gerd Binnig, Heinrich Rohrer and collaborators [1], revolutionized the field of surface science by providing, for the first time, real-space imaging of conductive surfaces with true atomic-scale resolution [2]. Its invention opened an entirely new era in nanoscience, ultimately earning the authors the Nobel Prize in Physics in 1986. The significance of STM stems not only from its impressive spatial resolution, but also from the fundamental quantum–mechanical principle on which it operates: electron tunneling through a classically forbidden vacuum barrier.

Metallic Tip

In a typical STM experiment, a metallic tip—usually made of tungsten, platinum–iridium, or gold—is electrochemically etched to produce an apex that ideally terminates in a single atom. This extremely sharp apex is crucial because the tunneling current is dominated by the outermost atom of the tip, ensuring that the technique is intrinsically sensitive to local atomic environments. The tip is mounted on a piezoelectric scanner capable of sub-ångström precision in all three spatial directions. When the tip is positioned within approximately 0.5–1 nm of a conductive surface, and a bias voltage of a few millivolts to a few volts is applied, electrons tunnel between the tip and the sample despite the presence of a seemingly impenetrable vacuum gap. This counterintuitive process is possible because the electron wavefunction does not abruptly vanish outside the material but instead decays exponentially into the vacuum [3]. The finite overlap of the tip and sample wavefunctions results in a measurable tunneling current.

Features

A defining feature of STM is the exponential dependence of the tunneling current on the tip–sample separation. A displacement of merely 1 Å can modify the current by approximately an order of magnitude. This sensitivity makes STM one of the most precise distance probes available, with vertical resolution routinely reaching the scale of a few picometers. Consequently, even minute variations in atomic positions or local electronic density of states are readily detectable. The feedback mechanism of STM exploits this behavior. During imaging in constant-current mode, the tunneling current is continuously monitored, and the vertical position of the tip is actively adjusted to maintain a fixed current value. The recorded height signal thus directly reflects the surface topography, producing images where individual atoms appear as protrusions or depressions depending on the bias of polarity. STM can also operate in constant-height mode, where the tip height is held fixed, and variations in current reflect changes in the local electronic environment. This mode is capable of very fast imaging, although it requires exceptionally flat surfaces because the tip otherwise crashes into the sample.

Scanning Tunneling Spectroscopy

Beyond topography, STM provides a wealth of spectroscopic information. By recording the tunneling current as a function of applied bias voltage at a fixed tip–sample distance, one obtains scanning tunneling spectroscopy (STS). STS directly probes the local density of electronic states (LDOS), enabling measurement of band gaps and detailed electronic structure due to its theoretical foundation described by Tersoff and Hamann [4]. STS has been widely used to investigate surface states [5], Kondo resonance [6, 7], superconducting energy gaps [8], quantized energy levels in confined systems [9, 10], and molecular orbitals [11], topological surface states [12]. Inelastic electron tunneling spectroscopy (IETS) extends the capability to vibrational and spin excitations of individual molecules [13, 14], enabling chemical identification at the single-molecule level. The ability to map LDOS with atomic resolution makes STM an indispensable tool in condensed-matter physics, particularly for studying low-dimensional materials such as graphene and other 2D materials [15–17], topological insulators [18], and strongly correlated electron systems [19]. In addition to imaging and spectroscopy, STM also enables atom and molecule manipulation, as demonstrated famously by the construction of the IBM logo from individual xenon atoms at the beginning of the 1990s [20]. The tip can be used to reposition atoms across the surface, induce chemical reactions by injecting electrons at specific energies, or locally modify adsorption configurations.

Experimental Environment and Accessible Physical Information

The operating environment of STM has a significant influence on its performance. While it can function in ambient conditions, the highest spatial and spectroscopic resolution is achieved under ultra-high vacuum (UHV) and low-temperature conditions, where thermal drift and vibrations are minimized. At cryogenic temperatures—often below 10 K—STM can resolve subtle electronic phenomena such as quasiparticle interference, inelastic tunneling signatures corresponding to vibrational or spin excitations, and coherent quantum states. Because STM probes the wavefunction overlap between the tip and sample, the contrast in images does not always correspond strictly to geometric height. Instead, the contrast can reflect variations in electronic density, orbital symmetry, bond order, or charge distribution. This dual sensitivity makes STM both a structural and an electronic probe. For example, STM can distinguish between different adsorption sites, image standing waves formed by surface-state electrons, and resolve intramolecular features such as nodal patterns of the highest occupied and lowest unoccupied molecular orbitals. Overall, STM remains one of the most powerful and versatile techniques in nanoscience. Its unique combination of atomic precision, electronic sensitivity, and manipulation capability has made it indispensable in disciplines ranging from surface chemistry and catalysis to molecular electronics, quantum materials, and plasmonics. Figure 15.1 illustrates the basic operating principle and the tunneling mechanism that enables atomic-scale resolution.

15.1.2 Theoretical Model of the Tunneling Current

In the following, we examine the characteristics of the tunneling current using a simple one-dimensional model of a metallic tip and a conductive sample separated by a small vacuum gap. Let the distance between the tip and the sample be s, the

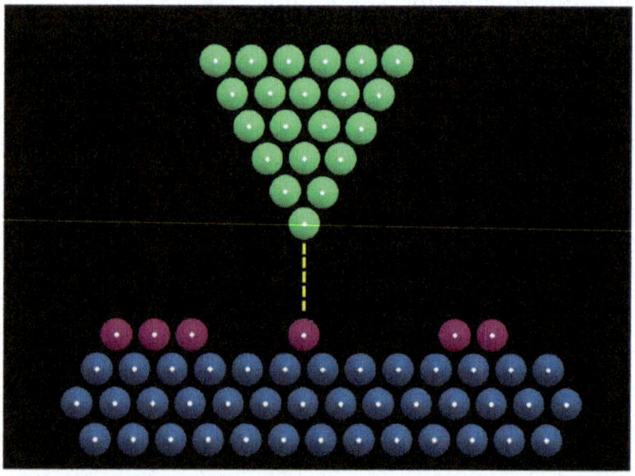

Fig. 15.1 Schematic illustration of the basic operating principle of scanning tunneling microscopy (STM), highlighting the quantum–mechanical tunneling process between a sharp metallic tip and a conductive sample that enables atomic-scale spatial resolution

applied voltage be V, and the work functions of the tip and the sample be ϕ_1 and ϕ_2, respectively. Denote the local density of electronic states at an energy level E, measured with respect to the Fermi level of the sample and the tip, as $\rho_s(E)$ and $\rho_t(E)$, respectively. Here, we assume that the energy distribution of electrons follows the Fermi distribution at 0 K, which is valid at sufficiently low temperature.

Let $T(E, eV)$ be the probability that an electron at energy level E tunnels to the other side when a potential difference V is applied across the tunneling barrier. If we consider only elastic tunneling processes, the tunneling current flowing between the two is given by the process in which an electron at energy E on the sample side tunnels into the energy level $-eV + E$ on the tip side and can thus be expressed as shown in Eq. (15.1).

$$I \propto \int_0^{eV} \rho_s(E)\rho_t(-eV + E)T(E, eV)dE \tag{15.1}$$

If the wavefunction of electrons in the vacuum barrier is denoted by $\Psi(z)$, the Schrödinger equation is given as shown in Eq. (15.2).

$$\left(-\frac{\hbar^2}{2m}\frac{d^2}{dz^2} + \phi(z)\right)\Psi(z) = E\Psi(z) \tag{15.2}$$

Here, z denotes the distance from the sample surface. We assume a stationary-state wavefunction as given in Eq. (15.3).

$$\Psi(z) = exp\left(\frac{i}{\hbar}S(z)\right) \tag{15.3}$$

Substituting Eq. (15.3) into Eq. (15.2) gives:

$$S'^2 - i\hbar S'' - 2m(E - \phi(z)) = 0 \tag{15.4}$$

When $\hbar$ is considered sufficiently small, S(z) can be approximated by expanding it in a power series with respect to $\hbar$ and retaining only the first few terms. Using this approach—known as the WKB approximation—we can obtain an approximate solution for the wavefunction.

If we expand $S(z)$ in powers of $\hbar$ and neglect terms of third order and higher, it can be expressed as shown in Eq. (15.5).

$$S(z) \approx S_0 + \frac{\hbar}{i}S_1 \tag{15.5}$$

Substituting Eq. (15.5) into Eq. (15.4) and comparing the coefficients of each power of $\hbar$, we obtain the following set of differential equations.

$$S_0'^2 = 2m(E - \phi(z)) \tag{15.6}$$

$$2S_1{}' = -\frac{S_0''}{S_0'} \tag{15.7}$$

We now consider the case in which the energy of electrons is smaller than the work function. Approximating the tilted rectangular potential of the vacuum tunneling barrier with a flat rectangular potential, the work function can be expressed as shown in Eq. (15.8).

$$\begin{aligned} \phi(z) &= \langle\phi\rangle + \frac{eV}{2} \\ \langle\phi\rangle &= \frac{\phi_1 + \phi_2}{2} \end{aligned} \tag{15.8}$$

Substituting Eq. (15.8) into Eq. (15.6) and solving the differential equation yields:

$$S_0 = \pm iz\sqrt{2m\left(\langle\phi\rangle - E + \frac{eV}{2}\right)} \tag{15.9}$$

Also, from Eqs. (15.6), (15.7), and (15.8), we have:

$$S_1 = -\frac{1}{2}\ln\left(\pm i\sqrt{2m\left(\langle\phi\rangle - E + \frac{eV}{2}\right)}\right) + \text{const} \tag{15.10}$$

From Eqs. (15.3), (15.5), (15.9), and (15.10), the general solution of the wavefunction is:

$$\Psi(z) = A_1\exp\left(\frac{z}{\hbar}\sqrt{2m\left(\langle\phi\rangle - E + \frac{eV}{2}\right)}\right) + A_2\exp\left(-\frac{z}{\hbar}\sqrt{2m\left(\langle\phi\rangle - E + \frac{eV}{2}\right)}\right) \tag{15.11}$$

Here, A_1 and A_2 are constants.

If $A_1 \neq 0$, then $\Psi(z) \to \infty$ as $z \to \infty$, and the wavefunction diverges, which is unphysical.

Therefore, $A_1 = 0$, and the wavefunction can be expressed as shown in Eq. (15.12).

$$\Psi(z) \propto \exp\left(-z\sqrt{\frac{2m}{\hbar^2}\left(\langle\phi\rangle - E + \frac{eV}{2}\right)}\right) \tag{15.12}$$

From this equation, it is evident that the wavefunction $\Psi(z)$ decays as the distance from the sample surface increases, with a decay constant κ.

$$\kappa = \sqrt{\frac{2m}{\hbar^2}\left(\langle\phi\rangle - E + \frac{eV}{2}\right)} \tag{15.13}$$

It follows that the wavefunction decays with decay constant κ, and that the larger the average work function ⟨φ⟩, the more rapid the decay. Furthermore, if we consider the wave vector component $K_{\parallel}$ (parallel to the surface) in the one-dimensional model, the decay constant becomes as shown in Eq. (15.14):

$$\kappa = \sqrt{\frac{2m}{\hbar^2}\left(\langle\phi\rangle - E + \frac{eV}{2}\right) + K_{\parallel}^2} \tag{15.14}$$

From this equation, it is evident that for electrons with the same energy E, the larger the in-plane wave vector $K_{\parallel}$, the more rapidly the wavefunction decays away from the surface. The tunneling probability $T(E, \text{eV})$ can also be expressed using the WKB approximation, in the same manner as the wavefunction, yielding the following Eq. (15.15).

$$T(E, eV) \cong \exp\left(-2s\sqrt{\frac{2m}{\hbar^2}\left(\langle\phi\rangle - E + \frac{eV}{2}\right)}\right) \tag{15.15}$$

Therefore, the tunneling probability of an electron becomes smaller when the energy of the tunneling barrier is higher relative to the electron energy, and when the distance between the tip and the sample is larger. In addition, as with the wavefunction, electrons with smaller in-plane wave vectors contribute more significantly to the tunneling current. Next, to calculate the tunneling current, we substitute Eq. (15.15) into Eq. (15.1). By integrating over the range of electron energy levels that contribute to the tunneling current, namely from 0 to eV, and taking into account the dominant exponential term, we obtain Eq. (15.16).

$$I \cong f(V)exp\left(-2s\sqrt{\frac{2m}{\hbar^2}\left(\langle\phi\rangle - \frac{eV}{2}\right)}\right) \tag{15.16}$$

Here, for simplicity of calculation, it is assumed that $\rho_s(E)$ and $\rho_t(E)$ are constant with respect to the energy E. In this context, $f(V)$ is a function that varies slowly with the applied voltage. When the physical quantities in this equation are expressed in units of electron volts (eV) and angstroms (Å), the equation takes the form shown in Eq. (15.17).

$$I \cong f(V)exp\left(-1.02 \times s\sqrt{\langle\phi\rangle - \frac{eV}{2}}\right) \tag{15.17}$$

For example, in Eq. (15.17), if the barrier width s is 10 Å and $\langle\phi\rangle - \frac{eV}{2}$ is 5 eV, the current changes by approximately one order of magnitude for every 1 Å change in distance.

15.1.3 Raman Spectroscopy

Basic Concept of Raman Spectroscopy and Vibrational Information

When light interacts with a material, a fraction of the incident photons undergoes inelastic scattering, resulting in a shift in their frequency. This effect was first discovered by C. V. Raman and his collaborators in 1928 [21]. In a typical light–matter interaction, the overwhelming majority of scattered photons retain the same frequency as the incident light. This elastic process is called Rayleigh scattering. However, a very small fraction—typically on the order of one part in 10^6–10^8—undergoes a change in frequency due to energy exchange with the vibrational states of the material. Scattered photons with lower frequency than the excitation light correspond to Stokes scattering, while those with higher frequency correspond to anti-Stokes scattering. These frequency shifts provide direct insight into the vibrational energy levels of the system. As depicted in Fig. 15.2, Raman scattering proceeds through the formation of a short-lived virtual state. When a photon interacts with a molecule, the electronic cloud is momentarily distorted, promoting the system into this non-stationary state rather than a real electronic excited level. If the molecule transitions to a higher vibrational level of the electronic ground state, a Stokes-shifted photon is emitted. Conversely, if the molecule initially occupies an excited vibrational level and relaxes to a lower one, the emitted photon has higher energy, producing an anti-Stokes shift. The relative intensities of Stokes and anti-Stokes lines therefore reflect the vibrational population distribution and, by extension, the temperature of the system. Because vibrational modes are sensitive to molecular geometry, bonding, and the surrounding environment, Raman spectroscopy provides a unique fingerprint of the material under investigation. Its ability to probe local symmetry, chemical composition, crystallinity, strain, and intermolecular interactions has made it indispensable for fields ranging from fundamental spectroscopy to nanotechnology and biosensing.

Classical Theory of Raman Scattering

Raman scattering can also be described using classical theory [22]. When light is incident on a molecule, the electric field of the incident light exerts a force on the electrons within the molecule. As a result, the electron cloud surrounding the atomic nuclei experiences a displacement due to this force, shifting from its original position centered around the nucleus. This displacement induces an electric dipole moment μ,

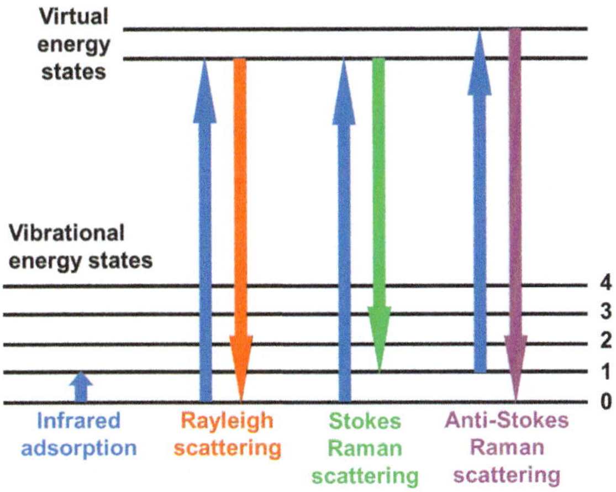

Fig. 15.2 Energy diagram of Raman scattering

leading to polarization within the molecule. The induced dipole moment μ is proportional to the electric field E. Letting α represent the polarizability, the relationship can be expressed as shown in Eqs. (15.18) and (15.19).

$$\mu = \alpha E \tag{15.18}$$

$$\begin{pmatrix} \mu_x \\ \mu_y \\ \mu_z \end{pmatrix} = \begin{pmatrix} \alpha_{xx} & \alpha_{xy} & \alpha_{xz} \\ \alpha_{yx} & \alpha_{yy} & \alpha_{yz} \\ \alpha_{zx} & \alpha_{zy} & \alpha_{zz} \end{pmatrix} \begin{pmatrix} E_x \\ E_y \\ E_z \end{pmatrix} \tag{15.19}$$

To simplify the discussion, we now consider the case, where both the dipole moment $\boldsymbol{\mu}$ and the electric field $\boldsymbol{E}$ are one-dimensional. Let the frequency of the incident light be ν_0, and the amplitude of the electric field be E_0. Then, the magnitude of the electric field can be expressed as shown in Eq. (15.20).

$$E = E_0 \cos 2\pi \nu_0 t \tag{15.20}$$

Substituting this into Eq. (15.18), the induced dipole moment can be expressed as shown in Eq. (15.21), indicating that the dipole moment oscillates with the same frequency as the incident light.

$$\mu = \alpha E_0 \cos 2\pi \nu_0 t \tag{15.21}$$

Next, we consider the case in which the polarizability changes due to molecular vibrations. Let ν be the vibrational frequency of the molecule and let α_0 be the polarizability when the nuclei are at their equilibrium positions. When the nuclei are displaced by an amount q from equilibrium due to molecular vibration, the polarizability α can be expanded in a Taylor series as shown in Eq. (15.22).

$$\alpha = \alpha_0 + \left(\frac{\partial \alpha}{\partial q}\right)_0 q + \frac{1}{2}\left(\frac{\partial^2 \alpha}{\partial q^2}\right)_0 q^2 + \frac{1}{3!}\left(\frac{\partial^3 \alpha}{\partial q^3}\right)_0 q^3 + \cdots \tag{15.22}$$

By neglecting terms of third order and higher, the expression can be simplified as shown in Eq. (15.23).

$$\alpha = \alpha_0 + \left(\frac{\partial \alpha}{\partial q}\right)_0 q \tag{15.23}$$

Here, $\left(\frac{\partial \alpha}{\partial q}\right)_0$ denotes the first derivative of the polarizability with respect to the displacement q, evaluated at the equilibrium position of the nuclei.

When the molecule vibrates at frequency ν, the nuclear displacement q_0 can be expressed as shown in Eq. (15.24), where q_0 is the maximum amplitude of the vibrational displacement.

$$q = q_0 \cos 2\pi \nu t \tag{15.24}$$

Substituting Eqs. (15.23) and (15.24) into Eq. (15.21), the induced dipole moment can be expressed as shown in Eq. (15.25).

$$\begin{aligned}
\mu &= \left[\alpha_0 + \left(\frac{\partial \alpha}{\partial q}\right)_0 q\right] E_0 \cos 2\pi \nu_0 t \\
&= \alpha_0 E_0 \cos 2\pi \nu_0 t + \left(\frac{\partial \alpha}{\partial q}\right)_0 q_0 E_0 \cos 2\pi \nu_0 t \cdot \cos 2\pi \nu t \\
&= \alpha_0 E_0 \cos 2\pi \nu_0 t + \frac{1}{2}\left(\frac{\partial \alpha}{\partial q}\right)_0 q_0 E_0 [\cos 2\pi (\nu_0 + \nu)t + \cos 2\pi (\nu_0 - \nu)t]
\end{aligned} \tag{15.25}$$

From Eq. (15.25), it is evident that the induced dipole moment contains not only a component oscillating at the frequency of the incident light ν_0 (Rayleigh scattered light), but also components oscillating at $\nu_0 + \nu$ (anti-Stokes scattered light) and $\nu_0 - \nu$ (Stokes scattered light).

Next, let us consider the coefficient of the second term in Eq. (15.25). Since $q_0 \neq 0$ and $E_0 \neq 0$, we focus on $\left(\frac{\partial \alpha}{\partial q}\right)_0$. If $\left(\frac{\partial \alpha}{\partial q}\right)_0 = 0$, the second term in Eq. (15.25) becomes zero, and Raman scattered light does not occur. Therefore, for Raman scattering to take place, it is necessary that $\left(\frac{\partial \alpha}{\partial q}\right)_0 \neq 0$, i.e., the molecular polarizability must change with molecular vibration.

15.1.4 *Surface- and Tip-Enhanced Raman Spectroscopy (SERS and TERS)*

Overview of SERS

Raman spectroscopy enables the analysis of molecular and crystal structures by obtaining vibrational information from a target sample. As described previously, Raman scattering originates from the inelastic interaction between incident photons and molecular vibrations. However, the intrinsic probability of Raman scattering is extremely low—only about 10^{-6} to 10^{-8} of the incident photons contribute to the Raman signal—because the induced dipole responsible for Raman scattering is much weaker than that associated with Rayleigh scattering. This very small scattering cross section fundamentally limits the sensitivity of conventional Raman spectroscopy, making it essentially impossible to detect the vibrational fingerprint of a single molecule or a small number of molecules under standard illumination conditions. To overcome this limitation, surface-enhanced Raman spectroscopy (SERS) was developed in the 1970s after the pioneering observations of Fleischmann et al. [23] and later quantitative interpretation by Jeanmaire and Van Duyne [24]. In SERS, Raman scattering is dramatically amplified by localized surface plasmon resonance (LSPR), which arises when a metallic nanostructure—typically composed of Au or Ag—is irradiated with laser light. LSPR corresponds to the collective oscillation of conduction electrons confined to the surface of metallic nanoparticles. When the frequency of the incident light matches the plasmon resonance frequency, a strongly enhanced and spatially confined electromagnetic (EM) field is generated near the metal surface. Two enhancement mechanisms contribute to the SERS effect:

(1) Electromagnetic enhancement (EM): The local field amplitude near plasmonic nanostructures can increase by several orders of magnitude. Since Raman scattering intensity scales with $|E|^4$ (product of excitation and scattered field enhancements), EM factors of 10^6–10^8 are commonly achieved, and enhancements exceeding 10^{10} have been reported under optimized "hotspot" conditions between closely spaced nanoparticles [25].
(2) Chemical enhancement (CE): A weaker (10–10^2) enhancement originating from charge-transfer interactions between the molecule and the metal surface, modifying the molecular polarizability [25].

Together, these mechanisms enable single-molecule Raman spectroscopy, which was first demonstrated convincingly in 1997 by Kneipp et al. [25]. Despite its remarkable sensitivity, SERS has a major limitation: because the enhanced field region extends only to $\approx$10–20 nm above the metal surface, but the illumination spot is still constrained by optical diffraction (~0.5–1 μm), the technique cannot achieve true nanoscale spatial resolution. Thus, SERS provides an averaged signal over many molecules present within the laser spot and cannot directly map the vibrational properties of individual nanoscale features or isolated molecules.

Overview of TERS

To overcome the spatial-resolution limit imposed by diffraction, Wessel proposed TERS in 1985 [26]. TERS combines scanning probe microscopy (SPM)—typically atomic force microscopy (AFM) or STM—with Raman spectroscopy to achieve simultaneous topographical and chemical imaging at the nanoscale. In TERS, a metallic tip (Ag or Au) acts as a nanoantenna: the laser excites localized surface plasmons specifically at the apex of the tip, generating an extremely confined ("nano-hotspot") EM field that enhances Raman scattering from molecules located directly beneath the tip apex. Because the enhancement is restricted to a nanometric volume (<10 nm lateral and ~1–2 nm vertical), TERS enables both high sensitivity and high spatial resolution [27]. Enhancement factors of 10^3–10^6 are commonly reported, with state-of-the-art techniques approaching single-molecule sensitivity [28]. Modern TERS experiments have achieved spatial resolutions in the range of 10–80 nm, depending on the tip geometry, excitation wavelength, and sample properties, and in some cases even below 3 nm, enabling submolecular vibrational imaging on surfaces [27, 29]. TERS therefore represents a powerful tool for nanoscale spectroscopy, capable of probing chemical composition, molecular orientation, local electronic structure, and intermolecular interactions far beyond the optical diffraction limit. The characteristics of TERS compared with other measurement techniques are shown in Fig. 15.3, and a schematic diagram of TERS is presented in Fig. 15.4.

15.1.5 Localized Surface Plasmon Resonance

In this section, we describe localized surface plasmons, which are the primary factor responsible for enhancement in TERS. A surface plasmon is a surface-bound eigenmode of plasma oscillation. Based on their characteristics, surface plasmons can be broadly categorized into two types: propagating surface plasmons, which are electromagnetic waves that propagate along the interface between a metal such as gold or silver and a dielectric; and localized surface plasmons, which are confined within metallic nanostructures such as those made of gold or silver and do not propagate.

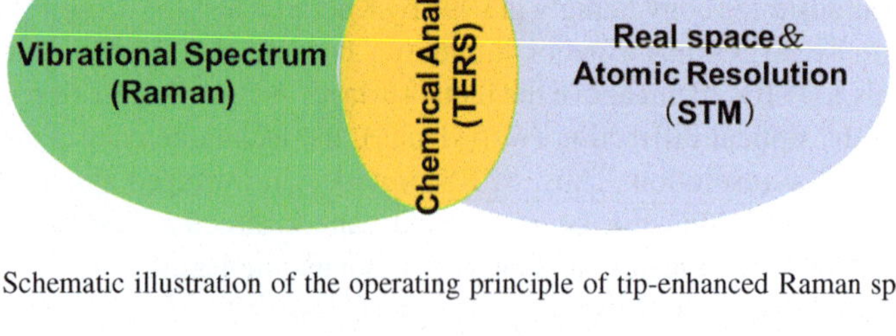

Fig. 15.3 Schematic illustration of the operating principle of tip-enhanced Raman spectroscopy (TERS)

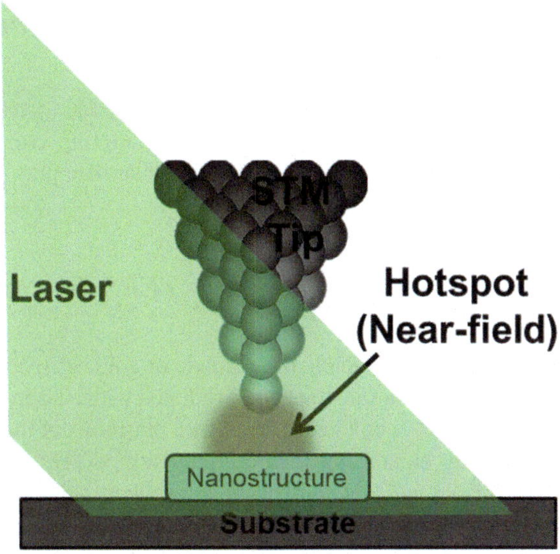

Fig. 15.4 Schematic diagram of TERS

Localized surface plasmons have the property of resonating with incident light when the frequency of the plasmon matches that of the light—this phenomenon is called localized surface plasmon resonance (LSPR). Let us now consider an isolated spherical metallic nanostructure embedded in a homogeneous medium. When a laser is irradiated onto this nanostructure, we can assume that the radius of the structure is much smaller than the wavelength of the incident light. Therefore, the electric field experienced by the nanostructure can be treated as a uniform electric field oscillating at angular frequency ω. By solving Laplace's equation for the potential and applying the appropriate boundary conditions, the metallic nanostructure can be regarded as a dipole with a polarizability given by Eq. (15.26). Here, R denotes the radius of the sphere, and ε_1 and ε_2 are the dielectric constants of the sphere and the surrounding medium, respectively.

$$\alpha = 4\pi R^3 \frac{\varepsilon_1(\omega) - \varepsilon_2}{\varepsilon_1(\omega) + 2\varepsilon_2} \tag{15.26}$$

When this dipole oscillates with time dependence e^{-iwt}, it generates dipole radiation. The resulting electric field, at a distance r from the center of the sphere, can be expressed as the sum of three components: a quasi-static field proportional to r^{-3}, an inductive field proportional to r^{-2}, and a radiative field proportional to r^{-1}. Among these, only the radiative field propagates to infinity, while the first two components are non-propagating and localized near the particle. The scattered field intensity, when integrated over a spherical surface of radius r centered on the particle, defines the generalized scattering cross section *C(r)*. Here k denotes the wave number; the generalized scattering cross section is expressed as shown in Eq. (15.27).

$$C(r) = \frac{|\alpha^2|}{6\pi}\left(\frac{3}{r^4} + \frac{k^2}{r^2} + k^4\right) \tag{15.27}$$

From Eq. (15.26), when $\mathrm{Re}[\varepsilon_1(\omega)] + 2\varepsilon_2 = 0$, the polarizability becomes extremely large, and the generalized scattering cross section also becomes large. This condition corresponds to localized surface plasmon resonance.

15.1.6 Tip Enhancement in TERS-Gap Mode

When a metal tip—such as one made of gold or silver—is brought within a distance of 1–2 nm from a substrate and irradiated with a laser, the localized surface plasmons at the tip apex couple with the surface plasmons of the substrate. This coupling leads to a further enhancement in the intensity of TERS, known as the *gap-mode effect* [30]. It is known that comparing cases where this coupling occurs and where it does not, the TERS intensity can vary by 2 to 3 orders of magnitude [31]. Consequently, STM-TERS, which does not require transparent samples, tends to exhibit stronger enhancement than AFM-TERS, which must use transparent samples due to setup constraints. Furthermore, if the distance between the tip and the substrate is denoted as d, the intensity of TERS in the gap mode has been found to be proportional to d^{-10} [32].

15.1.7 Spatial Resolution of TERS

The concept of spatial resolution Δx, defined as the minimum distance at which two objects can be clearly distinguished, was formulated by Abbe and Rayleigh in the late nineteenth century. It is given by Eq. (15.28).

$$\Delta x = \frac{0.61\lambda}{NA} \tag{15.28}$$

Here, λ denotes the wavelength of light, and NA represents the numerical aperture. This spatial resolution arises from the diffraction limit of light, and when using visible light, it is typically around 200–300 nm. Therefore, in conventional Raman spectroscopy or SERS, the spatial resolution of measurements is limited to this range. On the other hand, in measurements using TERS, only the Raman scattered light from molecules within the locally enhanced electromagnetic field is significantly amplified. As a result, the typical spatial resolution of TERS is approximately 10–80 nm, enabling measurements at the nanoscale that surpass the diffraction limit of light [33]. At this scale, the measurement resolution is generally considered to depend on the radius of curvature at the apex of the SPM probe tip. However, in the recent years, TERS measurements have been performed with resolutions much smaller than the curvature radius of the tip apex, achieving spatial resolutions below 1 nm. While the mechanism behind such high-resolution measurements has not yet been fully elucidated, all of these measurements have been carried out using gap-mode STM-TERS under ultra-high vacuum and low-temperature conditions [33–35]. Therefore, it could be considered that achieving higher spatial resolution requires measurement in the gap mode, where localized surface plasmons at the tip apex are coupled with surface plasmons on the substrate.

15.2 Chiral Molecular Systems

15.2.1 Molecular-Scale Observation of Helicene Derivatives

Overview of Helicene Molecules and Their Properties

Helicenes represent a captivating class of inherently chiral polycyclic aromatic compounds distinguished by their non-planar, screw-shaped (helical) molecular structure. This unique geometry arises from steric repulsion between ortho-fused aromatic rings, forcing the π-conjugated system into a helical twist that imparts both structural rigidity and pronounced chirality [36, 37]. The interplay between extended π-conjugation and helical distortion endows helicenes with exceptional chiroptical and electronic properties, making them a subject of intense research in materials science and supramolecular chemistry [38–40]. Helicenes are broadly categorized into carbohelicenes, composed exclusively of carbon atoms, and heterohelicenes, where one or more carbon atoms are replaced by heteroatoms such as nitrogen, sulfur, oxygen, boron, or phosphorus [36, 41, 42]. The introduction of heteroatoms significantly modulates their electronic structure, enabling precise control over optical properties, redox potentials, and intermolecular interactions. This tunability enhances their applicability in functional materials, ranging from asymmetric catalysis to optoelectronic devices. Owing to their strong chiroptical responses—including high dissymmetry factors (g-factors) and circularly polarized luminescence (CPL)—helicenes have emerged as key candidates for advanced technologies [43, 44]. They have been successfully incorporated into organic field-effect transistors (OFETs) capable of detecting circularly polarized light, as well as chiral organic light-emitting diodes (OLEDs) exhibiting polarized electroluminescence. Additionally, their stimuli-responsive behavior enables the development of tunable emissive materials for sensing and anti-counterfeiting applications. Beyond optoelectronics, helicenes show great promise in enantioselective recognition, where their inherent chirality facilitates selective molecular interactions for sensing and catalysis. Their multifunctionality and structural versatility position helicenes as promising building blocks for next-generation technologies, including optoelectronics, spintronics, chiral photonics, and enantioselective sensing. This broad applicability continues to drive innovation in molecular design and device engineering, solidifying their role as a cornerstone of modern materials chemistry.

STM and Atomic Force Microscopy (AFM) Studies of Carbohelicenes

Observation of helicene molecules at the single-molecule scale using STM is useful for understanding the molecular-scale processes of molecular recognition. One of the most important challenges is to clearly identify the chirality of adsorbed molecules and to clarify how chiral recognition progresses hierarchically in ordered structures. Numerous studies have reported the adsorption behavior of carbohelicene derivatives on metal surfaces such as Cu(111) [45–51], Ni(111) [52, 53], Au(111), and Ag(111) [54]. In addition, semiconductor surfaces such as TiO_2(110), TiO_2(011) [55], and InSb(001)-c(8 $\times$ 2) [56] can also be used as substrates for STM observation of helicenes. Chiral identification of individual helicene molecules adsorbed on surfaces can be further corroborated through non-contact atomic force microscopy (nc-AFM). There are many reports of nc-AFM using functionalized tips [57–59]. STM has emerged as an indispensable technique for probing helicene molecules at the single-molecule level, offering unprecedented insights into chiral recognition phenomena at the atomic scale. The technical ability to resolve molecular adsorption configurations and intermolecular interactions has significantly advanced our understanding of surface-mediated chirality transfer. However, a persistent challenge remains in the definitive determination of absolute molecular chirality and the mechanistic understanding of how chiral recognition evolves across different length scales—from single molecules to extended two-dimensional assemblies.

STM Studies of Thiaheterohelicene Derivatives

While STM has provided remarkable insights into carbohelicene systems, molecular-scale investigations of thiaheterohelicenes—characterized by their alternating benzene and thiophene units—remain relatively underdeveloped. A critical knowledge gap persists in understanding how the distinctive electronic properties of thiophene rings influence both molecular adsorption behavior and chiral recognition when these molecules interact with metal surfaces. This represents a significant gap in current understanding in the field, given the unique advantages offered by thiaheterohelicene systems. The sulfur atoms in the thiophene rings not only create natural anchoring points for functional group attachment but also modulate the π-electron density, which facilitates chemical modification by lowering the activation energy for functionalization reactions. Moreover, the specific interactions between thiophene moieties and metal substrates can lead to novel adsorption geometries and surface-mediated assembly patterns [60, 61]. The existing STM literature on thiaheterohelicenes, though limited, has revealed several important findings. Studies have successfully demonstrated the identification of molecular chirality in racemic [11]thiaheterohelicene (rac-[11]TH) at the single-molecule level [62], while investigations of [7]thiaheterohelicene derivatives on Au(111) surfaces have uncovered the formation of distinctive zigzag twin-row structures [63, 64]. Research on benzannulated tetrathia[7]helicene has provided insights into hierarchical self-assembly processes across different substrate interfaces [65], and work with dithia-[7]helicene systems has achieved unambiguous chirality determination [66]. Although reports of tunneling current-induced luminescence in these systems remain scarce compared to carbohelicenes, these collective findings demonstrate the valuable role of STM in probing chiral recognition phenomena at the molecular scale [67]. To fully understand these complex systems, our approach integrates STM with complementary analytical techniques. XPS provides crucial information about molecular interactions and bonding configurations, while Raman spectroscopy offers detailed vibrational fingerprints of surface-adsorbed species. This multimodal methodology enables comprehensive investigation of molecular interactions, molecule–substrate bonding, and the dynamic evolution of surface-ordered structures. By focusing on thiaheterohelicene derivatives as our primary chiral platform, we systematically evaluate how structural parameters—particularly molecular dimensions and functional group composition—influence chiral recognition processes. This structure–property relationship analysis provides fundamental insights for designing chiral interfaces with tailored recognition capabilities.

15.2.2 Rac-[5]Thiaheterohelicene and Rac-[7]Thiaheterohelicene

This section presents a comprehensive investigation of the adsorption behavior and stereochemical recognition in racemic mixtures of [5]thiaheterohelicene ([5]TH) and [7]thiaheterohelicene ([7]TH) on Ag(111), focusing on the structural dependence imposed by their differing numbers of aromatic rings. STM enables the unambiguous determination of absolute configuration for individual enantiomers while probing the extent of chiral recognition within periodic molecular assemblies. Complementary XPS studies reveal the nature of molecule–substrate interactions, with the absence of S–Ag bonding signatures confirming physisorption as the dominant adsorption mechanism. To further elucidate the molecular behavior at the interface, we employ SERS in conjunction with density functional theory (DFT) calculations. This combined approach provides vibrational fingerprints of the adsorbed species while enabling theoretical validation of the observed adsorption geometries and intermolecular interactions.

Figure 15.5 presents the molecular structures of (a) rac-[5]TH and (b) rac-[7]TH, illustrated as three-dimensional ball-and-stick models of both (M)- and (P)-type enantiomers after geometry optimization via DFT calculations. The rac-[5]TH molecule is composed of two benzene rings and three thiophene rings, while rac-[7]TH consists

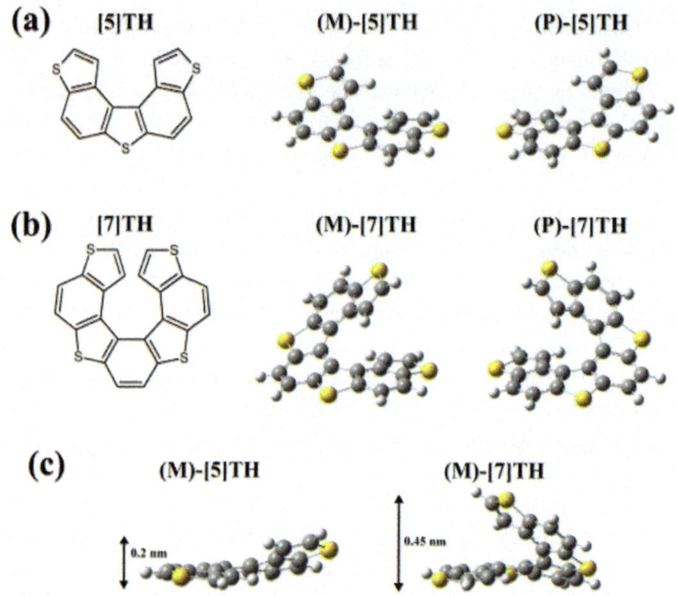

Fig. 15.5 Structures of **a** [5]Thiaheterohelicene and **b** [7]Thiaheterohelicene molecules with three-dimensional ball-and-stick models of (M)- and (P)-type enantiomers. **c** Side views of the molecules [68]

of three benzene and four thiophene rings. In both cases, the structures adopt helical conformations, forming pairs of enantiomers characterized by alternating, fused thiophene and benzene units, with thiophene rings capping both ends. At room temperature, rac-[5]TH rapidly racemizes, preventing the isolation of individual enantiomers, whereas rac-[7]TH is configurationally stable and can be resolved into its enantiomeric forms [69]. Figure 15.5c displays side views of both molecules, highlighting the vertical displacement of terminal sulfur atoms when the structures are aligned based on their central ring.

X-Ray Photoelectron Spectroscopy Analysis of Rac-[5]TH Molecules Deposited on Ag(111) Surface

To gain deeper insight into the interfacial interactions between *rac*-[5]TH molecules and the Ag(111) surface, XPS was employed, with particular attention given to the sulfur 2p core-level region. The goal was to assess whether the thiophene sulfur atoms engage in chemisorption—i.e., formation of S–Ag bonds—or whether they remain physisorbed via weak, non-covalent interactions.

Figure 15.6 presents the S 2p spectra for *rac*-[5]TH deposited on Ag(111) at two nominal film thicknesses: 50 nm (multilayer) and 1 nm (thin film). In the multilayer regime, the S $2p_{3/2}$ and S $2p_{1/2}$ peaks appear at 164.3 eV and 165.5 eV, respectively,

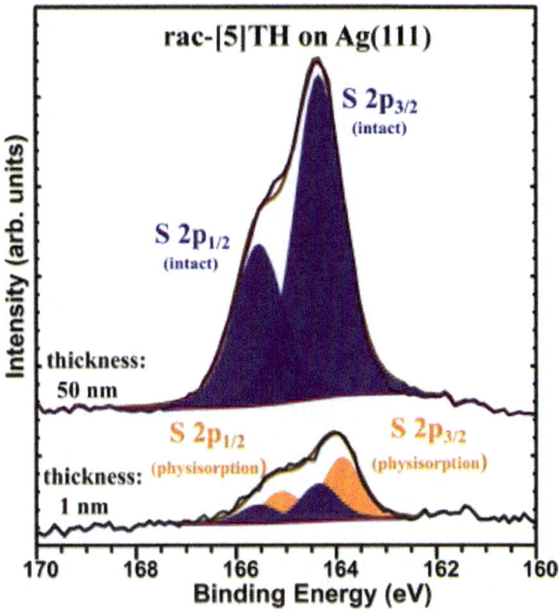

Fig. 15.6 X-ray Photoelectron Spectra (XPS) of the S 2p core level for rac-[5]TH molecules deposited on Ag(111) at 50 nm and 1 nm thicknesses indicates that the interaction between the molecules and the substrate is governed by physisorption, with no evidence of chemical bonding [68]

consistent with thiophene rings in chemically unperturbed environments [70, 71]. This confirms that the helicene backbone remains electronically and structurally intact in the absence of significant substrate interaction. In the 1 nm thin film, four peaks are observed, suggesting two chemically distinct sulfur environments. One set corresponds to unperturbed thiophene sulfur atoms, while the second shows a modest binding energy reduction of ~ 0.4 eV (S $2p_3/_2$ at 163.9 eV), indicating weak electronic interaction with the substrate. This shift is insufficient to indicate covalent bonding, as S–metal chemisorption typically produces much larger shifts of ~ 2.0 eV [70], suggesting that the interaction is dominated by van der Waals (physisorption) forces. The helical geometry of rac-[5]TH introduces vertical displacement of the sulfur atoms from the central plane, as confirmed by side-view molecular models (Fig. 15.5c). This spatial configuration effectively reduces orbital overlap between sulfur lone pairs and surface metal states, which in turn limits charge transfer and bond formation. Moreover, the presence of multiple fused rings leads to rigidity in the backbone, preventing the molecule from adopting a planar conformation favorable for strong adsorption. As such, both geometric and electronic factors combine to suppress chemisorption in favor of weak, dispersion-driven physisorption.

STM Observation of Rac-[5]Thiaheterohelicene Molecules Adsorbed on Ag(111) Surface

Under low coverage conditions, rac-[5]TH molecules could not be reliably imaged by STM on the Ag(111) surface, even when cooling the system from room temperature down to 80 K. This absence is attributed to the high surface mobility of rac-[5]TH molecules, which prevents aggregation or the formation of two-dimensional islands. These observations indicate a weak interaction between rac-[5]TH and the Ag(111) surface. Moreover, the strong interaction between the STM tip and the individual molecules leads to their displacement during scanning. Despite the presence of thiophene rings, rac-[5]TH does not appear to form strong chemical bonds with the silver substrate. The lack of island formation even at low temperatures further supports the presence of repulsive intermolecular interactions at sub-monolayer coverage [72]. Similar behavior has been previously observed for carbohelicene molecules, where weak molecule–substrate interactions hinder the formation of ordered STM images, even at temperatures as low as 60 K [66].

As the coverage increased, rac-[5]TH molecules began to form a variety of ordered structures. At sub-monolayer coverage, a low-density honeycomb-like arrangement was observed (Fig. 15.7). This structure is oriented along the Ag(111) $\langle 11\overline{2} \rangle$ direction, indicating a commensurate relationship between the molecular overlayer and the substrate lattice (Fig. 15.7a and b). Each unit cell of the honeycomb consists of six bright protrusions encircling a central dark void. The center-to-center distance between neighboring units is approximately 2.3 nm, while the spacing between adjacent bright protrusions is about 1.2 nm. Each protrusion corresponds to a single rac-[5]TH molecule. The formation of this structure is attributed to a balance between long-range, substrate-mediated repulsive interactions, and weak intermolecular attractions [73]. The estimated molecular density within the honeycomb structure is approximately 0.4 molecules/nm^2. As shown in Fig. 15.7b, point defects—interpreted as missing molecules—are occasionally observed. Due to the random orientations and rotational degrees of freedom of the molecules within this phase, obtaining high-resolution STM images of individual molecules is challenging. A proposed molecular arrangement model is shown in Fig. 15.7c, with the two enantiomers distinguished by color. Figure 15.7d and e present top and side views of the model, respectively, highlighting the differences in molecular dimensions and lattice constants, which are consistent with the presence of both enantiomers within the structural units.

As the coverage increased further, rac-[5]TH molecules self-assembled into several well-ordered structures on the Ag(111) surface, forming regular molecular rows. Figure 15.8a presents two representative structures observed at saturation coverage, hereafter referred to as Structure I and Structure II. As shown in Fig. 15.8b, these domains are clearly separated by distinct boundaries. Structure I consists of molecular rows oriented at an angle of $15 \pm 1°$ with respect to the $\langle 11\overline{2} \rangle$ crystallographic direction of Ag(111). High-resolution STM imaging (Fig. 15.8c) enables unambiguous identification of the absolute configuration (chirality) of individual rac-[5]TH molecules. Within each row, molecules organize into alternating homochiral dimers. In the STM images, individual molecules appear as crescent-shaped protrusions, with the bright ends extending above the surface. Molecules exhibiting clockwise-oriented protrusions correspond to the (P)-enantiomer, while those with counterclockwise features represent the (M)-enantiomer, as color-coded in Fig. 15.8c–e (blue and red, respectively). Each row is composed of repeating

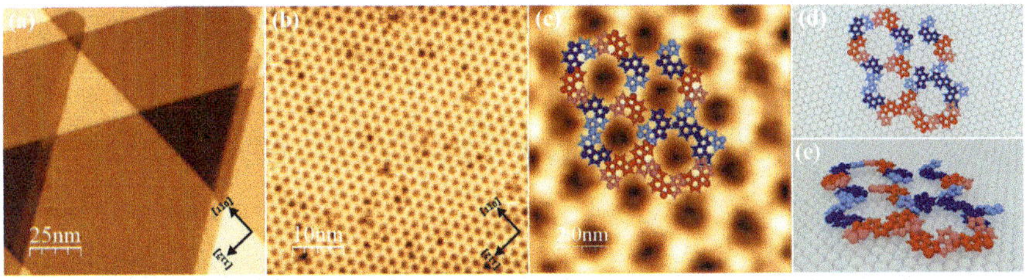

Fig. 15.7 STM images of rac-[5]TH molecules adsorbed on Ag(111) and proposed adsorption model at sub-monolayer coverage: **a** Formation of a low-density honeycomb-like structure aligned with the crystallographic directions of the Ag(111) surface (125×125 nm^2, 0.115 V, 50 pA); **b** Presence of point defects within the honeycomb lattice (50×50 nm^2, 0.15 V, 12 pA); **c** Proposed adsorption model of the honeycomb structure superimposed on the STM image, suggesting that each unit cell contains both (P)- and (M)-enantiomers, shown in blue and red, respectively; **d** Top view and **e** side view of the adsorption model based on the structure observed in (**c**) [68]

homochiral dimers, with adjacent dimers rotated by 180° relative to one another, producing a characteristic alternating pattern.

In contrast, Structure II is composed of molecular rows aligned along the $\langle 11\bar{2} \rangle$ crystallographic direction of the Ag(111) surface. Within this structure, adjacent molecules form heterochiral dimers (P–M or M–P), arranged in a back-to-side configuration with an approximate 90° rotational offset between neighboring molecules. Structure II can be further divided into two distinct substructures, denoted II-1 and II-2. As shown in Fig. 15.8f–k, these substructures differ in the relative orientation of the heterochiral pairs, highlighted by white ellipses. In substructure II-1, the ellipses are horizontally elongated, indicating a lateral alignment of the dimer axes, whereas in II-2, they are vertically elongated, reflecting a perpendicular arrangement. In addition to Structures I and II, several other ordered configurations were observed at saturation coverage, although these appeared less frequently. All identified arrangements can be classified into two categories based on dimer geometry: face-to-face and back-to-side types, depending on the relative orientation and chirality of the interacting molecules.

STM Observation of Rac-[7]Thiaheterohelicene Molecules Adsorbed on Ag(111) Surface

Figure 15.9a shows the STM image of rac-[7]TH molecules adsorbed at saturation coverage on the Ag(111) surface. In the ordered regions, a characteristic self-assembled zigzag twin-row structure is observed, with no detectable structural defects. These twin rows are oriented along one of the three equivalent close-packed directions of the Ag(111) surface, specifically the $\langle 1\bar{1}0 \rangle$ direction. This behavior contrasts with that of rac-[5]TH under similar conditions, where no ordered structures form at sub-monolayer coverage. In the STM images, each circular protrusion corresponds to a single rac-[7]TH molecule. However, distinguishing the chirality of

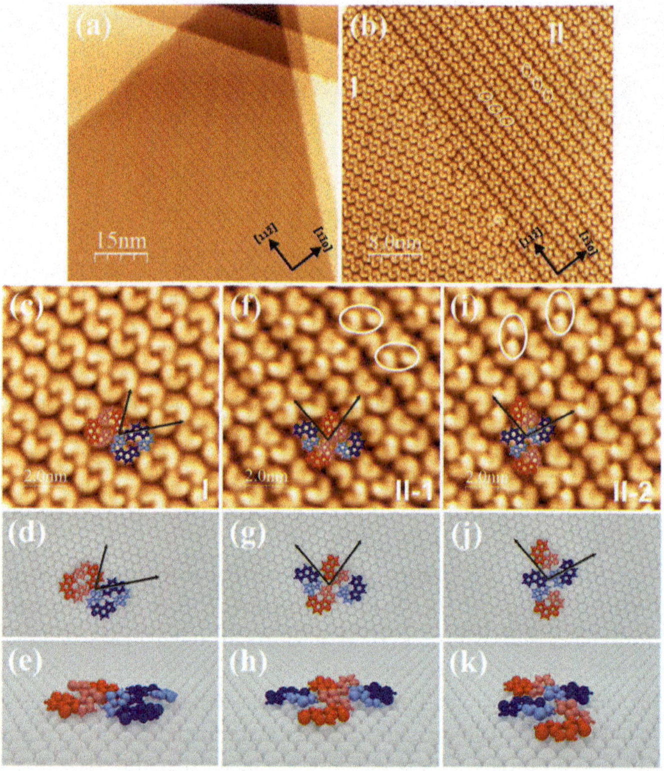

Fig. 15.8 STM images of the two predominant structures (I and II) formed by rac-[5]TH molecules on Ag(111) at saturation coverage and their corresponding adsorption models: **a** Structure I is characterized by parallel molecular rows oriented at an angle of 15 ± 1° with respect to the high-symmetry directions of the Ag(111) surface, whereas Structure II exhibits parallel rows aligned along the crystallographic directions of Ag(111) (75×75 nm^2, 0.5 V, 50 pA); **b** Coexistence of Structures I and II on the Ag(111) surface, separated by a distinct domain boundary (40×40 nm^2, 0.5 V, 50 pA); **c** High-resolution STM image of Structure I reveals the formation of homochiral dimers alternately aligned along the molecular rows (10×10 nm^2, 0.5 V, 50 pA); **d** Top view and **e** side view of the proposed adsorption model of Structure I based on the STM image in (**c**); **f** and **i** Substructures II-1 and II-2 observed within Structure II (10×10 nm^2, 0.5 V, 50 pA); **g** Top view and **h** side view of the adsorption model of the II-1 substructure, based on image (**f**); **j** Top view and **k** side view of the adsorption model of the II-2 substructure, based on image (**i**) [68]

individual molecules remains challenging. The uniformity and long-range order of the zigzag structure suggest that repulsive intermolecular interactions dominate over molecule–substrate interactions, which appear relatively weak. These weak interactions facilitate the formation of the observed highly ordered structures. Each molecule within the zigzag twin-row structure appears as a circular protrusion with an additional central bright spot, attributed to the topmost part of the rac-[7]TH molecule. The average diameter of the protrusions is 1.1 ± 0.1 nm, in good agreement with the theoretically predicted molecular size. STM analysis further reveals that the twin-row structure consists of alternating (M)-type and (P)-type enantiomers, and adjacent zigzag rows adopt distinct molecular conformations. The proposed adsorption model is shown in Fig. 15.9b and c. In this model, rac-[7]TH molecules are arranged in a unit cell with lattice parameters of 1.46 nm × 4.86 nm and an internal angle of 108°. This unit cell is overlaid on the STM image, and each molecule is labeled with its absolute configuration ((M) or (P)). The adsorption geometry assumes that the three thiophene and benzene rings lie nearly parallel to the Ag(111) surface, consistent with a physisorbed state. This adsorption behavior resembles that of benzohelicene molecules on Ag(111) and Au(111) surfaces in racemic mixtures [74]. The self-assembly of rac-[5]TH and rac-[7]TH molecules is governed by a delicate balance between intermolecular interactions—including stereochemical recognition—and molecule–substrate interactions. According to DFT and molecular dynamics (MD) simulations [75], thiaheterohelicene molecules interact with the Ag substrate primarily through van der Waals forces, confirming their physisorbed nature. Key parameters include an adsorption energy of less than 1.8 eV for a single [7]TH molecule, a minimum C–Ag atom distance of ~ 2.9 Å, and no significant dependence of adsorption energy on the adsorption site. MD simulations further indicate that the dominant intermolecular interaction is π–π stacking between terminal thiophene rings. These computational findings support the experimental observation that structural diversity in rac-[5]TH and rac-[7]TH assemblies arises primarily from differences in intermolecular interactions. The three-dimensional shape of helicene molecules leads to a limited contact area with the substrate, affecting both their adsorption and imaging. Notably, the internal structure of [5]TH is more clearly resolved in STM images than that of [7]TH, due to its flatter geometry, which allows better visualization of its helical shape. Additionally, the maximum surface coverage of [5]TH (~0.7 molecules/nm^2) is higher than that of [7]TH, consistent with its smaller molecular footprint. This suggests that [5]TH exhibits weaker stereochemical interactions, allowing for a greater variety of arrangements and enantiomer combinations. In contrast, [7]TH consistently forms the zigzag twin-row structure, reflecting stronger enantioselective interactions. To further understand the mechanism of chiral recognition and assembly, it is essential to compare STM-resolved structures obtained from both racemic and enantiomerically pure compounds. The results for the enantiopure systems are presented in the following sections.

Raman Spectroscopic Analysis of Rac-[7]Thiaheterohelicene Molecules Adsorbed on Ag(111) Surface

Figure 15.10a and b show the Raman spectra of rac-[5]TH and rac-[7]TH molecules, respectively, acquired using 532 nm laser excitation. These experimental spectra are compared with DFT-calculated Raman spectra of the corresponding (M)-type enantiomers. The red curve represent spectra obtained using a technique referred to as trench SERS, which utilizes Ag(111) substrates scratched to form microscale grooves that locally enhance plasmonic fields under laser excitation. This method, a variant of SERS, has the key advantage of enabling Raman measurements in vacuum directly

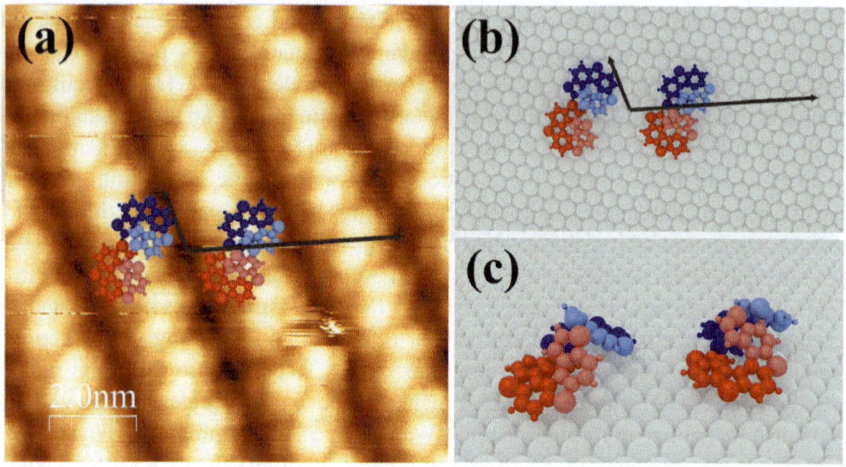

Fig. 15.9 **a** High-resolution STM image of self-assembled zigzagged twin rows, where individual rac-[5]TH molecules appear as circular protrusions with an additional bright spot positioned at the top (10 × 10 nm^2, − 2.6 V, 25 pA). **b** Top view and **c** side view of the proposed adsorption model based on the STM image in (**a**). The unit cell is indicated by black arrows in both (**a**) and (**b**) [68]

on the same samples used for STM imaging. For comparison, red spectra acquired under ambient conditions from thin molecular films are also shown. The black curves correspond to DFT-calculated Raman spectra of isolated (M)-[5]TH and (M)-[7]TH molecules, scaled by a factor of 0.99 o account for systematic frequency overestimation in the DFT calculations. In the experimental spectra, multiple peaks are observed in the 800–1800 cm^{-1} region, corresponding to the in-plane vibrational modes of the benzene and thiophene rings. As shown in Fig. 15.10a, both conventional SERS and trench SERS spectra of rac-[5]TH exhibit qualitatively similar features, indicating that the molecular vibrations are preserved across different measurement environments. Three prominent vibrational modes of (M)-[5]TH, identified in the DFT-calculated spectrum (see Fig. 15.10a), are consistent with the experimental observations. The first mode, near 1334 cm^{-1}, corresponds to C–C breathing (antisymmetric stretching) within the benzene and thiophene rings and is commonly assigned to the D mode, similar to that observed in polycyclic aromatic compounds such as coronene. The second mode, around 1577 cm^{-1}, represents C–C symmetric stretching, or the G mode. The third mode, near 1196 cm^{-1}, is a mixed vibration involving C–H wagging in both thiophene and benzene rings, along with C–C stretching within the benzene ring. A similar analysis applies to rac-[7]TH, as shown in Fig. 15.10b. The DFT-calculated spectrum for (M)-[7]TH (see Fig. 15.10b), when compared with experimental spectra from both conventional and trench SERS, shows only minor differences in intensity, while the key vibrational features are qualitatively consistent. The D mode appears at approximately 1334 cm^{-1}, the G mode near 1563 cm^{-1}, and the third prominent mode at 1194 cm^{-1}, again corresponding to a mixed C–H wagging and C–C stretching motion within the benzene ring. Notably, these SERS spectra differ significantly from the tip-enhanced Raman spectroscopy (TERS) spectra obtained from the well-ordered self-assembled structures of [7]TH-dial, presented in the next section. The observed SERS features suggest that the molecules contributing to the spectra adopt random orientations within localized electromagnetic fields generated by nanoscale surface roughness and defects. In such environments, local surface plasmon excitation during laser irradiation enables enhancement of Raman signals, while the flexible molecular orientations lead to spectra that closely resemble those of isolated, unbound molecules. Importantly, the good agreement between the experimental SERS spectra and DFT-calculated Raman modes confirms that the molecules remained chemically intact during measurement. This indicates that no thermal decomposition occurred under laser irradiation, despite the presence of strong localized fields. The trench SERS approach presented here provides valuable foundational insight for future TERS studies, particularly those aimed at resolving local vibrational signatures and chiral recognition mechanisms at the single-molecule level using plasmonic Ag(111) substrates.

15.2.3 [7]Thiaheterohelicene-2,13-Dicarboxaldehyde

Initial Adsorption and Self-Assembly Behavior on Au(111)

Figure 15.11 shows the molecular structure of rac-[7]hiaheterohelicene-2,13-dicarboxaldehyde ([7]TH-dial) and three-dimensional ball-and-stick models of its (M)- and (P)-enantiomers. When rac-[7]TH-dial molecules are thermally evaporated onto an Au(111) surface, different ordered molecular arrangements are observed depending on the surface coverage.

At low coverage, the adsorption behavior of rac-[7]TH-dial is strongly influenced by the intrinsic herringbone reconstruction of the Au(111) surface. STM reveals individual molecules as bright protrusions with an average apparent diameter of 1.1 ± 0.1 nm, consistent with the theoretical molecular size (~1.0 nm). Under these conditions, molecules preferentially adsorb at the elbow sites of the herringbone reconstruction as well as at step edges. In particular, adsorption predominantly occurs within the fcc regions of the reconstructed surface. This site-selective adsorption is consistent with the previous reports, which attribute the preferential binding of aromatic molecules to the fcc regions of Au(111) to their slightly higher adsorption energy compared to the hcp regions. These differences arise from subtle variations in atomic coordination and local electronic structure between the two domains [64, 76, 77]. As the surface coverage of rac-[7]TH-dial on Au(111) increases, the molecules self-assemble into zigzag-shaped twin-row structures preferentially aligned along the $\langle 11\overline{2} \rangle$ direction—coinciding with the orientation of the Au(111) herringbone reconstruction lines (see Fig. 15.12). Notably, these zigzag twin-row assemblies resemble those observed for racemic hydrocarbon-type heptahelicenes on Au(111). However, a key difference lies in the preferential adsorption direction: rac-[7]TH-dial aligns along $\langle 11\overline{2} \rangle$, whereas hydrocarbon heptahelicenes typically align along

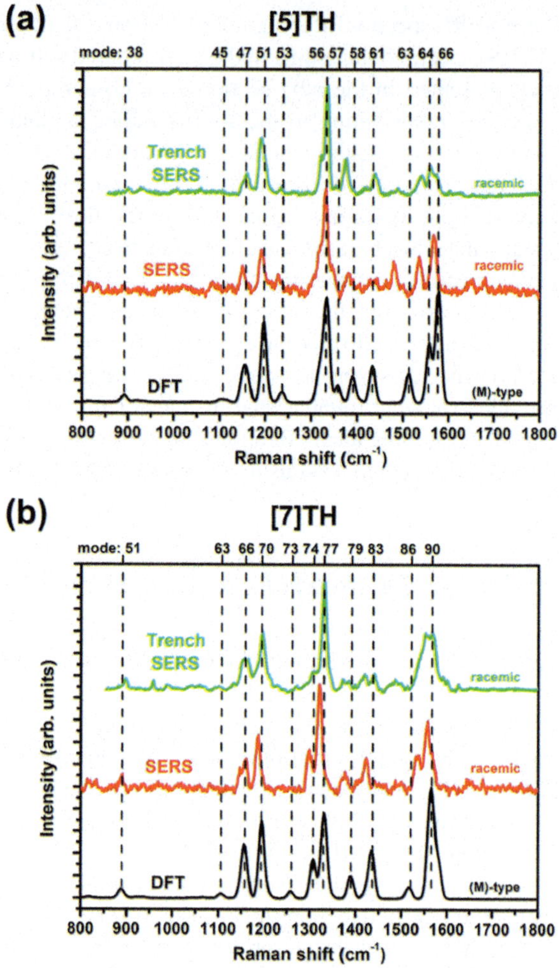

Fig. 15.10 **a** Conventional SERS and trench SERS spectra of [5]TH and **b** [7]TH molecules adsorbed on Ag(111), compared with DFT-calculated Raman spectra. The theoretical spectra were obtained using the B3LYP hybrid density functional in combination with the 6–311 + + G(d,p) basis set, and scaled by a factor of 0.99 to match the experimental data [68]

the $\langle 1\bar{1}0 \rangle$ direction. This distinction is attributed to the presence of thiophene rings and aldehyde functional groups in rac-[7]TH-dial, which play critical roles in modulating surface interactions and chiral recognition. Zigzag twin-row arrangements are commonly reported for helicene derivatives functionalized with polar groups on Au(111) surfaces [78].

Enantiomeric Arrangement and Zigzag Twin-Row Motifs

In the STM topography shown in Fig. 15.12c, transient noise spikes are observed between molecular rows. These are caused by strong tip–molecule interactions

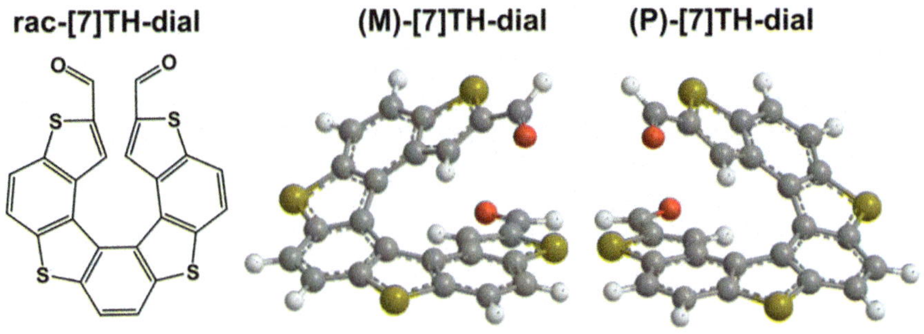

Fig. 15.11 Molecular structure of [7]thiaheterohelicene-2,13-dicarboxaldehyde ([7]TH-dial) and three-dimensional ball-and-stick models of its enantiomers. The chemical structure is shown alongside 3D representations of the (P)- and (M)-enantiomers, illustrating their helical geometry and stereochemistry [67]

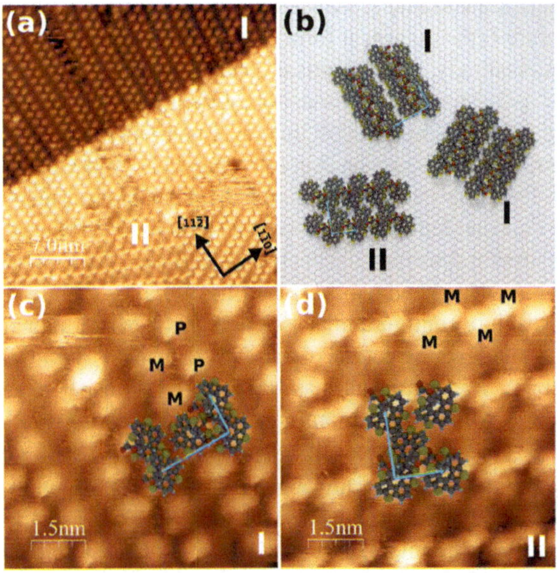

Fig. 15.12 STM images of well-ordered structures formed by rac-[7]TH-dial molecules on Au(111) at monolayer saturation coverage: **a** Dominant structure I consists of self-assembled zigzag twin rows preferentially aligned along the $\langle 11\overline{2}\rangle$ directions of the Au(111) surface. A less frequently observed structure, labeled II, is composed of unseparated twin rows tilted by 22 ± 2° from the $\langle 1\overline{1}0\rangle$ directions (35 × 35 nm^2, 1.0 V, 50 pA). **b** Proposed adsorption models for structures I and II, illustrating the molecular arrangement of rac-[7]TH-dial on Au(111). **c** High-resolution STM image of structure I, revealing zigzag twin rows composed of alternating (M)- and (P)-enantiomers (7.5 × 7.5 nm^2, 1.0 V, 50 pA). **d** High-resolution STM image of a homochiral domain of structure II, consisting exclusively of (M)-enantiomers (7.5 × 7.5 nm^2, 1.0 V, 50 pA) [67]

leading to local diffusion or reorientation of molecules during scanning. The relatively high mobility of rac-[7]TH-dial molecules on Au(111)—despite the presence of four thiophene rings capable of interacting with the gold surface—suggests that adsorption is primarily governed by physisorption, without significant chemical bonding [79]. Zigzag twin-row motifs are frequently observed for both racemic and enantiopure helicenes adsorbed on metal substrates [54]. Identifying the absolute configuration of surface-bound helicenes is essential for understanding enantiospecific intermolecular interactions and the mechanisms underlying chiral recognition. From high-resolution STM images (e.g., Fig. 15.12c), the absolute configuration of individual rac-[7]TH-dial molecules can be inferred: each molecule appears as a protrusion with a brighter lobe, corresponding to the topmost part of the helicene. If the protrusion curls clockwise from the bright end, the molecule is assigned as a (P)-enantiomer; if counterclockwise, as an (M)-enantiomer. STM analysis suggests that the dominant zigzag twin-row structure (Structure I) is composed of alternating (M)- and (P)-enantiomers. Two distinct types of twin-row configurations are observed: Structure I (the dominant form), and Structure II, which is less frequently encountered. Interestingly, even identical enantiomers can adopt different orientations within the same structure, demonstrating flexibility in packing arrangements. Racemic zigzag arrangements with alternating enantiomers have also been reported for hydrocarbon heptahelicenes on Au(111), Ag(111), and Cu(111) surfaces [51, 74]. Although STM allows identification of helical handedness, its resolution can be limited. Therefore, complementary non-contact atomic force microscopy (nc-AFM) is of interest for structural verification of chiral adsorbates. While nc-AFM is sensitive to instrumental parameters and requires careful calibration, it offers a powerful tool for confirming absolute configurations assigned via STM. As shown in Fig. 15.12a, a well-ordered monolayer structure of rac-[7]TH-dial on Au(111) is observed at saturation coverage. The dominant phase (Structure I) consists of zigzag twin rows aligned along $\langle 11\overline{2} \rangle$, while the less common Structure II consists of unresolved twin rows tilted by $22 \pm 2°$ from the $\langle 1\overline{1}0 \rangle$ direction. The proposed adsorption model for rac-[7]TH-dial on Au(111) is shown in Fig. 15.12b. In Structure I (Fig. 15.12c), molecules are arranged within a unit cell defined by lattice constants of 2.6 ± 0.1 nm and 1.4 ± 0.1 nm, with an internal angle of $90 \pm 1°$. This structure is composed of alternating (M)- and (P)-enantiomers. In contrast, Structure II (Fig. 15.12d) forms an enantiopure domain composed exclusively of (M)-enantiomers. The unit cell for Structure II is defined by lattice constants of 2.0 ± 0.1 nm and 1.5 ± 0.1 nm, with a similar internal angle of $90 \pm 1°$. These findings highlight the role of stereochemical interactions and functional group contributions in directing the self-assembly and enantioselective behavior of helicene-based molecules on metal surfaces. As the surface coverage of rac-[7]TH-dial on Au(111) increases, a zigzag-shaped twin-row self-assembled structure aligned along the $\langle 11\overline{2} \rangle$ direction—that is, along the lines of the herringbone reconstruction—is formed (see Fig. 15.12). Notably, such zigzag twin-row structures are similar to those observed upon deposition of racemic hydrocarbon-type heptahelicenes on Au(111). However, it should be emphasized that the rac-[7]TH-dial molecules preferentially adsorb along the $\langle 11\overline{2} \rangle$ direction, unlike hydrocarbon heptahelicenes, which typically align along the $\langle 1\bar{1}0 \rangle$ direction. This difference is attributed to the presence of thiophene rings and aldehyde functional groups in rac-[7]TH-dial, which play crucial roles in adsorption and molecular recognition on Au(111). Zigzag twin-row structures composed of racemic helicene molecules are generally observed for helicene derivatives with various functional groups on Au(111). In the STM topography image shown in Fig. 15.12c, noise spikes appear between molecular rows, due to strong interactions between the STM tip and molecules during measurement, resulting in lateral diffusion or reorientation of the molecules. The high molecular mobility on Au(111), despite the presence of four thiophene rings in the helical framework capable of chemical interaction with gold atoms, suggests the absence of chemical bonding between rac-[7]TH-dial and the Au(111) surface [79].

Saturation-Coverage Structures and High-Resolution STM Analysis

Figure 15.13 presents the highly ordered self-assembled structure (Structure III) of rac-[7]TH-dial molecules adsorbed on Au(111) under saturation coverage conditions. Structure III is characterized by zigzag-shaped twin rows tilted by $8 \pm 2°$ from the $\langle 1\bar{1}0 \rangle$ crystallographic direction, forming a monolayer with clearly defined domain boundaries. Except for occasional missing molecules, the domain exhibits nearly defect-free continuity. The underlying herringbone reconstruction of the Au(111) substrate remains visible, indicating that molecule–substrate interactions are relatively weak and dominated by van der Waals forces—sufficiently mild to preserve the surface reconstruction. The high degree of molecular ordering in Structure III enabled detailed high-resolution STM imaging, as shown in Fig. 15.13b and c. The observed molecular contrast and shape exhibit strong dependence on the applied bias voltage. At 0.5 V (Fig. 15.13b), each molecule appears as a circular protrusion with an additional bright feature located near one end, attributed to the aldehyde functional group. The average diameter of these protrusions is 1.1 ± 0.1 nm, with an apparent height of 0.25 ± 0.05 nm. Notably, molecules in adjacent rows display mirrored contrast features, indicating a heterochiral arrangement—specifically, alternating (M)- and (P)-enantiomers along the zigzag twin row. Such heterochiral molecular pairing is generally more energetically favorable and is the driving force behind the observed racemic twin-row assembly. Despite the high degree of structural order, the molecular rows do not exhibit lattice registry with the Au(111) substrate. A subtle spacing between the twin rows is consistently observed, further supporting the presence of weak intermolecular and molecule–substrate interactions. An adsorption model derived from STM image analysis is shown in Fig. 15.13d, with the molecules arranged in a rectangular unit cell defined by lattice parameters of 2.4 ± 0.1 nm and 1.5 ± 0.1 nm, and a lattice angle of $90 \pm 1°$. The overlaid model in Fig. 15.13e highlights the mirror-symmetric positioning of bright protrusions, facilitating the assignment of absolute configurations based on their handedness. Each molecule appears to adsorb with its helicene axis oriented perpendicular or slightly tilted relative to the Au(111) surface. When the coverage is further increased, the rac-[7]TH-dial layer transitions from a self-assembled zigzag twin-row to a single-row structure and is characterized by the formation of wide domains. Nonetheless, the herringbone reconstruction of the Au(111) substrate is maintained, and the interaction with the surface remains weak. Although the long-range ordered zigzag pattern disappears, the single-row structure is suggested to consist of dimers formed by two enantiomeric molecules.

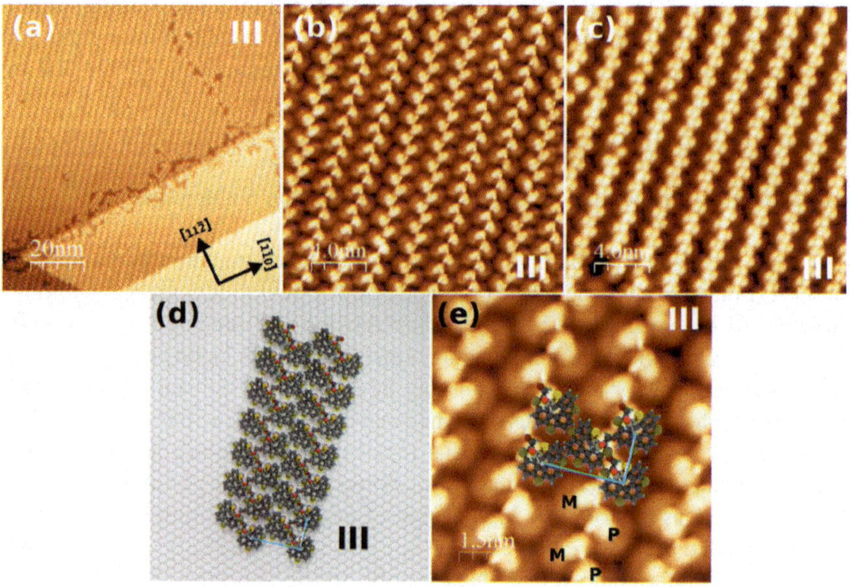

Fig. 15.13 STM images and adsorption model of the highly ordered monolayer of rac-[7]TH-dial molecules on Au(111) at saturation coverage (Structure III). **a** Large-area STM image showing self-assembled zigzagged twin rows tilted by $8 \pm 2°$ from the $\langle 1\bar{1}0 \rangle$ crystallographic direction, forming a well-ordered monolayer (100×100 nm^2, 0.5 V, 50 pA). The appearance of individual rac-[7]TH-dial molecules was strongly dependent on the applied bias voltage. **b, c** Bias-dependent STM images of the zigzag twin rows acquired at 0.5 V and 2.5 V, respectively (20×20 nm^2, 50 pA), illustrating the significant voltage influence on molecular contrast and apparent morphology. **d** Adsorption model for Structure III, showing rac-[7]TH-dial molecules forming heterochiral zigzag twin rows on Au(111). **e** High-resolution STM image revealing alternating (M)- and (P)-enantiomers within a single row, visualized as mirror-symmetric crescent-shaped features (7.5×7.5 nm^2, 0.5 V, 50 pA) [67]

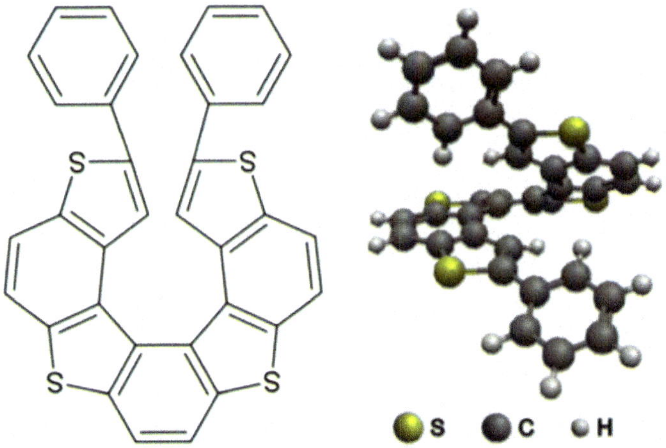

Fig. 15.14 Chemical structure of the (M)-enantiomer of 2,13-diphenyl[7]thiaheterohelicene ((M)-Ph-[7]TH) and its optimized gas-phase geometry shown as a ball-and-stick model [80]

15.2.4 (M)-Type 2,13-Diphenyl[7]Thiaheterohelicene

Formation of a Hexagonal Self-Assembled Monolayer on Ag(111)

Figure 15.14 shows the molecular structure of the (M)-type 2,13-diphenyl ([7]thiaheterohelicene (Ph-[7]TH), along with a ball-and-stick model of the optimized structure of the free molecule. When enantiopure (M)-Ph-[7]TH was deposited onto an Ag(111) substrate heated to 468 K, a hexagonal self-assembled monolayer covering the entire substrate surface was formed (Fig. 15.15a). In the STM image acquired under a bias voltage of + 3.0 V, six protrusions were observed inside each hexagonal unit. This structure is hereafter referred to as the "hexagonal structure." The hexagonal structures were aligned along the $\langle 11\overline{2} \rangle$ direction. Despite precise control over the amount of (M)-Ph-[7]TH deposited, no other self-assembled structures were observed. The distance between neighboring hexagonal structures was 3.48 ± 0.14 nm, and the molecular density was 0.58 ± 0.01 molecules nm^{-2}. The blue triangle connects the three brightest protrusions, with its edges aligned parallel to the $\langle 1\overline{1}0 \rangle$ direction. In the areas between hexagonal structures, faintly brighter regions were also observed, one of which is indicated by a white rectangle in the figure.

Electronic Structure and Voltage-Dependent STM Contrast

Meanwhile, the STS spectrum showed two peaks at − 2.6 V and + 1.3 V. The electronic structure of the free (M)-Ph-[7]TH molecule was evaluated by DFT calculations, where the highest occupied molecular orbital (HOMO) and lowest occupied molecular orbital (LUMO) energy levels were found to be − 5.667 eV and − 2.220

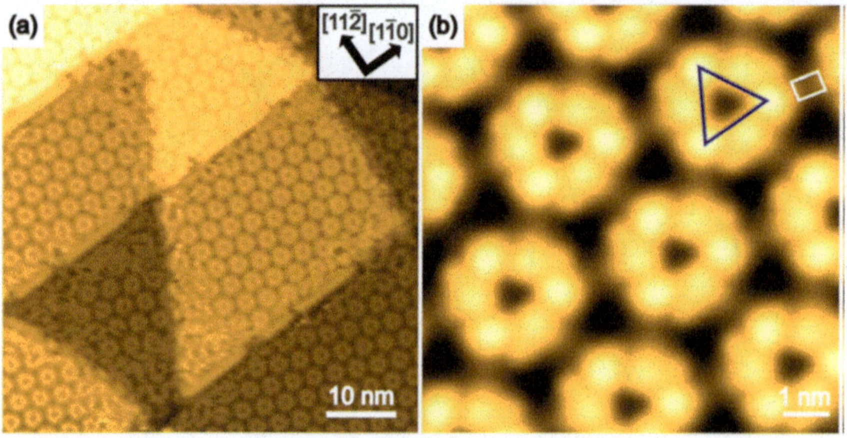

Fig. 15.15 STM images of the self-assembled (M)-Ph-[7]TH structure formed on the Ag(111) surface, acquired at a bias voltage of 3.0 V and a tunneling current of 50 pA: **a** overview image and **b** magnified view. The blue triangle in panel **b** marks the three brightest protrusions within a hexagonal unit, while the white rectangle highlights a bright region located between adjacent hexagonal assemblies [80]

eV, respectively. Based on these DFT results, the origin of the protrusions in the STM image is considered. The energy gap between the two STS peaks is 3.9 V, which is slightly larger than the HOMO–LUMO gap of 3.447 eV from DFT. It has been previously mentioned that DFT tends to underestimate the HOMO–LUMO gap compared to experimental values. The STS peaks at − 2.6 V and + 1.3 V correspond to the HOMO and LUMO, respectively. Due to the limited energy resolution of STS, multiple orbital energies overlap, making it difficult to resolve each orbital individually. To clarify the positions of the bright protrusions in (M)-Ph-[7]TH molecules within the STM images, the voltage dependence of the STM contrast was measured. STM images of the hexagonal structure obtained at + 3.0 V, + 2.5 V, + 2.0 V, and + 0.5 V are shown in Fig. 15.16. At + 3.0 V, six bright protrusions were observed, alternating in brightness. At + 2.5 V, three new protrusions appeared outside the hexagon, bringing the total to nine. The three brightest protrusions at this voltage matched the positions of those at + 3.0 V and are indicated by a blue triangle. At + 2.0 V, as shown in Fig. 15.16c, the three brightest protrusions at + 2.5 V split into two each, resulting in twelve observed protrusions. At this voltage, three new brightest protrusions appeared, not clearly visible at + 3.0 V. The triangle connecting these brightest points was tilted 31° counterclockwise with respect to the $\langle 1\bar{1}0 \rangle$ direction, indicated in Fig. 15.16c by a blue triangle. At + 1.5 V and + 1.0 V, twelve bright protrusions were similarly observed. At + 0.5 V, six elliptical protrusions were observed, but their positions differed from those at + 3.0 V. The triangle connecting the three brightest protrusions at this voltage, as shown in the blue triangle, was tilted 25° counterclockwise with respect to the $\langle 1\bar{1}0 \rangle$ direction, slightly different from that at + 2.0 V.

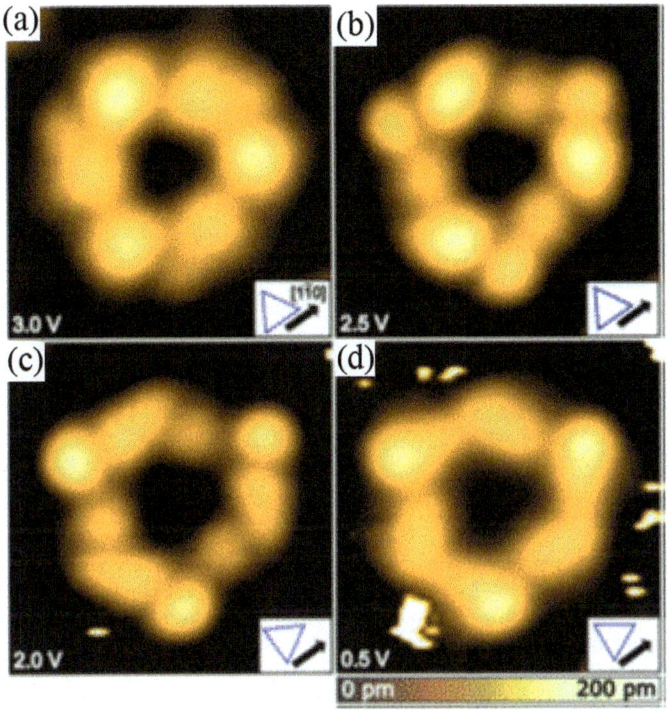

Fig. 15.16 Bias-dependent STM images of (M)-Ph-[7]TH acquired at sample bias voltages of **a** + 3.0 V, **b** + 2.5 V, **c** + 2.0 V, and **d** + 0.5 V over an area of 3.6 × 3.6 nm^2. All images were recorded at a tunneling current of 50 pA. The black arrow in the inset denote the $\langle 1\bar{1}0 \rangle$ crystallographic direction, while the blue triangles mark the orientations of the three brightest protrusions observed in each image [80]

This voltage dependence of the STM images can be explained as follows. According to DFT calculations, the energy difference between the LUMO and LUMO + 4 was 1.549 eV. When the STM image is acquired at + 3.0 V, which is 1.7 V higher than the LUMO peak (+1.3 V), the phenyl groups appear as protrusions. The previous section indicated that three aromatic rings of [7] helicene align parallel to the substrate and are physisorbed via van der Waals interactions, a conclusion supported by both DFT and molecular dynamics (MD) calculations. Accordingly, it is reasonable to assume that part of the helicene core and the phenyl groups in (M)-Ph-[7]TH are also oriented nearly parallel to the substrate and adsorbed via van der Waals forces. Based on this discussion, the surface adsorption model shown in Fig. 15.17 is proposed. In contrast, at + 0.5 V, the upper surface of the helicene core appears as protrusions. Thus, comparing theory and the voltage dependence of STM images is one of the most effective methods for constructing molecular models from STM data. In the model shown in Fig. 15.12b, the distance between phenyl groups and helicene cores of neighboring molecules alternates, resulting in brightness differences at + 3.0 V. The distance from the helicene core to adjacent phenyl groups was

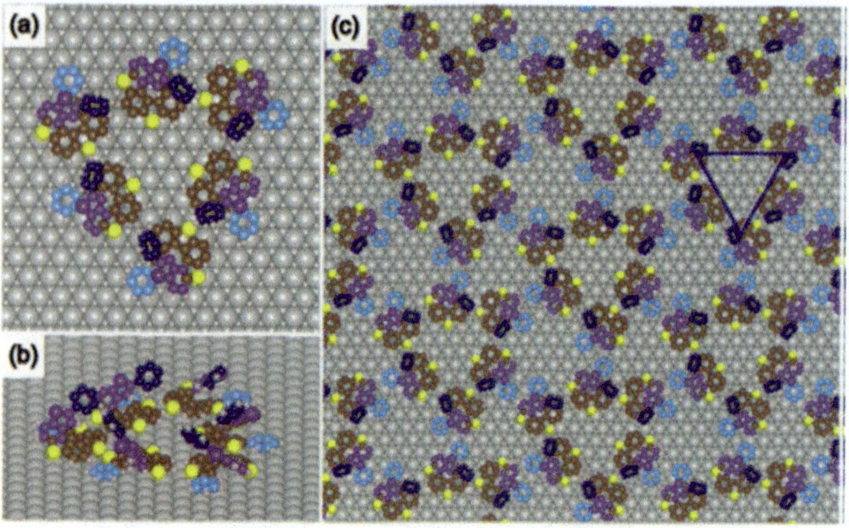

Fig. 15.17 Schematic representations of the distorted hexagonal arrangement of (M)-Ph-[7]TH viewed from **a** the top, **b** the side, and **c** a large-area perspective. Carbon, sulfur, hydrogen, and silver atoms are depicted as brown, yellow, white, and gray spheres, respectively. The blue triangle in panel **c** corresponds to the triangle defined by the three brightest protrusions observed in the STM image acquired at + 3.0 V (Fig. 15.4a). Dark blue and light blue colors denote the outer phenyl rings and those positioned closer to the Ag(111) surface, respectively, while the uppermost segment of the helicene backbone is highlighted in purple [80]

9.7 Å and 10.2 Å, respectively significant difference. The overlap of neighboring protrusions contributes to brightness variations, resulting in non-uniform hexagonal shapes.

Substrate-Driven Assembly and TERS Characterization

The schematic model of the self-assembled hexagonal structure shown in Fig. 15.17 can be described in matrix notation $\begin{pmatrix} 14 & 7 \\ 7 & -7 \end{pmatrix}$ with respect to the Ag(111) substrate. This suggests that (M)-Ph-[7]TH maintains commensurability with the Ag(111) surface.

As mentioned in the previous sections on thiaheterohelicene derivatives, various [7]carbohelicene derivatives have been reported to form zigzag arrangements where helicene molecules face each other. This arrangement arises from intermolecular interactions between the upper faces of adjacent molecules, as confirmed by DFT calculations. However, in Ph-[7]TH, which contains relatively bulky diphenyl groups, the phenyl groups close to the substrate prevent the helicene cores from approaching closely enough to enable π–π interactions, and therefore, no zigzag arrangement is formed. At low coverage, (M)-Ph-[7]TH molecules adsorbed randomly on the surface, and no hexagonal structure was formed. This indicates that the intermolecular interactions in (M)-Ph-[7]TH are weak and that the adsorption behavior is more strongly influenced by interactions with the substrate. In other words, van der Waals interactions between the terminal phenyl groups and the substrate, as well as between the helicene core and the substrate, dominate the adsorption behavior. As a result, a hexagonal structure reflecting the threefold symmetry of the substrate is formed—markedly different from other thiahelicene derivatives described earlier. Such substrate-symmetry-driven structures have also been reported for (M)-type [7]carbohelicenes, their derivatives, and oxygen-containing heterohelicenes. The variation in brightness of protrusions in the STM image at + 3.0 V indicates that the six protrusions in (M)-Ph-[7]TH form a distorted hexagon. The formation of such distortions is due to steric hindrance and van der Waals interactions, driven by a delicate balance between molecule–substrate and intermolecular interactions. Further, theoretical calculations are required to clarify this unique hexagonal structure.

To directly correlate this substrate-driven assembly with its local vibrational response, TERS investigation was performed. TERS spectra were obtained for (M)-Ph-[7]TH molecules that formed a self-assembled hexagonal structure, as shown in Fig. 15.18a. The blue line represents the near-field TERS spectrum acquired when the STM tip was positioned within the tunneling region, whereas the orange line corresponds to the far-field Raman scattering spectrum obtained when the distance between the tip and the substrate was 200–400 nm. Multiple peaks were observed only in the near-field spectrum within the range of 800–1800 cm^{-1}, indicating an enhancement of Raman scattering near the tip. It was also confirmed that the self-assembled structure remained intact before and after the TERS measurement. The Raman spectrum was reproduced using calculations based on DFT, as shown in Fig. 15.18a. For detailed assignments of vibrational modes, please refer to the supporting literature. Most of the Raman-active modes of the (M)-Ph-[7]TH molecule corresponds to in-plane vibrations of benzene rings, both within the helicene framework and in the phenyl substituents. The most intense peak, observed around 1320 cm^{-1}, originates from the C–C stretching mode of the side-chain benzene rings within the [7]TH framework. The identification of other peaks follows the same approach as in the previous section on [7]helicene. Additionally, peaks attributed to the phenyl functional group were also observed. Specifically, the peaks at approximately 996 cm^{-1}, 1489 cm^{-1}, and 1609 cm^{-1} correspond to the C–C stretching mode, the C–H rocking mode, and an in-plane vibrational mode of the phenyl group, respectively. These results strongly emphasize that vibrational spectra serve as molecular fingerprints. Furthermore, the TERS results showed excellent agreement with the DFT calculations in terms of both peak positions and relative intensities. In particular, the phenyl-derived peaks at 996 cm^{-1}, 1489 cm^{-1}, and 1609 cm^{-1} matched well with experimental observations. These findings indicate that (M)-Ph-[7]TH molecules, even under the intense local electromagnetic fields generated by localized surface plasmon excitations and strong laser irradiation, remain stably adsorbed on the Ag(111) substrate without undergoing thermal decomposition or detachment of the functional groups. Additionally, the position and intensity of peaks in the TERS spectrum of Ph-[7]TH are not significantly modulated by molecular adsorption or

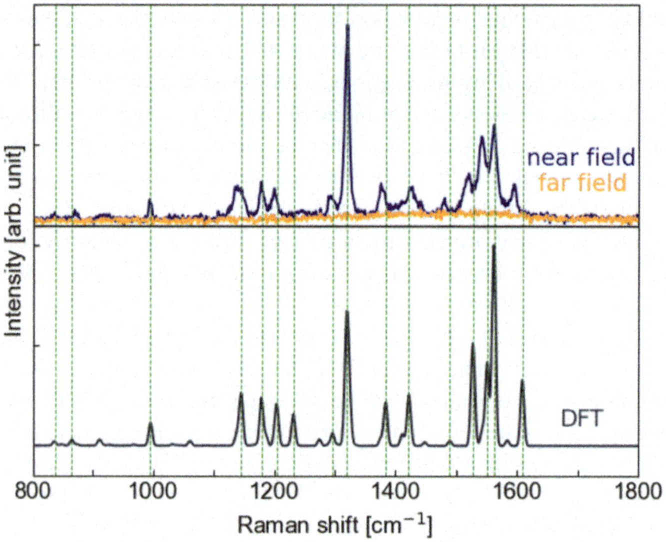

Fig. 15.18 (**a** TERS spectra acquired from the self-assembled hexagonal structure of (M)-Ph-[7]TH. Spectra recorded in the near-field and far-field configurations are shown in blue and orange, respectively. Near-field measurements were performed at a sample bias of + 50 mV and a tunneling current of 2 nA. The Raman spectrum of an isolated (M)-Ph-[7]TH molecule calculated using density functional theory (DFT) is displayed as a black curve. For comparison with the experimental data, the calculated vibrational frequencies were scaled by a factor of 0.982 to align the prominent peak at 1320 cm^{-1}. The green dotted lines indicate Raman peak positions obtained from the DFT calculations [80]

self-assembly. This result is consistent with the behavior of other helicene derivatives, suggesting that [7]TH molecules are physisorbed on the Ag(111) substrate entirely via van der Waals interactions.

15.3 Conclusions

This chapter has introduced the fundamental principles of STM, a technique that enables real-space observation of solid surfaces with atomic-scale resolution, together with two in situ optical methods integrated into STM operation: STM–TERS. While STM provides direct access to local electronic states at surfaces, it is often insufficient by itself to fully describe the physical and chemical properties of nanoscale structures and functional molecules adsorbed on solid substrates. STM–TERS address this limitation by synchronizing optical emission and vibrational analyses with STM probe scanning, thereby allowing spectroscopic information to be obtained with atomic and molecular spatial resolution. These techniques are now well established as powerful extensions of STM for nanoscale characterization.

As a representative example, this chapter has focused on the self-assembled structures of thiaheterohelicene derivatives and has shown that the combined STM-based approach is highly effective for analyzing chiral recognition at surfaces. Although helicenes are structurally simple molecules defined by their helical frameworks, they exhibit a surprisingly rich variety of periodic self-assembled structures, depending on molecular length and the presence or nature of functional groups. Chiral recognition in these systems arises from a delicate interplay of steric effects between neighboring molecules, intermolecular interactions such as van der Waals forces and Coulombic interactions including hydrogen bonding, and interactions between the molecules and the substrate. Together, these factors cooperatively give rise to distinct and characteristic molecular arrangements. It is important to emphasize that reliable determination of molecular packing and orientation cannot be achieved by STM imaging alone. A comprehensive understanding instead requires the integration of STM observations with complementary analyses of electronic states, vibrational spectroscopy, and theoretical modeling.

15.4 Outlook

Homochirality, a defining characteristic of biological systems, exhibits a hierarchical organization spanning length scales from individual molecules to entire organisms. Evaluating chirality at each level of this hierarchy naturally requires analytical approaches that are appropriate to the corresponding scale. Traditionally, the relationship between molecular chirality and optical activity has been investigated using optical rotation and circular dichroism measurements. However, because these methods rely on visible-light probes, the information obtained is fundamentally limited by the diffraction limit and therefore remains macroscopic in nature. In contrast, the identification of structural isomers and the analysis of chiral recognition mechanisms at the single-molecule level are most directly achieved through real-space observation of molecular conformations using probe microscopy techniques, as demonstrated throughout this chapter. Although the helicene derivatives discussed here are artificially synthesized chiral molecules and do not directly represent biomolecular systems, the chiral recognition processes analyzed provide a simple yet effective molecular model for gaining insight into chiral recognition phenomena in biological environments. From the perspective of optical asymmetry, STM-TERS, which exploit strong electromagnetic field enhancement in localized plasmonic environments, can be regarded as ultimate nanoscale tools for probing optical activity. Molecular-scale observation and evaluation of optical isomerism offer insightful perspectives on chiral recognition processes across different hierarchical levels. Looking ahead, continued advances in the theoretical description of light–matter interactions are expected to deepen our understanding of natural optical activity and spin-dependent optical phenomena. Such progress will also facilitate the discovery and interpretation of novel optical effects in which chirality and spin are intrinsically coupled, thereby opening new directions in chiral nanoscience and molecular optics.

References

1. Binnig G, Rohrer H, Gerber C, Weibel E. Surface studies by scanning tunneling microscopy. Phys Rev Lett. 1982;49(1):57–61. https://doi.org/10.1103/PhysRevLett.49.57.
2. Binnig G, Rohrer H. Scanning tunneling microscopy—from birth to adolescence. Rev Mod Phys. 1987;59(3):615–25. https://doi.org/10.1103/RevModPhys.59.615.
3. Tersoff J, Hamann DR. Theory and application for the scanning tunneling microscope. Phys Rev Lett. 1983;50(25):1998–2001. https://doi.org/10.1103/PhysRevLett.50.1998.
4. Tersoff J, Hamann DR. Theory of the scanning tunneling microscope. Phys Rev B. 1985;31(2):805–13. https://doi.org/10.1103/PhysRevB.31.805.
5. Kowalczyk PJ, Puchalski M, Kozłowski W, Dąbrowski P, Klusek Z, Olejniczak W. Investigation of the Shockley surface state on clean and air-exposed Au(111). Appl Surf Sci. 2008;254(15):4572–6. https://doi.org/10.1016/j.apsusc.2008.01.056.
6. Madhavan V, Chen W, Jamneala T, Crommie MF, Wingreen NS. Tunneling into a single magnetic atom: spectroscopic evidence of the Kondo resonance. Science. 1998;280(5363):567–9. https://doi.org/10.1126/science.280.5363.567.
7. Merino J, Gunnarsson O. Role of surface states in scanning tunneling spectroscopy of (111) metal surfaces with Kondo adsorbates. Phys Rev Lett. 2004;93(15): 156601. https://doi.org/10.1103/PhysRevLett.93.156601.
8. Fischer Ø, Kugler M, Maggio-Aprile I, Berthod C, Renner C. Scanning tunneling spectroscopy of high-temperature superconductors. Rev Mod Phys. 2007;79(1):353–419. https://doi.org/10.1103/RevModPhys.79.353.
9. Bakkers EPAM, Hens Z, Zunger A, Franceschetti A, Kouwenhoven LP, Gurevich L, Vanmaekelbergh D. Shell-tunneling spectroscopy of the single-particle energy levels of insulating quantum dots. Nano Lett. 2001;1(10):551–6. https://doi.org/10.1021/nl015572b.
10. Crommie MF, Lutz CP, Eigler DM. Confinement of electrons to quantum corrals on a metal surface. Science. 1993;262(5131):218–20. https://doi.org/10.1126/science.262.5131.218.
11. Teeter JD, Miller DP, Müllegger S. Stable π radical BDPA: adsorption on Cu(100) and survival of spin. ChemPhysChem. 2025;26(2): e202400852. https://doi.org/10.1002/cphc.202400852.
12. Roushan P, Seo J, Parker CV, Hor YS, Hsieh D, Qian D, Richardella A, Hasan MZ, Cava RJ, Yazdani A. Topological surface states protected from backscattering by chiral spin texture. Nature. 2009;460(7259):1106–9. https://doi.org/10.1038/nature08308.
13. Stipe BC, Rezaei MA, Ho W. Single-molecule vibrational spectroscopy and microscopy. Science. 1998;280(5370):1732–5. https://doi.org/10.1126/science.280.5370.1732.
14. Ho W. Single-molecule chemistry. J Chem Phys. 2002;117(24):11033–61. https://doi.org/10.1063/1.1521153.
15. Andrei EY, Li G, Du X. Electronic properties of graphene: a perspective from scanning tunneling microscopy and magnetotransport. Rep Prog Phys. 2012;75(5): 056501. https://doi.org/10.1088/0034-4885/75/5/056501.
16. George A, Fistul MV, Gruenewald M, Kaiser D, Lehnert T, Mupparapu R, Neumann C, Hübner U, Schaal M, Masurkar N, Arava LMR, Staude I, Kaiser U, Fritz T, Turchanin A. Giant persistent photoconductivity in monolayer MoS_2 field-effect transistors. NPJ 2D Mater Appl. 2021;5:15. https://doi.org/10.1038/s41699-020-00182-0.
17. Klusek Z, Dabrowski P, Kowalczyk P, Kozlowski W, Olejniczak W, Blake P, Szybowicz M, Runka T. Graphene on gold: electron density of states studies by scanning tunneling spectroscopy. Appl Phys Lett. 2009;95(11): 113114. https://doi.org/10.1063/1.3231440.
18. Zhang T, Levy N, Ha J, Kuk Y, Stroscio JA. Scanning tunneling microscopy of gate tunable topological insulator Bi_2Se_3 thin films. Phys Rev B. 2013;87(11): 115410. https://doi.org/10.1103/PhysRevB.87.115410.

19. Keimer B, Moore JE. The physics of quantum materials. Nat Phys. 2017;13(11):1045–55. https://doi.org/10.1038/nphys4302.
20. Morgenstern K, Lorente N, Rieder K-H. Controlled manipulation of single atoms and small molecules using the scanning tunnelling microscope. Phys Status Solidi B. 2013;250(9):1671–751. https://doi.org/10.1002/pssb.201248392.
21. Raman CV, Krishnan KS. A new type of secondary radiation. Nature. 1928;121(3048):501–2. https://doi.org/10.1038/121501c0.
22. Long DA. Classical theory of Rayleigh and Raman scattering. In: The raman effect: a unified treatment of the theory of Raman scattering by molecules. Chichester: Wiley; 2002. p. 31–48. https://doi.org/10.1002/0470845767.ch3.
23. Fleischmann M, Hendra PJ, McQuillan AJ. Raman spectra of pyridine adsorbed at a silver electrode. Chem Phys Lett. 1974;26(2):163–6. https://doi.org/10.1016/0009-2614(74)85388-1.
24. Jeanmaire DL, Van Duyne RP. Surface Raman spectroelectrochemistry: part I. heterocyclic, aromatic, and aliphatic amines adsorbed on the anodized silver electrode. J Electroanal Chem. 1977;84(1):1–20. https://doi.org/10.1016/S0022-0728(77)80224-6.
25. Kneipp K, Wang Y, Kneipp H, Perelman LT, Itzkan I, Dasari RR, Feld MS. Single molecule detection using surface-enhanced Raman scattering (SERS). Phys Rev Lett. 1997;78(9):1667–70. https://doi.org/10.1103/PhysRevLett.78.1667.
26. Wessel J. Surface-enhanced optical microscopy. J Opt Soc Am B. 1985;2(9):1538–41. https://doi.org/10.1364/JOSAB.2.001538.
27. Zhang Z, Sheng S, Wang R, Sun M. Tip-enhanced Raman spectroscopy. Anal Chem. 2016;88(19):9328–46. https://doi.org/10.1021/acs.analchem.6b02093.
28. Mauser N, Hartschuh A. Tip-enhanced near-field optical microscopy. Chem Soc Rev. 2014;43(4):1248–62. https://doi.org/10.1039/C3CS60258C.
29. Zhang R, Zhang Y, Dong ZC, Jiang S, Zhang C, Chen LG, Zhang L, Liao Y, Aizpurua J, Luo Y, Yang JL, Hou JG. Chemical mapping of a single molecule by Plasmon-enhanced Raman scattering. Nature. 2013;498(7452):82–6. https://doi.org/10.1038/nature12151.
30. Kazemi-Zanjani N, Vedraine S, Lagugné-Labarthet F. Localized enhancement of electric field in tip-enhanced Raman spectroscopy using radially and linearly polarized light. Opt Express. 2013;21(21):25271–6. https://doi.org/10.1364/OE.21.025271.
31. Stadler J, Oswald B, Schmid T, Zenobi R. Characterizing unusual metal substrates for gap-mode tip-enhanced Raman spectroscopy. J Raman Spectrosc. 2013;44(2):227–33. https://doi.org/10.1002/jrs.4169.
32. Pettinger B, Domke KF, Zhang D, Picardi G, Schuster R. Tip-enhanced Raman scattering: influence of the tip surface geometry on optical resonance and enhancement. Surf Sci. 2009;603(10):1335–41. https://doi.org/10.1016/j.susc.2008.08.033.
33. Richard-Lacroix M, Zhang Y, Dong Z, Deckert V. Mastering high resolution tip-enhanced Raman spectroscopy: towards a shift of perception. Chem Soc Rev. 2017;46(13):3922–44. https://doi.org/10.1039/C7CS00203C.
34. Chiang N, Chen X, Goubert G, Chulhai DV, Chen X, Pozzi EA, Jiang N, Hersam MC, Seideman T, Jensen L, Van Duyne RP. Conformational contrast of surface-mediated molecular switches yields Ångström-scale spatial resolution in ultrahigh vacuum tip-enhanced Raman spectroscopy. Nano Lett. 2016;16(12):7774–8. https://doi.org/10.1021/acs.nanolett.6b03958.
35. Tallarida N, Lee J, Apkarian VA. Tip-enhanced Raman Spectromicroscopy on the Ångström scale: bare and CO-terminated Ag tips. ACS Nano. 2017;11(11):11393–401. https://doi.org/10.1021/acsnano.7b06022.
36. Shen Y, Chen C-F. Helicenes: synthesis and applications. Chem Rev. 2012;112(3):1463–535. https://doi.org/10.1021/cr200087r.
37. Gingras M. One hundred years of helicene chemistry. Part 1: non-stereoselective syntheses of carbohelicenes. Chem Soc Rev. 2013;42(3):968–1006. https://doi.org/10.1039/C2CS35154D.
38. Jakubec M, Storch J. Recent advances in functionalizations of the helicene backbone. J Org Chem. 2020;85(21):13415–28. https://doi.org/10.1021/acs.joc.0c01837.
39. Mori T. Chiroptical properties of symmetric double, triple, and multiple helicenes. Chem Rev. 2021;121(4):2373–412. https://doi.org/10.1021/acs.chemrev.0c01017.

40. Nakai Y, Mori T, Inoue Y. Theoretical and experimental studies on circular dichroism of carbo[n]helicenes. J Phys Chem A. 2012;116(27):7372–85. https://doi.org/10.1021/jp304576g.
41. Gingras M. One hundred years of helicene chemistry. Part 3: applications and properties of carbohelicenes. Chem Soc Rev. 2013;42(3):1051–95. https://doi.org/10.1039/C2CS35134J.
42. Nowak-Król A, Geppert PT, Naveen KR. Boron-containing helicenes as a new generation of chiral materials: opportunities and challenges of leaving the flatland. Chem Sci. 2024;15(20):7408–40. https://doi.org/10.1039/D4SC01083C.
43. Sánchez-Carnerero EM, Agarrabeitia AR, Moreno F, Maroto BL, Muller G, Ortiz MJ, de la Moya S. Circularly polarized luminescence from simple organic molecules. Chem Eur J. 2015;21(39):13488–500. https://doi.org/10.1002/chem.201501178.
44. Qiu Z, Ju C-W, Frédéric L, Hu Y, Schollmeyer D, Pieters G, Müllen K, Narita A. Amplification of dissymmetry factors in π-extended [7]- and [9]Helicenes. J Am Chem Soc. 2021;143(12):4661–7. https://doi.org/10.1021/jacs.0c13197.
45. Fasel R, Parschau M, Ernst K-H. Chirality transfer from single molecules into self-assembled monolayers. Angew Chem Int Ed. 2003;42(42):5178–81. https://doi.org/10.1002/anie.200352232.
46. Ernst K-H, Parschau M, Fasel R. Differences in two-dimensional crystal structures: racemic and enantiopure heptahelicene on Cu(111). e-J Surf Sci Nanotechnol. 2004;2:136–140. https://doi.org/10.1380/ejssnt.2004.136.
47. Ernst K-H. Amplification of chirality in two-dimensional molecular lattices. Curr Opin Colloid Interface Sci. 2008;13(1):54–9. https://doi.org/10.1016/j.cocis.2007.08.011.
48. Parschau M, Ellerbeck U, Ernst K.-H. Chirality transfer by epitaxial mismatch in multilayered homochiral molecular films. Colloids Surf A. 2010;354(1):240–5. https://doi.org/10.1016/j.colsurfa.2009.06.008.
49. Stöhr M, Boz S, Schär M, Nguyen M-T, Pignedoli CA, Passerone D, Schweizer WB, Thilgen C, Jung TA, Diederich F. Self-assembly and two-dimensional spontaneous resolution of cyano-functionalized [7]Helicenes on Cu(111). Angew Chem, Int Ed. 2011;50(42):9982–6. https://doi.org/10.1002/anie.201102627.
50. Shchyrba A, Nguyen M-T, Wäckerlin C, Martens S, Nowakowska S, Ivas T, Roose J, Nijs T, Boz S, Schär M, Stöhr M, Pignedoli CA, Thilgen C, Diederich F, Passerone D, Jung TA. Chirality transfer in 1D self-assemblies: influence of H-bonding vs metal coordination between Dicyano[7]helicene enantiomers. J Am Chem Soc. 2013;135(41):15270–3. https://doi.org/10.1021/ja407315f.
51. Parschau M, Fasel R, Ernst K-H. Coverage- and enantiomeric excess-dependent enantiomorphism in two-dimensional molecular crystals. Cryst Growth Des. 2008;8(6):1890–6. https://doi.org/10.1021/cg701100r.
52. Ernst K-H, Böhringer M, McFadden CF, Hug P, Müller U, Ellerbeck U. Nanostructured chiral surfaces. Nanotechnology. 1999;10(3):355–61. https://doi.org/10.1088/0957-4484/10/3/320.
53. Ernst K-H, Kuster Y, Fasel R, McFadden CF, Ellerbeck U. Adsorption of helical aromatic molecules: heptahelicene on Ni(111). Surf Sci. 2003;530(3):195–202. https://doi.org/10.1016/S0039-6028(03)00489-8.
54. Ernst K-H. Stereochemical recognition of helicenes on metal surfaces. Acc Chem Res. 2016;49(6):1182–90. https://doi.org/10.1021/acs.accounts.6b00110.
55. Godlewski S, Prauzner-Bechcicki JS, Budzioch J, Walczak L, Stará IG, Starý I, Sehnal P, Szymoński M. [11]Anthrahelicene on TiO_2 Surfaces. Surf Sci. 2012;606(21):1600–7. https://doi.org/10.1016/j.susc.2012.06.015.
56. Prauzner-Bechcicki JS, Godlewski S, Budzioch J, Goryl G, Walczak L, Sehnal P, Stará IG, Starý I, Ample F, Joachim C, Szymoński M. [11]Anthrahelicene on InSb(001) c(8×2): a low-temperature scanning probe microscopy study. ChemPhysChem. 2010;11(16):3522–8. https://doi.org/10.1002/cphc.201000547.
57. Shiotari A, Tanaka K, Nakae T, Mori S, Okujima T, Uno H, Sakaguchi H, Sugimoto Y. Chiral discrimination and manipulation of individual heptahelicene molecules on Cu(001) by noncontact atomic force microscopy. J Phys Chem C. 2018;122(9):4997–5003. https://doi.org/10.1021/acs.jpcc.8b00487.

58. Irziqat B, Berger J, Mendieta-Moreno JI, Sundar MS, Bedekar AV, Ernst K-H. Transition from homochiral clusters to racemate monolayers during 2D crystallization of Trioxa[11]Helicene on Ag(100). ChemPhysChem. 2021;22(3):293–7. https://doi.org/10.1002/cphc.202000853.
59. Ishii A, Shiotari A, Sugimoto Y. Mechanically induced single-molecule helicity switching of graphene-nanoribbon-fused helicene on Au(111). Chem Sci. 2021;12:13301–6. https://doi.org/10.1039/D1SC03976H.
60. Jenkins SJ. Aromatic adsorption on metals via first-principles density functional theory. Proc R Soc A. 2009;465(2110):2949–76. https://doi.org/10.1098/rspa.2009.0119.
61. Adhikari S, Nepal NK, Tang H, Ruzsinszky A. Describing adsorption of benzene, thiophene, and xenon on coinage metals by using the Zaremba-Kohn theory-based model. J Chem Phys. 2021;154(12): 124705. https://doi.org/10.1063/5.0042719.
62. Taniguchi M, Nakagawa H, Yamagishi A, Yamada K. STM observation of Thia[11]Heterohelicene on Gold(111) and Gold(110) surfaces. Surf Sci. 2002;507–510:458–62. https://doi.org/10.1016/S0039-6028(02)01285-2.
63. Chaunchaiyakul S, Setiadi A, Krukowski P, Catalan FCI, Akai-Kasaya M, Saito A, Hayazawa N, Kim Y, Osuga H, Kuwahara Y. Nanoscale dehydrogenation observed by tip-enhanced raman spectroscopy. J Phys Chem C. 2017;121(33):18162–8. https://doi.org/10.1021/acs.jpcc.7b03352.
64. Chaunchaiyakul S, Krukowski P, Tsuzuki T, Minagawa Y, Akai-Kasaya M, Saito A, Osuga H, Kuwahara Y. Self-assembly formation of M-type enantiomer of 2,13-Bis(hydroxymethyl)[7]Thiaheterohelicene molecules on Au(111) surface investigated by STM/CITS. J Phys Chem C. 2015;119(37):21434–42. https://doi.org/10.1021/acs.jpcc.5b04961.
65. Cao H, Minoia A, De Cat I, Seibel J, Waghray D, Li Z, Cornil D, Mali KS, Lazzaroni R, Dehaen W, De Feyter S. Hierarchical self-assembly of Enantiopure and racemic Helicenes at the liquid/solid interface: from 2D to 3D. Nanoscale. 2017;9(45):18075–80. https://doi.org/10.1039/C7NR06700C.
66. Baciu BC, de Ara T, Sabater C, Untiedt C, Guijarro A. Helical nanostructures for organic electronics: the role of topological sulfur in Ad Hoc synthesized Dithia[7]Helicenes studied in the solid state and on a gold surface. Nanoscale Adv. 2020;2(5):1921–6. https://doi.org/10.1039/D0NA00045K.
67. Krukowski P, Chaunchaiyakul S, Akai-Kasaya M, Saito A, Osuga H, Kuwahara Y. Adsorption and light emission of a racemic mixture of [7]Thiaheterohelicene-2,13-carboxaldehyde on Au(111), Cu(001), and NiAl(110) surfaces investigated using a scanning tunneling microscope. J Phys Chem C. 2021;125(17):9419–27. https://doi.org/10.1021/acs.jpcc.1c01091.
68. Krukowski P, Hattori T, Okada M, Piskorski M, Lutsyk I, Saito A, Osuga H, Kuwahara Y. Study of stereochemical crystallization of racemic mixtures of [5] and [7]Thiaheterohelicene molecules on Ag(111) surface by scanning tunneling microscopy and Raman scattering spectroscopy. Appl Surf Sci. 2022;589: 152860. https://doi.org/10.1016/j.apsusc.2022.152860.
69. Arnaboldi S, Cauteruccio S, Grecchi S, Benincori T, Marcaccio M, Biroli AO, Longhi G, Licandro E, Mussini PR. Thiahelicene-based inherently chiral films for enantioselective electroanalysis. Chem Sci. 2019;10(5):1539–48. https://doi.org/10.1039/C8SC03337D.
70. Noh J, Ito E, Nakajima K, Kim J, Lee H, Hara M. High-resolution STM and XPS studies of thiophene self-assembled monolayers on Au(111). J Phys Chem B. 2002;106(29):7139–41. https://doi.org/10.1021/jp020482w.
71. Buckel F, Effenberger F, Yan C, Gölzhäuser A, Grunze M. Influence of aromatic groups incorporated in long-chain alkanethiol self-assembled monolayers on gold. Adv Mater. 2000;12(12):901–5. https://doi.org/10.1002/1521-4095(200006)12:12%3c901::AID-ADMA901%3e3.0.CO;2-B.
72. He Y, Cai Z, Shao J, Xu L, She L, Zheng Y, Zhong D. Quaterrylene molecules on Ag(111): self-assembly behavior and voltage pulse induced trimer formation. Phys Chem Chem Phys. 2018;20(17):12217–22. https://doi.org/10.1039/C8CP01005F.
73. Pawin G, Wong KL, Kwon K-Y, Bartels L. A homomolecular porous network at a Cu(111) surface. Science. 2006;313(5789):961–2. https://doi.org/10.1126/science.1129309.

74. Seibel J, Parschau M, Ernst K-H. Two-dimensional crystallization of enantiopure and racemic Heptahelicene on Ag(111) and Au(111). J Phys Chem C. 2014;118(50):29135–41. https://doi.org/10.1021/jp504673g.
75. Ye C, Hattori T, Hamamoto Y, Krukowski P, Inagaki K, Saito A, Osuga H, Morikawa Y, Kuwahara Y. Chiral recognition mechanism of two-dimensional self-assembly formed by [7]Thiaheterohelicene. J Phys Chem C. 2023;127(43):21305–12. https://doi.org/10.1021/acs.jpcc.3c04406.
76. Koslowski B, Tschetschetkin A, Maurer N, Mena-Osteritz E, Bäuerle P, Ziemann P. Terthiophene on Au(111): a scanning tunneling microscopy and spectroscopy study. Beilstein J Nanotechnol. 2011;2:561–8. https://doi.org/10.3762/bjnano.2.60.
77. Schimmel S, Sun Z, Baumann D, Krylov D, Samoylova N, Popov A, Büchner B, Hess C. Adsorption characteristics of $Er_3N@C_{80}$ on W(110) and Au(111) studied via scanning tunneling microscopy and spectroscopy. Beilstein J Nanotechnol. 2017;8:1127–34. https://doi.org/10.3762/bjnano.8.114.
78. Mairena A, Mendieta JI, Stetsovych O, Terfort A, Stará IG, Starý I, Jelínek P, Ernst K-H. Heterochiral recognition among functionalized heptahelicenes on noble metal surfaces. Chem Commun. 2019;55(71):10595–8. https://doi.org/10.1039/C9CC05317D.
79. Jiang T, Malone W, Tong Y, Dragoe D, Bendounan A, Kara A, Esaulov VA. Thiophene derivatives on gold and molecular dissociation processes. J Phys Chem C. 2017;121(50):27923–35. https://doi.org/10.1021/acs.jpcc.7b08006.
80. Kimura S, Hattori T, Ye C, Okada M, Kondo S, Sakurama Y, Saito A, Krukowski P, Osuga H, Kuwahara Y. STM/TERS observation of (M)-type Diphenyl[7]Thiaheterohelicene on Ag(111). Phys Chem Chem Phys. 2024;26(9):7658–63. https://doi.org/10.1039/D3CP05813A.

Chapter 16
Trends in Photocatalytic Properties of Three-Dimensionally Ordered Macroporous Metal Oxynitride Inverse Opals

Toshihiro Moriga

Abstract This chapter describes the synthesis of metal oxynitride inverse opal (IO) photonic crystals possessing a three-dimensionally ordered macroporous (3DOM) structure and discusses empirical design guidelines to maximize photonic band gap (PBG)-induced photocatalytic enhancement. In particular, it demonstrates the influence of the light localization effect of the slow photon phenomenon occurring at the PBG edges (blue edge and red edge) on photocatalytic properties. To maximize the benefits of light localization and the suppression of carrier recombination arising from this effect, a precise overlap between the photonic band gap and the optical absorption edge of the photocatalyst is a critical key point. Despite its importance, the position of the PBG in the previous studies has typically been discussed only in terms of theoretical values calculated from Bragg's law. However, a non-negligible discrepancy often arises between theoretical predictions and the actual PBG observed in real structures, representing a significant barrier to fully exploiting the PBG effect. Herein, we report a two-step process for the synthesis of 3DOM metal oxynitride inverse opal (IO) photocatalyst exhibiting brilliant structural colors, which combines colloidal crystal templating via vacuum-assisted impregnation with subsequent thermal nitridation under an ammonia atmosphere. This work, grounded in experimental approaches, provides a clear strategy to move beyond theoretical speculation and maximize the PBG effect, thus offering new paradigms for the rational design of high-performance photocatalysts.

Keywords Photocatalyst · Metal oxynitride · Photonic crystal · Inverse opal · Photonic bandgap · Slow photon · Hydrogen production

---

T. Moriga (✉)
Department of Applied Chemistry, Graduate School of Science and Technology for Innovation, Tokushima University, Tokushima, Japan
e-mail: moriga@tokushima-u.ac.jp

T. Onishi (ed.), *Advances in Quantum Science II*,
https://doi.org/10.1007/978-981-95-9721-5_16

16.1 Introduction

16.1.1 Energy and Environmental Pollution Issues and Metal Oxide Photocatalysts

The world confronts two critical challenges: global warming caused by anthropogenic CO_2 emissions and severe water contamination from industrial pollutants like organic dyes [1, 2]. Harnessing solar energy through photocatalysis offers a sustainable and powerful strategy to mitigate both issues. This technology utilizes semiconductor materials to convert light energy into chemical energy, driving two key processes: the mineralization of toxic organic compounds into benign products like CO_2 and H_2O, and the production of clean hydrogen (H_2) fuel via water splitting. The efficacy of a photocatalyst is dictated by its ability to absorb a broad range of light, which requires a narrow bandgap, and to efficiently separate the photogenerated electrons and holes to drive surface redox reactions [3]. While conventional oxide semiconductors like TiO_2 and $SrTiO_3$ have been widely studied due to high stability and low cost, their practical application is severely hampered by their large bandgaps. This inherent limitation restricts their photo response to the UV portion of the solar spectrum, which accounts for less than 5% of the total energy, leaving the vast visible and infrared regions untapped [4].

16.1.2 Visible-Light-Driven Metal Oxynitride Photocatalysts

Semiconductor metal oxynitrides, formed by substituting nitrogen into an oxide lattice, have recently emerged as a compelling class of materials for visible-light-driven photocatalysis. The introduction of N 2p orbitals, which are energetically higher than O 2p orbitals, elevates the valence band maximum [5]. This strategic modification narrows the band gap in materials such as TaON (Eg $\approx$ 2.4 eV) and $LaTiO_2N$ (Eg $\approx$ 2.0 eV), unlocking their ability to harness low-energy visible lights, unlike their wide-bandgap oxide counterparts [6, 7]. However, the conventional synthesis of these materials—typically involving high-temperature nitridation ($\geq$ 800 °C) of oxide precursors under flowing ammonia gas—presents significant drawbacks. The harsh reaction conditions inevitably create a high density of defects (e.g., reduced species of metal cations, oxygen, and nitrogen vacancies) that act as detrimental charge recombination sites, quenching photocatalytic activity [8–10]. Simultaneously, severe particle sintering occurs, drastically reducing the specific surface area and limiting the number of catalytically active sites. This creates a fundamental trade-off: enhancing visible-light absorption comes at the expense of poor charge separation and reduced surface area. Overcoming this challenge is one of the most critical hurdles preventing the widespread practical application of oxynitride photocatalysts.

16.1.3 Inverse Opal Photonic Crystals

Seeking to improve charge carrier utilization in photocatalysis, the scientific community is now focused on harnessing the slow photon effect within photonic crystal (PhC) architectures. PhCs, first reported in 1987, are periodic dielectric structures with visible-light wavelength that create a photonic band gap (PBG)—a forbidden energy range for photons which gives rise to structural color [11–13]. Furthermore, since the existence of light is forbidden at the PBG wavelength, luminescence due to radiative recombination within the photonic crystal is suppressed (i.e., the recombination process is suppressed). The difference between an optical bandgap and a photonic bandgap lies in whether their forbidden band acts in the electronic or optical wavelength region. In the case of an optical bandgap, it is formed due to the periodicity of the atom and differences in the electrostatic potential of atomic nuclei. Consequently, in the reflection spectrum, the optical bandgap creates a boundary where light is reflected and absorbed at a specific wavelength, as shown in Fig. 16.1a. In the case of photonic bandgaps, these are formed due to the periodicity of the microstructures constituting the photonic crystal being on the order of visible-light wavelengths, and the differences in refractive index within that period. Consequently, photonic bandgaps exhibit a spectrum such as that shown in Fig. 16.1b, diffracting and reflecting only light of specific wavelengths. Critically, at the PBG edges, the group velocity of light diminishes, leading to a "slow photon" effect. This effect dramatically enhances light-matter interactions by spatially localizing photon energy within the photocatalytic medium [11, 12]. The slow photon effect causes the optical state density to approach zero within the PBG, modulating the optical state density in its vicinity (at the PBG edge). The slow photon effect at the PBG edge theoretically causes the group velocity of light to approach zero, causing the propagating light to form standing waves within the crystal and localize at the high-dielectric and low-dielectric side materials. A facile and inexpensive route to fabricating these structures is colloidal crystal templating (Fig. 16.2) [14]. This bottom-up approach relies on the self-assembly of monodispersed polymer colloid spheres (e.g., polymethyl methacrylate (PMMA) or polystyrene (PS)) into an ordered opal template. After infiltration with a semiconductor precursor, the polymer scaffold is removed, yielding a highly ordered, macroporous inverse opal (IO) structure. The resulting IO architecture is nearly ideal for photocatalysis, providing a high surface area, enhanced light harvesting through multiple scattering, reduced charge carrier migration distances, and superior mass transport kinetics compared to conventional powdered materials [15].

16.1.4 Photonic Band Engineering with Blue and Red Edges

As discussed, the edges of the PBG are critical regions, where the "slow photon effect" localizes light energy. The short-wavelength side is termed the "blue edge," and the

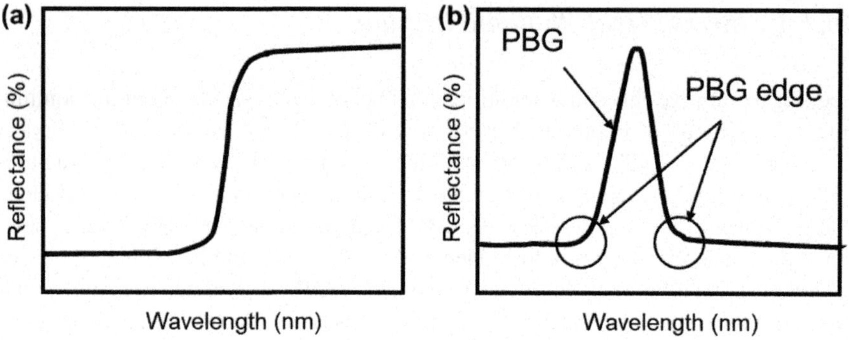

Fig. 16.1 Schematic diagram of the reflection spectrum illustrating the shape of the optical bandgap (**a**) and photonic bandgap (**b**)

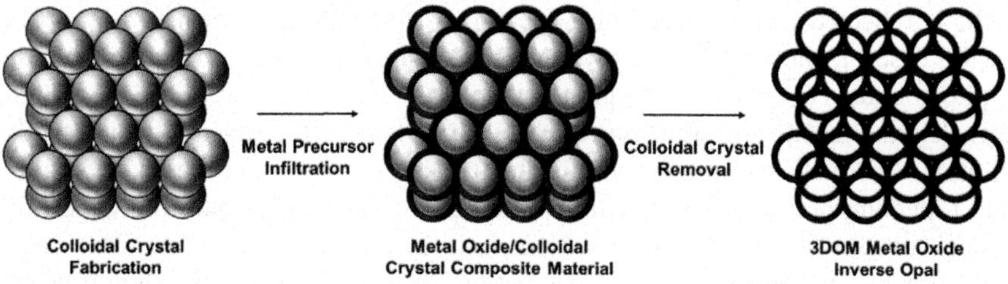

Fig. 16.2 Synthesis scheme for a three-dimensionally ordered macroporous (3DOM) metal oxide inverse opals using colloidal crystal. IO structure is synthesized by impregnating a colloidal crystal composed of monodisperse PMMA, PS, and SiO_2 colloidal spheres with a semiconductor precursor solution, followed by removal of the colloidal crystal via sintering or etching

long-wavelength side is the "red edge." The central requirement for harnessing these photonic enhancements is the precise spectral alignment of the optical absorption edge (i.e., optical energy bandgap) of semiconductor with the PBG edge (Fig. 16.3) [16]. The nature of this coupling is distinct for each edge: alignment with the blue edge concentrates the incident electric field in the low-dielectric voids, whereas alignment with the red edge concentrates it within the semiconductor scaffold [17]. Proper alignment has been shown to either amplify light absorption or inhibit charge recombination, both of which are pathways to superior photocatalytic performance. Essentially, these optical matching leverages the slow photon effect to trap photons, thereby increasing their effective path length and maximizing photon-semiconductor interactions. However, a critical gap exists in literature. Most prior work on inverse opal photocatalysts (e.g., TiO_2, $SrTiO_3$, CdS IO) has relied on theoretical PBG positions derived from Bragg's law [18–20]:

$$\lambda_{\max} = \frac{2d_{hkl}}{m}\sqrt{\left(n_{\mathrm{avg}}\right)^2 - (\sin\theta)^2} \tag{16.1}$$

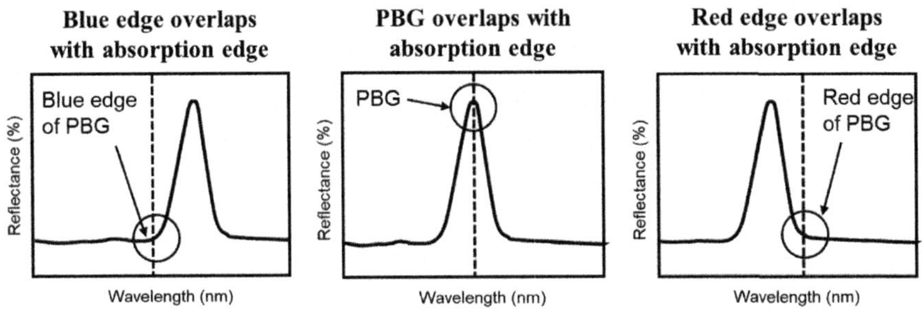

Fig. 16.3 Schematic diagram showing the overlap between the optical absorption edge of the photocatalyst and PBG and PBG edges (black solid line: PBG, black dot line: optical absorption edge)

where λ_{max} is the PBG position (in nm), d_{hkl} is the interplanar spacing (in nm) between the planes with Miller indices *hkl*, *m* is the order of diffraction, n_{avg} is the average refractive index of the IO, and θ is the incidence angle (in degrees) with respect to the surface normal of the *hkl* planes. However, estimating the PBG wavelength using this calculation is extremely difficult, as the PBG position is not only highly dependent on the angle of incidence but also redshifts significantly when moving from air to the aqueous media used in photocatalysis. Consequently, achieving and verifying the precise spectral alignment required to deconvolute these powerful photonic effects has remained a formidable experimental hurdle.

16.2 Enhancement of Photocatalytic Properties in Three-Dimensionally Ordered Macroporous (3DOM) TaON Inverse Opal (IO) with Accurate Slow Photon Effects

This section describes the synthesis of a visible-light-driven tantalum oxynitride (TaON) IO photocatalyst, which exhibits brilliant structural colors. The material was prepared by the thermal nitridation of a Ta_2O_5 IO, originally synthesized via a vacuum-assisted impregnation method, under an ammonia atmosphere. In addition, we investigate the influence of light diffraction and the slow photon effect, which occur at the PBG and its edges (the "blue edge" and "red edge"), on the photocatalytic hydrogen evolution reaction. A critical requirement for leveraging photonic band engineering is the precise alignment of the photocatalyst optical absorption edge with the PBG. This alignment necessitates the fine-tuning of multiple parameters, such as the macropore diameter of the TaON IO, the refractive index of the solvent filling the voids, and the angle of light incidence. The synthesis of IOs powder samples with identifiable PBG wavelengths and the adjustment of these wavelengths have been challenging due to this multiparameter dependency, impeding the definitive demonstration of the slow photon effect. To overcome these issues, this work aimed to experimentally maximize the PBG-induced enhancement of photocatalytic performance by simultaneously considering and optimizing these various parameters.

16.2.1 Difference Between Drop Method and Vacuum-Assisted Immersion Impregnation Method

Conventional infiltration of colloidal templates via drop method is often inadequate for creating high-quality photonic crystals. This method typically yields IOs (e.g., TiO_2, $SrTiO_3$) with poor quality, where any PBG is obscured by strong light scattering. To overcome this limitation, our group has developed a vacuum-assisted impregnation process that drastically improves the structural quality of Ta_2O_5 IOs, enabling the synthesis of powders with brilliant structural color. The drop method Ta_2O_5 IO powder is simply white color (This white color is attributable to the characteristic optical absorption edge of Ta_2O_5), showing no structural color, whereas the vacuum-assisted sample is a vibrant blue [21]. Their reflectance spectra confirm this observation (Fig. 16.4): the drop method sample has a broad, weak diffraction peak at 420 nm, while the vacuum-assisted sample exhibits a sharp, intense PBG at 440 nm [21]. This divergence, even when using identical PMMA templates, stems from a fundamental flaw in the drop method: incomplete and non-uniform filling of the template voids. This leads to a low solid volume fraction and low quality. In contrast, vacuum assistance ensures the complete evacuation of air and, consequently, a perfectly uniform infiltration by the $TaCl_5$/ethanol precursor. This results in superior structural integrity with less lattice shrinkage upon calcination, preserving a larger macropore diameter and a higher solid volume fraction. Consequently, the PBG of the vacuum-impregnated sample is not only more intense but also red-shifted, reflecting its higher-quality photonic structure.

Fig. 16.4 UV-vis spectra for Ta_2O_5 IO with different synthesis method (dot line: drop method, solid line: vacuum- assisted impregnation method)

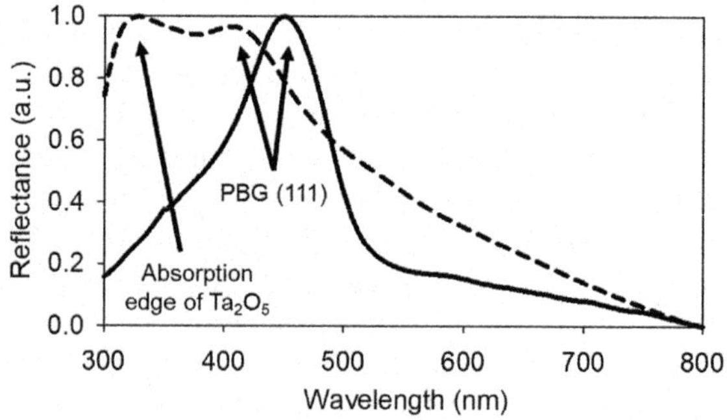

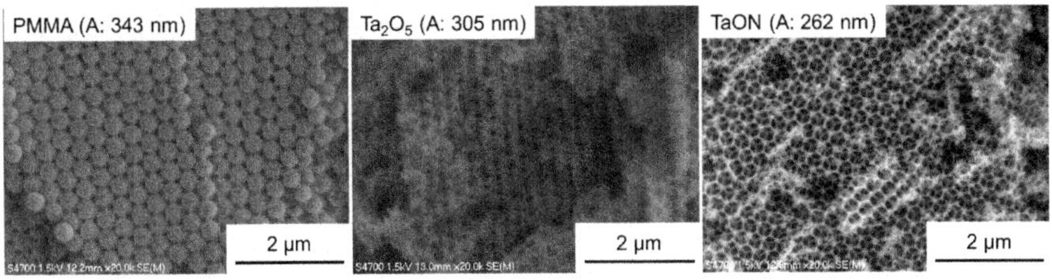

Fig. 16.5 Scanning electron microscope (SEM) images for the polymethyl methacrylate colloidal crystal (PMMA (A:343 nm)), Ta_2O_5 (A: 305 nm) and TaON (A:262 nm) samples. The inset numbers show the colloid diameter of polymethyl methacrylate colloidal sphere and macropore diameter of Ta_2O_5 (A), TaON (A)

16.2.2 Synthesis of the 3DOM TaON IO Photonic Crystals

TaON IOs photocatalyst samples with a uniform 3DOM structure were synthesized via a two-step process combining colloidal crystal templating and thermal nitridation. To control the alignment between the optical absorption edge of the TaON and its PBG or PBG edges, Ta_2O_5 IO precursor powders were synthesized using colloidal crystal templates with four different colloidal diameters. Subsequent TaON IO samples were synthesized by performing thermal nitridation at 800 °C for 5 h under an ammonia flow of 25–30 mL/min. As representative examples, SEM images of Ta_2O_5 (A:305 nm) synthesized using PMMA (A:343 nm) with a colloidal diameter of 343 nm as a sacrificial template, and TaON (A:262 nm) synthesized by thermal nitridation of Ta_2O_5 IO are shown (Fig. 16.5) [22]. It was confirmed that the PMMA colloidal spheres are arranged in a face-centered cubic structure with a close-packed arrangement. Furthermore, Ta_2O_5 IO sample synthesized via the template-inverted process using PMMA colloidal crystals exhibited a uniform inverse opal structure aligned with the sacrificial template. Additionally, TaON IO sample retained a uniform inverse opal, structure even after thermal nitridation process. Information on the colloidal particle diameter and macropore diameter of the four PMMA, Ta_2O_5, and TaON samples is given in Table 16.1. The intrinsic dark yellow-green color of TaON was significantly modulated to purple, blue, and bright green due to the structural color arising from PBG diffraction. Both scanning electron microscopy and optical images confirm that the synthesized TaON IOs possess a high-quality photonic crystal structure.

16.2.3 Optical Properties for the 3DOM TaON IO Photonic Crystals

A key advantage of vacuum-assisted impregnation method is the ability to precisely determine and tune the PBG position. Diffuse reflectance spectra of all samples,

Table 16.1 Information of colloid diameter of polymethyl methacrylate, macropore diameter of Ta_2O_5 IO and TaON IO samples

Sample	Colloid diameter of PMMA (nm)	Macropore diameter of Ta_2O_5 IO (nm)	Macropore diameter of TaON IO (nm)
(A)	343	305	262
(B)	297	250	231
(C)	270	238	213
(D)	256	228	199

both in air and when infiltrated with the 10 vol% aqueous methanol reaction solvent, revealed a characteristic reflectance peak from the FCC (111) plane (Fig. 16.6). Of particular importance are the spectra measured in the 10 vol% methanol solution, as they reveal the effective PBG wavelength present during the photocatalytic testing by accounting for the refractive index of solvent. Furthermore, by measuring across a wide range of incident angles (0–90°), we simulated the stirred conditions where light interacts with the powder isotopically. Contrary to the prediction from Bragg's law (Eq. 16.1), the PBG wavelength exhibited only a minor shift with increasing incidence angle. We attribute this to the polycrystalline nature of the IO powder; the random orientation of the FCC (111) planes throughout the powder averages out the angular-dependent diffraction. This result demonstrates that by considering both the solvent medium and the angularity of illumination, we successfully controlled the effective PBG position relative to the TaON optical absorption edge during the experiment.

To engineer the relationship between the electronic absorption edge of TaON and the PBG of the inverse opal structure, it was first essential to identify the intrinsic optical absorption edge of TaON material. Next, we identified the PBG wavelengths for the four TaON IO samples to clarify their positions relative to this optical absorption edge under realistic photocatalytic conditions. The optical absorption edge of TaON was determined to be located at 510 nm (corresponding to a band gap: 2.43 eV), a value that did not vary significantly with the macropore diameter [22]. It is important to compare the optical absorption edge at 510 nm with the PBG wavelength

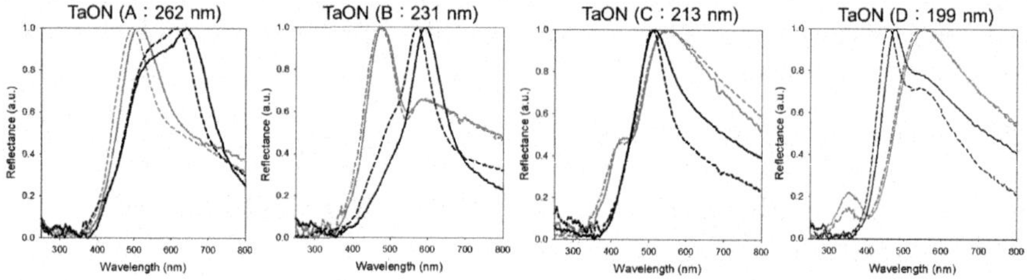

Fig. 16.6 Reflectance spectra for the 3DOM TaON IO powders with different macropore diameter showing the PBG wavelength (gray line: in air, black line: filled with 10vol% methanol solution, solid line: 0°, dot line: 90°)

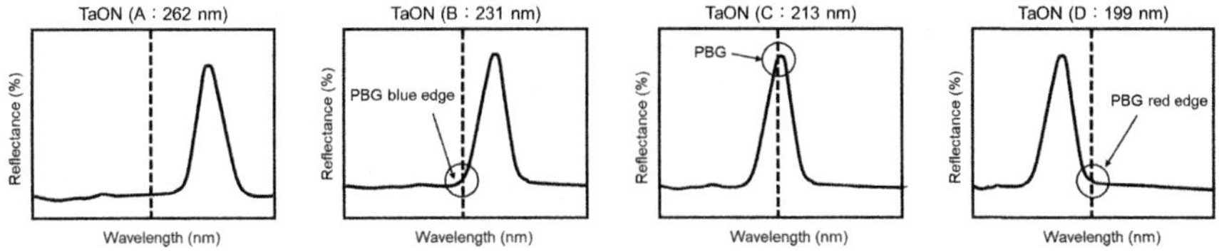

Fig. 16.7 Schematic diagram showing the relationship between the PBG wavelength and the optical absorption edge observed in photocatalytic testing (dot line: optical absorption edge of TaON, solid line: PBG wavelength)

(Fig. 16.6: black line) achieved by infiltrating a 10 vol% methanol aqueous solution, which is the solvent used in photocatalysis testing. The reason is that the relative position between the material optical absorption edge and the PBG is critically important for imparting the slow photon effect via photonic crystals. By comparing these two spectra, we can focus on the differences in photonic crystal effects that appear between samples during photocatalysis testing. The final alignment between the optical absorption edge of TaON photocatalyst and the effective PBG for each sample during photocatalysis is depicted in Fig. 16.7. A key point to note is that the optical absorption edge of TaON synthesized in this study shows no significant variation across the four samples with different macropore diameters, consistently occurring at 510 nm. Consequently, only the PBG wavelength shifts in the four samples with different macropore diameters.

- For **TaON (A:262 nm) sample**, the peak-top wavelength of PBG in a 10 vol% methanol aqueous solution is located at a wavelength (640 nm) significantly longer than the optical absorption edge; therefore, no photonic crystal effect is expected to occur.
- For **TaON (B:231 nm) sample**, the peak wavelength of PBG in a 10% methanol aqueous solution is located at 600 nm, and the short-wavelength edge (blue edge) of the PBG coincides precisely with the optical absorption edge of TaON.
- For **TaON (C:213 nm) sample**, the peak wavelength of PBG in a 10% methanol aqueous solution is located at 510 nm, precisely coinciding with the optical absorption edge of PBG and the optical absorption edge of TaON.
- For **TaON (D: 199 nm) sample**, the peak wavelength of PBG in a 10% methanol aqueous solution is located at 460 nm, and the long-wavelength edge (red edge) of the PBG coincides precisely with the optical absorption edge of TaON.

Consequently, the photocatalytic activities of TaON (B:231 nm), TaON (C:213 nm), and TaON (D:199 nm) samples directly affected by their distinct photonic band structures.

16.2.4 The Effect of Photonic Band Engineering on the Hydrogen Production of 3DOM TaON IOs

Hydrogen production was measured by irradiating a 10 vol% methanol aqueous solution with a 150 W xenon lamp ($\lambda > 380$ nm). In this work, H_2PtCl_6 was added as a precursor for the in situ photo-deposition method of platinum (Pt), which functions as a co-catalyst for hydrogen evolution [22]. Additionally, as a sample lacking the inverse opal structure of 3DOM (comparison sample), the NTaON sample was prepared by mechanically the TaON (A:262 nm) sample.

As given in Table 16.2, the order of H_2 production rates was as follows: TaON (B:231 nm) (0.3421 mmol g^{-1} h^{-1}) > TaON (D:199 nm) (0.3254 mmol g^{-1} h^{-1}) > TaON (A:262 nm) (0.275 mmol g^{-1} h^{-1}) $\approx$ TaON (C:213 nm) (0.2386 mmol g^{-1} h^{-1}) > NTaON (0.1054 mmol g^{-1} h^{-1}). This data clearly reveals that the two samples where a PBG edge was aligned with the optical absorption edge of TaON exhibited the highest photocatalytic activity. First, all TaON IO samples showed a marked improvement in H_2 evolution compared to the unstructured NTaON reference sample. This suggests that the inverse opal architecture provides superior light-harvesting efficiency compared to fine powder particles and/or facilitates the diffusion of reactants (methanol and protons) to active sites, thus mitigating mass-transfer limitations.

Among the IO samples, the differences in activity can be directly attributed to the PBG position. The H_2 production rate for TaON (A:262 nm) was modest because its PBG in the methanol solution was centered at 640 nm, far from the optical absorption edge of TaON, preventing any beneficial photonic effects. Conversely, the activity of TaON (C:213 nm) was significantly suppressed. For this sample, the PBG peak maximum directly overlapped with the optical absorption edge of TaON, causing PBG-induced diffraction to inhibit the fundamental light absorption by the photocatalyst itself. This result provides a clear experimental demonstration that a PBG can impede the intrinsic absorption of a photocatalyst when improperly aligned.

A further discussion is required to explain the exceptionally high photocatalytic H_2 evolution rates achieved by TaON (B:231 nm) and TaON (D:199 nm). Both samples appear to exhibit an enhanced slow photon effect and increased charge

Table 16.2 Summarized the relationship between optical absorption edge TaON and PBG wavelength in a 10 vol% methanol aqueous solution, hydrogen evolution rate and Photoluminescence (PL) intensity of TaON photocatalyst samples

TaON sample	Relationship between band gap and $PBG_{TaON,\ 10\ vol\%\ methanol\ aq.}$	H_2 evolution rates (mmol g^{-1} h^{-1})	Intensity at maximum PL
(A:262 nm)	No effect	0.2750	9.28
(B:231 nm)	Blue edge	0.3421	13.0
(C:213 nm)	PBG	0.2386	7.44
(D:199 nm)	Red edge	0.3254	10.3
NTaON	/	0.1054	6.06

carrier generation compared to the other inverse opal photocatalysts prepared in this study.

The photocatalytic enhancement observed for TaON (D: 199 nm), whose optical absorption edge aligns with the PBG red edge, can be explained based on established literature. At the red edge of a PBG, the group velocity of light is reduced, and the electric field of the resulting standing wave becomes highly localized within the inverse opal walls (in this case, the high-dielectric TaON). When the semiconductor optical absorption edge overlaps with this red edge, the intensified photon-semiconductor interaction boosts the generation of charge carriers—i.e., more valence band holes for methanol oxidation and more conduction band electrons for proton reduction. Consequently, the H_2 evolution rate is enhanced, even though the PBG partially suppresses light absorption at shorter wavelengths (< 510 nm). Thus, the slow photon effect in TaON (D:199 nm) is attributed to increased light absorption that leads to a higher rate of charge carrier generation.

Conversely, the mechanism for TaON (B:231 nm), whose absorption edge aligns with the PBG blue edge, is different. A key advantage here is that the PBG is positioned at slightly longer wavelengths than the absorption edge, meaning it does not negatively impact TaON fundamental light absorption. Furthermore, the presence of the PBG is expected to be beneficial for suppressing the radiative recombination of charge carriers. The inverse opal structure forbids emissive recombination processes within the PBG frequency band, effectively closing a key charge-loss pathway. This is supported by photoluminescence (PL) measurement, which shows that the radiative recombination for TaON is centered at 540 nm. For TaON (B:231 nm), this emission wavelength falls squarely within its PBG in the methanol solution, confirming the viability of this suppression mechanism. By inhibiting recombination, the populations of available valence band holes and conduction band electrons for the surface photoreaction increase, enhancing the H_2 production rate. Therefore, the effect observed for TaON (B:231 nm) is primarily attributed to a combination of strong, uninhibited light absorption and the suppression of charge carrier recombination. Ultimately, these results demonstrate that the slow photon effect, manifesting through distinct mechanisms at both the blue edge and the red edge of the PBG, can be effectively engineered to enhance the photocatalytic activity of TaON.

16.3 Enhancement of Photocatalytic Properties in 3DOM $LaTiO_2N$ Inverse Opal by Cation Stoichiometry Optimization

Perovskite lanthanum titanium oxynitride ($LaTiO_2N$) is being investigated as a promising photocatalyst for solar energy applications owing to its narrow bandgap (Bandgap $\approx$ 2.1 eV). A key approach to improving the performance of oxynitride photocatalysts involves suppressing the formation of defects that are generated during the high-temperature thermal nitridation process. In this section, we aimed to enhance the photocatalytic properties of $LaTiO_2N$ by leveraging a synergistic effect that combines two distinct strategies: the slow photon effect arising from the PBG edge and the suppression of defect formation. This will be achieved by precisely controlling the cation stoichiometry (La/Ti ratio) in a 3DOM $LaTiO_2N$ IO structure.

16.3.1 Microstructure of 3DOM $LaTiO_2N$ IO Photocatalysts

To synthesize the 3DOM $LaTiO_2N$ IO photocatalysts, a polystyrene colloidal crystal with a colloid diameter of 551 nm was prepared as a sacrificial template. It should be noted that for the synthesis of 3DOM IO powders containing multiple metal cations, the use of citric acid to stabilize the metal precursors via chelation was found to be necessary [15]. Without such stabilization, the sintering process during the thermal removal of the colloidal template is often unstable for multi-metal systems, which hinders the formation of the inverse opal structure. The $La_2Ti_2O_7$ IO sample synthesized by adding citric acid was confirmed to possess a uniform macroporous structure in the oxide state (Fig. 16.8). Furthermore, the $LaTiO_2N$ IO (La/Ti = 1.0) sample synthesized using $La_2Ti_2O_7$ IO (La/Ti = 1.0) as a precursor retained a uniform IO structure even after nitridation process at a high temperature of 850 °C (Fig. 16.8). $LaTiO_2N$ IO samples with varied cation ratios (La/Ti = 1.1–0.9) showed a similarly well-ordered 3DOM structure. A key feature of this design is that for the $LaTiO_2N$ IO synthesized from the colloidal crystal template with 551 nm colloid diameter, the blue edge of the PBG in water is aligned with the optical absorption edge of the $LaTiO_2N$ material [15]. Through this alignment, the $LaTiO_2N$ IO is endowed with a blue edge slow photon effect during photocatalytic testing, which is expected to enhance the utilization efficiency of photogenerated charge carriers.

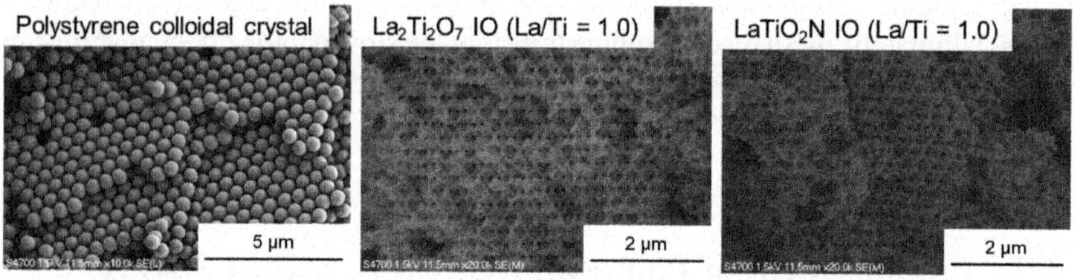

Fig. 16.8 Scanning electron microscope images (SEM) for the polystyrene colloidal crystal (colloid diameter: 551 nm), 3DOM $La_2Ti_2O_7$ (La/Ti = 1.0) and $LaTiO_2N$ IO (La/Ti = 1.0) powder. (Reproduced from Maekawa et al. 2025)

16.3.2 Optical and Composition Information of 3DOM $LaTiO_2N$ IO Photocatalysts

The crystal phase of the $LaTiO_2N$ IO samples with varying La/Ti ratios (1.1, 1.05, 1.0, 0.95, and 0.9) was primarily attributed to the perovskite-type $LaTiO_2N$ phase. However, it is noteworthy that in samples with La/Ti > 1, the excess lanthanum precipitated as lanthanum oxide (La_2O_3) [15]. As this precipitated phase could potentially have a negative impact on photocatalytic activity, it was subsequently removed by washing with HCl. Varying the La/Ti ratio in the $LaTiO_2N$ resulted in significant differences in the material optical properties, as well as its cation and anion composition. While there was no major change in the optical absorption edge of $LaTiO_2N$, the absorbance at wavelengths longer than 600 nm decreased significantly as the La/Ti ratio increased (Fig. 16.9). This long-wavelength absorbance is known to increase with higher defect concentrations, as defects introduce the new defects level between the valence and conduction bands. Furthermore, compositional analysis revealed that as the La/Ti ratio increased, the amount of surface Ti^{3+} decreased, while the oxygen content showed an increasing trend (Table 16.3).

These results directly indicate a reduction in defects, such as Ti^{3+} species and anion vacancies, that are formed during thermal nitridation. In summary, the optical and compositional analyses demonstrate that the defect concentration decreased in samples with La/Ti > 1, whereas it increased in samples with La/Ti < 1. This confirms that the concentration of defects in $LaTiO_2N$ can be effectively controlled by carefully managing the cation stoichiometry.

The phenomenon whereby the quantity of defects in $LaTiO_2N$ varies as a function of the La/Ti ratio is of significant interest. We postulate that the mechanism for defect

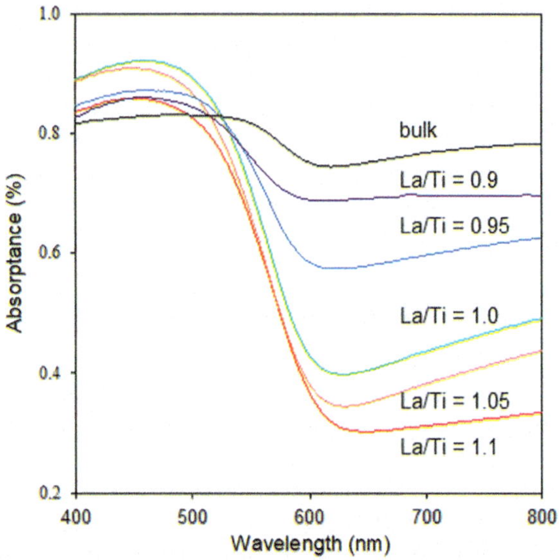

Fig. 16.9 Absorbance spectra for the 3DOM $LaTiO_2N$ IO powders with different La/Ti ratios and $LaTiO_2N$ bulk sample

Table 16.3 Summarized optical, composition, and surface area data for the $LaTiO_2N$ IO with different La/Ti ratios

La/Ti	$Ti^{3+}/(Ti^{3+}+Ti^{4+})$ (%)	O/N ratio	Anion composition	Surface area (m^2/g)
1.1	39.38	3.62	$La_{1.1}Ti\ (O_{2.74}N_{0.774})$	27.5
1.05	41.64	4.50	$La_{1.05}Ti\ (O_{2.82}N_{0.619})$	28.1
1.0	47.75	3.39	$LaTi\ (O_{2.57}N_{0.768})$	28.8
0.95	45.40	3.17	$La_{0.95}Ti\ (O_{2.45}N_{0.773})$	30.7
0.9	54.03	3.06	$La_{0.9}Ti\ (O_{2.01}N_{0.670})$	28.1

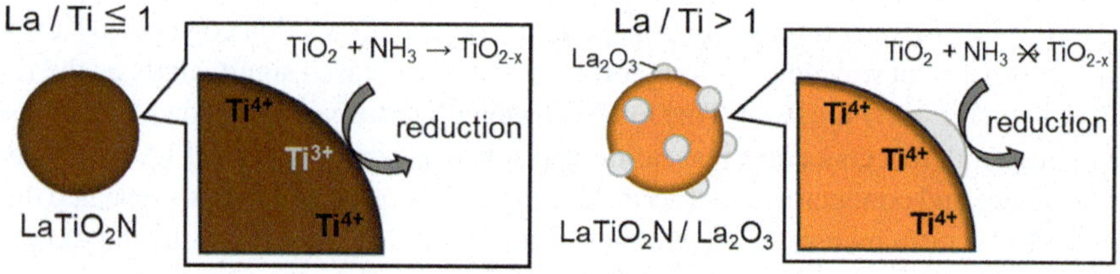

Fig. 16.10 Schematic diagrams of the mechanism suppressing defect generation. Typically, nitrogen is introduced after oxygen is removed by hydrogen gas generated from the thermal pyrolysis of ammonium gas. However, without nitrogen introduction, highly reducible titanium atoms are reduced ($Ti^{4+} \rightarrow Ti^{3+}$) to ensure charge balance, creating defects. In the La/Ti $\leq$ 1 sample (left side), the titanium reduction process dominates over the nitrogen introduction process due to the presence of excess titanium. In the La/Ti> 1 sample (right side), the precipitated La_2O_3 functioned as a protective agent by reducing excess titanium through interactions with titanium and oxygen on the $LaTiO_2N$ surface, thereby suppressing defect formation

reduction during thermal nitridation is strongly influenced by the precipitation of lanthanum oxide. In titanium-excess samples (La/Ti < 1), titanium is readily reduced during the thermal nitridation process. To maintain charge neutrality, this reduction is accompanied by the formation of anion vacancies, leading to a higher overall defect concentration. In contrast, for lanthanum-excess samples (La/Ti > 1), the excess lanthanum is considered to precipitate as lanthanum oxide [15]. This precipitated phase then could function as a protective agent, interacting with surface titanium and oxygen to suppress the excessive reduction of titanium (Fig. 16.10).

16.3.3 Tetracycline Degradation Using 3DOM $LaTiO_2N$ IOs

Tetracycline (TC) is used as antibiotic medication, but its persistence in the environment makes it a significant source of water pollution. Decomposing this persistent substance through oxidation-reduction reactions using photocatalysts and light energy offers one solution to the environmental pollution problem. For this reason, we dispersed $LaTiO_2N$ photocatalyst in a tetracycline aqueous solution containing a concentration of 10 ppm, irradiated it with a xenon lamp, and investigated its

Table 16.4 Decomposition rate constants of tetracycline using $LaTiO_2N$ photocatalyst samples obtained by pseudo-first-order rate plots

La/Ti	IO-1.1	IO-1.05	IO-1.0	IO-0.95	IO-0.9	bulk-1.0	Only light
Decomposition rates (min^{-1})	0.0038	0.0043	0.0031	0.0024	0.0023	0.0011	0.00009

decomposition rate. The photocatalytic decomposition rate of the pharmaceutical tetracycline using 3DOM $LaTiO_2N$ IO samples with different cation ratios is given in Table 16.4 [15].

Control experiments confirmed that the photolytic degradation of TC under light irradiation without a catalyst was negligibly slow, whereas the introduction of the photocatalysts resulted in significant photocatalytic decomposition. When compared to a non-IO bulk sample with the same La/Ti ratio, the $LaTiO_2N$ IO (La/Ti = 1.0) sample exhibited a TC degradation rate that was approximately three times higher. This enhancement is attributed to the synergistic benefits of the 3DOM architecture, including its increased surface area, facilitated diffusion of the dye molecules, and the slow photon effect occurring at the blue edge of the PBG. This result reaffirms that the 3DOM structure can substantially improve the performance of oxynitride photocatalysts compared to bulk state powders.

Furthermore, the variation of the cation ratio within the $LaTiO_2N$ IOs and—thus the concentration of defects—had a profound impact on the TC degradation rate. The rate of degradation increased as the La/Ti ratio increased, while it decreased as the La/Ti ratio was reduced. In samples with a high defect concentration (La/Ti < 1), photo-excited carriers are trapped by numerous deep defect states and are quickly deactivated via recombination (Fig. 16.11). Conversely, in samples with a low defect concentration (La/Ti > 1), the reduced number of bulk trap states leads to an increased carrier lifetime (Fig. 16.11). As a result, a significantly larger number of carriers are available to participate in the degradation of TC.

This result demonstrates that combining the 3DOM structure with a defect-reduction approach is a highly effective strategy for enhancing the photocatalytic properties of $LaTiO_2N$.

16.4 Summary

16.4.1 Enhancement of TaON Inverse Opal Photocatalyst Properties Via Photonic Band Engineering

We successfully synthesized TaON IO photocatalysts with a uniform 3DOM structure by utilizing a two-step process that combined colloidal crystal templating via vacuum-assisted impregnation with subsequent thermal nitridation. The resulting TaON IOs exhibit brilliant structural color and possess accurately identifiable PBG

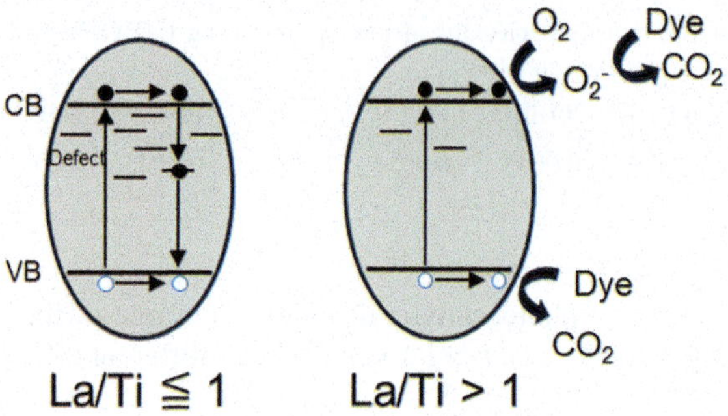

Fig. 16.11 Schematic diagram showing photocatalytic reactions based on differences in La/Ti ratio. In samples with many defects (La/Ti ≤ 1), photo-excited carriers are deeply trapped in numerous defect levels, causing many to deactivate (left). For samples with La/Ti > 1, fewer defect levels are generated, allowing charge carriers to be utilized in photocatalytic reactions (right). Holes decompose the tetracycline due to their strong oxidation power, while electrons reduce oxygen into reactive oxygen species. These reactive oxygen species, possessing strong oxidation power, then decompose the tetracycline

wavelengths, which enabled precise control over the alignment between the PBG and the photocatalyst optical absorption edge—a task that was previously intractable due to its dependence on multiple parameters. The 3DOM IO architecture led to a significant enhancement in hydrogen evolution compared to fine powder particles, owing to its unique properties that improve light-harvesting efficiency, promote charge carrier separation, and facilitate mass transport. More specifically, we confirmed clear evidence of a moderate photocatalytic activity enhancement induced by the slow photon effect in TaON IO samples, where the PBG edge was aligned with the material optical absorption edge. Aligning the red edge of PBG with the absorption edge promoted light absorption and electron-hole pair generation, thereby increasing the H_2 evolution rate. Meanwhile, aligning the blue edge of PBG with the absorption edge suppressed electron-hole pair recombination while still permitting strong light absorption, which also resulted in an enhanced H_2 production rate. These findings provide valuable new insights into the phenomenon of photocatalytic activity amplification via the slow photon effect in inverse opal photonic crystals and demonstrate how this effect can be strategically exploited to engineer superior photocatalyst performance.

16.4.2 Enhancement of $LaTiO_2N$ Inverse Opal Photocatalyst Properties Via Cation Ratio Optimization

We have introduced a synergistic approach to significantly enhance the properties of oxynitride photocatalysts, which combines the light localization effects of a 3DOM photonic crystal structure with a defect-reduction strategy based on cation stoichiometry control. The formation of the 3DOM structure for this multi-cation system, using a sacrificial template method, was successfully achieved by stabilizing the metal precursors via chelation with citric acid.

By slightly varying the La/Ti ratio in $LaTiO_2N$, we could effectively control the concentration of defects generated during thermal nitridation. In lanthanum-excess samples, a precipitated lanthanum oxide phase protects the $LaTiO_2N$ surface, thereby suppressing the excessive reduction of titanium and the associated formation of anion vacancies. The key to this protective function is likely the high stability of the La^{3+} ion; unlike titanium, which has multiple stable valence states, lanthanum is exceptionally stable, and its oxide is not readily reduced, allowing it to act as an effective passivation layer.

As a result of this combined strategy, the photocatalytic activity of the La/Ti > 1 samples was enhanced by a factor of more than four compared to the bulk, fine powder morphology. These findings demonstrate that a combination of 3DOM photonic crystal engineering and cation ratio control is a powerful and promising strategy for significantly improving the photocatalytic activity of oxynitride materials.

Acknowledgements The author is especially grateful to Mr. Taiki Maekawa, a PhD-course student of Graduate School of Science and Technology for Innovation, Tokushima University.

References

1. Nemiwal M, Gosu V, Zhang TC, Kumar D. Metal organic frameworks as electrocatalysts: hydrogen evolution reactions and overall water splitting. Int J Hydrog Energy. 2021;46(41):10216–38. https://doi.org/10.1016/j.ijhydene.2020.12.146.
2. Wang S, Li H, Xu L. Application of zeolite MCM-22 for basic dye removal from wastewater. J Colloid Interface Sci. 2006;295(1):71–8. https://doi.org/10.1016/j.jcis.2005.08.006.
3. Chen L, Chuang Y, Nguyen TB, Chang J-H, Lam SS, Chen C-W, Dong C-D. Novel molybdenum disulfide heterostructure nanohybrids with enhanced visible-light-induced photocatalytic activity towards organic dyes. J Alloys Compd. 2020;848(25):156448. https://doi.org/10.1016/j.jallcom.2020.156448.
4. Liu J, Zhao H, Wu M, Van der Schueren B, Li Y, Deparis O, Ye J, Ozin GA, Hasan T, Su B-L. Slow photons for Photocatalysis and photovoltaics. Adv Mater. 2017;29(17):1605349. https://doi.org/10.1002/adma.201605349.
5. Maeda K, Takata T, Hara M, Saito N, Inoue Y, Kobayashi H, Domen K. GaN:ZnO solid solution as a Photocatalyst for visible-light-driven overall water splitting. J Am Chem Soc. 2005;127(23):8286–7. https://doi.org/10.1021/ja0518777.
6. Kasahara A, Nukumizu K, Takata T, Kondo JN, Hara M, Kobayashi H, Domen K. LaTiO2N as a visible-light ($\leq$600 nm)-driven Photocatalyst (2). J Phys Chem B. 2003;107(3):791–7. https://doi.org/10.1021/jp026767q.
7. Hitoki G, Takata T, Kondo JN, Hara M, Kobayashi H, Domen K. An oxynitride, TaON, as an efficient water oxidation photocatalyst under visible light irradiation ($\lambda \leq$ 500 nm). Chem Commun. 2002;16:1698–9. https://doi.org/10.1039/B202393H.
8. Chen K, Xiao J, Hisatomi T, Domen K. Transition-metal (oxy)nitride photocatalysts for water splitting. Chem Sci. 2023;14:9248–57. https://doi.org/10.1039/D3SC03198E.

9. Matsukawa M, Ishikawa R, Hisatomi T, Moriya Y, Shibata N, Kubota J, Ikuhara Y, Domen K. Enhancing photocatalytic activity of LaTiO2N by removal of surface reconstruction layer. Nano Lett. 2014;14(2):1038–41. https://doi.org/10.1021/nl404688h.
10. Dong Y, Ai F, Sun-Waterhouse D, Murai K, Moriga T, Waterhouse GIN. Optical and photocatalytic properties of three-dimensionally ordered macroporous Ta_2O_5 and Ta_3N_5 inverse opals. Chem Mater. 2023;35(19):8281–300. https://doi.org/10.1021/acs.chemmater.3c01903.
11. Yablonovitch E. Inhibited spontaneous emission in solid-state physics and electronics. Phys Rev. 1987;58:2059. https://doi.org/10.1103/PhysRevLett.58.2059.
12. Yablonovitch E. Photonic band-gap structures. J Opt Soc Am B. 1993;10(2):283–95. https://doi.org/10.1364/JOSAB.10.000283.
13. Maekawa T, Ikeda Y, Fukuma H, Horikawa T, Waterhouse GIN, Murai K, Moriga T. Optical properties of photonic crystal balls of different sizes synthesized using a microfluidic device. Mater Sci Semicond Process. 2025;197:109716. https://doi.org/10.1016/j.mssp.2025.109716.
14. Waterhouse GIN, Waterland MR. Opal and inverse opal photonic crystals: fabrication and characterization. Polyhedron. 2007;26(2):356–8. https://doi.org/10.1016/j.poly.2006.06.024.
15. Maekawa T, Tateishi N, Ikeda Y, Maekawa H, Waterhouse GIN, Murai K, Moriga T. Enhancing the photocatalytic activity of $LaTiO_2N$ inverse opal photonic crystals by La/Ti stoichiometry optimization to suppress Ti^{3+} defects. Chem Mater. 2025;37(18):7439–49. https://doi.org/10.1021/acs.chemmater.5c01864.
16. Wu M, Liu J, Wang C, Huang S, Deng Z, Li Y, Su B-L. Probing significant light absorption enhancement of titania inverse opal films for highly exalted photocatalytic degradation of dye pollutants. Appl Catal B: Environ. 2014;150:411–20. https://doi.org/10.1016/j.apcatb.2013.12.037.
17. Deparis O, Mouchet SR, Su B-L. Light harvesting in photonic crystals revisited: why do slow photons at the blue edge enhance absorption? Phys Chem Chem Phys. 2015;17:30525–32. https://doi.org/10.1039/C5CP04983K.
18. Wan Y, Wang J, Wang X, Xu H, Yuan S, Zhang Q, Zhang M. Preparation of inverse opal titanium dioxide for photocatalytic performance research. Opt Mater. 2019;96:109287. https://doi.org/10.1016/j.optmat.2019.109287.
19. Chang Y, Yu K, Zhang C, Yang Z, Feng Y, Hao H, Jiang Y, Lou L-L, Zhou W, Liu S. Ternary CdS/au/3DOM-SrTiO3 composites with synergistic enhancement for hydrogen production from visible-light photocatalytic water splitting. Appl Catal B. 2017;215:74–84. https://doi.org/10.1016/j.apcatb.2017.05.054.
20. Liu Q, Liang S, Wang J, Zhang R, Wang X. Cadmium Sulfide 3D photonic crystal with hierarchically ordered macropores for highly efficient photocatalytic hydrogen generation. Inorg Chem. 2022;61(6):2920–8. https://doi.org/10.1021/acs.inorgchem.1c03798.
21. Maekawa T, Maekawa H, Ikeda Y, Onoe T, Waterhouse GIN, Murai K, Moriga T. Synthesis of polycrystalline Ta_2O_5 inverse opal photonic crystal powders and their optical characterization. Open Ceram. 2024;20:100688. https://doi.org/10.1016/j.oceram.2024.100688.
22. Maekawa T, Huang Y-S, Tateishi N, Nakanishi A, Onoe T, Dong Y, Waterhouse GIN, Murai K, Moriga T. Slow photon photocatalytic enhancement of H_2 production in TaON inverse opal photonic crystals. J Solid State Chem. 2024;329:124404. https://doi.org/10.1016/j.jssc.2023.124404.

Chapter 17
Computational and Experimental Approaches for Protein–Metal Complex Hybrid Materials

Takashiro Akitsu

Abstract In this chapter, the author presents their experimental and computational studies on complex systems of metal complexes and proteins. For a long time, the authors have been performing single-crystal X-ray structure analysis and obtaining (polarized) UV–vis (ultraviolet–visible) absorption, circular dichroism, infrared absorption, and fluorescence spectra of chiral metal complexes in crystal-packed, in-solution, and soft-matter (synthetic polymers, biopolymers, colloids, etc.) systems to elucidate new phenomena and mechanisms of intermolecular interactions. In particular, we have been conducting experimental and computational studies on artificial metalloproteins, i.e., protein–protein complexes and metal (ligand) complexes, including their building blocks, as well as experimental and computational evaluations of protein–ligand complexes to elucidate their functions. For metal complexes, we have chosen mainly those with analogs of the typical type of "Schiff base ligand." Schiff base is a general term for compounds formed by dehydration of aldehydes or ketones with primary amines. Schiff base salt complexes are rich in coordinating environment changes around the central metal, such as steric effects of substituents, and various studies, including spectroscopic, magnetic, and redox studies, have been conducted on the coordination environment. The proteins selected include the immunoprotein CD19, the viral protein JaGAr, human serum albumin, egg white lysozyme, and laccase, many of which are commercially available and easy to handle. Various examples, including the formation of supramolecular assemblies through intermolecular interactions with biomolecules (proteins) as well as metal complex molecules alone, the physical properties due to chirality, molecular recognition through noncovalent bonding, supramolecular arrangements in soft matter, and chemical reactions, are presented.

Keywords Metal complexes · Schiff base · Protein · Hybrid materials · Computational chemistry · Spectroscopy · Lysozyme · Artificial metalloprotein · Laccase · Electron transfer · Photoisomerization · Chirality

---

T. Akitsu (✉)
Department of Chemistry, Faculty of Science, Tokyo University of Science, Tokyo, Japan
e-mail: akitsu2@rs.tus.ac.jp

T. Onishi (ed.), *Advances in Quantum Science II*,
https://doi.org/10.1007/978-981-95-9721-5_17

17.1 Introduction Computational and Mathematical Viewpoint for Organic/Inorganic Hybrid Complexes

17.1.1 Characterization

Organic/inorganic hybrid materials and metal complexes are characterized not only by their structural hierarchy, which endows them with properties and functions that cannot be possessed by a single component, but also according to computational chemistry and mathematical considerations. Examples include artificial metalloproteins, i.e., protein–protein complexes and metal (ligand) complexes as hybrid materials [1]. These functional materials have attracted attention in terms of their reactions, physical properties, and structures. The author would like to introduce some examples from recently published forewords and proceedings papers [2].

For example, the author's group has studied suitable metal complexes as mediators of native metalloenzymes to reduce oxygen, namely to ensure smooth electron transfer from the cathode to laccase from *Trametes versicolor* within biofuel cells (Fig. 17.1). During the study, various physicochemical aspects of these systems were considered [3]. The topics of interest and techniques used included electron transfer associated with chiral-induced spin selectivity (without magnetic fields), the Weigert effect (selective photoisomerization by electric dipole transitions) of azobenzene (AZ), polarized photoemission spectroscopy, ligand–protein density functional theory (DFT) calculations based on chiral molecular recognition, and spectroscopic interpretation considering the spatial factors of molecules. Electrochemistry, electron spin resonance, and magnetic measurements were also used to study the redox reactions of these paramagnetic metal species. Physical chemistry is not clearly separated from solid-state physics or biophysics. For example, in a published study on laccase, the author investigated reactions and their mechanisms using chemical experiments and computational chemistry or electromagnetic and quantum mechanical considerations [4].

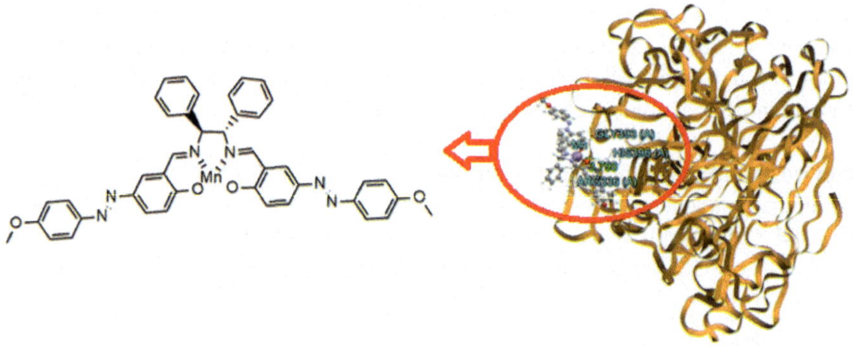

Fig. 17.1 Docking feature of a Schiff base complex (left) in laccase (right) as mediator

17.1.2 Schiff Base Ligand

Metal complexes of Schiff bases (imine or azomethine-N=CH−) are typical types and have various optical, magnetic, and redox (reaction) functions. These complexes are an evergreen subject in terms of both chemical experiments and calculations and mainly include the following targets [5]: basic (physical and chemical) properties of Schiff bases; formation mechanism or principles in the synthesis of Schiff bases; reactions with a Schiff base or its metal complex; Schiff base ligands in metal complexes; analysis or classification of Schiff base compounds; medical or biological applications of Schiff bases; and the use of Schiff bases in materials science and engineering.

17.1.3 Metal Complexes

There are compounds with photoelectron transfer (Ru) and redox (V=O) functions involving metal ions or metal sites that are characteristic of molecular design, such as for catalysts, for example, the electron spins of Ru and V=O complexes. The spin state of an electron can play an important role in these functions [6]. Lanthanide ions in optically and magnetically functional molecules have high coordination numbers due to the coordination bonding of ions with large ionic radii with metal elements [7]. According to the hard and soft acid and base (HSAB) principle, lanthanide ions prefer to coordinate with O atoms. Similar to other metal ions, lanthanide ions are stabilized by the chelating effect of the ligand. However, relativistic effects make computational chemistry of compounds containing lanthanides difficult [8].

To discuss the structure–property relationships of metal complexes alone and the docking of metal complexes with proteins, the three-dimensional steric structures of the metal complexes must be known. Crystal structure analysis is a powerful tool for experimental studies of both proteins and metal complexes. The recent rise of data science has broadened the possibilities for integration with computational chemistry and the use of mathematical considerations. For example, the crystal structure of a Cu complex was rereported from another point of view by using new computational methods to evaluate experimental data. Synthesis, identification, physical property functions, their confirmation, and explanations are discussed from the perspective of structure–physical property relationships. Crystal structure analysis has now become an everyday tool. Computational chemistry that promotes secondary uses, such as Hirshfeld surface analysis, is also enriched and is important for reanalysis of crystals [9]. Intermolecular interactions and protein–ligand docking in crystals should not be unrelated (Fig. 17.2). Because although the situation is different during crystal packing of homogeneous molecules and during docking between heterogeneous molecules, potential intermolecular interaction sites of a certain molecule can at least be identified.

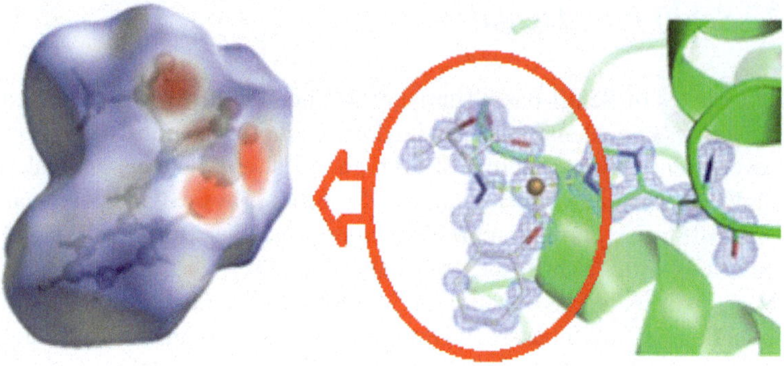

Fig. 17.2 Hirshfeld surface of a Schiff base complex (left) binding to lysozyme (right). Both are based on each result from experimental X-ray crystal structure analysis

17.1.4 Crystal Symmetry

Space group symmetries beyond those of crystalline states are applicable also in chemistry. When predicting a crystal structure or analyzing an unknown powder crystal, the crystal system and space group that can logically exist in a single crystal must be determined by trial and error. In addition, the crystal symmetry is useless for soft-matter materials [10]. The mathematical geometry and groups for low-symmetry metal complex systems would be useful in this context. Not only chiral molecules but also complex and low-symmetry supramolecular metal complexes and aggregates can be classified into one of the 230 space groups, at least in terms of the crystal symmetry. Arguments based on classical point groups cannot be applied to metal complexes with low symmetry, and computational chemistry becomes essential [11]. Specifically, the crystallographic result (even if it belongs to $P1$ symmetry) belongs to a space group. However, not all real molecular 3D structures have such high symmetry as the "hexahedron octahedron = O_h point group." Computational chemical methods are commonly used to attribute d–d transitions to real molecules (Fig. 17.3).

17.1.5 Artificial Intelligence and Computational Chemistry

Recently, artificial intelligence (AI) or category theory has potentially become important in any chemistry field. "Mathematical chemistry," as an academic field, was proposed in the twentieth century as a way of thinking that abstracts and expresses "variables," "symbols," and "functions." Mathematics in the sense of arithmetic and statistics used to calculate chemical data is excluded here for now. However, "group theory," which is well known as the mathematics that describes the three-dimensional shape of compounds, the electronic structure of atoms and molecules, and the geometric structure and symmetry of crystals, is widely used in various fields

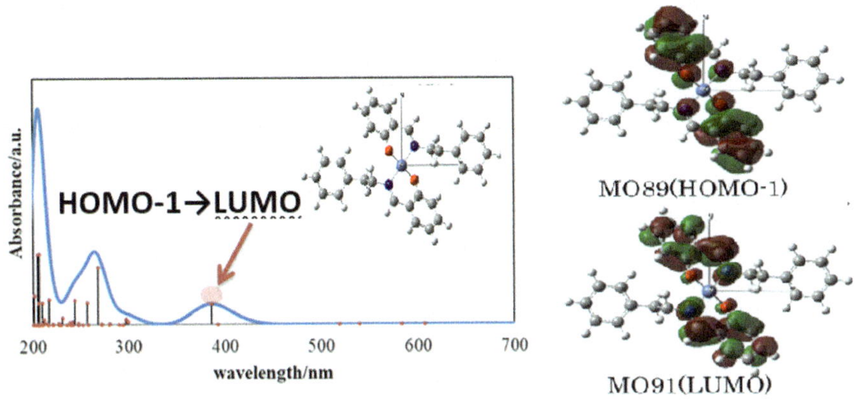

Fig. 17.3 Simulated ultraviolet–visible (UV–vis) spectrum of a Schiff base metal complex (left) assuming the experimental crystal structure of low-symmetry (inserted) and orbitals associated with the peak around 380 nm (right). *Highest occupied molecular orbital (HOMO), lowest unoccupied molecular orbital (LUMO)

of material science. While achievements have been made in the application of artificial intelligence to various sciences (such as computing and social networks), the application of "category theory" to chemistry is regarded as being worthy of attention [12].

In view of this interest, any topic beyond the scope of each computational chemistry method should be treated with the aid of experimental results of hybrid materials of metal complexes and functional materials of particular proteins. Examples of the authors' research are presented later in this chapter. How to discuss information such as intermolecular interactions, molecular assembly states, and packing, which is beyond the scope of application of each computational chemistry program and theory, is still a subject of research [13].

17.2 Spectroscopic Approaches for the Immune Protein CD19 and Binuclear Complexes

17.2.1 Immune Protein CD19 and Fluorescence Spectroscopy

The immune protein CD19 is expressed on B cells and is often used as a biomarker in immunology and oncology, especially in flow cytometry. It is a useful tool for diagnosis of primary immunodeficiency diseases, but traditionally, the analysis of this intracellular protein has often been difficult for technical reasons, partly because of loss of fluorescence emission or dissociation from the protein molecule. CD19 functions as a coreceptor for B-cell receptor signaling. It forms a complex with CD21,

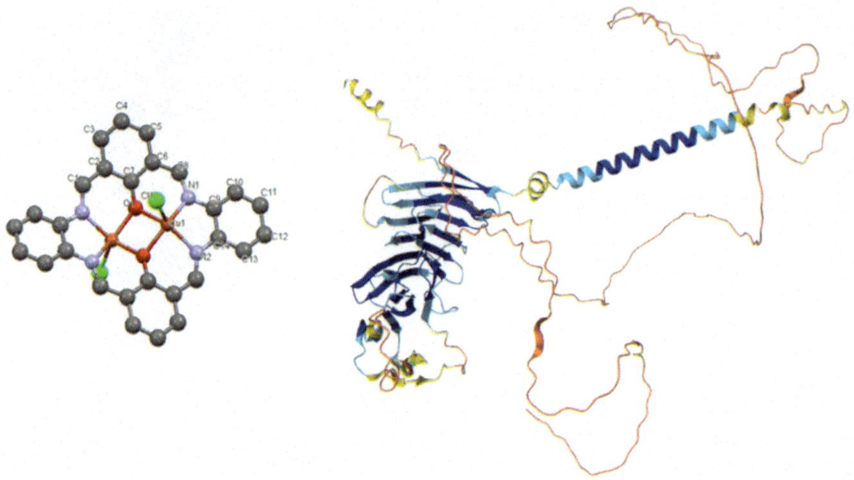

Fig. 17.4 (Left) crystal structure of a Zn(II) complex for binding to CD19 [16]. (Right) Computational three-dimensional structure of CD19 based on PDB

CD81, and CD225 to amplify antigen-induced signaling and lower the threshold for B-cell activation. This process involves phosphorylation by kinases, activation of the phosphatidylinositol 3-kinase pathway and promotion of B-cell survival and proliferation. According to a practical example, a novel Schiff base binuclear Zn(II) complex can significantly increase the fluorescence intensity of the CD19 antibody, even though it does not specifically bind to CD19. This situation will be explained using an easy-to-understand schematic diagram (Fig. 17.4). X-ray crystallography and in silico docking simulations were used to investigate the interaction of the complex with the CD19 antibody. Docking simulations suggested that the complex does not bind directly to CD19 but is included in the center of the CD19 trimer structure, leading to an increase in fluorescence emission [14].

17.2.2 Experimental Spectroscopic Approaches

Needless to say, understanding the three-dimensional structure (crystallography) and electronic states (quantum chemistry) is the key to research on metal complexes and metalloproteins [15]. UV–vis (ultraviolet–visible), circular dichroism (CD), and fluorescence (FL) spectroscopies are essential tools for analyzing the molecular structure, electronic transitions, and optical properties of materials. These techniques are widely used to investigate the behavior of metal complexes, organic molecules, and hybrid materials under different conditions. In addition, polarization experiments can also be used to evaluate the molecular orientation in thin films or polymer matrices [16].

The chiral or achiral Ni(II) and Zn(II) complexes identified by these spectroscopic methods showed an increase in optical anisotropy under UV light irradiation after dispersion in polymer films. Transition dipole moments along the molecular long axis aligned perpendicular to the polarized UV light electric vector were deduced [17].

Ultraviolet–Visible (UV) Spectroscopy

UV–vis spectroscopy is used to study electronic (d–d, charge transfer (CT), intraligand) transitions, particularly in metal complexes (Fig. 17.5). How do orientations change UV–vis spectra? When molecules align parallel to the electric field vector of the incident light, absorption is maximized because of the improved alignment of the electronic transition dipoles. When molecules align perpendicularly, absorption is reduced because of the lower overlap with the electric field. In thin films, polarized UV–vis spectra reveal the dichroic ratio ($R = A_{\text{parallel}}/A_{\text{perpendicular}}$), which indicates the strength of the molecular alignment. UV–vis spectra can reveal the alignment. For an aligned molecular structure, a high dichroic ratio ($R > 1$) suggests the presence of aligned molecules. For randomly oriented molecules, a ratio close to 1 ($R \approx 1$) indicates an isotropic (random) molecular distribution. The circular diagram of a polarized light spectrum is a diagram in which the angle of the polarizer is plotted as the circumference and the absorbance (Abs.) at that angle as the radius (Fig. 17.5). It can be seen that the absorption is stronger in the direction of angles with larger Abs., i.e., the dipole moment of the molecule is oriented. The approximate Abs. ratio at a specific perpendicular angle is taken as the R value. A discussion of the molecular orientation using computational chemistry (transition dipole moments) and electromagnetism (dipole–dipole interactions and electric and magnetic field vectors) is also possible potentially for ideal data [18].

Circular Dichroism (CD) Spectroscopy

CD spectroscopy is used to study chiral molecules and optically active systems by measuring the differential absorption of left and right circularly polarized light. The following results and interpretations are expected: Cotton effects in d–d, charge transfer (CT), and π–π^* transitions of chiral molecules and induced chirality, in which CD signals appear when achiral molecules acquire supramolecular conformations from chiral molecules. However, discussing the CD spectra of oriented samples is often difficult because of the problem of artifact peaks. When molecules preferentially align in one direction (to the extent that molecular motion is hindered and isotropy is lost), the intensity of the CD signal is "falsely" altered. In liquid crystals and oriented polymer films, care must be taken to avoid such changes in the CD peak due to structural chirality and molecular packing symmetry (although there are experimentally simple or essential ways to avoid this) [19]. A theoretical framework explaining the intensity manifestation of linear dichroism (LD) and CD

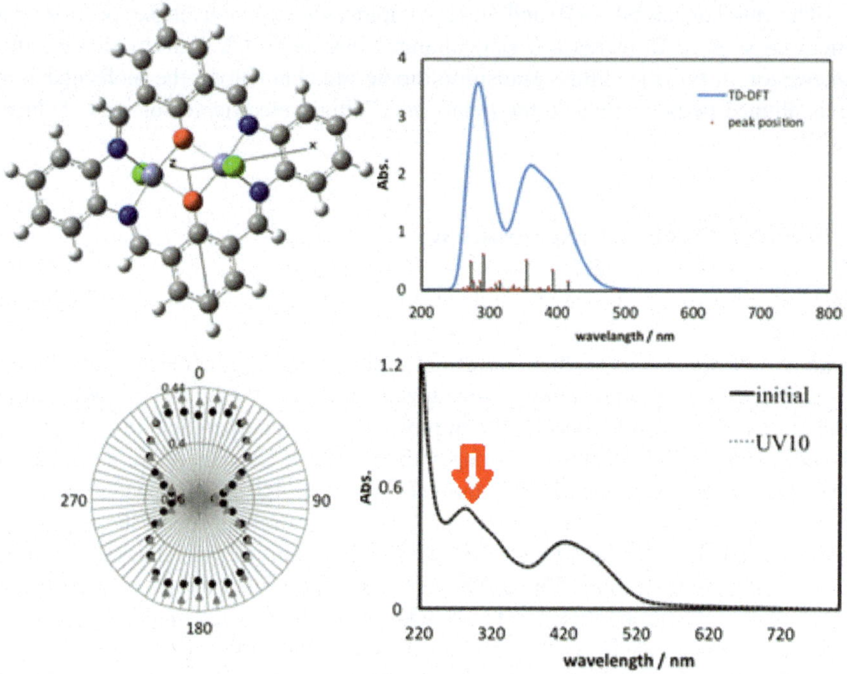

Fig. 17.5 (Above) Density functional theory (DFT)-optimized structure (left) and simulated ultraviolet–visible (UV–vis) spectrum with transitions for assignments (right). (Below) Plot of peak absorbance of polarized spectra (left) and polarized spectra with an arrow of the peak after UV light irradiation for 10 min (right)

exists, which was the basis for the development of an experimental apparatus that can be used to experimentally demonstrate the intensity variation of artifact peaks that depend on the surrounding environment. However, incorporating the factors caused by an isotropic molecular orientation and the surrounding environment (e.g., solvent viscosity) into the energies and transition probabilities of the ground and excited states in computational chemistry is currently difficult.

Fluorescence (FL) Spectroscopy

One of the characteristics of FL spectroscopy is that it provides both static and dynamic information. Static information (e.g., the energy difference between excited and ground states and the transition probabilities) can be the subject of ordinary computational chemistry, whereas dynamic information (e.g., the FL lifetime, quantum yield, and energy transfer to other molecules) is still difficult to discuss. Therefore, the experimental data are sometimes interpreted by empirical rules. FL spectroscopy is used to investigate the luminescence properties of molecules that are typically excited by UV or visible light. It can also be used to estimate the extent of intermolecular interactions and contacts based on the decrease in the fluorescence intensity (quenching) that accompanies energy transfer. The degree of quenching is sometimes analyzed using Stern–Volmer plots to provide kinetic parameters generally. In addition, FL spectroscopy provides the basis for the flow cytometry method [20].

How the FL spectrum changes with the molecular orientation can be discussed as follows: (1) Dipole orientation: The emission intensity depends on whether the fluorescent molecules are aligned parallel or perpendicular to the excitation source. (2) Excimer formation: Strong molecular stacking increases the emission of the excimer and redshifts the fluorescence. (3) Polarization of emission: Aligned molecules exhibit polarized fluorescence, which is used in optoelectronic device applications.

Therefore, the following information will be obtained with FL: (1) Molecular interactions: The fluorescence lifetime of molecules changes as they experience local environmental changes. (2) Energy transfer efficiency: Förster resonance energy transfer (FRET) more efficiently occurs in aligned structures.

17.2.3 Computational Spectroscopic Approach

Since metal atoms have many electrons and protein molecules have many atoms, both the computational volume and the accuracy in the simulation of spectra related to electronic transitions must be considered. The author has experience in using semiempirical molecular orbital methods under conditions of insufficient computer resources. For other studies, theoretical calculations of chiral Schiff base Zn(II) complexes were used to qualitatively analyze the UV–vis, CD, and fluorescence spectra. They were performed by the use of Gaussian 09, Revision A.02. In semiempirical calculations, structure optimization and reference vibration analysis were performed using PM6 or the improved Hamiltonian [21], and excited state calculations were performed using the ZINDO Hamiltonian, which was applied for the spectra of metal complexes [22]. DFT calculations were then used, but the agreement of the experimental spectra with the semiempirical calculations was often better (Fig. 17.6).

The selection of the calculation accuracy and cost is always a vexing issue. The theoretical absorption spectra calculated by ZINDO were compared with the measured spectra, and the molecular orbital diagrams involved in the transitions were discussed. The characteristics of the transitions of organic ligands were well discussed. For one chiral Schiff base Zn(II) complex, the calculations revealed that the absorption bands corresponded to π–π^* transitions in the salicylaldehyde moiety, with the amine moiety contributing little to the electronic state. This is consistent with why the absorption intensities of the absorption bands of the four complexes shown below are almost identical. Therefore, it is considered that the change in the fluorescence intensity at the excitation wavelength is due to the contribution of the excited state of the salicylaldehyde moiety. The absorption bands are composed of several transitions, of which the transition with the highest absorption intensity is

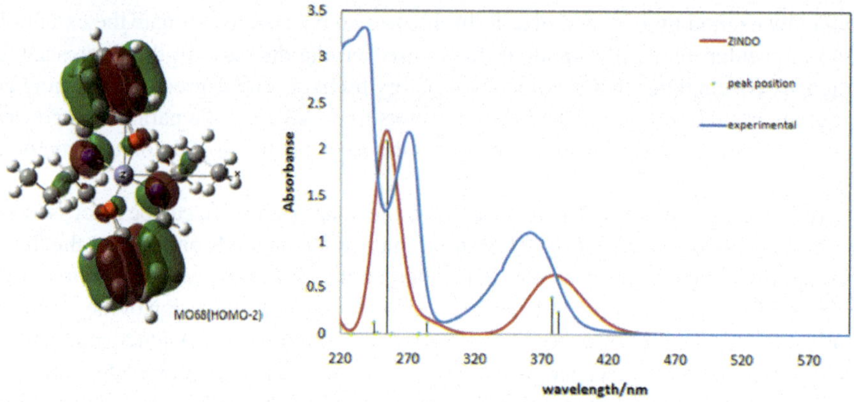

Fig. 17.6 Schiff base Zn(II) complexes showing an orbital and experimental (blue) and ZINDO simulated (red) UV–vis spectra for this complex

the n-π^* transition from the N atom to the salicylaldehyde moiety, which has a high transition contribution [23].

17.3 Docking Proteins with Copper Complexes of Amino Acid Schiff Bases

17.3.1 Antiviral Activity of Copper Complexes

The author has studied experimental protein–ligand (metal complex) docking, which is of biological interest, with the aid of various computational methods. Schiff bases derived from amino acids are known to form stable complexes with transition metals, and various effects can be exerted by changing the central metal ion and substituents. When the Cu(II) ion is used as the central metal ion, the complex shows excellent antimicrobial activity. Using microwave synthesis, our laboratory succeeded in systematically synthesizing Cu(II) complexes of amino acid Schiff bases with different substituents in a short time and demonstrated that they exhibit effective antibacterial activity against several fungi, including *Escherichia coli* [24].

With the frequent occurrence of global pandemics over the past several years, countermeasures against unknown pathogens have become essential. These diseases can be caused by any pathogenic microorganism, including bacteria, fungi, and viruses. Therefore, there is a need for compounds that exhibit inactivating or growth-inhibiting activity against a wide range of pathogenic microorganisms. Compounds with antimicrobial activity are considered to tend to inactivate enveloped viruses. In a previous study [25], the antiviral activity of 20 amino acid Schiff base Cu(II) complexes with different substituents synthesized in our laboratory was investigated mainly by computational chemistry to search for compounds that exhibit inactivation and growth inhibitory effects against a wide range of pathogenic microbes.

17.3.2 Three-Dimensional Structure of Viral Proteins

In such cases, if the three-dimensional structures (crystal structures) of the proteins are unknown, docking calculations with the ligand are impossible. However, in recent years, when the three-dimensional structure of a protein can be calculated with high precision, there have been situations in which the barrier to this research seems low. First, protein structure prediction using AlphaFold2 [26] and ColabFold allowed us to estimate the three-dimensional structure of the Japanese encephalitis virus (JEV) protein with high accuracy [27]. By using this method, reliable structures can be obtained even when experimental crystallographic analysis is not feasible (Fig. 17.7). In addition, based on the obtained structural data, we could prepare a model suitable for docking simulations by removing unnecessary elements such as water molecules.

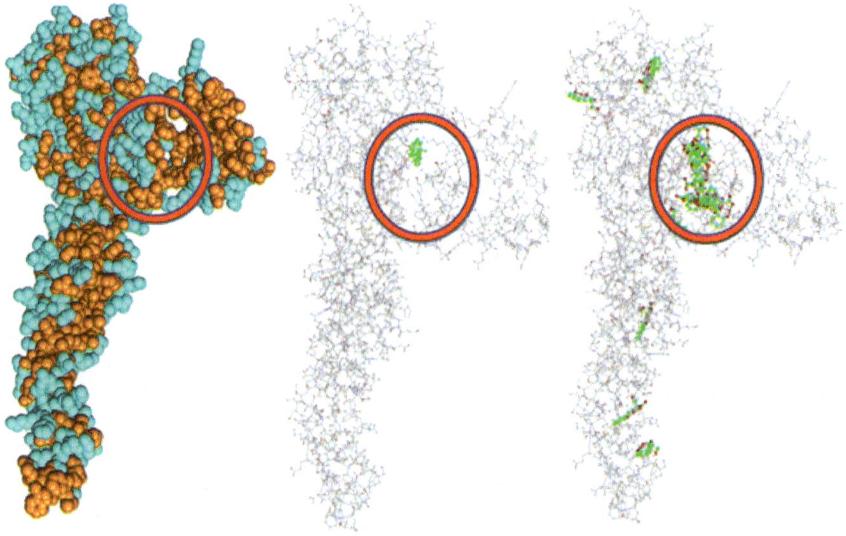

Fig. 17.7 Three-dimensional structure of E protein of Japanese encephalitis virus: JEV (JaGAr01 strain) predicted by AlphaFold2 (left), with hydrophobic groups colored orange and hydrophilic groups colored light blue. The optimal docking structure of JEV with the Ser complex where (red circles) showed the highest GoldScore (center). The process of searching for the docking structure of JEV with the Ser complex, which examined about 30 candidates (right)

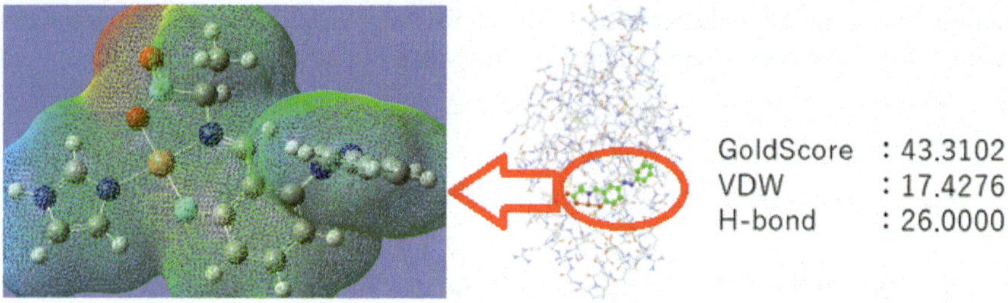

Fig. 17.8 Example of electron density distribution of amino acid Schiff base Cu(II) complex (left) and protein–ligand docking feature and scores for the complex (right)

17.3.3 Optimal Geometry and Electronic Structure of Copper Complex

Next, conventional density functional theory (DFT) calculations using Gaussian 09 [28] were used to determine the optimal geometry and electronic structure of the Cu(II) complex used as the ligand. In this analysis, the electronic distribution of the Cu(II) complex was evaluated (Fig. 17.8). This result suggests that the Cu atom is expected to interact with the electrophilic site of the JEV protein. We also evaluated how the difference in the electronic state of the complexes with and without the Cl– group, which is experimentally more antimicrobially active, affects the binding affinity.

17.3.4 Docking Simulations

To further analyze the interaction between the ligand and the JEV protein, docking simulations were performed using the GOLD program with a genetic algorithm for flexible docking [29]. The results revealed that the amino acid Schiff base Cu(II) complex that showed the highest binding affinity (GoldScore: 51.1385) had features and sites of interaction with important amino acid residues involved in JEV function. It is considered that their proximity to certain residues reduce the activity of the virus, which has led to the selection of candidate complexes that are promising inhibitors of JEV. Furthermore, the binding affinities of the Cu(II) complexes without Cl groups were generally lower than those of the complexes with Cl groups, confirming that the molecular shape (quantification of this feature is a future issue) and hydrophobic interactions significantly influence the binding affinity.

The drug-like properties of the candidate complexes were evaluated, and all of Lipinski's rules (molecular weight, log *P* (n-octanol–water partition coefficient), number of hydrogen bond donors/acceptors, etc.) were satisfied [30], and sophisticated ADMET analysis [31] confirmed that these complexes exhibited good pharmacokinetics in vivo. This study demonstrated that by using computational chemistry approaches, effective candidate compounds can be screened while avoiding experimental costs. This study is expected to accelerate the development of new therapeutic agents against the Japanese encephalitis virus.

17.3.5 Combined Computational and Experimental Approaches

A combined computational and experimental approach has been employed to analyze how metal complexes bind to proteins. In this study, amino acid Schiff base Cu(II) complexes were systematically synthesized, and their structures and properties were analyzed by experimental and computational methods. Specifically, Fourier transform infrared (FT-IR), UV–vis, CD, FL, and electron spin resonance (ESR) spectroscopies and cyclic voltammetry (CV) measurements were used to investigate the complex properties in detail. In addition, DFT calculations were used to optimize the geometry and electronic structure of the complexes, and the results were compared with experimental data [32]. DFT calculation under various functional and basis conditions [33] were performed (Table 17.1), and the result was benchmarked with experimental values (d–d bands of UV–vis spectra). Higher accuracy requires longer calculation times, but this does not necessarily result in better agreement between the calculated and experimental values.

In addition, when discussing the reaction mechanism of superoxide dismutase (SOD) [34], natural bond orbital (NBO) analysis, which is usually used [35], was performed to obtain bond information and calculate the bond order. Energy calculations of amino acid Schiff base Cu(II) complexes were performed using the counterpoise (CP) method [36]. All calculations were performed within the DFT framework. Cam-B3LYP was used as the functional, and D3 dispersion force correction was applied. cc pvdz was applied as the basis set, and the Gaussian 09 program was employed.

The interaction with lysozyme was analyzed through a combination of crystallization experiments and protein–ligand docking simulations using DFT-optimized

Table 17.1 Comparison of the d–d peak wavelengths of an amino acid Schiff base Cu(II) complex under different DFT conditions

d–d peak/nm	B3LYP	CAM-B3LYP	ωB97YP
STO-3G	452.20	429.20	429.75
3-21G	470.20	444.65	440.28
6-31G	470.86	446.13	526.88
6-31G(d)	469.13	445.25	438.30
6-31G(d,p)	469.99	446.17	439.27
6-31+G(d,p)	473.03	479.80	477.36
6-311G(d,p)	479.87	474.51	482.75
6-31+G(d,p)	475.08	472.35	479.99

structures. Lysozyme is a protein whose crystal structure has been analyzed in detail, and the His residues exposed on its surface often serve as binding sites for metal complexes. Cocrystallization of lysozyme and Cu(II) complexes was attempted using the hanging drop vapor diffusion method, and the results suggested that complexes with specific amino acids, especially those containing Thr, are suitable for crystallization. However, crystal structure analysis has not been performed, and further studies are needed [37].

For protein–ligand docking simulations, the interactions between lysozyme and Cu(II) complexes were evaluated using GOLD software. The calculations revealed two patterns in the binding mode of the ligands, with some complexes adsorbing on the protein surface and others fitting into pockets of the protein. However, GOLD does not properly handle the coordination bonding of metal complexes, suggesting that there are still issues in predicting binding considering the contribution of Cu(II) ions. Further analysis using other docking algorithms and force fields (for molecular dynamics) is needed to improve the validity of the calculation results [38].

Additionally, the relationship of structural changes on the protein side (i.e., not the substrate side), rather than on the small-molecule ligand side, with reactions and functions can be investigated via dynamical computations. Lysozyme conformational changes due to the binding of amino acid Schiff base Cu(II) complexes were investigated using quantum mechanics/molecular dynamics (QM/MM) methods [39], but for more precise analysis, molecular dynamics (MD) calculations (QM/MM-MD) with the reactive force field (ReaxFF) were performed in addition to QM/MM calculations [40]. Specifically, the SOD activity of complexes with Ser and Val derivative ligands in the bound state of lysozyme was experimentally measured, and the superoxide $\left(O_2^-\right)$ binding and reaction processes were investigated in detail via calculations. The calculations revealed that in the Ser complex, after the substrate (superoxide) was bound to the Cu moiety, the electron density was changed by Arg14, His15, and Thr89 of lysozyme, and super oxide dismutase (SOD) activity was expressed. In contrast, in the Val complex, although superoxide was bound to the Cu moiety, the complex separated from the lysozyme over time, and the SOD activity decreased. These calculations indicate that the appropriate amino acid-derived ligand facilitates binding of superoxide and cooperates with specific residues of lysozyme to exert SOD activity [41].

Furthermore, QM/MM-MD simulations using ReaxFF potential revealed structural changes not observed by conventional QM/MM calculations (Fig. 17.9), particularly the electron-deficient properties of Arg14 involved in SOD activity. In this calculation, the total energy tends to remain constant at the initial calculated value, which means that there is no significant change over time. Furthermore, it is considered that the guanidine group of Arg14 is converted to hydrogen peroxide by hydrogen elimination. These results indicate that the difference in SOD activity between the Ser complex and the Val complex can be explained by structural changes after superoxide binding and interactions with the surrounding amino acid residues. This discovery will contribute to the design of artificial metalloproteins that target superoxide and the development of new catalysts for controlling reactive oxygen species (ROS) [42].

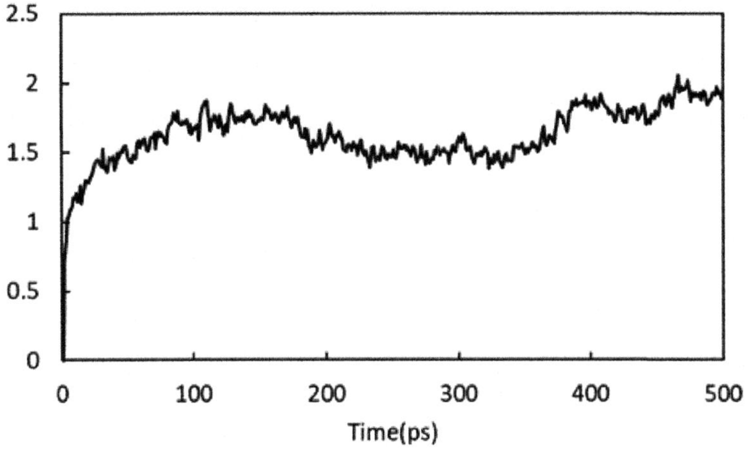

Fig. 17.9 Total energy in molecular dynamics (MD) calculations of the Val complex-lysozyme hybrid

17.4 Infrared Free Electron Laser Decomposition of Human Serum Albumin and Proteins with Schiff Base Complexes

17.4.1 Human Serum Albumin (HSA)

Human serum albumin (HSA) is a protein found in plasma that is known to be involved in the transport of metal ions and metal complexes associated with drugs in the blood. It fluoresces and is commonly evaluated by fluorescence spectroscopy. It is also an interesting enzyme that exhibits esterase-like activity due to its substrate diversity.

17.4.2 Infrared Free Electron Laser

When HSA is complexed with a Schiff salt-based Zn(II) complex and irradiated by an infrared free electron laser (IR-FEL) [43], the conformational change of the protein is accelerated by the change in the folding structure of the protein molecules. Medical applications are also expected, for example, to target the accumulation of heavy metal ions that cause disease or proteins with abnormal structures and destroy them by binding appropriately designed metal complexes (Fig. 17.10).

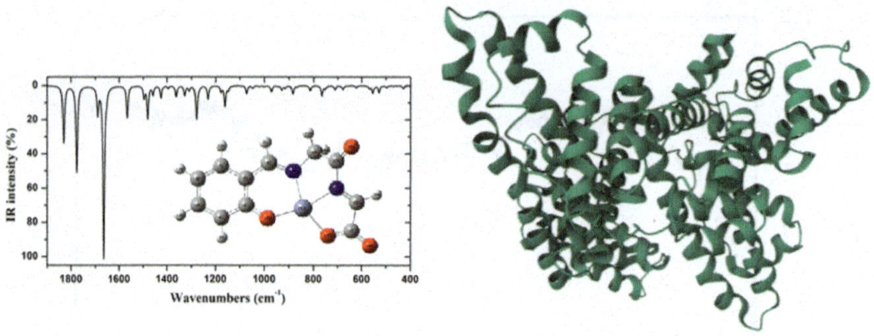

Fig. 17.10 DFT simulated infrared free electron laser (IR) spectra (left) from optimized structure (inserted) and crystal structure of Human serum albumin (HSA) from PDB (right)

17.4.3 Long-Ultraviolet (UVA) Absorption Effects

Similar to the Schiff salt-based complexes with long-ultraviolet (UVA) absorption effects previously reported, as infrared rays are harmful to skin collagen depending on their intensity, the creation of metal complexes that absorb infrared rays would be useful. We attempted IR-FEL irradiation of proteins complexed with metal complexes that clearly absorb infrared rays. IR-FEL irradiation of proteins complexed with metal complexes has rarely been studied [44].

For longer-wavelength UVA light up to approximately 400 nm, in contrast to Zn(II) oxide, certain Schiff base Zn(II) complexes incorporating amino acid derivative moieties can absorb UVA light, thus preventing UVA light from reaching the skin and causing many spillover effects, with potential applications in sunscreen. In contrast, absorption of IR light by sunscreen can be simply assumed to protect skin in the IR region.

17.4.4 Stern–Volmer Plot

Docking measurements of small molecules of complex rice rape, proteins, and DNA in the field of biochemistry can be verified using absorption and fluorescence measurements, as well as simulations. The obtained peak intensities of the fluorescence quenching spectra were used to create Stern–Volmer plots (Fig. 17.11). Stern–Volmer equation:

$$F_0/F_1 = 1 + K_{SV}[Q] = 1 + K_q K_0[Q]$$

where F_0: fluorescence intensity in the absence of quencher, F: fluorescence intensity in the absence of quencher, K_{SV}: quenching constant: t_0: fluorescence lifetime, k_q: two-molecule quenching rate constant, and $[Q]$: solution concentration, provides

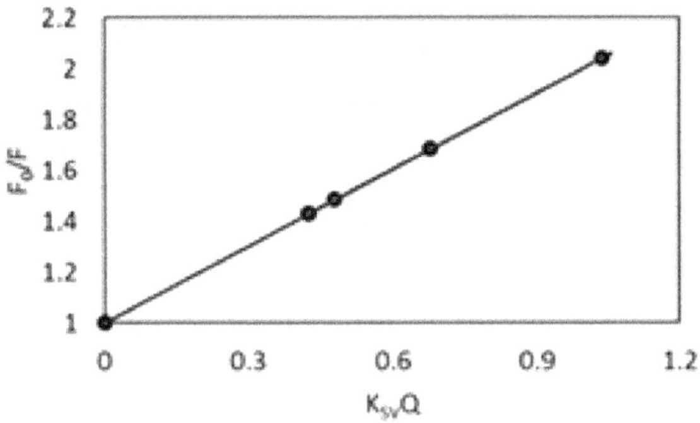

Fig. 17.11 Decreasing fluorescence (FL) intensity along the Stern–Volmer plot of a protein due to binding of a metal complex

some quenching parameters from decreasing of fluorescence intensity. The vertical axis represents the ratio of the fluorescence intensity to the fluorescence intensity without the addition of the quencher complex, the horizontal axis represents the concentration of the complex, and the two axes in the graph are on logarithmic scales.

17.4.5 Complexation of HSA with Zinc Complex

An IR-FEL can degrade aggregating proteins via excitation of vibrational bands. In a previous study, hybrid materials of HSA containing several new Schiff base Zn(II) complexes incorporating amino acids or dipeptide derivative moieties were prepared, synthesized, and characterized by UV–vis, CD, and IR spectroscopy. Time-dependent or conventional DFT calculations were also performed to investigate the vibrational modes [45] of the Zn(II) complexes corresponding to the imine C=N, amide I, and amide II bands in the IR spectra obtained under IR irradiation. Analysis of the secondary structure suggested that the inclusion of the Zn(II) complex in HSA altered the structure of HSA, making it more brittle than the original HSA. This result was one of the features of vibrational excitation by an IR-FEL, as opposed to electronic excitation by UV or visible light.

The complexation of HSA with the synthesized Zn(II) complexes was confirmed by measuring the UV–vis, CD, and fluorescence spectra of the mixed solutions. Specifically, the UV–vis spectra revealed that some of the electronic transition peaks that appeared in the case of the complex alone disappeared. The CD spectra revealed a change in the spectral shape of HSA alone, suggesting a change in the secondary structure of HSA (due to the adsorption of the complex), and the fluorescence spectra revealed a change in molecular bonding, called fluorescence resonance energy transfer (FRET), which is a phenomenon in which molecules bond to each other. In the case of (TD-)DFT calculations for FRET [46], FRET should be ensured to not occur unless the molecules are very close to each other, as in experiments. In Fig. 17.12,

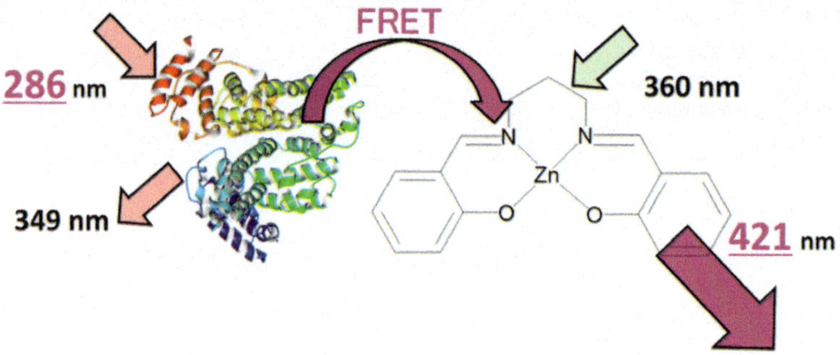

Fig. 17.12 Quenching of HSA due to a Schiff base Zn(II) complex under Förster resonance energy transfer (FRET) mechanism

straight arrows denote excitation or emission light and their wave lengths (predominant ones are 286 and 421 nm, respectively) and a round arrow denotes energy transfer of FRET. In addition, we also measured the fluorescence lifetime of proteins by transient absorption spectroscopy. (Very short) Time-dependent phenomena and phenomena involving two or more molecules are difficult to handle via computational chemistry.

17.4.6 Infrared Free Electron Laser (IR-FEL) Irradiation

For IR-FEL irradiation at the FELTUS Institute, a cast film of the complex was prepared by evaporating water, i.e., the solvent, from the solution of the complex, and this film was irradiated. The wavelengths of the IR irradiation light were the wavelength absorbed by the carbon–nitrogen double (imine) bond (C=N) in the Schiff base Zn(II) complex (wavenumber = 1622 cm^{-1}), the wavelength absorbed by the carbon–oxygen double bond (C=O) in the amide bond in HSA (wavenumber = 1622 cm^{-1}), the wavelength absorbed by the carbon–oxygen double bond (C=O) of the amide bond in HSA (wavenumber = 1652 cm^{-1}), and the wavelength absorbed by the nitrogen–hydrogen single bond (N–H) of the amide bond in HSA (wavenumber = 1537 cm^{-1}). The irradiation time was set to 0, 5, 10, 20, and 30 min, and FT-IR measurements were performed after each irradiation to examine the percentage of α-helix and β-sheet structures through waveform separation of the C=O and N–H spectral peaks of the amide bonds. Infrared irradiation at 1622 cm^{-1} had no effect. In contrast, irradiation at 1652 and 1537 cm^{-1}, which energizes the amide bonds in HSA, did not generally change the secondary structure of HSA alone, whereas the complex with the Zn(II) complex showed a change in the proportion of α-helix and β-sheet structures with the duration of infrared irradiation [47]. The ratio of α-helix and β-sheet structures in the complex with the Zn(II) complex changed with the irradiation time. In other words, complexation with Zn(II) complexes resulted in an

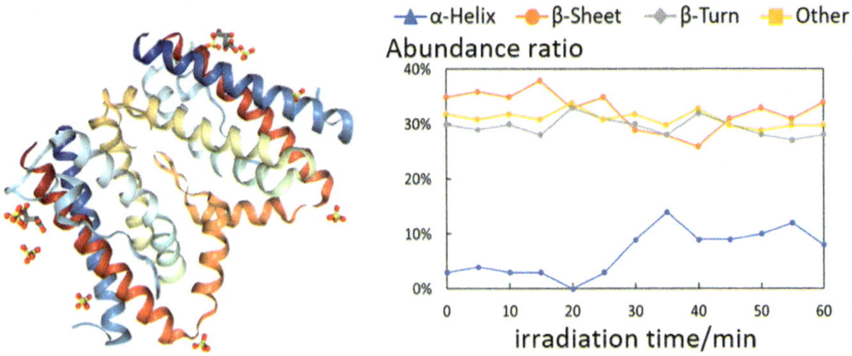

Fig. 17.13 Crystal structure of IL-6 (left) and infrared free electron laser (IR-FEL) irradiation time and the abundance ratio of remaining secondary structures (right)

enhancement of the secondary structure change of HSA caused by IR-FEL irradiation [48, 49]. Similarly, IR-FEL irradiation was performed on the immune protein interleukin IL-6 (Fig. 17.13), *D*- or *L*-amino acids, and dipeptides to investigate differences in chirality [50]. As there is controversy, however, the energy difference of the chirality theoretically predicted [51] under these conditions could not be experimentally clarified.

17.4.7 Infrared (IR) Spectroscopy of Organic/Inorganic Hybrid Materials

Infrared free electron laser (IR-FEL) irradiation is a technique that uses a high-power, high-intensity laser with a tunable wavelength in the infrared region. This laser can irradiate in the mid-infrared region (5–14 μm), has linear polarization, and produces picosecond pulses of laser light. An example is a reported enhancement effect of metal complexes on damaged protein molecules.

The docking properties due to weak intermolecular interactions were also investigated using fluorescence spectroscopy (quenching, temperature dependence, and decay lifetime) with the help of computational docking simulations. DFT calculations support the reliable assignment of experimental IR bands for Zn(II) complexes and can provide stable conformations of Zn(II) complexes (i.e., several possible optimized structures) to suggest docking functions. The HSA + Zn(II) complex is much more damaged than HSA based on the residual a-helix of HSA, suggesting that the inclusion of the Zn(II) complex in HSA leads to a conformational change in HSA, resulting in a more fragile structure. This is a feature of the excitation induced by the IR-FEL, which is a vibrational excitation, as opposed to electronic excitation by UV or visible light.

In addition to breaking chemical bonds, the advantage of the IR spectrum is that each peak is wavelength-selective, i.e., sharper bands can be obtained than in the UV–vis spectrum. By utilizing this property, experiments and computational chemistry can be combined to study the improvement in the anisotropy of molecular orientation induced by polarized light irradiation. This is how a series of studies evolved from synthetic polymers to the development of other systems for polymer matrices, such as proteins, for new photofunctional materials. To attribute peaks in the IR spectra or correlate DFT results to molecular orientation experiments, for example, we synthesized organic/inorganic hybrid materials consisting of chiral Schiff base Co(II), Ni(II), Cu(II), or Zn(II) complexes and azobenzene (AZ) in synthetic polymer polymethylmethacrylate (PMMA) cast films. Organic/inorganic hybrid materials were synthesized and observed via polarized electron (UV–vis) or IR spectroscopy to determine the molecular orientation photoinduced by linear or circular polarization.

In the polarized UV–vis spectra [52], each component cannot be separately observed for the following reasons. First, the π–π' bands of AZ and the complex overlap. In addition, the Zn(II) complex has no d–d bands. Therefore, IR spectroscopy was employed to selectively observe the C–N band of the complex and the N–N band of AZ as a complex in the PMMA film; when using TD-DFT calculations, a reliable assignment of the IR peaks requires a theoretical interpretation of the IR spectra based on normal vibrations.

Therefore, new chiral Schiff base Ni(II), Cu(II), and Zn(II) complexes were synthesized. −OH groups were incorporated to enhance intermolecular interactions with the PMMA matrix. This molecular design aims to induce a supramolecular orientation under linearly or circularly polarized UV light irradiation [53]. In addition, normal mode AZ analysis of the PMMA film (not the gas phase) and simulated IR spectra of the complexes from TD-DFT calculations were used to separately observe the components within the hybrid material. To the best of our knowledge, this is the first attempt to selectively interpret the molecular orientations induced after polarized UV light irradiation.

The light-induced supramolecular orientation was investigated, and in all cases, the optical anisotropy increased, with the most pronounced increase in anisotropy for the Ni(II) complex + AZ + PMMA. Upon irradiation with circularly polarized light, new CD peaks appeared, which were attributed to the induced helical orientation. Polarized IR measurements allowed us to separately observe the molecular orientation of AZ and the complexes. Parallel alignments of the azo N=N (AZ) and imine C=N (complex) moieties perpendicular to the electrical vector of the irradiated light were found. This is evidence of the supramolecular orientation of the complexes, with transmission from the reorientation of AZ directly rotated by the so-called Weigert effect (Fig. 17.14) [54].

The transition moment at a certain absorption wavelength is related to the atomic coordinates of the molecule, so if the atomic coordinates of the molecule are rotated around the origin, the transition moment accordingly rotates. Therefore, the coordinate system of the atomic coordinates of a molecule with a three-dimensional structure used in UFF force field calculations is based on the definition of the coordinates for the transition moment [18]. The UFF calculation [55] was adopted for lightweight

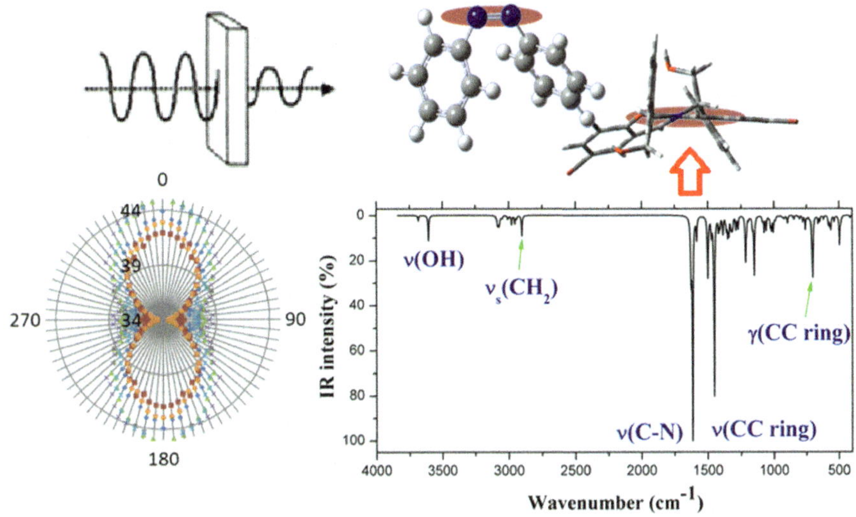

Fig. 17.14 (Above) Polarized UV light irradiation (left) and anisotropically oriented azobenzene (AZ) and Schiff base complex (right). (Below) Circular plot of intensity of polarized IR spectrum (C=N band) (left) and DFT simulated IR spectrum with marked C=N peak (right)

calculation of the positional relationship (due to rotation) of a two-molecule system and the corresponding energy change.

An example in which the transition moment is located on the molecular skeleton and the atomic coordinates and transition moment of the molecule rotate around the origin owing to the coordinate transformation is given [56]. By rotating the atomic coordinates 90° around the origin in the xy-plane, we obtained a suitable orientation for the transition moment after rotation of the atomic coordinates (Fig. 17.15).

17.5 Investigation of Lysozyme and Metal Complexes Using Crystal Structures

17.5.1 Artificial Metalloproteins

Hen egg white lysozyme (Fig. 17.16) is one of the most easily obtained proteins (some of which were mentioned in the previous section, as there are many research examples). It is a catalytic enzyme that hydrolyzes the mucopolysaccharides that make up bacterial cell walls and has a lytic effect on bacteria. In addition, as artificial metalloproteins or biomaterials, the introduction of nonnatural metal ions and metal-containing complementary molecular families into proteins has dramatically expanded the diversity of protein functions and range of applications. Reported artificial metalloproteins have been utilized in a wide range of fields, including as tumor

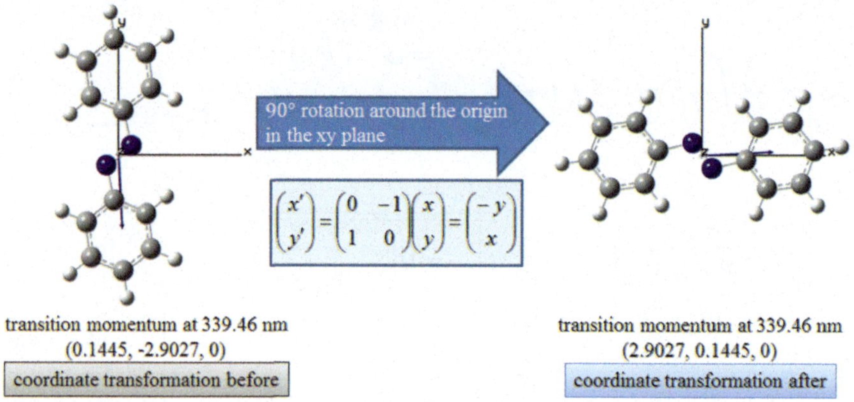

Fig. 17.15 Rotation of the transition moment resulting from rotation of the atomic coordinates of azobenzene (AZ) by 90° around the origin in the xy-plane (blue arrows indicate the magnitude and direction of the transition moment)

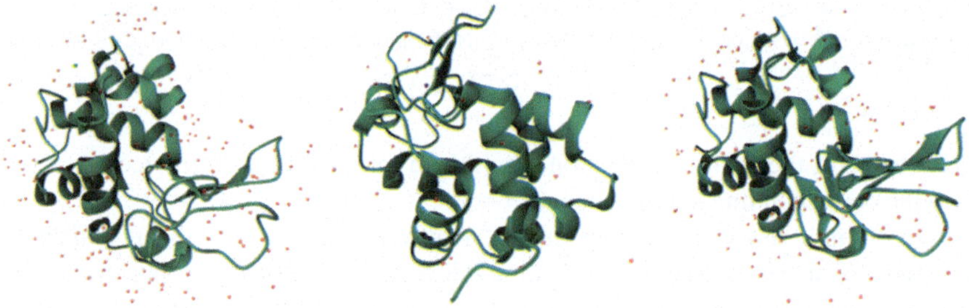

Fig. 17.16 Some crystal structures of hen egg white lysozyme from PDB 1DPX (left), 3WUN (middle), and 2LYM (right)

markers. The author and colleagues previously designed many artificial metalloprotein systems. However, this was our first attempt, as we narrowed down promising candidate compounds using databases such as the Cambridge Structural Database (CSD) [57] and computational chemistry (protein–ligand docking based on crystal structure databases in this case) before conducting experiments.

17.5.2 Lysozyme-Copper Complex

Focusing on lysozyme, we created complex systems of lysozyme with approximately 20 types of salen-type Cu(II) complexes registered in the CSD and investigated their behavior by virtually screening their behavior using the molecular docking method. To screen the complex systems, structural information for data-driven studies was obtained from the Protein Data Bank (PDB) [58] for lysozyme and from CSD for metal complexes and was optimized via DFT using Gaussian 09 and GaussView 5.0.9w with B3LYP and 6-31G basis functions. The optimized structures were subjected to CSD investigations and docking simulations with 20 different complexes and egg white lysozyme using the GOLD program.

Among the complexes with lysozyme, the electrochemical effects of the four complexes with the highest docking scores were measured; cyclic voltammetry (CV) measurements revealed that there was little difference between the N_2 and O_2 loading results for lysozyme alone, but in the complex system of the four selected metal complexes and lysozyme, for O_2 loading, an increase in the reduction peak was observed, revealing a synergistic effect between the metal complexes and lysozyme. The highest docking scores corresponded to the largest increase in CV measurements among the four complexes. These results suggest that the increased affinity with lysozyme established an electron transfer pathway. This was confirmed in smoother electron transfer. In other words, four salen Cu(II) complexes with good docking (GOLD) scores were selected as promising candidates, and were actually incorporated into lysozyme. The redox experiments were performed to confirm increasing reduction peaks [59].

17.5.3 Hirshfeld Surface Analysis

While analysis of individual data of crystallographic results has been used to investigate the intermolecular interactions of crystallographically characteristic molecules, statistical investigations can also be performed by analyzing a large amount of data and extracting information. This is based on the hypothesis that the results from Hirshfeld surface analysis [60] are data that contain all the information such as the position and type of atoms, bond lengths, and electrostatic potentials, and they are output in terms of molecular interactions. This analysis illustrates areas of long and short distances between atoms and can quantify the contribution of intermolecular interactions between some atoms. The method used in this statistical investigation was a type of unsupervised learning called principal component analysis (PCA) [61]. In this learning method, various interactions are evaluated and aggregated, and the two components with the highest contribution rates are set as principal components and plotted on a graph on a two-dimensional plane with the two principal components as axes to investigate correlations. In a previous study, Hirshfeld analysis was performed on each of approximately 1000 datasets to determine the proportions of H…H, C…H, C…C, and O…H interactions in the overall interaction of the molecule. The C…H interaction includes H inside the Hirshfeld surface and C outside the surface, and similarly, the O…H interaction includes H inside the surface and O outside the surface. By performing unsupervised learning on the data collected in this way, some correlations between the molecular features and the type and proportion of interactions could be found.

17.5.4 *Presence of Crystalline Water Molecules*

A survey of the complexes obtained in our laboratory revealed that the greater the number of crystalline water molecules present, the greater the contribution of hydrogen bonding tended to be. Therefore, we investigated the correlation between the number of crystalline water molecules and the contribution of intermolecular interactions and performed Hirshfeld surface analysis on approximately 1000 files collected from the Cambridge Structure Database to determine the contributions of four types of intermolecular interactions [62].

17.5.5 *Intermolecular Interaction Between Zinc Complex and Lysozyme*

In a previous study of catalytic reactions, the formation of Zn(II) complexes with lysozyme was investigated. Spectroscopic measurements were used to examine the weak noncovalent intermolecular interaction properties between two Zn(II) complexes and lysozyme. After confirming the synthesis of the Zn(II) complexes, the CD and corresponding UV–vis spectra were measured for the hybrid system with lysozyme. No significant changes in the low-wavelength region reflecting changes in the secondary structure were observed (in the case of the lysozyme + Zn(II) complex, there was a slight change at approximately 350–410 nm). This result suggests that the association of the Zn(II) complex has little effect on the three-dimensional structure of lysozyme, in contrast to HSA. The fluorescence spectra of lysozyme and the two Zn(II) complexes revealed quenching of the emission from lysozyme in both systems after the association of the Zn(II) complex. However, only relatively weak interactions were observed between the Zn(II) complex and lysozyme, as these interactions were not confirmed in all the peaks of both complexes. The variable temperature spectra also revealed that heating to room temperature decreased the degree of quenching by lysozyme. It is considered to be accompanied by a change in the energy transfer pathway due to a conformational change in lysozyme caused by the external temperature.

Furthermore, lacking of Zn(II) complex in lysozyme crystals were confirmed after soaking solutions of complexes (Fig. 17.17). These results support the existence of a weak bond between the complex and lysozyme, which is bound to His, Trp, and Glu residues at the active site. Docking calculations to estimate the best binding site indicated that both Zn(II) complexes are stable in a noncovalent or noncoordinated conformation bound to a gap on the protein surface [63].

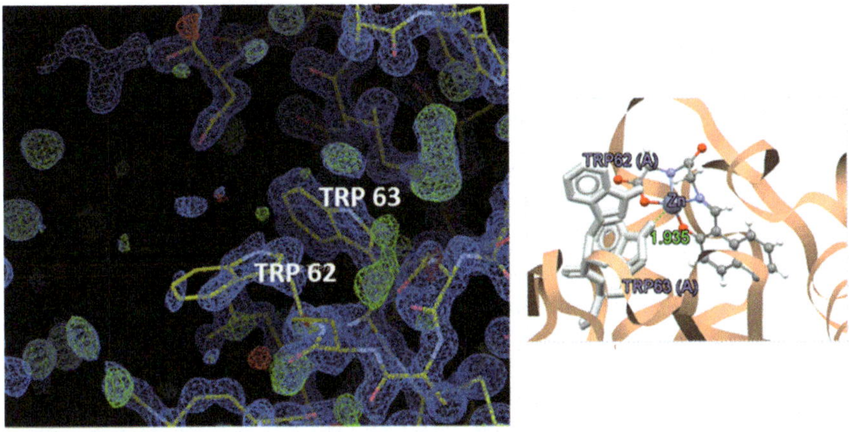

Fig. 17.17 Experimental electron density of lysozyme without the dipeptide Zn(II) complex (left) and computational docking feature around there (right)

17.5.6 Binding Between Copper Complex and Lysozyme

Steric structures pf protein in Protein Data Bank (PDB) and electronic structures of metal complexes from DFT is often helpful to rational design ligand–protein binding materials. Based on the above findings, hydrophobic Val derivative Schiff base Cu(II) complexes and dipeptide (AlaAla, GlyGly) derivative Schiff base Cu(II) complexes were introduced into egg white lysozyme, and X-ray crystallography was used to confirm the presence of Cu ions and dissociative ligands and the binding of Cu(II) ions in these Cu complexes. The results showed that the Val derivative, a Schiff base Cu(II) complex, was weakly bound at Arg114 of egg white lysozyme. In other Cu(II) complexes, binding of Cu(II) ions and dissociated ligands to various residues was observed.

GOLD docking simulations for lysozyme and Schiff base Cu complexes revealed high docking scores for Ala and His in previous reports, but this trend was not shown in experiments. This is because experiments do not always reproduce binding sites, and GOLD calculations are not good at discussing the binding between metal ions and proteins. Therefore, AlphaFold2 program was used to steric predict structures of metal ion docking, and it was also used to expect structure around binding residues [64]. Recent developments in machine learning-based metal binding site prediction have also been reported [65]. The results showed that, in some cases, Cu(II) ions dissociate from the ligand and are incorporated into lysozyme molecules. Ion dissociation from the ligand was facilitated by protein crystals rather than by bare ions due to the dissolution of some $CuCl_2$ [66].

When crystal structures are not always available, however, computational chemistry can be relied upon to aid in understanding experimental results. Here, cis–trans photoisomerization of the azobenzene moieties of azo-arginine Schiff base Cu(II) and azo-threonine Schiff base Cu(II) complexes was performed via UV irradiation.

The experimental results of FT-IR, fluorescence, UV–vis, electron spin resonance (ESR), CD, and CV measurements were compared with those of DFT calculations of metal complexes via Gaussian and docking simulations with lysozyme by GOLD.

The FT-IR measurements yielded vibrational spectra of the imine and carbonyl moieties of the ligand. The same results were given in the DFT calculations. The fluorescence measurements revealed the binding of the Cu(II) complex to lysozyme, with a possible quenching mechanism based on the wavelength shift to longer wavelengths due to the binding of the Cu(II) complex to lysozyme. The UV–vis measurements revealed a decrease in the absorbance in the cis form due to photoisomerization of the azobenzene moiety, with a peak shift from 666 nm. The DFT calculations confirmed a shift of the peak from trans to cis toward shorter wavelengths. The ESR measurements of the g values and A values of the Cu(II) complex alone and the Cu(II) complex–lysozyme were compared, and the results suggested that the bond between the Cu atom and the ligand was stronger after docking. The CV measurements revealed a decrease in the current after lysozyme binding compared with the Cu(II) complex alone. The redox potential of the Cu(II) complex with an arginine ligand showed a larger change after lysozyme binding than that of the threonine complex. Electron density analysis [67] and molecular orbital calculations of the Cu(II) complexes via DFT and docking simulations with lysozyme by GOLD suggested that the arginine ligand was more adsorbed in the hydrophobic pocket of lysozyme than the threonine ligand.

Furthermore, GOLD docking simulations suggested that there are sites where the Cu(II) complex enters lysozyme and sites where it does not. It is attributed to the bulky structure of the azobenzene (AZ) moiety [68]. In this way, even if the binding between lysozyme and a metal complex is spectroscopically confirmed, experimentally proving it via crystallography is not necessarily possible. In this case, computational chemistry often provides useful supporting data about AZ [69].

17.6 Laccase Metal Complexes for Promising Biofuel Cells

17.6.1 Electron Transfer in Laccase

Laccase is a multicopper oxidase (containing Cu active sites type 1 (T1), type 2 (T2), and type 3(T3)) that accepts electrons from a substrate near T1 (catalyzing the oxidation reaction) and catalyzes the four-electron reduction of oxygen to water near T2 and T3 after electron transfer in a protein molecule (Fig. 17.18). Biofuel cells are attracting attention as enzyme-based energy conversion devices and are expected to be environmentally friendly energy sources. Efficient promotion of the oxygen reduction reaction is an important property of a cathode catalyst, but the efficiency of the electron transfer between the electrode and the enzyme remains an issue in laccase-based biofuel cells. To solve this potential problem of electron transfer involving tunneling, hopping, and coupling [70], mediators (mediators) that

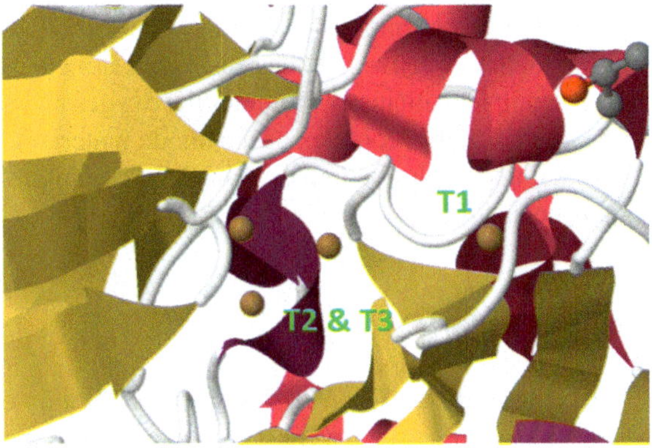

Fig. 17.18 Crystal structure of laccase showing mononuclear T1 site and trinuclear T2 and T3 sites

mediate electron transfer must be developed. To address this issue, we present five studies on electron transfer systems using photoresponsive and chiral complexes and discuss how they can contribute to improving performance. The electron transfer performance and the relationship (distance and direction) between metal sites were clarified, but it seems difficult to discuss the reaction kinetics using conventional quantum chemical calculations.

17.6.2 Photoisomerization

In the early stage of study, the design and evaluation of photoresponsive chiral complexes with azobenzene (AZ)-containing Schiff base ligands were carried out (for example, Fig. 17.1). Chiral Schiff base ligands containing azo groups are attracting attention as photoresponsive electron transfer mediators, because their molecular recognition and conformation change upon photoisomerization of AZ. Photoisomerization has a significant effect on the electron transfer properties of Cu(II) complexes, particularly the cis form, which has the highest current values. This suggests that the cis form more efficiently transfers electrons, contributing to its performance as a cathode catalyst in biofuel cells. From these preliminary experimental results, it is considered that dynamic control of electron transfer through photoisomerization play an important role in new energy conversion technologies [71].

Mn(III/II) Complexes

Mn(III/II) complexes containing azo-anthraquinones were designed, and computational chemistry was used to select the best complexes and evaluate their electron transfer properties. After screening candidate Mn complexes using DFT, a Schiff base ligand that combines anthraquinone with redox activity and azobenzene with a photoisomerization property was found to be the best candidate, and the complex was synthesized. Furthermore, docking simulations confirmed that the complex has a high affinity for the T1 active site of laccase, indicating that the interaction between laccase and the complex efficiently promotes the oxygen reduction reaction. It was concluded the potential of this complex as a cathode material for biofuel cells and establishes a rational approach to complex design by combining computational chemistry and experiments [72].

Cu(II) Complexes

Cu(II) complexes of AZ and redox-active anthraquinone were synthesized, and their interactions with laccase were investigated. Since the AZ moiety exhibits photoisomerization and the anthraquinone moiety has redox activity, these complexes have the properties of electron transfer mediators. From experiments, it was confirmed that the cis–trans isomerization directly affects their electron-transfer ability, since the cis form exhibits efficient electron transfer superior to the trans form. In short, photoismerization due to UV irradiation resulted in tuning of electron transfer efficiency. This enables dynamic control of electron transfer using photoisomerization, which is expected to improve the performance of cathode catalysts for biofuel cells. Photoisomerization can contribute to the tuning of the electronic structure of a molecule, i.e., the redox potential. In addition, the change in the molecular orientation of azo groups under polarized UV light irradiation (Weigert effect) is expected to contribute to a change in the molecular orientation, i.e., to a consequent improvement in the interaction with hydrophobic pockets on the protein surface (three-dimensional structural fitting), which is expected to lead to a decrease in the electron transfer distance between the mediator complex and the active site of the protein that receives the electron [73].

17.6.3 Enzymes as Electrocatalysts

In biofuel cells, the use of enzymes as electrocatalysts is called a direct electron transfer (DET)-type enzyme-functional electrode reaction. To increase the electron transfer rate, the distance between the redox center in the enzyme and the electrode should be short, the electrons should be easily transferred, and the redox-related reorientation energy should be small. However, controlling these factors only with enzymes is very difficult. To externally control these factors, gold nanoparticles, carbon nanotubes, and even anthracene have been proposed to intervene between the electrode and the enzyme as conductive nanowires to increase the electron transfer rate between the enzyme and electrode to improve DET-type catalytic reactions [74].

Electron Transfer Mediators

It is difficult to calculate the intermolecular electron transfer pathways in proteins [75]. A method to smoothly facilitate electron transfer involves the introduction of an electron transfer mediator (a low-molecular-weight redox substance relaying electron in an electron transfer pathway) between the enzyme and electrode. Examples of simple electron transfer mediators include $K_3[Fe(CN)_6]$ and $[Cu(phen)_2(NO_3)_2]$, but a variety of other metal complexes have been reported. The corresponding reaction is called a mediator electron transfer (MET)-type enzyme-functional electrode reaction and is applicable to most redox enzyme systems. In this method, the selection of the mediator and the modification method is important. In general, when similar compounds are used as mediators, the smaller the difference between the redox potential of the mediator and the redox potential of the enzyme is, the lower the reaction rate between the enzyme and the mediator. Since the reaction rate between the enzyme and mediator is related to the current density, increasing the power density requires the appropriate redox potentials. Therefore, the selection or development of an appropriate mediator is necessary to increase the output density. Notably, however, the current density is not only a function of the reaction rate between the enzyme and mediator but also of several other factors, including the rate of electron transfer between mediators, the density of electrochemically active enzymes per unit area, the rate of proton conduction in the electrodes and between the anode and cathode, and the diffusion rate of fuel or oxidant in the electrodes [73].

The most common method of facilitating electron transfer through an intermediate mediator between the electrode and the enzyme is to immobilize the mediator on the electrode surface. Immobilization increases the concentration of mediators and enzymes on the electrode surface, and the current from the electrochemical reaction between the enzyme and mediator is expected to be higher [76].

Chiral-Induced Spin Selectivity (CISS) Effect

Naaman et al. [77] reported that electrons transmitted through a helical organic molecule with chirality have chirality-dependent spin selectivity, which they named the chiral-induced spin selectivity (CISS) effect. In many cases, electron transfer in vivo occurs through proteins, which are organic molecules. In this case, electron transfer was considered to occur with reduced heat generation such that the proteins do not undergo thermal denaturation. As has become recently clear, this phenomenon highly depends on the electron spin, and the link between the spin selectivity of proteins (helical materials) and heatless electron transfer, as well as its mechanism, has attracted much attention in recent years. For example, if spin control becomes possible, then the memory capacity of a device can be doubled according to a simple calculation. In this context, chiral and redox-active (potentially paramagnetic) complexes with AZ for molecular alignment must be good strategies for mediators (Fig. 17.19).

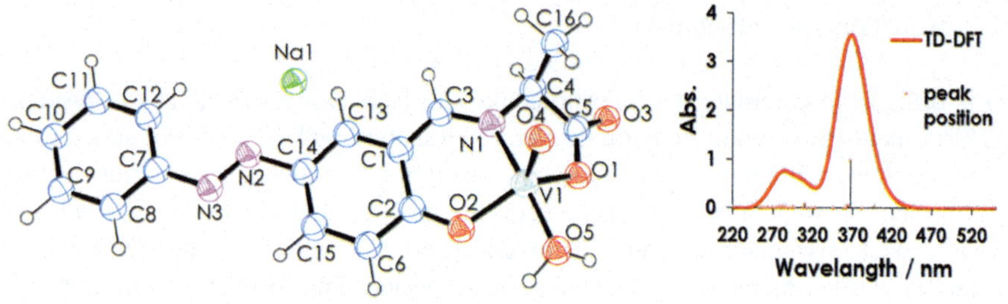

Fig. 17.19 Crystal structure (left) and DFT-simulated UV–vis spectra of azobenzene-containing chiral amino acid paramagnetic metal complex

To date, the CISS effect has been observed in oligopeptides, amino acids (Cys), double-stranded deoxyribonucleic acid (DNA), and membrane proteins. However, since this research has only recently begun, there are still few broad cross-disciplinary applications, i.e., optical switching. Therefore, we considered the application of spin-polarized electrons induced by the CISS effect to the transfer of electrons to laccase (Fig. 17.20).

A new electron transfer system was developed by combining chiral oligopeptides with Schiff base Cu(II) complexes. The chiral oligopeptides were shown to be able to induce CISS effects through their interaction with the complexes, and the use of Cu complexes containing photoisomerizable AZ was found to alter the electron transfer efficiency. A marked change in the electron transfer efficiency was observed between the L- and D-form oligopeptides, suggesting that the CISS effect is involved in electron transfer. This study indicates that electron transfer using chiral-induced spin control is a novel approach in cathode catalysts for realizing biofuel cells with high efficiency.

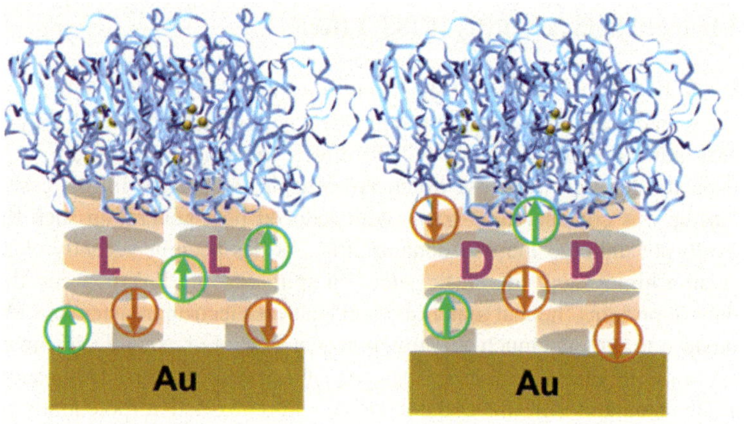

Fig. 17.20 Cathode design (Au substrate—chiral oligopeptide—laccase with mediator like Fig. 17.17) for CISS effect

These results suggest that photoresponsive and chiral complexes have the potential to significantly improve the efficiency of electron transfer in biofuel cells. Notably, only intramolecular DFT calculations of the complex and docking calculations of the laccase complex are currently possible. Dynamic control of electron transfer utilizing photoisomerization and spin-selective control is expected to play a very important role in new energy conversion technologies. Further studies are needed to incorporate these systems into actual fuel cell devices to achieve more efficient energy conversion and to commercialize biofuel cells [78].

17.6.4 Molecular Docking Simulations

Molecular Interactions Between Cu(II) Complex and Laccase

To predict the molecular interactions between a newly synthesized Cu(II) Schiff base and amino acid residues of laccase, molecular docking simulations were performed using the AutoDock 4.2.6 program on a Windows platform with 8 GB of RAM and an Intel I 5 processor (usually equipped in the laboratory). The protein structure of laccase was obtained from PDB 1GYC. A chain of laccase was modified by removing the water-bound ligand. Missing amino acids were checked, and hydrogen bonds were added to the protein structure. The center grid box was x: 16.895, y: 5.931, z: 37.151, where the number of points in the x, y, and z directions were $30 \times 30 \times 30$ Å^3, and the grid spacing was 0.3750 Å. The three-dimensional (3D) structure of the complex, which is essential for docking, was taken from the DFT-optimized structure. The docking procedure of AutoDock Tools was performed using the Lamarck genetic algorithm [79]. The conformation of the best fitting ligand was selected based on the minimum binding energy. One of our Cu(II) complexes exhibited a docking score of −9.9 kcal/mol [72, 80].

Reactivity of Fe/Mn Complex with Laccase

Now, let us turn to a different topic and introduce some preliminary research aimed at utilizing the multiple properties of magnetism, light, and redox. One idea is to use a complex as a molecular-based magnet as a redox mediator. Docking of the protein and the complex can be confirmed by photochemical quenching. The reaction of a heterotrinuclear Fe/Mn complex with the T1 moiety of laccase was investigated: Mn(III) is coordinated in-plane by the O and N atoms of the ligand and axially by the O and N atoms of water and the bridging cyanide. The reactivity with laccase was evaluated by spectroscopic and CV measurements and docking simulations. Experimental studies with different concentrations revealed that there is an optimal mediator concentration that depends on the temperature and electrode type. Docking analysis using computer simulations confirmed that the complexes bind to hydrophobic pockets on the laccase surface, but no improvement in electron transfer was observed after docking compared to laccase alone. The results suggest that the smaller complex, which properly enters the hydrophobic pocket and has a shorter electron transfer distance, performs better than the present cyanide-bridged trinuclear Mn(III)-Fe(III)–Mn(III) complex. Fluorescence measurements revealed that laccase activity was quenched with increasing concentration of this complex [80].

Assessing Mediators in Combination with DFT Calculations

With biofuel cells, we have studied the use of redox complexes as mediators to improve electron transfer between the cathode and laccase enzyme, which is an oxygen-reducing catalytic process. Herein, we systematically investigated "cyanide (bridged) metal complexes" (such as $[Fe(CN)_6]^{3-/4-}$) as suitable electrochemical and steric mediators. For example, we referred to a trinuclear cyanide-bridged Au–Cu complex ($[Cu(N\text{-ethylenediamine})_2][Ag(CN)_2]_2$) as a mediator and its experimental data (IR: C≡N 2124 cm^{-1}, ESR: $g = 2.01$ and 2.064 with laccase; CV: $E^0 = 0.406$, -0.237, and -0.821 V (Ag/AgCl)) previously reported to apply new computational study methods such as the Fukui function for DFT and AlphaFold2 for discussion [81].

DFT calculations (considering their characteristics) helped to interpret the experimental results of the chemical reactivity [81]. In general, biofuel cells have the advantages of being able to generate electricity at room temperature and having a low environmental impact. Mediators such as cyanide complexes are used to improve electron transfer between the electrode and laccase. In this electron transfer, not only the redox potential but also the electronic structure and steric factors are important. Because electron transfer must occur in one direction, and electron transfer must be induced in the opposite direction for different enzymes. Spectroscopic and electrochemical measurements, DFT calculations using Fukui functions [82], and docking simulations were performed to investigate efficient cathode designs for biofuel cells, using several cyanometallic complexes as mediators. In the optimized structures, the hardness, electrophilicity, and local properties of the molecules were evaluated. The results revealed that the metal center of the complexes has both nucleophilic and electrophilic properties. The Fukui function provides detailed information about the (redox) reactivity. In other words, the Cu complexes were found to have a strong affinity for laccase (Fig. 17.21).

17.7 Summary

Thus, in the study of metal complex-protein hybrid systems, the authors and their collaborators have used both experimental and computational methods, combining experimental spectroscopy, photochemistry and electronic structure, hybridization and functional non-essential contributions, and utilizing concepts from condensed matter chemistry for this purpose. What the authors particularly want to emphasize is

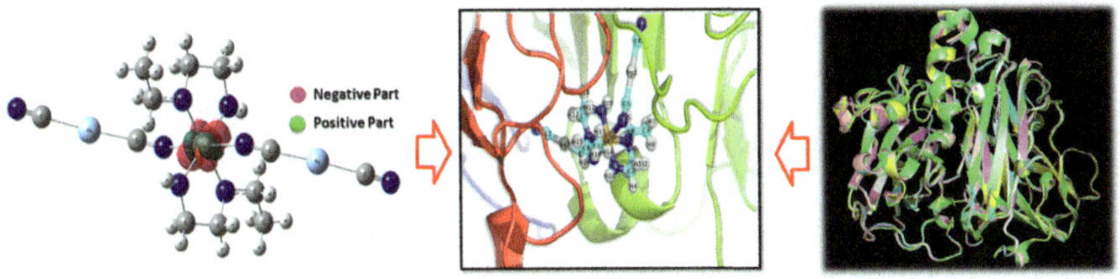

Fig. 17.21 Cyanide-bridged Ag–Cu–Ag complex calculated with Fukui function (left), laccase-Ag–Cu–Ag complex docking feature using AutoDock (middle), and three-dimensional structure of laccase estimated by Alphafold2 (right)

that if used appropriately, computational chemistry can be useful not only for understanding experimental results, but also for proposing efficient experiments. What the author particularly wants to emphasize is that if used appropriately, computational chemistry can be useful not only for understanding experimental results, but also for proposing efficient experiments.

17.8 Perspective

17.8.1 Toward More Complicated Hybrid Systems

Quantum Dots

In the future, more functions can be linearly combined by the introduction of new components. Quantum dots (metal nanoparticles and metal compound semiconductors) are tools that are often coupled with proteins. Introducing composite materials and metal complexes, we can examine the possibility of using multiple redox, optical, and magnetic properties of other-component composite materials of metal complexes, quantum dots, and proteins.

Gold Nanoparticle

Chiral salen-type Schiff base complexes were synthesized. The adsorption structures of gold nanoparticle (AuNP) and synthesized complexes were predicted from the observed CD spectra and the transition dipole moments calculated by quantum chemical calculations. The structures were observed via scanning tunneling microscopy (STM). AuNP showed optical activity due to surface plasmon resonance upon irradiation with light at approximately 500 nm, which was observed in the induced CD spectrum. In addition, for gold nanoparticles complexed by metal complexes through dipole–dipole interactions, the induced CD spectra increase (decrease) when the dipole moments are perpendicular (parallel) to each other according to theory [83]. Analysis of the intensity of the induced CD spectra revealed that the chiral metal complexes had a strong peak at approximately 600 nm, which was different from the induced CD peak of AuNPs at 500 nm, suggesting that an adsorption interaction occurred.

We calculated the optimized structures and dipole moments of the chiral metal complexes that presented strong peaks via DFT calculations. In the image diagram of the adsorption with AuNPs (Fig. 17.22), the CH_3O- moiety of the ligand was adsorbed on the surface of the AuNPs, and the dipole moment had an axis of symmetry with the chiral metal complex. The surface structure analysis via STM measurements revealed that a 1.194 nm molecule was adsorbed on the surface of the AuNPs, which is close to the length of the chiral metal complex (1.266 nm); thus, the chiral metal complex was standing on the surface of the AuNPs. The length of the chiral metal complex was close to 1.266 nm [84].

Next, by using AZ-containing analogous complexes, the changes in the optimized structure and dipole moment after cis–trans photoisomerization were considered via DFT calculations. The stability of the trans complexes was greater than that of the cis complexes for all three metal complexes. The coordination environments of the optimized structures differed depending on the central metals: (1) a planar structure in Ni(II), (2) a distorted planar structure in Cu(II), (3) a distorted planar structure in Ni(II) more than Cu(II), (4) the most compressed tetra-coordination

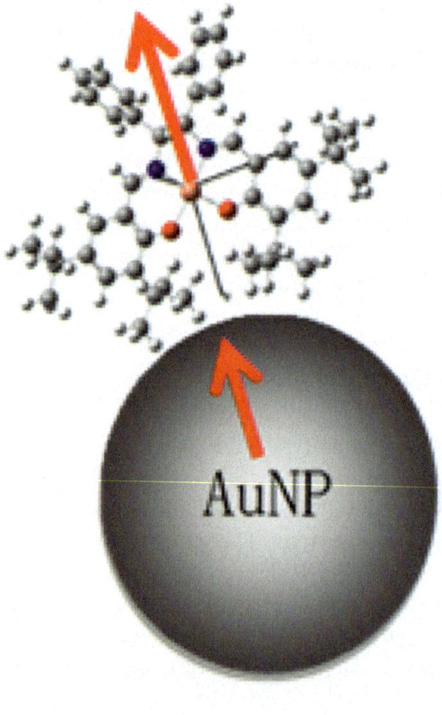

Fig. 17.22 Chiral complex (red arrow depicts dipole moment) adsorbed appropriately on the surface of AuNP (red arrow depicts dipole moment) exhibiting induced dichroism (CD)

structure in Zn(II). The dipole moments were greater in the cis form than in the trans form for all the metal complexes. From the CD measurements, the Ni(II) and Cu(II) complexes with Au nanoparticles showed peaks at approximately 500 nm and 600 nm, respectively. The peak of the cis complex was larger than that of the trans complex, which was derived from the induced CD spectrum. In contrast, in the Zn(II) complex–AuNP composite, there was little change in the magnitude of the induced CD spectrum between the trans and cis forms. The reason why the induced CD spectra changes were observed after cis–trans photoisomerization in the Ni(II) and Cu(II) complexes with AuNPs but not in the Zn(II) complex is that a change in the dipole moment must occur in the molecule, and the dipole moment of the chiral metal complex with the optimized structure determined via DFT calculations was not affected by photoisomerization. This finding was consistent with the prediction of the dipole moment of the complex. Therefore, the induced CD spectra of AuNPs obtained by quantum chemistry calculations were helpful [85, 86].

Semiconducting TiO_2

For semiconductors, metal complex–TiO_2 composites were synthesized and characterized by UV–vis, fluorescence, and CV measurements to investigate the utility of AZ-containing chiral salen-type metal complexes as a model dye-sensitized solar cell (DSSC) via complexation with TiO_2. The metal complexes were synthesized using Fe(II), Cu(II), and Zn(II) as the central metal ion and salen-type chiral ligands. The UV–vis spectra of the synthesized azo-containing chiral salen-type metal complexes revealed a long wavelength shift in the peak corresponding to the π–$\pi\star$ transition in comparison with that of similar chiral salen-type metal complexes without the AZ moiety. The energy gaps of the AZ chiral salen-type metal complexes were found to be 2.234 eV for the Fe(II) complex, 1.527 eV for the Cu(II) complex, and 2.282 eV for the Zn(II) complex by CV measurements. The highest occupied molecular orbital (HOMO) and lowest unoccupied molecular orbital (LUMO) energy levels of each azo-containing chiral salen-type metal complex were calculated via DFT calculations (Fig. 17.22); the (HOMO, LUMO) energy levels of the Fe(II) complex were (−5.483, −3.423 eV), those of the Cu(II) complex were (−5.312, −3.785 eV), and those of the Zn(II) complex were (−5.579, −3.297 eV). The LUMO energy levels of the azo-containing chiral salen-type metal complexes are higher than the conduction band of TiO_2 (−0.50 V), indicating that these complexes can be used as DSSCs. Compared with that before UV irradiation, the adsorption of Fe(II) complexes on TiO_2 increased by approximately 47% after UV irradiation, and by approximately 61% after polarized UV irradiation. The increase in adsorption on TiO_2 due to UV irradiation can be attributed to the fact that the molecular structure of the cis complex is more curved than that of the trans complex, making penetration of the porous structure of TiO_2 easier. The change in the dipole moment calculated by DFT due to photoisomerization affects the adsorption interaction, and the Weigert effect, in which the molecular orientation is restricted by polarized light, is considered to have acted [87, 88] (Fig. 17.23).

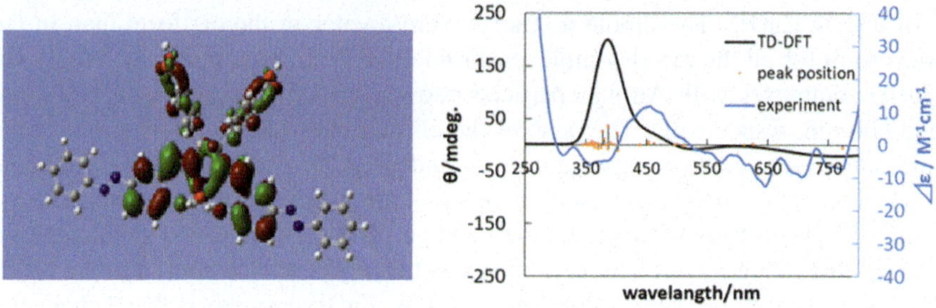

Fig. 17.23 DFT calculated LUMO of the azobenzene (AZ) chiral salen-type Fe(II) complexes (left) and the resulting DFT simulated and experimental circular dichroism (CD) spectra (right)

17.8.2 Our Future Goal

Our future goals are to understand the unique interactions and functions of metal complex–protein complexes both experimentally and computationally. Artificial metalloproteins have two properties (homogeneous metal catalysis and enzymatic catalysis), and the incorporation of metal-containing portions into protein scaffolds can impart new catalytic functions to polymers. There are four possible strategies for synthesizing metalloproteins: covalent bonding, supramolecular assembly, bonding, and metal substitution; this study falls into the supramolecular assembly category. The acquisition of experimental data on metal complexes alone and their interpretation and prediction with the aid of computational chemistry have been getting easier in recent years. Proteins, in contrast, are relatively more difficult than small-molecule metal complexes, both in terms of experimental data acquisition and available computational chemistry methods, owing to their high computational complexity and system complexity.

Acknowledgements This work was partly supported by a Grant-in-Aid for Scientific Research (B) KAKENHI (24K00912).

References

1. Vornholt T, Leiss-Maier F, Jeong WJ, Zeymer C, Song WJ, Roelfes G, Ward TR. Artificial metalloenzymes. Nat Rev Methods Primers. 2024;4:78.
2. Akitsu T. Symmetry in organic/inorganic hybrid materials. Symmetry. 2022;14:1624.
3. Akitsu T. Hybrid or component?—Schiff base complexes and laccase. Compounds. 2022;2:307.
4. Kim S, Nakane D, Akitsu T. Physical approaches with DFT for metal complex-based photofunctional materials. AIP Conf Proc. 2022;2611:020004.
5. Takeda C, Nakane D, Akitsu T. Recent advances in Chiral Schiff base compounds in 2023. Molecules. 2023; 28:7990
6. Akitsu T. Electron spin aspects of Ru and V=O complexes for drugs. Int J Clin Images Med Rev. 2022;1:1.

7. Akitsu T. Lanthanide complexes in recent molecules. Molecules. 2022;27:6019.
8. Kawahara K, Okumura Y, Takiguchi Y, Nakane D, Akitsu T. Lightening calculations for Schiff base lanthanide complexes. AIP Conf Proc. 2024;3030:020006.
9. Akitsu T, Suda S, Katsuumi N. Re-reporting the crystal structure of copper complex from another point of view. Edelweiss Chem Sci J. 2021;4:27.
10. Akitsu T, Higashi Y, Tsuchiya R, Imae T, Komatsu K, Nakane D, Moon D. Chemistry for space group symmetry beyond crystals. Symmetry. 2024;16:319.
11. Akitsu T. Mathematical geometry and groups for low-symmetry metal complex systems. Molecules. 2023;28:4509.
12. Akitsu T. Category theory in chemistry. Compounds. 2023;2:334.
13. Akitsu T, Suda S, Katuumi N. Beyond the scope of each computational chemistry. Edelweiss Chem Sci J. 2021;4:25.
14. Yasuda S, Takase M, Sano A, Akitsu T. Schiff base dinuclear Zn(II) complexes as assisting fluorescent probe for CD19 antibody protein. Int J Chem Eng Appl. 2017;8:290.
15. Solomon EI, Leve ABP. Inorganic electronic structure and spectroscopy;1999. NY, USA: Weily-International.
16. Nakatori H, Haraguchi T, Akitsu T. Polarized light induced molecular orientation control of rigid Schiff base Ni(II), Cu(II), Zn(II) binuclear complexes as polymer composites. Symmetry. 2018;10:147.
17. Takase M, Yagi S, Haraguchi T, Noor S, Akitsu T. Linearly polarized UV light-induced optical anisotropy of PVA films and flexible macrocycle Schiff Base Ni(II), Cu(II), Zn(II) Dinuclear complexes. Symmetry. 2018;10:760.
18. Yamazaki A, Akitsu T. Polarized spectroscopy and polarized UV light-induced molecular orientation of chiral diphenyl Schiff base Ni(II) and Cu(II) complexes and azobenzene in a PMMA film. RSC Adv. 2012;2:2975.
19. Kuroda R, Harada T, Takahashi H. Fast and artifact-free circular dichroism measurement of solid-state structural changes. Chirality. 2023;36:e3622.
20. Schmidt W. Optical spectroscopy in chemistry and life sciences an introduction. WeinHeim: Wiley-VCH; 2005.
21. Nováček M, Řezáč J. PM6-ML: the synergy of semiempirical quantum chemistry and machine learning transformed into a practical computational method. J Chem Theory Comput. 2025;21:678.
22. Mantz YA, Musselman RL. ZINDO calculations of the ground state and electronic transitions in the Tetracyanonickelate Ion, $Ni(CN)_4{}^{2-}$. Inorg Chem. 2002;41:5770.
23. Miura T, Onodera T, Endo S, Yamazaki A, Akitsu T. Some hybrid systems of chiral Schiff base Zn(II) complexes and photochrmic spiropyrans for envitonmental ion sensing. Am J Anal Chem. 2014;5:751.
24. Otani N, Fayeulle A, Nakane D, Léonard E, Akitsu T. Synthesis, identification and antibacterial activities of amino acid Schiff Base Cu(II) complexes with chlorinated aromatic moieties. Appl Microbiol. 2022;2:438.
25. Miyazaki R, Pradhan S, Roymahapatra G, Nakane D, Akitsu T. Design, synthesis and characterization of copper(II) Schiff Base complexes with Chloro Groups as antivirus drug candidates against Japanese Encephalitis Virus (JaGAr strain). In: Taylor JC (ed) Advances in chemistry research, vol 84. NY, USA: Nova Science Publishers, Inc.;2024, p 119.
26. Jumper J, Evans R, Pritzel A, Green T, Figurnov M, Ronneberger O, Tunyasuvunakool K, Bates R, Žídek A, Potapenko A, Bridgland A, Meyer C, Kohl SAA, Ballard AJ, Cowie A, Romera-Paredes B, Nikolov S, Jain R, Adler J, Back T, Petersen S, Reiman D, Clancy E, Zielinski M, Steinegger M, Pacholska M, Berghammer T, Bodenstein S, Silver D, Vinyals O, Senior AW, Kavukcuoglu K, Kohli P, Hassabis D. Highly accurate protein structure prediction with AlphaFold. Nature. 2021;596:583.
27. Yang Z, Zeng X, Zhao Y, Chen R. AlphaFold2 and its applications in the fields of biology and medicine. Sig Transduct Target Ther. 2023;8:115.
28. Kumar CRS. Approach of density functional theory to molecules using Gaussian. Int J Pure Appl Phys. 2019;15:1.

29. Jones G, Willett P, Glen RC, Leach AR, Taylor R. Development and validation of a genetic algorithm for flexible docking. J Mol Bio. 1997;267:727.
30. Lipinski CA, Lombardo F, Dominy BW, Feeney PJ. Experimental and computational approaches to estimate solubility and permeability in drug discovery and development settings. Adv Drug Deliv Rev. 2001;46:3.
31. Honma T. ADMET prediction and its applications to new structure proposal AI. Medchem News. 2018;28:167.
32. Takeshita Y, Takakura K, Akitsu T. Multifunctional composites of Chiral Valine derivative Schiff Base Cu(II) complexes and TiO_2. Int J Mol Sci. 2015;16:3955.
33. Bursch M, Mewes J-M, Hansen A, Grimme S. Best-practice DFT protocols for basic molecular computational chemistry. Agnew Chem Int Ed. 2022;61:e202205735.
34. Furuya T, Nakane D, Kitanishi K, Katsuumi N, Tsaturyan A, Shcherbakov IN, Unno M, Akitsu T. A novel hybrid protein composed of superoxide-dismutase-active Cu(II) complex and lysozyme. Sci Rep. 2023;13:6892.
35. Halim SA, El-Meligy AB, El-Nahas AM, El-Demerdash SH. DFT study, and natural bond orbital (NBO) population analysis of 2-(2-Hydroxyphenyl)-1-azaazulene tautomers and their mercapto analouges. Sci Rep. 2024;14:219.
36. Nguyen ALP, Izgorodina EI. Behavior of counterpoise correction in many-body molecular clusters of organic compounds: Hartree-Fock interaction energy perspective. J Comput Chem. 2022;43:568.
37. Akitsu T, Miyazaki R, Hasegawa K, Senoo K, Nakane D. Docking simulation of (anionic) Schiff base metal complexes and proteins. AIP Conf Proc. 2025;3269:020007.
38. Miyazaki R, Tamagawa T, Takeda C, Takiguchi Y, Kuroda Y, Nakane D, Akitsu T. Systematic preparation and computation of hybrid materials composed of amino acid Schiff Base copper(II) complexes and lysozyme. In: Taylor JC, editor. Advances in Chemistry Research, 78, 57. NY USA: Nova Science Publishers Inc.; 2023.
39. Lin H. QM/MM methods. Am Chem Soc. 2024. https://doi.org/10.1021/acsinfocus.7e7030.
40. Senftle TP, Hong S, Islam MM, Kylasa SB, Zheng Y, Shin YK, Junkermeier C, Engel-Herbert R, Janik MJ, Aktulga HM, Verstraelen T, Grama A, van Duin ACT (2016) The ReaxFF reactive force-field: development, applications and future directions. NPJ Comput Mater 2:15011.
41. Nakane D, Akiyama T, Suzuki S, Miyazaki R, Akitsu T. Improvement of the SOD activity of the Cu^{2+} complexes by hybridization with lysozyme and its hydrogen bond effect on the activity enhancement. Front Chem. 2024;11:1330833.
42. Miyazaki R, Nakane D, Akitsu T. Analysis of superoxide dismutase reaction mechanism of lysozyme containing amino acid Schiff base copper(II) complexes using molecular dynamics calculations. AIP Conf Proc. 2025;3269:020008.
43. Kawasaki T, Zen H, Ozaki K, Yamada H, Wakamatsu K, Ito S. Application of mid-infrared free-electron laser for structural analysis of biological materials. J Synchrotron Radiat. 2021;28:28.
44. Yoshida N, Shimada T, Hiroki H, Takase M, Akitsu T. L-valine and salicylaldehyde derivative Schiff base Zn(II) complexes as UVA sunscreen. Jacobs J Inorg Chem. 2016;1:6.
45. Singh JS, Khan MS, Uddin S. A DFT study of vibrational spectra of 5-chlorouracil with molecular structure, HOMO–LUMO, MEPs/ESPs and thermodynamic properties. Polym Bull. 2023;80:3055.
46. Kitoh-Nishioka H, Yokogawa D, Irle S. Förster resonance energy transfer between fluorescent proteins: efficient transition charge-based study. J Phys Chem C. 2017;121:4220.
47. Yang H, Yang S, Kong J, Dong A, Yu S. Obtaining information about protein secondary structures in aqueous solution using Fourier transform IR spectroscopy. Nat Protoc. 2015;10:382.
48. Onami Y, Koya R, Kawasaki T, Aizawa H, Nakagame R, Miyagawa Y, Haraguchi T, Akitsu T, Tsukiyama K, Palafox MA. Investigation by DFT methods of the damage of human serum albumin including amino acid derivative Schiff Base Zn(II) complexes by IR-FEL irradiation. Int J Mol Sci. 2019;20:2846.
49. Onami Y, Kawasaki T, Aizawa H, Haraguchi T, Akitsu T, Tsukiyama K, Palafox MA. Degradation of human serum albumin by infrared free electron laser enhanced by inclusion of a salen-type Schiff base Zn (II) complex. Int J Mol Sci. 2020;21:874.

50. Fujisawa N, Onami Y, Kawasaki T, Haraguchi T, Akitsu T, Tsukiyama K. Infrared free electron laser or polarized ultraviolet photolysis of hierarchical and Chiral components of interleukin-6, AlanylAlanine and Alanine. Trends Peptide Protein Sci. 2020;5:e8.
51. Salam A. The role of chirality in the origin of life. J Mol Evol. 1991;33:105.
52. Michl J, Thulstrup EW. Spectroscopy with polarized light. NY, USA: Wiley-VCH; 1995.
53. He Z, Cheng X, Wang Z, Zhang W. Chiral structures in azobenzene-containing systems: construction, regulation, and application. Responsive Mater. 2024;2:e20240010.
54. Ito M, Akitsu T, Palafox MA. Theoretical interpretation of polarized light-induced supramolecular orientation based on normal mode analysis of azobenzene as hybrid materials in PMMA with chiral Schiff base Ni(II), Cu(II), and Zn(II) complexes. J Appl Solut Chem Model. 2016;5:30.
55. Rappe AK, Casewit CJ, Colwell KS, Goddard WA III, Skiff WM. UFF, a full periodic table force field for molecular mechanics and molecular dynamics simulations. J Am Chem Soc. 1992;114:10024.
56. Khoroshyy P, Martinez-Seara H, Myšková J, Lazar J. Dynamics of transition dipole moment orientation in representative fluorescent proteins. Phys Chem Chem Phys. 2023;25:22117.
57. Pallikara I, Skelton JM, Hatcher LE, Pallipurath AR. Going beyond the ordered bulk: a perspective on the use of the cambridge structural database for predictive materials design. Cryst Growth Des. 2024;24:6911.
58. Rosenberg AA, Marx A, Bronstein AM. A dataset of alternately located segments in protein crystal structures. Sci Data. 2024;11:783.
59. Kashiwagi K, Pradhan S, Haraguchi T, Akitsu T. Structural and physical properties of salen-type copper(II) complexes in lysozyme for oxygen reduction. J Indian Chem Soc. 2019;96:593.
60. Spackman MA, Jayatilaka D. Hirshfeld surface analysis. CrystEngComm. 2009;11:19.
61. Greenacre M, Groenen PJF, Hastie T, D'Enza AI, Markos A, Tuzhilina E. Principal component analysis. Nat Rev Methods Primers. 2002;2:100.
62. Suda S, Tateno A, Nakane D, Akitsu T. Hirshfeld surface analysis for investigation of intermolecular interaction of molecular crystals. Int J Org Chem. 2023;13:57.
63. Akitsu T, Kuroda Y, Suda S, Furuya T, Haraguchi T, Unno M. Weakly non-covalent docking of amino-acid Schiff base Zn(II) complex to lysozyme. Key Eng Mater. 2021;888:105.
64. Lin Y-F, Cheng C-W, Shih C-S, Hwang J-K, Yu C-S, Lu C-H. MIB: metal ion-binding site prediction and docking server. J Chem Inf Model. 2016;56:2287.
65. Dürr SL, Levy A, Rothlisberger U. Metal3D: a general deep learning framework for accurate metal ion location prediction in proteins. Nat Commun. 2023;14:2713.
66. Takahashi K, Miyazaki R, Nakane D, Aiyelabola TO, Akitsu T. Preliminary investigation of copper(II) Ion binding or complex coordination in lysozeme molecules. J Mater Sci Chem Eng. 2024;12:98.
67. Koch D, Pavanello M, Shao X, Ihara M, Ayers PW, Matta CF, Jenkins S, Manzhos S. The analysis of electron densities: from basics to emergent applications. Chem Rev. 2024;124:12661.
68. Kuroda Y, Miyazaki R, Shimonishi D, Nakane D, Akitsu T. Coordination and photoisomerization of azobenzene-amino acid Schiff base copper(II) complexes to lysozyme. J Mater Sci Chem Eng. 2023;11:34.
69. Aleotti F, Petropoulos V, Hannah VO, Pettini M, Mancinelli M, Pecorari D, Maiuri M, Medri R, Mazzanti A, Preda F, Perri A, Polli D, Conti I, Cerullo G, Garavelli M. Engineering azobenzene derivatives to control the photoisomerization process. J Phys Chem A. 2023;127:10435.
70. Cahen D, Pecht I, Sheves M. What can we learn from protein-based electron transport junctions? J Phys Chem Lett. 2021;12:11598.
71. Kunitake F, Jong-Yeon K, Yagi S, Yamzaki S, Haraguchi T, Akitsu T. Chiral recognition of azo-schiff base ligands, their Cu(II) complexes, and their docking to laccase as mediators. Symmetry. 2019;11:666.
72. Kajiwara K, Yamane S, Haraguchi T, Pradhan S, Sinha C, Parida R, Giri S, Roymahaptra G, Moon D, Akitsu T. Computational design of Azo-anthraquinone Schiff Base Mn complexes as mediators for biofuel cell cathode. Chem Eng. 2019;13:23.

73. Kajiwara K, Pradhan S, Haraguchi T, Sinha C, Parida R, Giri S, Roymahaptra G, Akitsu T. Photo-tunable azobenzene-anthraquinone schiff base copper complexes as mediators for laccase in biofuel cell cathode. Symmetry. 2020;12:797.
74. Haouas B, Zouaghi MO, Aissa MAB, Batanero B, Arfaoui Y. Synthesis, characterization and computational study of the conjugated thiol-systems for self-assembled monolayers on gold electrode. J Mol Struct. 2024;1296:136832.
75. Saen-Oon S, Lucas MF, Guallar V. Electron transfer in proteins: theory, applications and future perspectives. Phys Chem Chem Phys. 2013;15:15271.
76. Orlando C, Rizzo IC, Arrigoni F, Zampolli F, Mangiagalli M, Di Gennaro P, Lotti M, De Gioia L, Marino T, Greco C, Bertini L. Mechanism of non-phenolic substrate oxidation by the fungal laccase Type 1 copper site from Trametes versicolor: the case of benzo[a]pyrene and anthracene. Dalton Trans. 2024;53:12152.
77. Bloom BP, Paltiel Y, Naaman R, Waldeck DH. Chiral induced spin selectivity. Chem Rev. 2024;124:1950.
78. Kashiwagi K, Tassinari F, Haraguchi T, Banerjee-Gosh K, Akitsu T, Naaman R. Electron transfer via helical oligopeptide to laccase including chiral schiff base copper mediators. Symmetry. 2020;12:808.
79. Morris GM, Goodsell DS, Halliday RS, Huey R, Hart WE, Belew RK, Olson AJ. Automated docking using a lamarckian genetic algorithm and an empirical binding free energy function. J Comput Chem. 1998;19:1639.
80. Ishizuka A, Mishima K, Nakane D, Azam M, Akitsu T. Heteronuclear Fe(III)/Mn(III) salen complex as electron transfer mediator for laccase-based biocathodes: structural investigations, molecular docking, and electrochemical studies. J Mol Struct. 2023;1291:136040.
81. Roymahapatra G, Pradhan S, Sato S, Nakane D, Hait M, Bhattacharyya S, Saha R, Akitsu T. Computational study on docking of laccase and cyanide-bridged Ag–Cu complex for designing the improved biofuel cell cathode. ES General. 2023;21:957.
82. Pucci R, Angilella GGN. Density functional theory, chemical reactivity, and the Fukui functions. Found Chem. 2022;24:59.
83. Govorov AO, Fan Z, Hermandez P, Slocik JM, Naik RR. Theory of circular dichroism of nanomaterials comprising chiral molecules and nanocrystals: plasmon enhancement, dipole interactions, and dielectric effects. Nano Lett. 2010;10:1374.
84. Oshima M, Matsuno M, Tsutsumi Y, Sunaga N, Haraguchi T, Akitsu T. Synthesis of chiral Schiff base metal complex inducing CD and elucidation of structure of adsorption on surface of gold nanoparticles. Int J Org Chem. 2017;7:153.
85. Tsutsumi Y, Sunaga N, Haraguchi T, Akitsu T. Induced CD from chiral Schiff base metal complexes involving azo-dye groups to gold nanoparticles. J Indian Chem Soc. 2017;94:1163.
86. Sunaga N, Haraguchi T, Akitsu T. Orientation of chiral schiff base metal complexes involving azo-groups for induced CD on gold nanoparticles by polarized UV light irradiation. Symmetry. 2019;11:1094.
87. Tanaka S, Sato H, Ishida Y, Deng Y, Haraguchi T, Akitsu T, Sugiyama M, Hara M, Moon D. Photo-control of adsorption of dye metal complexes incorporating chiral Schiff base ligands containing azo-groups on TiO_2. J Korean Chem Soc. 2018;62:328.
88. Yamane S, Hiyoshi T, Tanaka S, Ikenomoto S, Numata T, Takakura K, Haraguchi T, Palafox MA, Hara M, Sugiyama M, Akitsu T. Substituent effect of Chiraldiphenyl Salen Metal (M = Fe(II), Co(II), Ni(II), Cu(II), Zn(II)) complexes for new conceptual DSSC dyes. J Chem Chem Eng. 2018;11:135.

Chapter 18
Basic Technique for Oriented Molecular Beam

Dock-Chil Che

Abstract This chapter reviews the basic technique for the oriented molecular beam using the hexapole electric field, which is composed of the six-pole electric field. The hexapole electric field is used for both selections of the rotational state and the orientation of molecules in the laboratory frame by the interaction of electric field with the molecular dipole moment. This technique is widely applicable not only to the symmetric-top molecules but also to the radicals or the asymmetric molecules. The studies using the oriented molecular beam have been clarified that the chemical reaction considerably depends on the molecular orientation. We introduce the basic experimental technique for the oriented molecular beam and the theoretical calculation of the orientational distribution after passing through the hexapole state selector.

Keywords Quantum dynamics · Stereodynamics · Hexapole electric field · State selection · Oriented molecular beam · Symmetric-top molecule · OH radical · Asymmetric-top molecule · Orientation dependence in reaction · Orientational distribution

18.1 General Introduction

The idea that chemical reaction depends on the mutual orientation of reactant molecules is one of the most intuitive concepts. Measurements of the steric effect in atom-molecule reactions can provide insight into the anisotropies of interest potential and the reaction dynamics [1–6]. However, due to the difficulty for controlling molecular orientation prior to the chemical reaction, only a handful of experiments have been reported [1–3]. This chapter aims to describe the technique of molecular orientation control and the stereodynamics study using the oriented molecular beams.

D.-C. Che (✉)
Department of Chemistry, Graduate School of Science, Osaka University, Toyonaka, Osaka, Japan
e-mail: che@chem.sci.osaka-u.ac.jp

T. Onishi (ed.), *Advances in Quantum Science II*,
https://doi.org/10.1007/978-981-95-9721-5_18

18.1.1 History of Stereodynamics Research

Direct experimental studies on steric effects have become possible with the development of modern vacuum systems and electronics. One of the powerful techniques for elucidating the steric effect in chemical reaction is the application of the hexapole electric field, which is composed of the six-pole electric field [4–13]. By using this technique, molecular orientation of symmetric-top molecules such as CF_3H, NH_3 and so on can be easily selected in the laboratory frame by the first-order Stark effect between the interaction of electric filed and the molecular dipole moment [14–17]. The pioneering work using the hexapole state selection has been studied by two groups of Brooks and of Bernstein. The reaction of oriented CH_3I with either K or Rb atom to produce KI or RbI demonstrated the existence of steric effects in the gas phase for the first time [4, 12].

$$K + CH_3I \rightarrow KI + CH_3 \tag{18.1}$$

$$Rb + CH_3I \rightarrow RbI + CH_3 \tag{18.2}$$

The angular distribution of product KI was found to be scattered in the backward direction. Large flux of KI was observed when the K atom was incident toward the I end of CH_3I molecule. For the Rb + CH_3I reaction, the large orientation dependence was observed. By the analysis of the experimental results, the existence of a cone of nonreactive area at the CH_3-end was suggested. Interestingly, no significant steric effect in the reaction:

$$K + CF_3I \rightarrow KI + CF_3 \tag{18.3}$$

has been reported [13]. This observation is well explained by the long-rage interaction of the two kinds of harpoon mechanism, which is an electron of alkali metal atom transfers to the CF_3I molecule in the first step. In the next step, the formed two ions of K^+ and CF_3I^- interact by the Coulomb force. This mechanism had not been observed in reaction with unoriented molecules. The orientation dependences of the chemiluminescent channel in the reaction of metastable argon atom with oriented molecules have been studied in Kuwata's groups [14–23]. The considerable orientation dependence of the formation of the excited radicals have been observed in the reactions [14–22]:

$$CF_3H + Ar(^3P) \rightarrow CF_3^* + H + Ar \tag{18.4}$$

$$NH_3 + Ar(^3P) \rightarrow NH_2^* + H + Ar \tag{18.5}$$

$$CH_3CN + Ar(^3P) \rightarrow CH_3 + CN^* + Ar \tag{18.6}$$

These experimental results tell us that the chemical reaction considerably depends on the molecular orientations. Here, the basic technique for the oriented molecular beam using the hexapole electric field will be presented.

18.2 Experimental Technique Using Hexapole Electric Field

18.2.1 Quantum Numbers for Rotation in Symmetric-Top Molecule

Oriented molecular beam was first applied to the symmetric-top molecules because of the simply expressed molecular motion in the hexapole electric field. Figure 18.1 shows the illustrated symmetric-top molecule in the electric field. A symmetric-top molecule is a molecule which has three-fold axis of symmetry. Two of the three principal moments of inertia are equal to each other. Examples of such molecules are methyl iodide (CH_3I), acetonitrile (CH_3CN), and so on. The total angular momentum, J, of the molecules is represented by two precession motions. One is the molecular axis, K, which rotates around J. The other is the rotation of the J around the direction of the electric field, M. In classical picture, the orientation, that is the angle between the K vector (molecular axis) and M vector (direction of electric field), is expressed by the use of three quantum numbers for rotation of a symmetric-top molecule: J, K, and M.

$$\cos\theta = \frac{KM}{J(J+1)} \tag{18.7}$$

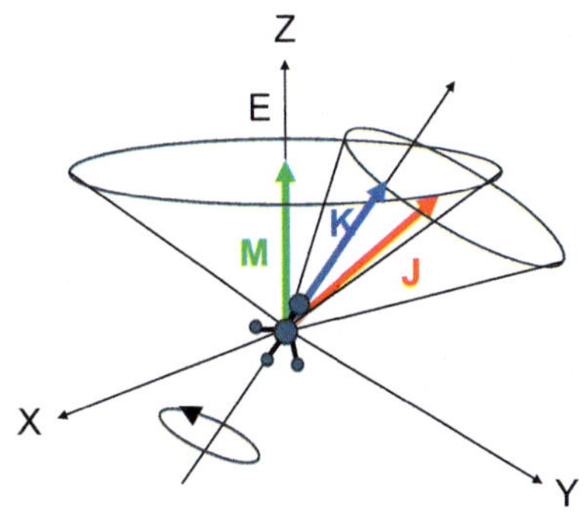

Fig. 18.1 Rotational behavior of symmetric-top molecule in an electric field. The magnitude of total angular momentum and its projection on to the molecular axis are denoted by J and K, respectively. The projection of J on to z axis is denoted by M

18.2.2 Rotational Energy

A statistical problem to be considered is an ensemble average over all orientation of molecules in the molecular beam as an orientational distribution. Orientational distribution can be calculated by taking the weighted sum of the distributions for each rotational state for all possible combinations of J, K, and M. They can be obtained from the trajectory simulation in the hexapole field by taking care about the interaction of molecules with the electric field. (See in Sect. 18.2.3 for the trajectory simulation.) When a molecule with a permanent dipole moment is placed in an electric field, the rotational energy of the molecule is perturbed, and the (2 J + 1)-fold degeneracy arising from different M quantum numbers are removed [24]. This perturbation of rotational energy is known as the Stark effect. Using the first-order perturbation theory [25], this energy (W) can be expressed by,

$$W = \int \mu E \sin\theta d\theta = -\mu E \cos\theta \tag{18.8}$$

where μ is the magnitude of the dipole moment of molecules, E is the field strength, and θ is defined by the angle between the dipole moment and the direction of electric field.

18.2.3 Trajectory Simulation in Hexapole Electric Field

Figure 18.2 shows the experimental setup of a hexapole electric field in a molecular beam apparatus. The advantage of using the hexapole field is that the symmetric molecules can be focused on a space after passing through the electric field. This indicates that the beam intensity does not decrease as compared with a conventional beam. Hexapole electric field consists of six rods alternatively charged and equally spaced in a circle centered about the beam axis as shown in the inset of figure. In this setup, the strength of electric field (E) is given by

$$E = 3V_0 \frac{r^2}{R^3} \tag{18.9}$$

where V_0 is the applied hexapole voltage, r is the distance of the molecule from the center of hexapole, and R is the radius of the hexapole field. Using Eqs. (18.8) and (18.9), the motion of molecule can be solved by the force exerted to a molecule in the field (F):

$$F = \frac{\partial W}{\partial r} = -\mu \cos\theta \frac{\partial E}{\partial r} = -kr\cos\theta \tag{18.10}$$

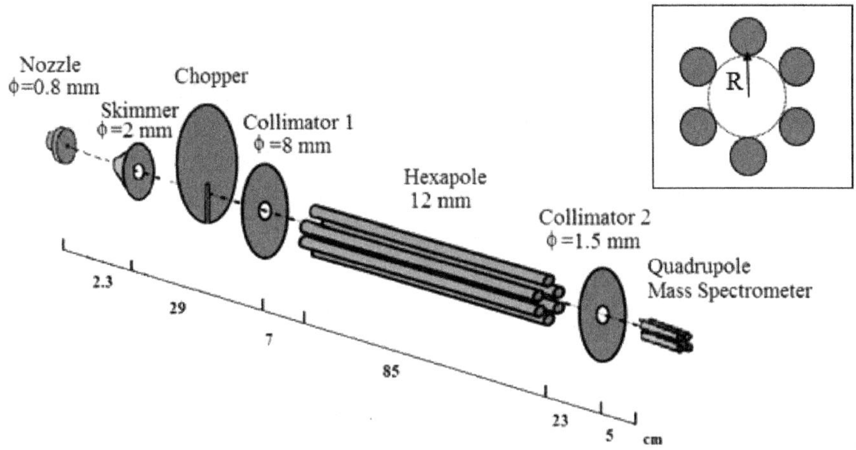

Fig. 18.2 Experimental apparatus for the focused beam. The dimensions are shown in figures: distances are expressed in cm and diameters in mm. The inner panel shows a view of the hexapole along its longitudinal axis. Radius of hexapole rod is 6 mm. Inlet radius (R) of hexapole electric field is 6 mm

where $k = \left(\frac{6V_0\mu}{R^3}\right)$. The sign of F depends upon whether cosθ is positive or negative. For a symmetric-top molecule, the sign of F is determined by the KM value in Eq. (18.7). If molecules have a positive sign, the trajectory shows the simple harmonic motion and focuses toward the beam axis. On the other hand, if molecules have negative sign, then molecules deflect from the beam axis. The molecules with $< \cos\theta > = 0$ are not affected from the electric field. From the transmitted molecule of hexapole field with the $|JKM >$ state, one can calculate the orientational distribution, respectively. By the overall summation of the orientational distribution, we can determine the orientational distribution in the molecular beam.

18.2.4 Quantal Orientational Distribution of Molecules

In order to calculate the quantal orientational distribution of molecules after passing through the hexapole state selection, we consider the distribution of a specific $|JKM >$ state in laboratory frame. For a molecule in the state of $|JKM >$, Choi and Bernstein have evaluated the quantal probability distribution, P_{JKM}, of a specific orientation as.

$$P_{JKM}(\cos\theta) = \frac{(2J+1)}{2}\sum_{n=0}^{2J} C_n(JKM)P_n(\cos\theta) \tag{18.11}$$

where the expansion coefficients of $C_n(JKM)$ are determined as a function of J, K, and M values using the following 3j-symbol [26].

$$C_n(JKM) = (2n+1)(-1)^{M-K}\begin{pmatrix} J & J & n \\ K & -K & 0 \end{pmatrix}\begin{pmatrix} J & J & n \\ M & -M & 0 \end{pmatrix} \tag{18.12}$$

The 3j-symbol originates from the coupling of the angular momentum, and it can be expressed by

$$\begin{pmatrix} j_1 & j_2 & j_3 \\ m_1 & m_2 & m_{13} \end{pmatrix} = \left[\frac{(j_1+j_2-j_3)!(j_1-j_2+j_3)!(-j_1+j_2+j_3)!}{(j_1+j_2+j_3)!}\right]^{\frac{1}{2}} \times \begin{bmatrix} (j_1+m_1)!(j_1-m_1)! \\ (j_2+m_2)!(j_2-m_2)! \\ (j_1+m_1)!(j_3-m_3)! \end{bmatrix} \times \sum_k \left[\frac{(-1)^{k+j_1+j_2-m_3}}{\begin{matrix} k!(j_1+j_2-j_3+k)! \\ (j_1-m_1-k)!(j_2+m_2-k)! \\ (j_3-j_2+m_1+k)!((j_3-j_1-m_2+k)! \end{matrix}}\right] \tag{18.13}$$

$P_n(\cos\theta)$ in Eq. (18.11) is the Legendre polynomials expressed as follows:

$$P_n(\cos\theta) = \frac{1}{2^n n!}\frac{d^n}{dx^n}(x^2-1)^n \tag{18.14}$$

The orientational distribution of molecular beam is a sum over the individual states weighted by the population of each state. Finally, the orientational distribution of rotational energy, $W_v(\cos\theta)$, of molecular beam after through the hexapole field at the applied voltage (V) is obtained by.

$$\begin{aligned} W_V(\cos\theta) &= \sum_{JKM} P_{JKM}(\cos\theta)W_n(JKM) \\ &= \sum_{JKM}\frac{2J+1}{2}\sum_n C_n(JKM)P_n(\cos\theta)W_n(JKM) \\ &= \sum_{JKM}\frac{2J+1}{2}\sum_n C_n(JKM)W_n(JKM)P_n(\cos\theta) \end{aligned} \tag{18.15}$$

W_n (JKM) is a weight for each rotational state directly reflecting the population after the state selection and can be obtained the trajectory simulation. As an example, the orientational distribution for the acetonitrile molecular beam is shown in Fig. 18.3. The classical distribution is shown by the histogram, and the smooth curve shows the quantal distribution obtained from Eq. (18.7). Using this orientational distribution of the molecular beam, the orientation dependences in the reaction of the metastable Ar

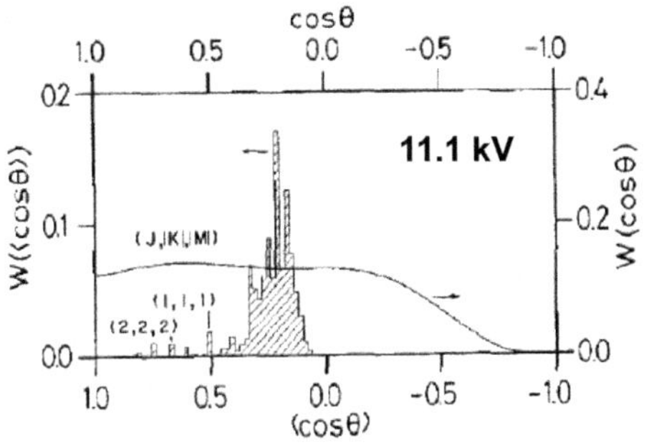

Fig. 18.3 Classical, W (<cosθ >), and quantal, W(cosθ), distributions of CH_3CN molecular beam after through the hexapole field at 11.1 kV. The set of three rotational quantum numbers is given as $|JKM>$ state. The classical distribution is shown by histograms. Smooth curve shows the quantal distribution transformed from the classical distribution

with the acetonitrile (CH_3CN), ammonia (NH_3), and trifluoromethane (CF_3H) have been studied in our group [14–23].

18.3 Single State Selection of Hydroxyl Radicals

18.3.1 Orientational Quantum State

The hydroxyl radical (OH) plays a key role in combustion, atmospheric reactions [27–29]. The open-shell structure of OH radical is unstable and reactive. This leads to the large orientational dependence in reaction due to the anisotropic distribution of the molecular orbital for the radical.

The molecular orientation of OH radical can be characterized by the quantum number of J, Ω, and M_J, where J is the total angular momentum, Ω is the spin–orbit momentum, and M_J is the projection of J on to the electric field, respectively. Ter Meulen and coworkers have reported the first state selection of OH radicals using the short length (23 cm) of hexapole electric field [30–36]. The formed OH radicals were detected by means of the laser-induced fluorescence (LIF) technique as indicated the energy diagram of detection in Fig. 18.4. Using the oriented OH beam, the nonreactive collision systems of OH with At and N_2 have been studied. Unfortunately, the observed focusing curve, which is the beam intensity upon the applied voltage to the hexapole field, is increasing monotonously due to the short hexapole electric field. So far, no focusing curves for the higher rotational states have been reported as well. Single quantum state selection of OH radicals has been provided by Curtiss's group using the continuous molecular beam detected by the mass spectrometer [37]. The observed focusing curve presented clear structures originating from the quantum states: $|J\Omega M_J = |\frac{3}{2} \pm \frac{3}{2} \mp \frac{3}{2}$ and $|\frac{3}{2} \pm \frac{3}{2} \mp \frac{1}{2}$. The trajectory simulations were in good agreement with the experimental results, suggesting the successful selection

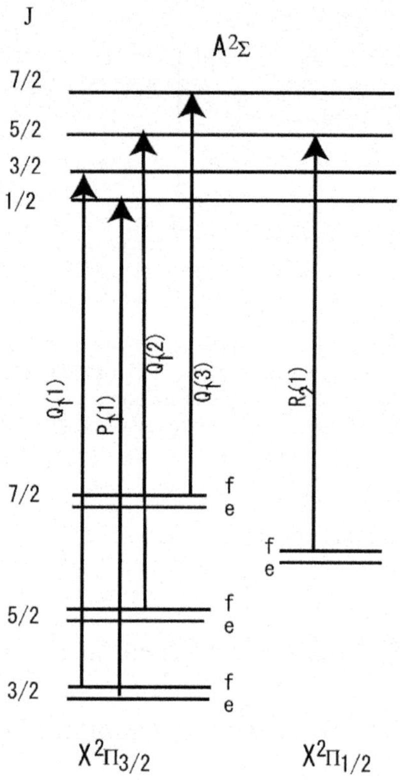

Fig. 18.4 Schematic energy diagram of the $OH(A^2\Sigma \leftarrow X^2\Pi)$ LIF detection. The energy difference between the components of each Λ-doublet is exaggerated for reasons of clearness

of the single quantum state selection. However, the focusing curves for the higher rotational state could not be resolved because of the mass spectrometry detection.

18.3.2 Probed Transitions of OH Radicals

We also succeeded in the single quantum state selection of OH radical using the 1-m hexapole field combined with the LIF detection [38]. Using this detection system, we can measure the focusing curves consisting of a single $|J\Omega M_J>$ quantum state, enabling reliable comparison with trajectory simulations of individual rotational states. As a result, one can extract the effects of Λ-type doubling and the spin–orbit interaction of OH in an electric field. We performed trajectory simulations considering the effect of Λ-type doubling and successfully calculated the orientation distribution.

Figure 18.4 shows the energy diagram of $OH(A^2\Sigma \leftarrow X^2\Pi)$ LIF detection. The energy levels are labeled with J and Ω. The lowest state of $OH(X^2\Pi_{3/2})$ rotation is $|J\Omega> = |\frac{3}{2}\frac{3}{2}>$. The electronic ground states, such as $\Pi_{3/2}$ and $\Pi_{1/2}$ define the value of Ω. Each rotational level is split into e (lower) and f (upper) components due to Λ-type doubling. In the LIF spectrum, the e and f components are probed via P (or R) and Q branch transitions, respectively. The observed transitions are indicated by arrows in the figure.

18.3.3 Experimental Apparatus for State Selection of OH Radicals

Figure 18.5 shows the experimental apparatus for the state selection of OH radical beam combined with the LIF detection system. Radical source of OH is composed of a pulsed valve nozzle and a ring electrode. The ring electrode is kept at negative high voltages of − 1.5 kV. H_2O vapor seeded in Ar carrier gas is expanded into the vacuum chamber through the pulsed valve. A discharge takes place between the nozzle and the ring electrode. Due to supersonic expansion, rotational states of OH radical beam are cooled down. After skimming by a skimmer, the OH radical beam enters the hexapole inhomogeneous field, and is selected the rotational state of OH by the Stark effect. Then OH radical beam is detected by LIF technique using the A ← X transition at 308 nm as shown in Fig. 18.4, which laser wavelength generated by a YAG-Dye laser system. Under low laser power conditions, the relative population N of OH in the ground state can be derived from the fluorescence intensity I according to $N \propto I/(BP)$, where B is the Einstein absorption coefficient and P denotes the laser power [39]. At sufficiently high laser power where all transitions are saturated, the population becomes directly proportional to the fluorescence intensity, $N \propto I$ [40]. Two experimental configurations were employed. In the first, LIF signals were recorded in counting mode under low laser power. In the second, signals were integrated using a boxcar system under saturation conditions. The latter configuration was used to evaluate the beam intensity of the molecular beam. Saturation was verified by measuring the dependence of the fluorescence signal on laser power. Under confirmed saturation conditions, the population of the probed OH state is directly proportional to the detected fluorescence intensity for all transitions considered. The UV laser beam was unfocused and had a diameter of 5 mm, exceeding the diameter of the molecular beam at the interaction region.

18.3.4 Trajectory Simulation Based on First-Order Stark Effect

Trajectory simulation is based on the first order of Stark effect [37]. The first-order Stark energy including the Λ-type doubling can be expressed as follows.

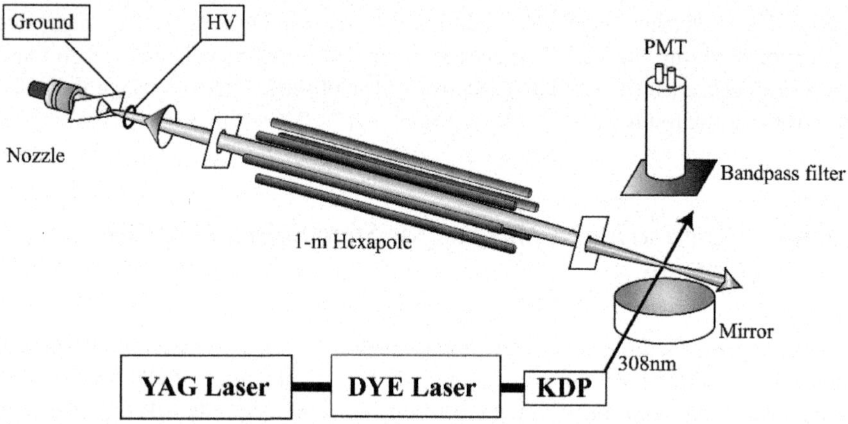

Fig. 18.5 Experimental apparatus for the state selection of OH radicals with the LIF detection system

$$W_E = \frac{W_1 + W_2}{2} \pm \left[\frac{(W_1 - W_2)^2}{4} + \mu^2 E^2 \left(\frac{\Omega M_J}{J(J+1)} \right)^2 \right]^{1/2} \quad (18.16)$$

whereas W_1 and W_2 indicate energy levels of the Λ-states at given J state, μ is the permanent dipole moment of OH, and E is the field strength of electric field.

$$\begin{aligned} F &= -\frac{\partial W_E}{\partial r} = -\frac{\partial W_E}{\partial E} \cdot \frac{\partial E}{\partial r} \\ &= \pm \left[\frac{(W_1 - W_2)^2}{4} + \mu^2 E^2 \left(\frac{\Omega M_J}{J(J+1)} \right)^2 \right]^{-1/2} \mu^2 \left(\frac{\Omega M_J}{J(J+1)} \right)^2 E \frac{\partial E}{\partial r} \end{aligned} \quad (18.17)$$

The force acting on the OH radical in the hexapole electric field is described by Eq. (18.16). In Eq. (18.8), the negative sign corresponds to a trajectory that is focused by the hexapole field and is associated with the upper Λ-doubling component. The positive sign denotes a trajectory that diverges from the beam axis, corresponding to the lower Λ-doubling component. Radial motion of OH radical in the hexapole field was numerically calculated with a time-step of 1 μs. Number of successful trajectories through the detector aperture were counted as an intensity of beam.

18.3.5 Signal Enhancement by Hexapole Electric Field

Figure 18.6a displays the LIF spectrum of the OH (A ← X) transition obtained with the hexapole voltage is off. A comparison of the $P_1(1)$ and $Q_1(1)$ transition intensities reveals that the upper Λ-doubling component is more populated than

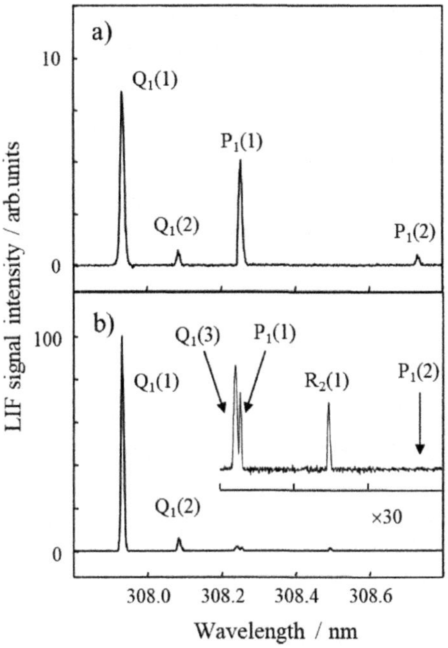

Fig. 18.6 Laser-induced fluorescence (LIF) spectrum of the OH radical. **a** Hexapole voltage is turned off **b** Hexapole voltage set to 13.2 kV

the lower component. This population preference is presumably related to the OH production mechanism in the electrical discharge source employed in the present study. The rotational temperature of the OH(X) beam was evaluated from the intensity ratio of the $Q_1(1)$ and $Q_1(2)$ transitions, corresponding to the two lowest rotational levels, yielding a value of 42 K.

Figure 18.6b presents the LIF spectrum recorded with a hexapole voltage of 13.2 kV. As described in the preceding section, activation of the hexapole field results in focusing of the upper Λ-doubling component, while the lower component is defocused. Accordingly, an enhancement in the signal intensities of the Q ($\Omega = 3/2$) and R ($\Omega = 1/2$) transitions is observed when the hexapole voltage is applied. However, the signal intensity decreases for P ($\Omega = 3/2$) transitions. The signal intensities for $Q_1(1)$, $Q_1(2)$, $Q_1(3)$, and $R_2(1)$ transitions originating from $|\frac{33}{22} >$, $|\frac{53}{22} >$, $|\frac{73}{22} >$, *and* $|\frac{11}{22} >$ rotational states dramatically increase when the hexapole voltage is turned on, whereas the intensities for $P_1(1)$ and $P_1(2)$ transitions originating from $|\frac{33}{22} >$ *and* $|\frac{53}{22} >$ decrease.

18.3.6 Beam Intensity Depending on Hexapole Voltage for Each Rotational State

Figure 18.7 shows the focusing curve of ground-state $OH\,|X^2\Pi_{1/2}, J = \frac{1}{2}>$, probed by the $R_2(1)$ transition. The experimental data are represented by open circles. A pronounced peak is observed at an applied voltage of 4.0 kV. The solid line in figure shows the results of the trajectory simulations, which is in good agreement with the experimental data. From the simulation analysis, the translational temperature of the molecular beam was determined to be 19 K.

The simulated focusing profile displays a noticeable asymmetry around the peak position, which arises from the influence of Λ-type doubling. The calculated maximum at 4.0 kV corresponds to a half-wave-period trajectory, whereas the full-wave-period trajectory is predicted to occur at 14.1 kV. Calculations neglecting Λ-type doubling yield peak positions at 3.5 and 13.2 kV, respectively, and one cannot reproduce the experimental observations. These findings clearly demonstrate the significant role of Λ-type doubling in determining the OH trajectories within the hexapole field.

To further assess the validity of the trajectory simulations, the focusing characteristics of additional rotational states were examined. Figure 18.8a presents the focusing curve for the ground-state $OH\,|X^2\Pi_{3/2}, J = \frac{3}{2}, f>$, probed via the $Q_1(1)$ transition. This state constitutes the dominant component (93%) of the focused molecular beam. The experimental data are indicated by open circles. The focusing profile exhibits two distinct peaks at 1.6 and 6.0 kV. In the LIF detection of the $J = \frac{3}{2}$ state, two

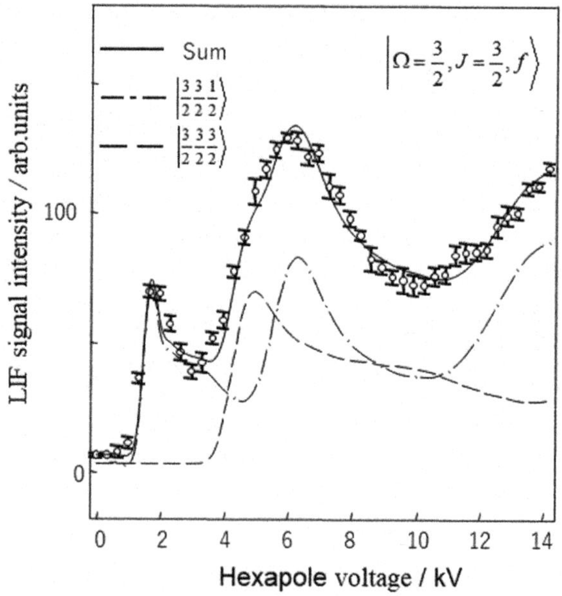

Fig. 18.7 Focusing curve of $|X^2\Pi_{1/2}, J = \frac{1}{2}, f>$ of the OH radical. Experimental results are shown as open circles, result of the trajectory simulation as solid line

$|J\Omega M_J>$ components of $|\frac{333}{222}>$ and $|\frac{331}{222}>$ contribute to the observed signal. The simulated focusing curves for these two components, calculated under the assumption of equal populations, are shown as dash-dotted and dashed lines, respectively, while the solid line represents their sum. The combined simulation reproduces the experimental data with good accuracy. From the measured signal intensity, the beam intensity was estimated to be approximately ~ 3×10^{16} molecules / sr s at a hexapole voltage of 6.0 kV.

Figures 18.8b and c show the focusing curves of $OH|X^2\Pi_{3/2}, J = \frac{5}{2}, f>$ and $|X^2\Pi_{3/2}, J = \frac{7}{2}, f>$ probing the Q_1 (2) and Q_1 (3) LIF signals. Experimental results are presented as open circles in the figure. For the $J = \frac{5}{2}$ state, the focusing curve includes contributions from three $|M_J|$ states of $\frac{1}{2}$, $\frac{3}{2}$, and $\frac{5}{2}$. As an initial approximation, the trajectories were calculated under the assumption of equal populations for these M_J states; the resulting simulation is shown as a solid line. While satisfactory agreement with experiment is obtained in the low-voltage region, the simulation overestimates at higher hexapole voltages. It is considered that this discrepancy reflects the inadequacy of the equal-population assumption, and improved agreement is often achieved by adopting a non-uniform M_J distribution. For the $J = \frac{7}{2}$, state, five M_J components contribute to the focusing behavior, rendering the trajectory calculations

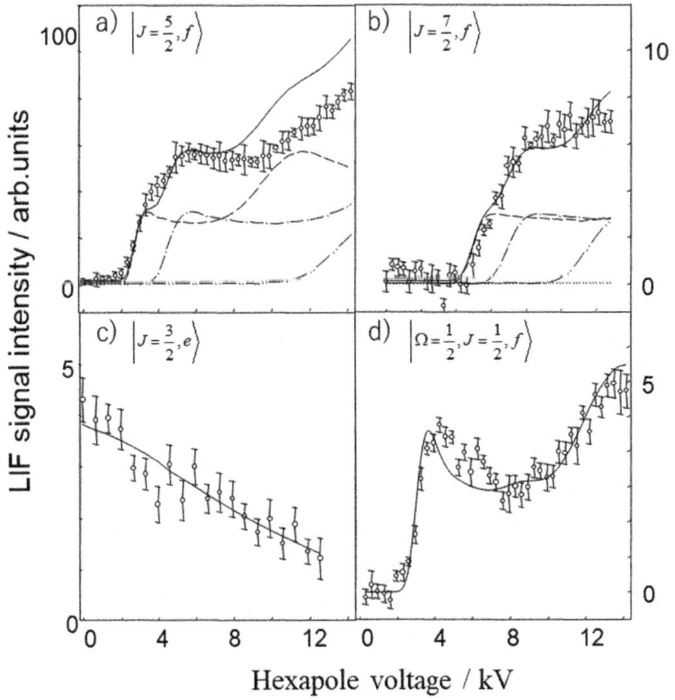

Fig. 18.8 Focusing curves of the OH radical **a** $|J = \frac{3}{2}, f>$ **b** $|J = \frac{5}{2}, f>$ **c** $|J = \frac{3}{2}, f>$ **d** $|J = \frac{3}{2}, e>$

more involved. Nevertheless, the simulated curve reproduces the experimental data with good accuracy. It should be noted that the influence of Λ-type doubling becomes increasingly pronounced for higher J states.

Figure 18.8d displays the focusing curve for $OH|X^2\Pi_{3/2}, J = \frac{3}{2}, e >$, corresponding to the lower Λ-doubling component of the $J = \frac{3}{2}$ level. As anticipated, the signal intensity decreases monotonically with increasing hexapole voltage. The solid line represents the trajectory simulation, which agrees well with the experimental results. At a hexapole voltage of 6 kV, the intensity ratio between the upper and lower Λ-doubling components for $J = \frac{3}{2}$ is approximately 250. This pronounced difference demonstrates that the hexapole electric field provides an effective means of selectively enhancing a specific Λ-doubling component.

18.4 Stereodynamics Study in the Reaction of Oriented OH with HBr

Four-atom reactions of the type AB + CD have attracted substantial experimental and theoretical attention in the field of reaction dynamics [27]. Among these, reactions of OH with hydrogen halide molecules serve as prototypical systems for investigating the fundamental mechanisms of four-atom processes [28, 29]. In particular, the OH + HCl and OH + HBr reactions are of significant importance in atmospheric chemistry, as they generate halogen atoms that play a critical role in catalytic ozone depletion [41]. Owing to its open-shell electronic structure, the OH radical exhibits high chemical reactivity. Consequently, a pronounced dependence of the reaction dynamics on the spatial orientation of OH is anticipated for the OH + HBr system [42–44].

$$OH + HBr \rightarrow H_2O + Br \tag{18.18}$$

Here, we present about orientation dependence in the reaction of HBr with the oriented OH radicals.

18.4.1 Experimental Setup of Hydroxyl Radical Beam

State selection of the OH radical beam using a 1-m hexapole state selector has been described in the previous section [38]. An intense OH beam was generated by the discharge source. Figure 18.9 shows the set up for the study of the oriented OH radical with HBr molecule. The He-seeded HBr supersonic beam was generated by a pulsed valve. Signal intensity depending upon the hexapole voltage is shown in Fig. 18.10a, suggesting the successful single state selection of $|J\Omega M_J> = |\frac{3}{2}\frac{3}{2}\frac{3}{2}>$ state. The HBr beam intersects the oriented OH beam at right angles at the interaction region. After traversing a 1-m electrostatic hexapole, the OH radicals are oriented in the laboratory frame by an external orienting field in combination with a guiding field. The applied orienting field has a strength of 8.5 kV cm^{-1} and is aligned parallel to the relative velocity vector of the two reactants (61°). This field strength is sufficient to achieve complete orientation of the $OH\ (X^2\Pi)$ Λ-doublet states [37, 45, 46]. To investigate the orientation dependence of the OH + HBr reaction, the polarity of the orienting field was reversed to generate three distinct configurations corresponding to O-end attack, H-end attack, and a randomly oriented distribution of the OH radical. The resulting orientational distributions are illustrated in Fig. 18.10b. The Br atoms produced in the reaction were detected by vacuum ultraviolet laser-induced fluorescence (VUV-LIF) [47–50]. This method was employed in place of REMPI to avoid significant absorption interference from the primary HBr beam. The VUV radiation was generated by four-wave mixing in Mg vapor and tuned to the Br $\left(^2P_3/2 \rightarrow ^4 P_3/2\right)$ resonance transition at 154 nm. In the present study, Br atoms formed in the ground $^2P_3/_2$ state were monitored. The LIF signals were recorded using a photon-counting system equipped with a solar-blind photomultiplier tube (Hamamatsu R6835). During data acquisition, signals obtained with the hexapole voltage turned off were subtracted from those recorded with the voltage applied, thereby eliminating background contributions arising from the direct OH beam.

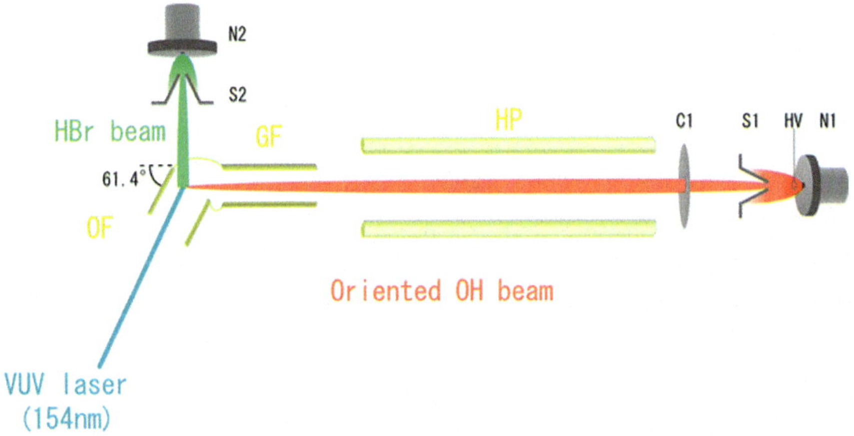

Fig. 18.9 Schematic representation of the experimental arrangement for the reaction between oriented OH radicals and HBr. After transmission through the guiding field, the OH radicals are aligned in the laboratory frame by an applied orienting electric field. The Br atoms generated in the reaction are detected by vacuum ultraviolet laser-induced fluorescence (VUV-LIF)

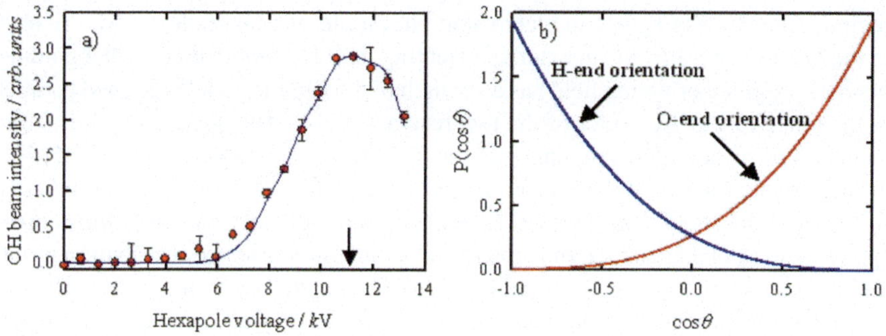

Fig. 18.10 **a** Focusing curve OH molecular beam seeded in He. The OH beam intensities are represented by circles, while the solid line denotes the results of the trajectory simulation. The crossed-beam experiments were carried out at a hexapole voltage of 11 kV, as indicated by the arrow **b** Orientational distribution of the $OH\left(X^2\Pi 3/2\right)$ rotational state. The blue curve, P(cosθ), represents the H-end attack configuration, whereas the red curve, P(−cosθ), corresponds to the O-end attack configuration

18.4.2 Reaction Cross Section

Reaction cross section $S(E_c)$ of the ground-state Br formation in the OH + HBr reaction can be expressed as

$$S(E_c) \propto \frac{I_{SIG} v_{Br}}{I_{VUV}[OD][HBr]v_{rel}} \tag{18.19}$$

where I_{SIG} and I_{VUV} denote LIF signal intensity and VUV laser beam intensity, respectively. V_{rel} is the relative velocity of reactants in the laboratory frame, and v_{Br} corresponds to the diffusion motion of Br atoms away from the detection region. $[OH]$ and $[HBr]$ stand for the beam densities of OH and HBr at the crossing volume. Given angular and speed distribution of Br products, one can obtain the relative reaction cross section.

The relative reaction cross section for each orientation was defined as the ratio of the measured signal intensity to that obtained under randomly oriented conditions. The results indicate that the O-end attack is favored over the H-end attack by a factor of 3.4 ± 2.3. To quantitatively interpret the observed orientation dependence, a simple hard-sphere model was employed to estimate the acceptance cone for Br atom formation [10, 51]. Within this framework, the reactivity ratio, R, between O-end and H-end attacks can be formulated in terms of the orientational probability distribution function P(cosθ) [52–55], as given below:

$$R = \frac{\int_{\cos\theta_A}^{1} P(-\cos\theta)d\cos\theta}{\int_{\cos\theta_A}^{1} P(\cos\theta)d\cos\theta} \tag{18.20}$$

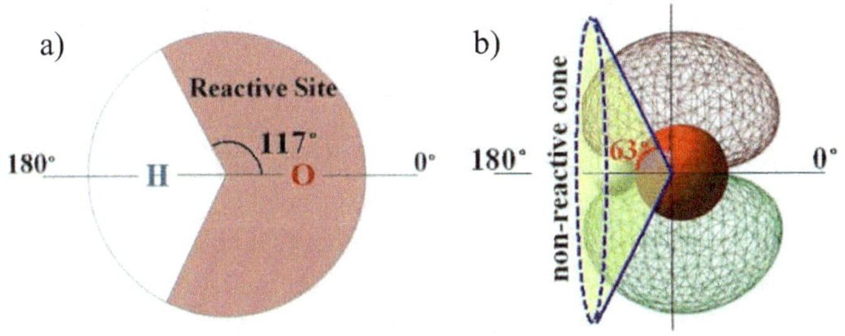

Fig. 18.11 **a** Experimental results for the acceptable angle in reaction are shown by the shaded area **b** Spatial distribution of electrons in the highest occupied molecular orbital (HOMO) from ab initio calculation. Nonreactive area by hydrogen atom was displayed by the cone of steric hindrance

In this model, the OH molecule is divided by a critical angle θ_A into two regions: the O-end is assigned unit reactivity ($\sigma_O = 1$), whereas the H-end is assumed to be nonreactive ($\sigma_H = 0$). The OH dipole moment vector is defined as pointing toward the H-end, and the orientation angle θ is given by the angle between the orienting field and the dipole moment vector. Solving Eq. (18.19) with the experimentally determined orientational probability distribution for the OH beam yields a reactivity ratio $R = 3.4 \pm 2.3$. Figure 18.11a shows a polar representation of the model, where the shaded area indicates the acceptance cone defined by θ_A, determined to be 117 $\pm$ 13°.

18.4.3 Spatial Distribution of Highest Occupied Molecular Orbital (HOMO) Electron

For comparison with the hard-sphere model described above, the spatial distribution of the highest occupied molecular orbital (HOMO) of the OH radical was calculated using density functional theory (DFT) at the B3LYP/6–311 + + G(d,p) basis set. The resulting orbital is depicted in Fig. 18.11b. The HOMO electron density is predominantly localized on the oxygen atom and exhibits a nonbonding character. This result provides a visual rationale for the experimental findings, suggesting that the direct attractive interaction between the HOMO electron of OH and the HBr molecule plays a decisive role in the reaction. Such an interpretation is consistent with chemical intuition. For comparison, the nonreactive cone derived from the model is also shown in Fig. 18.11b, overlaid with the HOMO electron distribution. The extent of this nonreactive cone exceeds the atomic radius of hydrogen, indicating substantial steric hindrance imposed by the hydrogen atom.

18.5 State Selection of the Asymmetric-Top Molecule Propylene Oxide

Although the hexapole electric field has been widely employed to achieve the selection of molecular orientations, state selections have been limited to the symmetric-top molecules or some linear molecules such as N_2O or OH [1–3]. The reason is that the Stark energy can be well described by the first order of perturbation theory for these molecules. It enables us the trajectory simulation of molecules in the hexapole electric field, leading to calculation of the orientational distribution of molecular beam. More recently, we have succeeded in the state selection and the trajectory simulation for the asymmetric molecules using the hexapole field [56]. In order to fully characterize the rotational state distribution, complexity of the asymmetric molecule is higher than most cases previously studied. The energies of the rotational states have been calculated depending on the electric field by diagonalizing the complete Stark Hamiltonian:

$$H_s = -\sum_F \sum_g E_F \mu_g \Phi_{Fg} = E_z \sum_g \Phi_{Zg} \tag{18.21}$$

where F is the three laboratory frame components, g is the three momentum axes components of molecules for asymmetric rotor, E is the electric field, and Φ is the direction cosine. For a more detailed explanation, please refer to the references [24, 58]. This allows us to calculate the trajectories of the molecules in the hexapole electric field. In this section, we introduce the experimental technique for the state selection of the asymmetric molecules and the calculation methods of the trajectory in the hexapole electric field. The propylene oxide is used as an example of the asymmetric molecule in this study.

18.5.1 Rotational Energy of Asymmetric-Top Molecules

Molecular orientation of asymmetric-top molecules can be characterized by three unequal components of the moment of inertia along the three axes denoted as a, b, and c as shown in Fig. 18.12 [58]. Molecular orientation of asymmetric-top molecules can be characterized by three distinct principal moments of inertia about the a, b, and c axes, as illustrated in Fig. 18.12 [58]. Their structure lies intermediate between the two limiting cases of symmetric-top rotors, namely the prolate and oblate tops. The rotational states of an asymmetric-top molecule are specified by the total angular momentum quantum number J and its projection onto a space-fixed quantization axis, M. In contrast to symmetric-top molecules, the projection of J onto the molecular symmetry axis, denoted by K, is not a good quantum number for asymmetric tops. Nevertheless, the limiting-case quantum numbers corresponding to the prolate-top limit (K_{-1}) and the oblate-top limit (K_1) can be combined to define a pseudo-quantum

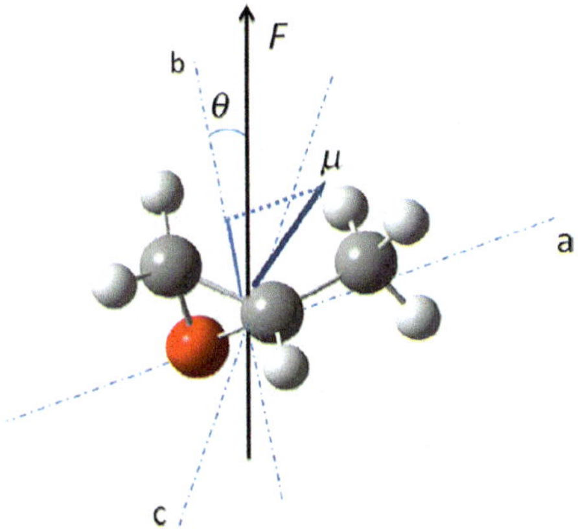

Fig. 18.12 Orientation of the propylene oxide molecule relative to the applied orienting field F. The three principal axes, **a**, **b**, and **c**, as well as the dipole moment μ, are indicated. The angle θ is defined as the angle between the component of the dipole moment μ along the principal b-axis and the direction of the orienting field F

number τ, given by $\tau = K_{-1} - K_1$. This quantity provides a convenient labeling scheme for the rotational states of asymmetric-top molecules. Under zero-field conditions, the rotational energy of an asymmetric-top molecule can thus be expressed as follows [58]:

$$W_{J\tau M} = \frac{1}{2}(A + C)J(J + 1) + \frac{1}{2}(A - C)W_{J\tau K}(\kappa) \tag{18.22}$$

where κ is the Ray's asymmetry parameter and $W_{J\tau K}(\kappa)$ is the reduced energy obtained from the diagonalization of the reduced Hamiltonian matrix. A, B, and C are the rotational constants, given in [59] for propylene oxide.

18.5.2 Field-Dependent Rotational Energy of Asymmetric-Top Molecules

In the absence of an external electric field, the projections of the total angular momentum J onto the space-fixed quantization axis, labeled by M, are energetically degenerate. When an electric field is applied, this degeneracy is partially lifted. The field-induced shifts in the rotational energy levels give rise to the Stark effect, and the corresponding energy gradients generate the Stark force, which is often alternatively interpreted in terms of an effective dipole moment. Field-dependent rotational energy levels $W^M_{J\tau}(E)$ and wave functions $C^M_{J\tau}(E)$ are calculated as eigenvalues and eigenvectors of the Stark matrix $H_S(E)$ for the construction of the Stark matrix [58]. In matrix notation:

$$W_S(E) = CH_SC^T \tag{18.23}$$

where C is an orthogonal matrix and C^T its transpose. This large matrix diagonalization is a computational bottleneck in problems of this complexity.

18.5.3 Trajectory Simulation for Asymmetric-Top Molecules

Let us describe the trajectory simulation for the asymmetric-top molecules in the hexapole field. The hexapole electric field E is given by Eq. (18.9). The magnitude of the force $F(r)$ applied to a molecule in the hexapole electric field is written by

$$F(r) = -\frac{dW_{J\tau}^{M}(E)}{dr} = -\frac{dW_{J\tau}^{M}(E)}{dE}\frac{dE}{dr} \tag{18.24}$$

By solving Eq. (18.14), one can calculate the trajectory of the symmetric molecule in the hexapole field. The number of rotational states $W_{J\tau}^{M}(E)$, which govern the trajectories, depends on the rotational temperature of the molecular beam. This parameter serves as a fundamental quantity for determining the rotational state population distribution according to the Boltzmann equation. For the pure propylene oxide beam and the Ar-seeded beam, with rotational temperatures of 30 K and 35 K, respectively, trajectory simulations were performed for rotational states up to a maximum quantum number of $J = 22$, corresponding to a total of 8371 states. In contrast, for the He-seeded beam with a rotational temperature of 10 K, simulations were carried out up to $J = 12$, resulting in 1546 rotational states included in the trajectory calculations. It is notable that we have compared both cases of the avoided crossing and diabatic passage due to the high electric field condition in the simulation.

By applying an orienting electric field, it is possible to achieve control over the molecular orientation in the laboratory frame. To preserve this orientation, the transition from the hexapole field to the orienting field must proceed under adiabatic conditions, such that the molecules retain their alignment. The adiabatic criterion is satisfied when the direction of the electric field experienced by the molecules varies sufficiently slowly during their passage between the two field regions. Accordingly, we examine the influence of introducing a guiding (buffer) field on the preservation of molecular orientation, as discussed, for example, by Bulthuis et al. [60].

Figures 18.13a shows the focusing curves for the pure propylene oxide beam and for the He-seeded beam. The beam intensities were measured using a quadrupole mass spectrometer. In each case, the experimental focusing curves are compared with simulated results obtained under the diabatic and adiabatic limiting conditions. The rotational temperature was determined from the best fit between the simulated and experimental focusing curves. For both limiting cases, a rotational temperature of 30 K was obtained, which is somewhat higher than the previously estimated value of $T_{rot} = 20$ K. The simulated focusing curves for the adiabatic and diabatic limits coincide at low and high hexapole voltages, whereas noticeable differences appear in the intermediate voltage range. For the propylene oxide beam seeded in He, the velocity distribution measured in previous time-of-flight experiments exhibited

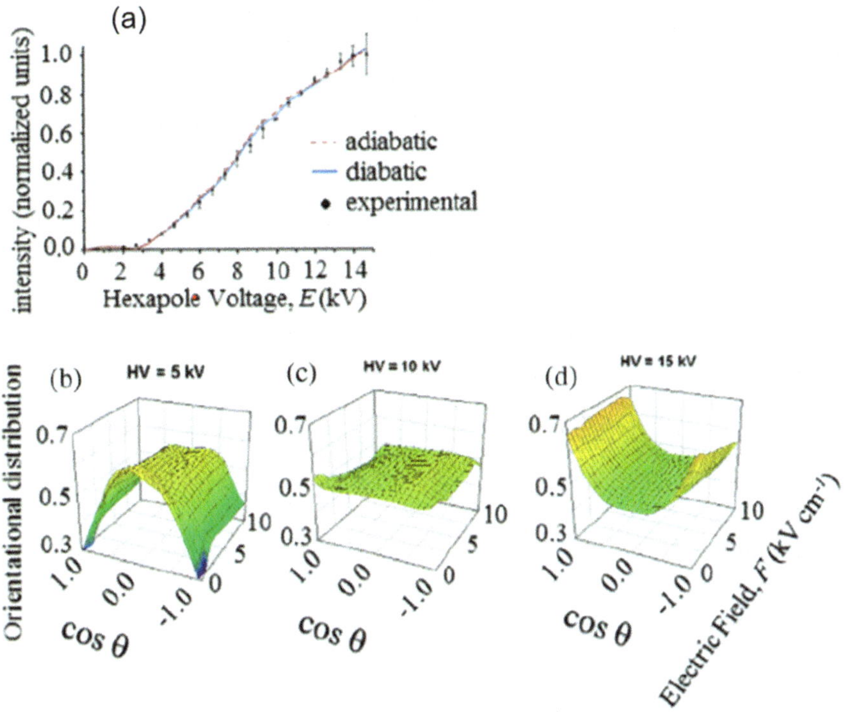

Fig. 18.13 Focusing curves of the pure propylene oxide beam obtained experimentally (dots) and calculated by trajectory simulations under the diabatic (blue solid line) and adiabatic (red dashed line) limiting conditions. Figures 19.13b–d present the orientational distributions in the adiabatic limit for selected rotational states at hexapole voltages of 5, 10, and 15 kV, respectively

a peak velocity of 1000 m/s and a full width at half maximum of 20 m/s. The rotational temperature estimated from the present simulations is $T_{rot} = 10$ K for both the diabatic and adiabatic limits, consistent with the previous work [59]. In this case, the simulated focusing curves under the two limiting conditions are nearly identical and provide good agreement with the experimental data. The corresponding orientational distributions are also shown in the figure.

18.6 Conclusions

In this chapter, we aimed to describe the basic technique for the oriented molecular beam using the hexapole electric field. In Sect. 18.2, we described the experimental technique for state selection using a hexapole field, the calculation of molecular orbitals in the hexapole field, and the calculation of the orientation distribution function in the molecular beam after passing through the hexapole field. In

Sect. 18.3, state selection of hydroxyl radical was described. By detecting the laser spectroscopy, it enables us the individual state selection using the hexapole electric field. In Sect. 18.4, the application of the reaction dynamics in the OH + HBr was described. Here, the orientation dependence in the reaction has been clarified. In Sect. 18.5, we demonstrated that the research technique, which was previously limited to the symmetric molecules, can be applied to the selection of orientation states of more general asymmetric molecules such as the chiral molecules.

Over the years, electrostatic hexapole field demonstrated to be a powerful tool for the study of the stereodynamics such as the collisional and photodissociation process. Recent developments of the hexapole electrostatic field enable us to be rotated in the rotational conformer selection [57]. Additionally, this technique can be used for the orientational selection of chiral molecules. Therefore, the possibility arises that the photofragment scattering distribution of enantiomers could be discriminated in those angular distributions if the parent chiral molecules were oriented [61–63].

References

1. Levine RD, Bernstein RB. Molecular reaction dynamics and chemical reactivity. New York: Oxford University Press; 1987.
2. Scoles G. Atomic and molecular beam methods, vol. 1. Oxford: Oxford Inuv; 1988.
3. Liu K, Wagner A. The chemical dynamics and kinetics of small radicals advanced series in physical chemistry, Vol. 6. World Scientific; 1995.
4. Brooks RB. Science. 1976;196:11.
5. Bernstein RB, Herschbach DR, Levine RD. 1987;91:5365: Parker DH. Bernstein RB Ann Rev Phys Chem. 1989;40:561.
6. Stolte S. Ber Bunsenges Ohys Chem. 1982;86:413.
7. Kramer KH, Bernstem RB. J Chem Phys. 1965;42:767.
8. Brooks RB, Jones EM. J Chem Phys. 1966;45:3449.
9. Beuhler RJ Jr, Bemstein RB, Kramer KH. J Am Chem Soc. 1966;88:5331.
10. Parker DH, Chakravarty KK, Bernstein RB. J Phys Chem. 1981;85:466.
11. Brooks PR, Jones EM, Smrth K. J Chem Phys. 1969;51:3073.
12. Parker DH, Chakravarty KK, Bernstein RB. Chem Phys Lett. 1982;86:113.
13. Carman HS, Harland PW, Brooks PR. J Phys Chem. 1986;90:944.
14. Ohoyama H, Kasai T, Ohashi K, Kuwata K. Chem Phys Lett. 1986;131:20.
15. Ohoyama H, Kasai T, Ohashi K, Kuwata K. Chem Phys Lett. 1987;136:236.
16. Ohoyama H, Igro T, Kasai T, Kuwata K. Chem Phys Lett. 1993;209:361.
17. Ohashi K, Kasai T, Che DC, Kuwata K. J Phys Chem. 1989;93:5484.
18. Ohashi K, Kasai T, Kuwata K. J Phys Chem. 1988;92:5954.
19. Kasai T, Che DC, Ohashi K, Kuwata K. Chem Phys Lett. 1989;163:246.
20. Che DC, Kasai T, Ohoyama H, Ohashi K, Fukawa T, Kuwata K. J Phys Chem. 1991;98:8159.
21. Che DC, Kasai T, Ohoyama K, Kuwata K. J Phys Chem. 1994;98:132.
22. Che DC, Kasai T, Ohoyama K, Kuwata K, Kono M, Tabayashi K, Shobatake K. Chem Lett. 1994; 133.
23. Kasai T, Matsunami T, Fukawa T, Che DC, Ohashi K, Fukunishi H, Ohoyama H, Kuwata K. Phys Rev Lett. 1993;70:3864.
24. Townes CH, Schawlow AL. Microwave spectroscopy. New York: Dover Publication, Inc.; 1975.
25. Landau LD, Lifshits EM. Quantum Mechnics. Oxford: Pergamon Press; 1977.

26. Chio SE, Bernstein RB. J Phys Chem. 1986;85:150.
27. Miller JA, Fisk GA. Chem Eng News. 1987;65:22.
28. Melissas VS, Truhlar DG. J Chem Phys. 1993;99:1013.
29. Miyoshi A, Yamauchi N, Matsui H. J Phys Chem. 1996;100:4893.
30. Beek MC, ter Meulen JJ. Chem Phys Lett. 2001;337:237.
31. ter Meulen JJ, Meerts WL, van Mierlo GWM, Dymanus A. Phys Rev Lett. 1976;36:1031.
32. Schreel K, Schleipen J, Eppink A, ter Meulen JJ. J Chem Phys. 1993;99:8713.
33. Schreel K, ter Meulen JJ. J Chem Phys. 1996;313:491.
34. van Beek MC, Schreel K, ter Meulen JJ. J Chem Phys. 1998;109:1302.
35. Schreel K, ter Meulen JJ, Alexander MH. J Chem Phys. 2000;113:628.
36. Schreel K, ter Meulen JJ, Alexander MH. J Chem Phys. 2000;113:637.
37. Hain TD, Weibel MA, Backstrand KM, Curtiss TJ. J Phys Chem A. 1997;101:7674.
38. Hashinokuchi M, Che DC, Watanabe D, Fukuyama T, Koyano I, Shimizu Y, Woelke A, Kasai T. Phys Chem Chem Phys. 2003;5:3911.
39. Chidsey IL, Crosley DR. J Quant Spectrosc Radiat Transf. 1980;23:187.
40. Andresen P, Ondrey GS, Titze B, Rothe EW. J Chem Phys. 1984;80:2548.
41. Wayne RP. Chemistry of the atmospheres. Oxford: Oxford University; 1991.
42. Ravishankara AR, Wine PH, Wells JR. J Chem Phys. 1985;83:447.
43. Sims R, Smith IW, Clary DC, Bocherel P, Rowe BR. J Chem Phys. 1994;101:1748.
44. Clary DC, Nyman G, Hernandez R. J Phys Chem. 1994;101:3704.
45. Moise A, Parker DH, ter Meulen JJ. J Chem Phys. 2007;126: 124302.
46. Schreel K, ter Meulen JJ. J Phys Chem A. 1997;101:7639.
47. Wallace SC, Zdasiuk G. Appl Phys Lett. 1976;28:449.
48. Hepburn JW, Klimek D, Liu K, Polanyi JC, Wallace SC. J Phys Chem. 1978;69:4311.
49. Hepburn JW, Klimek D, Liu K, Macdonald RG, Northrup F, Polanyi JC. J Chem Phys. 1981;74:6226.
50. Hepburn JW, Klimek D, Liu K, Macdonald RG, Northrup F, Polanyi JC. J Chem Phys. 1981;75:3353.
51. Stolte S, Chakravorty KK, Bernstein RB, Parker DH. Chem Phys. 1982;71:353.
52. Che DC, Matsuo T, Yano Y, Bonnetb L, Kasai T. Phys Chem Chem Phys. 2008;10:1419.
53. Che DC, Doi A, Yamamoto Y, Okuno Y, Kasai T. Phys Scr. 2009;80:48110.
54. Tsai PY, Che DC, Nakamura M, Lin KC, Kasai T. Phys Chem Chem Phys. 2010;12:2532.
55. Tsai PY, Che DC, Nakamura M, Lin KC, Kasai T. Phys Chem Chem Phys. 2011;13:1419.
56. Dc C. Kanda K, Palazzetti F, Aquilanti V, Kasai T. Chem Phys. 2012;399:180.
57. Caglioti C, Masaaki N, Che DC, Tsai PY, Palazzetti F. Symmetry. 2022;14:317.
58. Hain TD, Moison RM, Curtiss TJ. J Chem Phys. 1999;111:6797.
59. Che DC, Palazzetti F, Okuno Y, Aquilanti V, Kasai T. J Phys Chem A. 2010;114:3280.
60. Bulthuis J, Milan JB, Janssen MHM, Stolte S. J Chem Phys. 1991;94:7181.
61. Che DC, Kawamata H, Nakamura M, Tso CJ. Muthiah B Kasai T. J Chin Chem Soc. 2023;70:1297.
62. Palazzetti F, Tsai PY, Lombardi A, Nakamura M, Che DC, Kasai T, Lin KC, Aquilanti V. Rend Fis Acc Lincei. 2013;24:299.
63. Kawamata H, Che DC, Nakamura M, Kasai T. J Phys Chem A. 2022;126:8844.

The manufacturer's authorised representative in the EU is Springer Nature Customer Service Centre GmbH, Europaplatz 3, 69115 Heidelberg, Germany. If you have any concerns regarding our products, please contact ProductSafety@springernature.com

Printed and bound by CPI Group (UK) Ltd, Croydon, CR0 4YY
14/10/2026
02233569-0002